# Lecture Notes

T0238058

## The Lecture Notes in Physics

The series Lecture Notes in Physics (LNP), founded in 1969, reports new developments in physics research and teaching – quickly and informally, but with a high quality and the explicit aim to summarize and communicate current knowledge in an accessible way. Books published in this series are conceived as bridging material between advanced graduate textbooks and the forefront of research and to serve three purposes:

- to be a compact and modern up-to-date source of reference on a well-defined topic

- to serve as an accessible introduction to the field to postgraduate students and nonspecialist researchers from related areas

- to be a source of advanced teaching material for specialized seminars, courses and schools

Both monographs and multi-author volumes will be considered for publication. Edited volumes should, however, consist of a very limited number of contributions only. Proceedings will not be considered for LNP.

Volumes published in LNP are disseminated both in print and in electronic formats, the electronic archive being available at springerlink.com. The series content is indexed, abstracted and referenced by many abstracting and information services, bibliographic networks, subscription agencies, library networks, and consortia.

Proposals should be sent to a member of the Editorial Board, or directly to the managing editor at Springer:

Christian Caron
Springer Heidelberg
Physics Editorial Department I
Tiergartenstrasse 17
69121 Heidelberg / Germany
christian.caron@springer.com

L. Socha

# Linearization Methods for Stochastic Dynamic Systems

 Springer

Author

Lesław Socha
Cardinal Stefan Wyszyński University in Warsaw
Faculty of Mathematics and Natural Sciences
College of Sciences
ul.Dewajtis 5
01-815 Warsaw, Poland
and
University of Silesia
Department of Mathematics, Physics and Chemistry
ul.Akademicka 4
40-007 Katowice, Poland
leslawsocha@poczta.onet.pl

L. Socha, *Linearization Methods for Stochastic Dynamic Systems*, Lect. Notes Phys. 730
(Springer, Berlin Heidelberg 2008), DOI 10.1007/978-3-540-72997-6

ISSN 0075-8450
ISBN 978-3-642-09210-7          e-ISBN 978-3-540-72997-6

Springer is a part of Springer Science+Business Media
springer.com
© Springer-Verlag Berlin Heidelberg 2008
Softcover reprint of the hardcover 1st edition 2008

Cover design: eStudio Calamar S.L., F. Steinen-Broo, Pau/Girona, Spain

*To my wife Ewa and my sons Jaroslaw, Marcin and Wojtek*

# Preface

To describe real observed processes in many fields such as physics, chemistry, biology, economics and engineering, researchers use mathematical models. For dynamic processes, these models contain many different types of equations such as ordinary or partial differential equations, difference equations, and algebraic equations. Because of the nature of the considered problems, these equations are usually nonlinear. In general case, it is not possible to find the exact solution for such equations. Except for some special cases, the solutions are approximate. They can be obtained, for instance, by simple and popular methods called linearization methods.

The second important problem of mathematical modeling is the fact that mathematical models of the considered processes contain uncertainty terms that represent the researcher's or designer's lack of knowledge about parameter values and excitations. In this case, the quantities are assumed to be deterministic, time-dependent, or stochastic processes. Therefore, nonlinear stochastic differential equations are the natural mathematical tools in the analysis of stochastic dynamic systems. Unfortunately, only a narrow class of these equations are exactly solvable. To obtain approximate solutions, two groups of methods were developed in the literature. The first group contains numerical algorithms and simulation methods. The second group consists of analytical approximate methods. As in the deterministic case, the simplest and popular of these methods are linearization methods. They have been intensively developed since 50 years and still new publications appear introducing new linearization approaches and its applications in many fields.

The objective of this book is to give a discussion about the existing linearization methods in an ordered form. A wide spectrum of these methods from theoretical point of view and applications is presented. Although the material is limited to models with continuous external and parametric excitations, it covers over 100 different approaches. I apologize to those whose publications were omitted. The book is addressed to researchers and graduate students dealing with mathematical modeling of uncertain dynamic systems, mainly in response analysis or control synthesis.

Some parts of the material presented in this book were used as notes in my graduated courses on random vibration in State University of Sao Paulo in Guaratingueta in Brazil in 1988, State University of New York in Buffalo, USA, in 1999, University of Silesia in Katowice, Poland, in 2000, and in the International Centre For Mechanical Sciences in Udine, Italy, in 2005.

I am grateful to my teachers and numerous colleagues, friends, and graduate students with whom I have sought collaboration and coauthored research papers. I feel myself privileged to have had the opportunity to be taught by the outstanding teachers of random vibration in Poland Professors B. Skalmierski and A. Tylikowski who first initiated my research interest in linearization methods. I appreciate my research collaboration and discussions with colleagues and friends K. Popp from Hannover University, J.L. Willems from Gent University, N. Nascimento from SUSP in Guaratingueta, I. Elishakoff from Florida University and C. Proppe from Karlsruhe University. I am specifically thankful to my coauthors and friends Professor T.T. Soong from SUNY in Buffalo and my graduate student M. Pawleta from Silesian Technical University. I also would like to express my appreciation to several of my students M. Święty, A. Wojak, and J. Czekaj from University of Silesia and P. Kaczyński and E. Seroka from Cardinal Stefan Wyszyński University who helped me in the preparation of the manuscript.

Much of the material in this book was generated from sponsored research during the past 20 years. I am indebted to the sponsors of my research project including the State Committee of Scientific Research in Poland, Alexander von Humboldt Foundation in Germany, National Center for Earthquake Engineering Research in the USA and Cardinal Stefan Wyszyński University in Warsaw, Poland.

# Contents

# 1

# Introduction

Linearization methods are the oldest and the most popular methods of approximation. First, the Taylor formula was used to approximate nonlinear deterministic functions by linear terms, i.e., by limiting the Taylor expansion of a nonlinear function to first two terms, $f(x) \cong f(x_0) + f'(x_0)(x - x_0)$. Also other approximations were introduced, for instance, minimization of square error. Unfortunately, Taylor expansion could not be applied to nonlinear dynamic systems, because in many mathematical models, nondifferentiable functions or hysteresis appears. These models were developed and studied mainly by control engineers. They proposed a linearization in frequency domain called *harmonic balance* or *describing function* method. The mathematical background for this approach was given by Russian researchers Krylov (1934) and Bogoluboff (1937) partially presented in [7] and developed by many authors (see for details and references, for instance, [10, 18]). The natural generalization of *describing function* for nonlinear function of stochastic process was an approach called *statistical linearization* that was carried out virtually simultaneously by Booton [1] and Kazakov [6]. The objective of this method is to replace nonlinear elements in a model by corresponding linear forms where the linearization coefficients can be found based on a specified criterion of linearization. These methods were extended and developed mainly by Russian authors in the 1950s and 1960s in connection with the modeling of automatic control systems and next developed in statistical physics and in mechanical and structural engineering.

A different philosophy of the replacement of a nonlinear oscillator under Gaussian excitation by a linear one under the same excitation for which the coefficients of linearization can be found from a mean-square criterion was proposed by Caughey [2, 3]. Similarly to Krylov and Bogoluboff who studied deterministic vibration systems by asymptotic methods [7], Caughey called his approach by *equivalent linearization*. Since Caughey has used the same name of method for a few versions of linearization methods with different criteria some misunderstandings regarding the derivation of the formulas for linearization coefficients appeared in the literature. These misunderstandings

L. Socha: *Introduction*, Lect. Notes Phys. **730**, 1–5 (2008)
DOI 10.1007/978-3-540-72997-6_1                    © Springer-Verlag Berlin Heidelberg 2008

caused that the statistical linearization proposed by Kazakov and equivalent linearization are mainly treated in the literature as the same methods. However, some authors in their papers or books introduce different names for these methods. For instance, in the book of Roberts and Spanos, [12] statistical linearization in "Kazakov's sense" is described in the section entitled *Nonlinear elements without memory*. Similarly, in the book [14], statistical linearization is introduced in the section entitled *Memoryless Transformation* while equivalent linearization is presented in the section entitled *Transformation with memory*. Also, in the survey papers by Socha and Soong [17] and Socha [15], both approaches were separately reviewed. Therefore, we will discuss statistical and equivalent linearization approaches in Chaps. 4 and 5 separately. Both approaches are used to approximate nonlinear systems by equivalent linearized ones for a given type of excitation (see Fig. 1.1).

All methods of statistical and equivalent linearization can be considered in different fields such as state space, frequency domain, distribution space, characteristic functions space, and entropy field. Usually they consist of two main steps. In the first one based on the type of field and a linearization criterion, we find explicit or implicit analytical formulas for linearization coefficients. They depend on unknown response characteristics (mean value, variance, and higher-order moments). In the second step, we replace the unknown characteristics by the corresponding ones determined for linearized systems. These two steps are repeated in an iterative procedure until converge. It is illustrated in the case of moment criteria in Fig. 1.2.

It is clear that both linearization coefficients and approximate response characteristics of dynamic system depend not only on nonlinear elements and system parameters, but also on the type of the excitation, for instance, Gaussian or non-Gaussian, external or parametric, stationary or nonstationary, continuous or impulsive. Therefore, the block "Excitation" in Fig. 1.1 can be expanded to the one presented in Fig. 1.3.

The number of possible applications of linearization methods defined by different fields of consideration, types of criteria, and types of excitations to stochastic dynamic systems is very large (over 1000). However, not all possible cases were considered in the literature. Also, a very wide area of applications of linearization methods in many cases required specific modifications of these methods.

The material till 1990 was reviewed, for instance, in the book [12] and in the survey paper [17]. The development of linearization methods in the study of stochastic models of dynamic systems in theoretical aspects as well as in application fields over the last decade was intensive. Over 200 papers

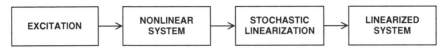

**Fig. 1.1.** Schematic of the replacement of a nonlinear system by linearized one

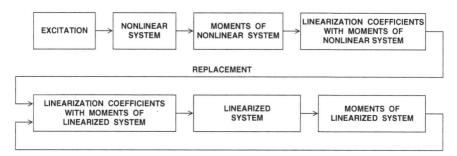

**Fig. 1.2.** Schematic of the determination of linearization coefficient for moment criteria

in journals and conference proceedings have been appeared during last 15 years. In several books and texts, for instance [8, 9, 13, 14], one can find that linearization methods are already treated as standard mathematical tools in the analysis of stochastic dynamic systems. They have been partially reviewed in [5, 4, 11, 15, 16].

The objective of this book is to give a discussion of the existing linearization methods in an ordered form. A general classification of the main linearization methods for stochastic dynamic systems is presented in Fig. 1.4.

Since the material is very wide, we restrict our considerations to continuous stochastic processes and in order to give the readers an idea of development of linearization methods, in Chaps. 5–8 we start the presentation of each group of methods with approaches proposed by pioneers and next we show only the

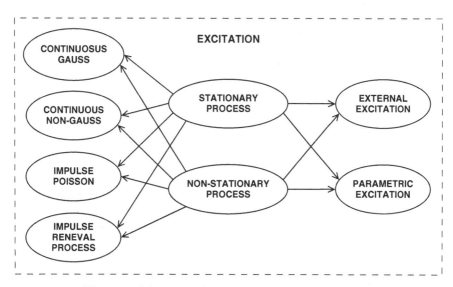

**Fig. 1.3.** Schematic of the different types of excitation

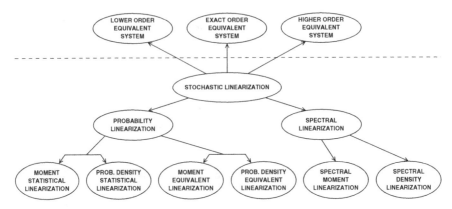

**Fig. 1.4.** Schematic of the classification of linearization methods for stochastic dynamic systems

basic ones. At the end of each chapter in "Bibliography notes" we shortly review other important methods.

The book is organized as follows. In Chap. 2 we quote some basic results of continuous stochastic processes connected with random vibration including informations about stochastic differential equations, stochastic stability, and the method of Fokker–Planck–Kolmogorov equation. The problem of the determination of moment equations for linear and nonlinear systems with external and parametric excitation is discussed in Chaps. 3 and 4, respectively. Theoretical results connected with derivation of linearization coefficients, linearization approaches with moment criteria including energy criteria and with criteria in probability density functions space and spectral density domain for dynamic systems under external excitations are presented in Chaps. 5 and 6. The problem of the determination of an equivalent nonlinear system is discussed in Chap. 7 and the linearization of dynamic systems with parametric excitation in Chap. 8. Applications of stochastic linearization in response analysis of hysteresis models, structures under earthquake, sea waves and wind excitation, and in control of nonlinear stochastic systems are shown in Chap. 9. In particular, we show an application of the classical technique for linear systems with quadratic criterion under Gaussian excitation (LQG) and linearization methods with different criteria in joint iterative procedures that determine quasioptimal controls for nonlinear stochastic dynamic systems. The discussion of the accuracy of linearization methods using theoretical estimation methods and simulations is given in Chap. 10. There are also reported literature examples of verification of linearization methods in modeling and response analysis by experiments. At the end of this chapter, a summary discussion about advantages and disadvantages of linearization methods from theoretical point of view and applications is given.

# References

1. Booton, R.C.: The analysis of nonlinear central systems with random inputs. IRE Trans. Circuit Theory **1**, 32–34, 1954.
2. Caughey, T.: Response of Van der Pol's oscillator to random excitations. Trans. ASME J. Appl. Mech. **26**, 345–348, 1959.
3. Caughey, T.: Random excitation of a system with bilinear hysteresis. Trans. ASME J. Appl. Mech. **27**, 649–652, 1960.
4. Elishakoff, I.: Random vibration of structures: A personal perspective. Appl. Mech. Rev. **48**, 809–825, 1995.
5. Elishakoff, I.: Stochastic linearization method: A new interpretation and a selective review. Shock Vib. Dig. **32**, 179–188, 2000.
6. Kazakov, I.E.: An approximate method for the statistical investigation for nonlinear systems. Trudy VVIA im Prof. N. E. Zhukovskogo **394**, 1–52, 1954 (in Russian).
7. Krilov, N. and Bogoliuboff, N.: Introduction to Non-linear Mechanics. Princeton University Press, Princeton, 1943.
8. Lin, Y. and Cai, G.: Probabilistic Structural Dynamics. McGraw Hill, New York, 1995.
9. Lutes, L. and Sarkani, S.: Stochastic Analysis of Structural and Mechanical Vibrations. Prentice Hall, Upper Saddle River, New Jersey, 1997.
10. Naumov, B.: The Theory of Nonlinear Automatic Control Systems. Frequency Methods. Nauka, Moscow, 1972 (in Russian).
11. Proppe, C., Pradlwarter, H., and Schueller, G.: Equivalent linearization and Monte Carlo simulation in stochastic dynamics. Probab. Eng. Mech. **18**, 1–15, 2003.
12. Roberts, J. and Spanos, P.: Random Vibration and Statistical Linearization. John Wiley and Sons, Chichester, 1990.
13. Solnes, J.: Stochastic Processes and Random Vibrations Theory and Practice. John Wiley and Sons, Chichester, 1997.
14. Soong, T. and Grigoriu, M.: Random Vibration of Mechanical and Structural Systems. PTR Prentice Hall, Englewood Cliffs, NJ, USA, 1993.
15. Socha, L.: Linearization in analysis of nonlinear stochastic systems – recent results. Part I. Theory. ASME Appl. Mech. Rev. **58**, 178–205, 2005.
16. Socha, L.: Linearization methods in stochastic dynamic systems. In: I. Elishakoff (ed.), Mechanical Vibration: Where Do We Stand. CISM Lecture Notes, pp. 249–319. Springer, Wien, 2007.
17. Socha, L. and Soong, T.: Linearization in analysis of nonlinear stochastic systems. Appl. Mech. Rev. **44**, 399–422, 1991.
18. Thaler, G. and Pastel, M.: Analysis and Design of Nonlinear Feedback Control Systems. MacGraw-Hill, New York, 1962.

# 2

# Mathematical Preliminaries

In this chapter the basic concepts and results concerning stochastic calculus of continuous stochastic processes in Euclidean spaces are presented. Therefore we omit some introductory facts from probability theory sending the readers to well-known text books. We start our considerations with some information about moments and cumulants of random variables.

## 2.1 Moments and Cumulants of Random Variables

To obtain formulas for moments and cumulants of random variables we use the characteristic function defined by

**Definition 2.1.** *Let $g(x)$ be the probability density function of random variable $X$, and $\Theta \in R$ be a real variable, then*

$$\Phi(\Theta) = E[\exp\{i\Theta X\}] = \int_{-\infty}^{+\infty} \exp\{i\Theta x\}g(x)dx \qquad (2.1)$$

*is called characteristic function of random variable $X$, where $E[.]$ is the operation of averaging, $i = \sqrt{-1}$.*

For $n$ random variables $X_1, \ldots, X_n$ the definition formula of joint characteristic function is given by

$$\Phi(\Theta_1, \ldots, \Theta_n) = E[\exp\{i\sum_{j=1}^{n}\Theta_j x_j\}]$$

$$= \int_{-\infty}^{+\infty} \ldots \int_{-\infty}^{+\infty} \exp\left\{i\sum_{j=1}^{n}\Theta_j x_j\right\} g(x_1, \ldots, x_n)dx_1, \ldots, dx_n$$

$$(2.2)$$

L. Socha: *Mathematical Preliminaries*, Lect. Notes Phys. **730**, 7–58 (2008)
DOI 10.1007/978-3-540-72997-6_2 &copy; Springer-Verlag Berlin Heidelberg 2008

or in vector notation

$$\Phi(\boldsymbol{\Theta}) = E[\exp\{i\boldsymbol{\Theta}^T\mathbf{X}\}] = \int_{R^n} \exp\{i\boldsymbol{\Theta}^T\mathbf{x}\}\, g(\mathbf{x})dx\,, \qquad (2.3)$$

where $\boldsymbol{\Theta} = [\Theta_1, \ldots, \Theta_n]^T$, $\mathbf{X} = [X_1, \ldots, X_n]^T$, $\mathbf{x} = [x_1, \ldots, x_n]^T$.

One can find the joint characteristic function of $m$ random variables $X_1, \ldots, X_m$, $m < n$, by substituting into equality (2.2)

$$\Theta_{m+1} = \Theta_{m+2} = \cdots = \Theta_n = 0.$$

The function $\Phi(\boldsymbol{\Theta})$ can be expanded by the Maclaurin series

$$\Phi(\boldsymbol{\Theta}) = \Phi_1(\boldsymbol{\Theta}) = 1 + i\sum_{j=1}^{n}\Theta_j E[X_j] + \frac{i^2}{2!}\sum_{j=1}^{n}\sum_{k=1}^{n}\Theta_j\Theta_k E[X_j X_k] + \cdots \quad (2.4)$$

or presented in an equivalent form

$$\Phi(\boldsymbol{\Theta}) = \Phi_2(\boldsymbol{\Theta}) = \exp\left\{i\sum_{j=1}^{n}\Theta_j\kappa_1[X_j] + \frac{i^2}{2!}\sum_{j=1}^{n}\sum_{k=1}^{n}\Theta_j\Theta_k\kappa_2[X_j X_k] + \ldots\right\}.$$
$$(2.5)$$

Formula (2.5) is equivalent to the expansion of logarithm of characteristic function by Maclaurin series

$$ln(\Phi_2(\boldsymbol{\Theta})) = i\sum_{j=1}^{n}\Theta_j\kappa_1[X_j] + \frac{i^2}{2!}\sum_{j=1}^{n}\sum_{k=1}^{n}\Theta_j\Theta_k\kappa_2[X_j X_k] + \ldots\,, \qquad (2.6)$$

where $E[X_1\ldots X_n]$ are *moments of random variables* $X_1\ldots X_n$, and $\kappa_n[X_1\ldots X_n]$ are called *cumulants* or *semi invariants* of $n$ order. They can be found by differentiation of (2.4) and (2.6), respectively.

$$E[X_1\ldots X_p] = \frac{1}{i^p}\frac{\partial^p\Phi_1(\boldsymbol{\Theta})}{\partial\Theta_1\ldots\partial\Theta_p}\bigg|_{\boldsymbol{\Theta}=0}\,, \qquad p \leq n\,, \qquad (2.7)$$

$$\kappa_p[X_1\ldots X_p] = \frac{1}{i^p}\frac{\partial^p ln(\Phi_2(\boldsymbol{\Theta}))}{\partial\Theta_1\ldots\partial\Theta_p}\bigg|_{\boldsymbol{\Theta}=0}\,, \qquad p \leq n\,. \qquad (2.8)$$

Function $ln(\Phi_2(\boldsymbol{\Theta}))$ is called *generating function for cumulants*. The relationship between moments and cumulants can be found by the replacement of the differentiated functions $\Phi_j(\boldsymbol{\Theta})$, $j = 1, 2.$, i.e.,

$$E[X_1\ldots X_p] = \frac{1}{i^p}\frac{\partial^p\Phi_2(\boldsymbol{\Theta})}{\partial\Theta_1\ldots\partial\Theta_p}\bigg|_{\boldsymbol{\Theta}=0}\,, \qquad p \leq n\,, \qquad (2.9)$$

$$\kappa_p[X_1\ldots X_p] = \frac{1}{i^p}\frac{\partial^p ln(\Phi_1(\boldsymbol{\Theta}))}{\partial\Theta_1\ldots\partial\Theta_p}\bigg|_{\boldsymbol{\Theta}=0}\,, \qquad p \leq n\,. \qquad (2.10)$$

For instance,

$$E[X_j] = \kappa_1[X_j], \quad E[X_j X_k] = \kappa_2[X_j X_k] - \kappa_1[X_j]\kappa_1[X_k],$$

$$E[X_j X_k X_l] = \kappa_3[X_j X_k X_l] + 3\{\kappa_1[X_j]\kappa_2[X_k X_l]\}_s$$

$$+\kappa_1[X_j]\kappa_1[X_k]\kappa_1[X_l] , \tag{2.11}$$

where $\{.\}_s$ denotes a symmetrizing operation with respect to all its arguments, i.e., the arithmetic mean of all different permuted terms similar to the one within the braces. It is illustrated by the following example:

$$\{\kappa_1[X_j]\kappa_2[X_k X_l]\}_s = \frac{1}{3}\{\kappa_1[X_j]\kappa_2[X_k X_l] + \kappa_1[X_k]\kappa_2[X_j X_l] + \kappa_1[X_l]\kappa_2[X_j X_k]\}.$$

Cumulants have properties similar to independent random variables, i.e., a cumulant of sum of random variables is equal to sum of cumulants of separated random variables. Similarly, one can generate the characteristic function of a central random variable $X - E[X]$ and next by differentiating it with respect to $\Theta$ one can obtain central moments

$$\mu_p = E[(X_1 - m_1)\ldots(X_p - m_p)] = \frac{1}{i^p} \frac{\partial^p [\exp\{-i\Theta^{\mathrm{T}}\mathbf{m}\}\Phi(\Theta)]}{\partial\Theta_1\ldots\partial\Theta_p}\bigg|_{\Theta=0} , \quad p \le n , \tag{2.12}$$

where $\mathbf{m} = E[\mathbf{X}] = [m_1,\ldots,m_n]^T$.

## 2.2 Gaussian and Non-Gaussian Distributions

Gaussian random variable (normal variable) plays a particular role in probability theory and statistics.

**Definition 2.2.** *Vector random variable* $\mathbf{X}$ *is called r-dimensional Gaussian if its characteristic function has the form*

$$\Phi(\Theta) = \exp\left\{i\Theta^{\mathrm{T}}\mathbf{m} - \frac{1}{2}\Theta^T K\Theta\right\} , \tag{2.13}$$

*where* $\mathbf{m} = [m_1,\ldots,m_r]^T$ *is the vector of mean values and* $\mathbf{K} = [K_{jl}]$, $j$, $l = 1,\ldots,r$ *is the covariance matrix*

$$\mathbf{m} = E[\mathbf{X}], \quad \mathbf{K} = E[(\mathbf{X} - E[\mathbf{X}])(\mathbf{X} - E[\mathbf{X}])^T] . \tag{2.14}$$

If the covariance matrix $\mathbf{K}$ is an invertible matrix, then the probability density function of the Gaussian random variable $\mathbf{X}$ has the form

$$g_G(\mathbf{x}) = \frac{1}{\sqrt{(2\pi)^r|\mathbf{K}|}} \exp\left\{-\frac{1}{2}(\mathbf{x} - \mathbf{m})^T \mathbf{K}^{-1}(\mathbf{x} - \mathbf{m})\right\} , \tag{2.15}$$

where $|\mathbf{K}|$ is the determinant of the matrix $\mathbf{K}$. This probability density function has some important analytical properties. One of them is that the Hermite polynomials $\{H_\nu(\mathbf{x}), G_\nu(\mathbf{x})\}$ constitute a biorthonormal system of polynomials with Gaussian probability density function (treated as a weighting function), i.e.,

$$\int_{-\infty}^{+\infty} \ldots \int_{-\infty}^{+\infty} g_G(\mathbf{x}) H_\nu(\mathbf{x}) G_\mu(\mathbf{x}) dx_1 \ldots dx_r = c\delta_{\nu\mu} = \begin{cases} 0 \ for \ \mu \neq \nu, \\ c \ for \ \mu = \nu, \end{cases}$$

$$(2.16)$$

where $g_G(\mathbf{x})$ is determined by (2.15), while Hermite polynomial $H_\nu(\mathbf{x})$, $G_\nu(\mathbf{x})$ can be determined as follows [43]

$$H_\nu(\mathbf{x}) = (-1)^{\sigma(\nu)} \exp\left\{\frac{1}{2}\mathbf{x}^\mathbf{T}\mathbf{K}^{-1}\mathbf{x}\right\}$$

$$\frac{\partial^{\sigma(\nu)}}{\partial x_1^{\nu_1} \ldots \partial x_n^{\nu_r}} \exp\left\{-\frac{1}{2}\mathbf{x}^\mathbf{T}\mathbf{K}^{-1}\mathbf{x}\right\} , \qquad (2.17)$$

$$G_\nu(\mathbf{x}) = (-1)^{\sigma(\nu)} \exp\left\{\frac{1}{2}\mathbf{x}^\mathbf{T}\mathbf{K}^{-1}\mathbf{x}\right\}$$

$$\left[\frac{\partial^{\sigma(\nu)}}{\partial y_1^{\nu_1} \ldots \partial y_n^{\nu_r}} \exp\left\{-\frac{1}{2}\mathbf{y}^\mathbf{T}\mathbf{K}^{-1}\mathbf{y}\right\}\right]_{\mathbf{y}=\mathbf{K}^{-1}\mathbf{x}} , \qquad (2.18)$$

where $\nu$ is the multi index, $\nu = [\nu_1, \ldots, \nu_r]^T$, $\sigma(\nu) = \sum_{i=1}^r \nu_i$, $c = \prod_{i=0}^r \nu_i!$.

This important property gives an opportunity for the presentation of the probability density function of any continuous vector random variable in the form of a series, under condition that there exist moments of the considered random variable of any order

$$g_N(\mathbf{x}) = g_G(\mathbf{x}) \left[1 + \sum_{k=3}^{+\infty} \sum_{\sigma(\nu)=k} \frac{c_\nu H_\nu(\mathbf{x} - \mathbf{m})}{\nu_1! \ldots \nu_r!}\right] , \qquad (2.19)$$

where $g_G(\mathbf{x})$ is determined by (2.15) and the coefficients $c_\nu$ can be found from the following condition

$$c_\nu = \int_{-\infty}^{+\infty} \ldots \int_{-\infty}^{+\infty} g_G(\mathbf{x}) G_\nu(\mathbf{x} - \mathbf{m}) dx_1 \ldots dx_r = E[G_\nu(\mathbf{X} - \mathbf{m})] = G_\nu(\boldsymbol{\mu}) ,$$

$$(2.20)$$

where $G_\nu(\boldsymbol{\mu})$ is a linear combination of central moments of variable $X$ obtained as a result of the replacement of each term $(x_1 - m_1)^{\nu_1} \ldots (x_r - m_r)^{\nu_r}$ by the corresponding moment $E[(x_1 - m_1)^{\nu_1} \ldots (x_r - m_r)^{\nu_r}]$.

The coefficients $c_\nu$ are called *quasi-moments*, and the number $\sigma(\nu)$ *order of quasi-moments* $c_\nu$. From relation (2.20) it follows that quasi-moment of order $k$ is a linear combination of central moments of up to the $k$th order.

To obtain an approximate probability density function of a non-Gaussian variable the first sum can be limited to $k = N$. Then a normalized constant must be introduced. This approximation can be used in both cases: when the vector random variable $\mathbf{X}$ has moments of higher order than $N$ or not. However, there are some problems with direct truncation of the approximation series. It may often appear meaningless negative values at the probability density function. We will discuss this problem in Sect. 2.5.

In order to obtain approximate probability density functions of scalar nonlinear random variables $Y_j = \psi_j(x_j)$, $j = 1, \ldots, n$ one can use, for instance, the Gram–Charlier expansion [43],

$$
g_{Y_j}(y_j) = \frac{1}{\sqrt{2\pi}c_j\sigma_{y_j}} \exp\left\{ -\frac{(y_j - m_{y_j})^2}{2\sigma_{x_j}^2} \right\}
$$
$$
\left[ 1 + \sum_{\nu=3}^{N} \frac{c_{\nu j}}{\nu!} H_\nu\left( \frac{y_j - m_{y_j}}{\sigma_{y_j}} \right) \right], \tag{2.21}
$$

where $m_{y_j} = E[y_j]$, $\sigma_{y_j}^2 = E[(y_j - m_{y_j})^2]$; $c_{\nu j} = E[G_\nu(y_j - m_{y_j})]$, $\nu = 3, 4, \ldots, N$ are quasi-moments, $c_j$ are normalized constants, $H_\nu(x)$ and $G_\nu(x)$ are Hermite polynomials of one variable. For instance, a few first polynomials are given below:

$$
H_3(x) = \frac{1}{\sigma_{x_j}^6}(x^3 - 3\sigma_{x_j}^2 x), \quad H_4(x) = \frac{1}{\sigma_{x_j}^8}(x^4 - 6\sigma_{x_j}^2 x^2 + 3\sigma_{x_j}^4),
$$

$$
H_5(x) = \frac{1}{\sigma_{x_j}^{10}}(x^5 - 10\sigma_{x_j}^2 x^3 + 15\sigma_{x_j}^4 x),
$$

$$
H_6(x) = \frac{1}{\sigma_{x_j}^{12}}(x^6 - 15\sigma_{x_j}^2 x^4 + 45\sigma_{x_j}^4 x^2 - 15\sigma_{x_j}^6), \tag{2.22}
$$

$$
G_3(x) = x^3 - 3\sigma_{x_j}^2 x, \quad G_4(x) = x^4 - 6\sigma_{x_j}^2 x^2 + 3\sigma_{x_j}^4,
$$
$$
G_5(x) = x^5 - 10\sigma_{x_j}^2 x^3 + 15\sigma_{x_j}^4 x,
$$
$$
G_6(x) = x^6 - 15\sigma_{x_j}^2 x^4 + 45\sigma_{x_j}^4 x^2 - 15\sigma_{x_j}^6. \tag{2.23}
$$

## 2.3 Moments and Cumulants of Stochastic Processes

The theory of stochastic processes was created as a generalization of the concept of random variables. In the case of random variable, each elementary event was assigned a real number. Such model in many cases was not precise enough. In fact, in many real processes (physical, biological, economical, each elementary event was assigned a real function, for instance, a function of time or defined on a set of real parameters. It leads to the following definition of stochastic process.

**Definition 2.3.** *Let $(\Omega, F, P)$ be the probability space and $R^+ = [0, +\infty)$. The family $X = (\xi_t), t \in R^+$ of random variables $\xi_t = \xi_t(\omega)$ is called (real) stochastic process with continuous time $t \in R^+$. In the case when time parameter $t$ belongs to the set of natural numbers $N = \{0, 1, \ldots\}$, the family $X = \{\xi_t\}, t \in N$, is called random sequence or stochastic process with discrete time.*

Usually we will use the notation $\xi(t, \omega)$ for processes with continuous time and $\xi_n(\omega)$ for processes with discrete time, i.e., $\xi_n(\omega) = \xi(n, \omega)$. For fixed $\omega \in \Omega$ the function of time $\xi(t, .), t \in R^+$ or $t \in N$, is called *trajectory* or *sample path* corresponding to the elementary event $\omega$. Sometimes for convenience we will omit parameter $\omega$ in notation of stochastic processes, i.e., $\xi(t) = \xi(t, \omega), t \in R^+, \xi_n = \xi_n(\omega), n \in N$, which should not lead to misunderstanding.

In contrast to random variables denoted by capital letters, for instance, $X_1, X_2, Y_1, Y_2$, stochastic processes will be denoted by small letters, for instance, $x_1(t), x_2(t), y_1(t), y_2(t)$.

For fixed times $t = t_1, t_2, \ldots, t_n$ stochastic process $\xi(t)$ reduces to a finite number of random variables $\xi(t_1), \ldots, \xi(t_n)$; these are characterized by joint probability distribution

$$F_{t_1, \ldots, t_n}(x_1, \ldots, x_n) = P\{\xi(t_1) < x_1, \ldots, \xi(t_n) < x_n\}, \qquad (2.24)$$

or for continuous processes by joint probability density function $g(t_1, x_1, \ldots, t_n, x_n)$ or by joint characteristic function

$$\Phi(\Theta_1, t_1, \ldots, \Theta_n, t_n) = E\left[\exp\left\{\sum_{j=1}^{n} i\Theta_j \xi(t_j)\right\}\right]. \qquad (2.25)$$

The natural generalization of the joint characteristic function is the characteristic functional defined by

$$\Phi(\Theta(t)) = E\left[\exp\left\{i \int_{R^+} \Theta(t)\xi(t)dt\right\}\right], \qquad (2.26)$$

where the function $\Theta(t)$ belongs to the class of functions for which the integration under exponent is well defined.

The transformation from (2.26) to (2.25) can be done by substitution

$$\Theta(t) = \sum_j \Theta_j \delta(t - t_j), \qquad (2.27)$$

where $\delta(t)$ is the Dirac generalized function.

Like random variables, moments and cumulants for stochastic processes are derived by differentiation of the corresponding characteristic functionals

$$\Phi(\Theta(t)) = \Phi_1(\Theta(t)) = 1 + i \sum_{j=1}^{n} \Theta_j(t) E[x_j(t)]$$

$$+ \frac{i^2}{2!} \sum_{j=1}^{n} \sum_{k=1}^{n} \Theta_j(t) \Theta_k(t) E[x_j(t) x_k(t)] + \dots \qquad (2.28)$$

or their equivalent form

$$\Phi(\Theta(t)) = \Phi_2(\Theta(t)) = \exp \left\{ i \sum_{j=1}^{n} \Theta_j(t) \kappa[x_j(t)] \right.$$

$$\left. + \frac{i^2}{2!} \sum_{j=1}^{n} \sum_{k=1}^{n} \Theta_j(t) \Theta_k(t) \kappa[x_j(t) x_k(t)] + \dots \right\} . \qquad (2.29)$$

Then formulas (2.7–2.12) are also satisfied for stochastic processes, where the terms $X_j, X_k, X_l, \dots$ are replaced by stochastic processes $x_j(t), x_k(t), x_l(t), \dots$ or by one stochastic process for various times $t_j, t_k, t_l, \dots$, i.e., $X_j = x(t_j)$. Using the joint probability density function one can calculate the higher-order moments for continuous processes

$$E[x_1^{p_1}(t_1) \dots x_n^{p_n}(t_n)] = \int_{-\infty}^{+\infty} \dots \int_{-\infty}^{+\infty} [x_1^{p_1}(t_1) \dots x_n^{p_n}(t_n)]$$

$$g(x_1, t_1, \dots, x_n, t_n) dx_1 \dots dx_n . \qquad (2.30)$$

## 2.4 Second-Order Stochastic Processes

The particular important class of stochastic processes with respect to various applications are second-order stochastic processes, i.e., complex-valued processes with bounded second-order moments,

$$E[|x(t, \omega)|^2] < \infty , \quad t \in R^+ . \qquad (2.31)$$

The quantities characterizing second-order stochastic processes are *auto-correlation functions and autocovariance functions* defined, respectively, by

$$R_{xx}(t_1, t_2) = E[x(t_1)\overline{x(t_2)}] \qquad (2.32)$$

and

$$K_{xx}(t_1, t_2) = E[(x(t_1) - E[x(t_1)])\overline{(x(t_2) - E[x(t_2)])}] . \qquad (2.33)$$

They are sometimes shortly called *correlation function and covariance function* and denoted by $R_x(t_1, t_2)$ and $K_x(t_1, t_2)$ or $R(t_1, t_2)$ and $K(t_1, t_2)$, respectively; an upper line $\overline{(.)}$ denotes the complex conjugate.

For $t_1 = t_2 = t$

$$K_{xx}(t,t) = E[(x(t) - E[x(t)])^2] = \sigma_x^2(t) , \qquad (2.34)$$

where $\sigma_x(t)$ is the standard deviation of process $x(t)$.

For two different processes one can define *cross correlation function* and *cross covariance function*

$$R_{xy}(t_1, t_2) = E[x(t_1)\overline{y(t_2)}] , \qquad (2.35)$$

$$K_{xy}(t_1, t_2) = E[(x(t_1) - E[x(t_1)])\overline{(y(t_2) - E[y(t_2)])}] \qquad (2.36)$$

and for complex-valued vector process $\mathbf{x}(t)$ *the matrix correlation function* and *the matrix covariance function* are defined by

$$\mathbf{R_{xx}}(t_1, t_2) = E[\mathbf{x}(t_1)\mathbf{x}^*(t_2)] , \qquad (2.37)$$

$$\mathbf{K_{xx}}(t_1, t_2) = E[(\mathbf{x}(t_1) - E[\mathbf{x}(t_1)])(\mathbf{x}(t_2) - E[\mathbf{x}(t_2)])^*] , \qquad (2.38)$$

respectively, where an asterisk denotes the transposition and the complex conjugate.

Similarly, one can define *cross correlation matrix function* and *cross covariance matrix function*

$$\mathbf{R_{xy}}(t_1, t_2) = E[\mathbf{x}(t_1)\mathbf{y}^*(t_2)] , \qquad (2.39)$$

$$\mathbf{K_{xy}}(t_1, t_2) = E[(\mathbf{x}(t_1) - E[\mathbf{x}(t_1)])(\mathbf{y}(t_2) - E[\mathbf{y}(t_2)])^*] , \qquad (2.40)$$

Since the properties of correlation function and covariance function are similar and there is the strict relationship between them, we present only the properties of the correlation function

The properties of the correlation function are as follows:

- *Nonnegative definiteness.* Each correlation function is nonnegatively defined, i.e., for any finite sequence $\{t_n\}$, $t_1, \ldots, t_n \in R^+$ and any complex numbers $\alpha_1, \ldots, \alpha_n$ the following inequality is satisfied

$$\sum_{j=1}^{n}\sum_{k=1}^{n}\alpha_j\bar{\alpha}_k R(t_j, t_k) = E\left[|\sum_{j=1}^{n}\alpha_j x(t_j)|^2\right] \geq 0 . \qquad (2.41)$$

The inverse theorem also satisfied; if $R(t, s)$ is a nonnegative by defined function, then one can always find a second-order process $x(t)$, that correlation function $R_x(t, s)$ is equal to the given one.

- *Symmetry*

$$R(t_1, t_2) = \overline{R(t_2, t_1)} . \qquad (2.42)$$

- *Schwarz Inequality*

$$|R(t_1, t_2)| \leq \sqrt{R(t_1, t_1)R(t_2, t_2)} . \qquad (2.43)$$

- *Closure property.* Sum and product of two correlation functions are also correlation functions.
- If $\{R_i(t_1, t_2), i = 1, 2, \ldots\}$ is a sequence of correlation functions defined on $R^+ \times R^+$ converging to $R(t_1, t_2)$ in each point $(t_1, t_2) \in R^+ \times R^+$, then $R(t_1, t_2)$ is a correlation function.
- *Bilinear forms.* The limit point of each converging sequence of bilinear forms

$$\sum_{j=1}^{n} \sigma_j(t_1) \bar{\sigma}_j(t_2) \tag{2.44}$$

is a correlation function, where $\sigma_1, \sigma_2, \ldots$ is a sequence of functions.

For second-order process one can define the continuity, differentiation, and integration in mean-square sense.

**Definition 2.4.** *(continuity in mean square). Stochastic process $x(t)$ is continuous in mean-square sense if*

$$l.i.m.(x(t + \Delta t) - x(t)) = \lim_{\Delta t \to 0} E[(x(t + \Delta t) - x(t))^2] = 0 , \tag{2.45}$$

*where l.i.m. denotes the limit in mean-square sense.*

**Theorem 2.1.** *The necessary and sufficient condition for continuity in mean-square sense of the process $x(t)$ is the existence of correlation function $R(t_1, t_2)$ continuous on the set $\{(t_1, t_2) \in R^+ \times R^+ : t_1 = t_2\}$.*

**Definition 2.5.** *(differentiation in mean-square sense). Differential in mean-square sense of stochastic process $x(t)$ is defined by*

$$\frac{d}{dt} x(t) = \dot{x}(t) = l.i.m._{\Delta t \to 0} \frac{x(t + \Delta t) - x(t)}{\Delta t} . \tag{2.46}$$

**Theorem 2.2.** *The necessary and sufficient condition for differentiation in mean-square sense of the process $x(t)$ is the existence of second-order derivative of correlation function $\frac{\partial^2 R(t_1, t_2)}{\partial t_1 \partial t_2}$ bounded and continuous on the set $\{(t_1, t_2) \in R^+ \times R^+ : t_1 = t_2\}$.*

**Definition 2.6.** *(integration). Let $f(t)$ be a complex function defined over an interval $[a, b]$, and let $\{T_n\}$ be a collection of all finite partitions of the interval $[a, b]$, i.e.,*

$$T_n = \{a = t_0^{(n)} < t_1^{(n)} < \ldots < t_n^{(n)} = b\} ,$$

$$\lim_{n \to +\infty} \max_{1 \leq i \leq n} (t_i^{(n)} - t_{i-1}^{(n)}) = 0 . \tag{2.47}$$

*Integral in mean-square sense* over the interval $[a, b]$ is the limit of Riemann's sums

$$\int_a^b f(t)x(t)dt = l.i.m._{n \to +\infty} \sum_{i=0}^{n-1} f(t'_{in})x(t'_{in})(t_{i+1}^{(n)} - t_i^{(n)}) , \qquad (2.48)$$

where $t'_{in}$ is any sequence satisfying the following inequalities

$$t_i^{(n)} \le t'_{in} \le t_{i+1}^{(n)} . \qquad (2.49)$$

**Theorem 2.3.** *Integral in mean-square sense $\int_a^b f(t)x(t)dt$ exists if and only if the ordinary double Riemann integral*

$$\int_a^b \int_a^b f(t_1)\overline{f(t_2)}R(t_1, t_2)dt_1 dt_2 \qquad (2.50)$$

*exists and is finite.*

Averaging operation is commutative with differentiation and integration in mean-square sense, i.e.,

$$\frac{dE[x(t)]}{dt} = E\left[\frac{dx(t)}{dt}\right] , \qquad (2.51)$$

$$E\left[\int_a^b f(t)x(t)dt\right] = \int_a^b f(t)E[x(t)]dt . \qquad (2.52)$$

This property is valid for any linear operation $L$, i.e., if $L_t$ is a linear operator transforming of a second-order stochastic process $x(t)$ into another second-order stochastic process $y(t)$, i.e.,

$$y(t) = L_t[x(t)] , \qquad (2.53)$$

then

$$E[y(t)] = L_t[E[x(t)]] , \qquad (2.54)$$

and the correlation function of process $y(t)$ is determined by

$$R_{yy}(t_1, t_2) = L_{t_1} L_{t_2} R_{xx}(t_1, t_2) . \qquad (2.55)$$

In the particular case when $L_t = L$,

$$R_{yy}(t_1, t_2) = L^2 R_{xx}(t_1, t_2) . \qquad (2.56)$$

In the analysis of second-order stochastic processes one can use the spectral representation of process $x(t)$, i.e., integral Fourier transform defined by

$$\hat{x}(\lambda) = \frac{1}{2\pi} \int_{-\infty}^{+\infty} x(t)e^{-i\lambda t}dt , \qquad (2.57)$$

where $\lambda \in R$; integral (2.57) and the next integrals should be treated in the sense of main value.

The inverse transform has the form

$$x(t) = \int_{-\infty}^{+\infty} \hat{x}(\lambda)e^{i\lambda t}d\lambda . \qquad (2.58)$$

We assume that the autocorrelation function $R_{xx}(t_1, t_2)$ of the stochastic process $x(t)$ is a piecewise continuous function and bounded function,

$$\int_{-\infty}^{+\infty} \int_{-\infty}^{+\infty} |R_{xx}(t_1, t_2)| dt_1 dt_2 < \infty . \qquad (2.59)$$

Then the necessary and sufficient condition for the existence of spectral representation $x(t)$ is the existence of the generalized power spectral density function of the nonstationary process $x(t)$:

$$S_{xx}(\lambda_1, \lambda_2) = \frac{1}{4\pi^2} \int_{-\infty}^{+\infty} \int_{-\infty}^{+\infty} e^{-i(\lambda_1 t_1 + \lambda_2 t_2)} R_{xx}(t_1, t_2) dt_1 dt_2 . \qquad (2.60)$$

The inverse transformation has the form

$$R_{xx}(t_1, t_2) = \int_{-\infty}^{+\infty} \int_{-\infty}^{+\infty} e^{i(\lambda_1 t_1 + \lambda_2 t_2)} S_{xx}(\lambda_1, \lambda_2) d\lambda_1 d\lambda_2 . \qquad (2.61)$$

Relations (2.60) and (2.61) are called *generalized Wiener–Chinczyn formula* [36].

Another representation of the mean-square continuous process $x(t)$ is its orthogonal expansion.

Let $K_x$ denote a family of all random variables obtained as linear transformations of a given continuous in mean-square stochastic process $\{x(t), t \in R^+\}$. The scalar product is defined by

$$< x, y >= E[xy] . \qquad (2.62)$$

Let us assume that $\{z_n, n = 1, 2, \ldots\}$ is a given complete orthonormal basis in $K_x$ and let $\sigma_n(t)$ be the coordinate of process $x(t)$ in this basis, i.e.,

$$\sigma_n(t) =< x(t), z_n > , \qquad (2.63)$$

then from general Karhunen theorem of representation [55] it follows that the following expansions

$$x(t, \omega) = \sum_{n=1}^{\infty} \sigma_n(t) z_n(\omega) , \qquad (2.64)$$

$$R_{xx}(t_1, t_2) = \sum_{n=1}^{\infty} \sigma_n(t_1) \bar{\sigma}_n(t_2) \qquad (2.65)$$

are equivalent. One can show [55] under additional assumptions that both series are convergent in mean-square sense and uniformly over any finite interval $[a, b]$, respectively. If we consider a finite sum in expansion (2.64) then we obtain an approximation of a real continuous in mean-square nonstationary process $x(t)$.

A wide class of second-order stochastic processes frequently used in applications are the so-called weakly stationary processes.

**Definition 2.7.** *Stochastic process* $\xi(t)$, $t \in R^+$, *is called weakly stationary process or stationary in a wide sense if for any real number* $\Delta \in R^+$ *and any* $t, s \in R^+$ *the following relations are satisfied:*

$$E[|\xi(t)|^2] < \infty \,,$$
$$E[\xi(t)] = E[\xi(t + \Delta, \omega)] \,,$$
$$E[\xi(t + \Delta)\overline{\xi(s + \Delta)}] = E[\xi(t)\overline{\xi(s)}] \,, \qquad (2.66)$$

*i.e., the first- and second-order moments are invariant with respect to variable* $t$. *For simplicity, we will omit the word "weakly" in further considerations.*

The direct corollary following from this definition is the fact that the mean value and the variance of stationary process are constants with respect to $t$ and correlation and covariance functions depend on one variable (difference of arguments) $t_2 - t_1$, i.e.,

$$E[\xi(t)] = m_\xi = \text{const} \,, \qquad (2.67)$$
$$E\left[(\xi(t) - E[\xi(t)])^2\right] = \sigma_\xi^2 = \text{const} \,, \qquad (2.68)$$
$$R_\xi(t_1, t_2) = R_\xi(t_2 - t_1) = R_\xi(\tau) \,, \qquad (2.69)$$
$$K_\xi(t_1, t_2) = K_\xi(t_2 - t_1) = K_\xi(\tau) \,, \qquad (2.70)$$

where $t_1 = t$, $t_2 = t + \tau$.

Relations (2.41–2.43) for stationary processes are reduced to the following ones

$$\sum_{j=1}^{n} \sum_{k=1}^{n} \alpha_j \bar{\alpha}_k R(t_j - t_k) \geq 0 \,, \qquad (2.71)$$

$$R(\tau) = \overline{R(-\tau)} \,, \qquad (2.72)$$

$$|R_{xy}(\tau)| \leq \sqrt{R_{xx}(0) R_{yy}(0)} \,, \qquad (2.73)$$

where $R(\tau)$ is a nonnegative definite function.

**Theorem 2.4.** *The necessary and sufficient condition for mean-square differentiation of a second-order stationary process* $x(t)$ *is the existence of the first- and second-order derivatives of* $R(\tau)$ *continuous and finite at* $\tau = 0$.

The following relations are also satisfied:

$$-\frac{\partial}{\partial s}R_{xx}(t-s) = R_{\dot{x}x}(t-s) \quad or \quad -\frac{d}{d\tau}R_{xx}(\tau) = R_{\dot{x}x}(\tau) , \quad (2.74)$$

$$\frac{\partial}{\partial t}R_{xx}(t-s) = R_{\dot{x}x}(t-s) \quad or \quad \frac{d}{d\tau}R_{xx}(\tau) = R_{\dot{x}x}(\tau) , \quad (2.75)$$

$$\frac{\partial^2}{\partial s\partial t}R_{xx}(t-s) = -R_{\dot{x}\dot{x}}(t-s) \quad or \quad \frac{d^2}{d\tau^2}R_{xx}(\tau) = R_{\dot{x}\dot{x}}(\tau) , \quad (2.76)$$

$$\frac{\partial^{p+r}}{\partial s^p\partial t^r}R_{xx}(t-s) = (-1)^p R_{x^{(p)}x^{(r)}}(t-s) \quad or \quad (-1)^p\frac{d^{p+r}}{d\tau^{p+r}}R_{xx}(\tau)$$
$$= R_{x^{(p)}x^{(r)}}(\tau) . \quad (2.77)$$

These relations can be applied to the determination of correlation function of the mean-square derivative of a stationary process using only the correlation function of the considered process (the knowledge of the probability distribution is not required).

If for a second-order stationary process $x(t)$ the condition $\frac{d}{d\tau}R_{xx}(\tau)|_{\tau=0}=0$ is satisfied, then

$$\dot{R}_{xx}(0) = R_{\dot{x}x}(0) = E[\dot{x}(t)x(t)] = 0 , \quad (2.78)$$

i.e., a second-order stationary process and its mean-square derivative are uncorrelated.

The power spectral density function $S_{xx}(\lambda)$ for a stationary process $x(t)$ is defined as the Fourier transform of the corresponding autocorrelation function $R_{xx}(\tau)$

$$S_{xx}(\lambda) = \frac{1}{2\pi}\int_{-\infty}^{+\infty} R_{xx}(\tau)e^{-i\lambda\tau}d\tau . \quad (2.79)$$

The inverse transform has the form

$$R_{xx}(\tau) = \int_{-\infty}^{+\infty} S_{xx}(\lambda)e^{i\lambda\tau}d\lambda . \quad (2.80)$$

In a particular case for a stationary zero-mean process, we find

$$\sigma_x^2 = R_{xx}(0) = \int_{-\infty}^{+\infty} S_{xx}(\lambda)d\lambda . \quad (2.81)$$

From properties of Fourier transform it follows that

$$S_{\dot{x}\dot{x}}(\lambda) = \lambda^2 S_{xx}(\lambda) . \quad (2.82)$$

It can be shown that condition (2.82) holds if and only if

$$\int_{-\infty}^{+\infty} \lambda^2 S_{xx}(\lambda)d\lambda < \infty . \quad (2.83)$$

Frequently in applications the following property of zero-mean stationary process $x(t)$ is used

$$S_{xx}(0) = \frac{1}{2\pi} \int_{-\infty}^{+\infty} R_{xx}(\tau)d\tau \qquad (2.84)$$

or

$$\frac{1}{4\pi^2} \int_{-\infty}^{+\infty} \int_{-\infty}^{+\infty} R_{xx}(t_2 - t_1)dt_1 dt_2 = S_{xx}(0)t + \Theta(t) , \qquad (2.85)$$

where $|\Theta(t)| < At^\varepsilon$, $A > 0$ and $0 \le \varepsilon < 1$.

## 2.5 Gaussian and Non-Gaussian Processes

A very important class of stochastic processes because of their applicability in the forming of mathematical models of real physical processes are Gaussian processes (or normal processes). There are a few definitions of Gaussian process in the literature. We use the following one [46].

**Definition 2.8.** *A vector stochastic process* $\mathbf{x}(t)$, $\mathbf{x} \in R^r$, $t \in R^+$ *is called Gaussian (or normal) process if for any integer* $n \in \mathbf{N}$ *and any subset* $\{t_1, \ldots, t_n\}, t_i \in R^+$, $n \ge 1$ *vector random variables* $\mathbf{x}(t_1), \ldots, \mathbf{x}(t_n)$ *have the joint Gaussian distribution, i.e., their characteristic function for any real vectors* $\mathbf{\Theta}_1, \ldots, \mathbf{\Theta}_n$ *is given by*

$$\Phi(\mathbf{\Theta}_1, t_1, \ldots, \mathbf{\Theta}_n, t_n) = E\left[\exp\left\{\sum_{j=1}^n i\mathbf{\Theta}_j^T \mathbf{x}(t_j)\right\}\right]$$

$$= \exp\left\{\sum_{j=1}^n i\mathbf{\Theta}_j^T \mathbf{m}(t_j) - \frac{1}{2}\sum_{j=1}^n \sum_{k=1}^n i\mathbf{\Theta}_j^T \mathbf{K}(t_j, t_k)\mathbf{\Theta}_k\right\} ,$$

$$(2.86)$$

*where* $\mathbf{m}(t)$ *and* $\mathbf{K}(t_1, t_2)$ *are the mean value and the covariance matrix of the vector process* $\mathbf{x}(t)$, $t \in R^+$, *respectively;*

$$\mathbf{\Theta} = [\mathbf{\Theta}_1^T, \ldots, \mathbf{\Theta}_n^T]^T, \quad \mathbf{m} = [\mathbf{m}_1^T, \ldots, \mathbf{m}_n^T]^T.$$

If the covariance matrix $\mathbf{K}(t_i, t_j)$, $i, j = 1, \ldots, n$, is not singular, then the joint probability density function of the vector random variables $\mathbf{x}(t_1), \ldots,$ $\mathbf{x}(t_n)$ has the form

$$g_G(\mathbf{x}_1, t_1, \ldots, \mathbf{x}_n, t_n) = [(2\pi)^{n^2}|\mathbf{K}|]^{-1/2} \exp\left\{-\frac{1}{2}(\mathbf{u} - \mathbf{m_x})^T \mathbf{K}^{-1}(\mathbf{u} - \mathbf{m_x})\right\} , \qquad (2.87)$$

where $\mathbf{u} = [\mathbf{x}_1^T, \ldots, \mathbf{x}_n^T]^T$, $\mathbf{m_x} = [\mathbf{m}_1^T, \ldots, \mathbf{m}_n^T]^T$, $|\mathbf{K}|$ is the determinant of $n^2 \times n^2$-dimensional block covariance matrix. It has the form $\mathbf{K} = [\mathbf{K}(t_i, t_j)]$,

$i, j = 1, \ldots, n$. In the particular case when the matrix elements are one dimensional $[\mathbf{K}(t_i, t_j)] = K(t_i, t_j)$, the covariance matrix $\mathbf{K}$ is $n \times n$ dimensional,

$$\mathbf{K} = \begin{bmatrix} K(t_1, t_1) & K(t_1, t_2) & \ldots & K(t_1, t_n) \\ K(t_2, t_1) & K(t_2, t_2) & \ldots & K(t_2, t_n) \\ \vdots & \vdots & \ddots & \vdots \\ K(t_n, t_1) & K(t_n, t_2) & \ldots & K(t_n, t_n) \end{bmatrix} . \tag{2.88}$$

The important place of the Gaussian process between second-order processes follows from the following theorem [46].

**Theorem 2.5.** *Let $T = (-\infty, +\infty)$, $m(t)$ and $K(s, t)$ be arbitrary complex-valued functions defined over $T$ and $T \times T$, respectively, such that*

*(a) $K(t, s) = \bar{K}(s, t)$*
*(b) if $\{t_1, \ldots, t_n\}$ is an arbitrary finite set of values of $T$, then the covariance matrix $K(t_i, t_j)$ is nonnegative definite.*

Then there exists a Gaussian stochastic process $x(t)$ for which

$$E[x(t)] = m(t) , \tag{2.89}$$

$$E[(x(s) - m(s)) \; \overline{(x(t) - m(t))}] = K(s, t) . \tag{2.90}$$

**Corollary 2.1.** *For an arbitrary stochastic process with finite moments of the first and second order there exists a Gaussian process $x(t)$ having the same mean value and covariance function.*

Here we quote some basic properties of a scalar Gaussian stochastic process [46]:

- A Gaussian process is entirely determined by its mean value $m_x(t)$ and correlation function $R_x(t_1, t_2)$.
- The necessary condition for the continuity of the realization of a real Gaussian process is the continuity of its mean and correlation function.
- If a Gaussian process $\xi(t)$, $t \in [a, b]$, is separable, $E[\xi(t)] = 0$ and if there exist such constants as $c > 0$, $\varepsilon > 0$ so that

$$|K_\xi(t_1, t_1) + K_\xi(t_2, t_2) - 2K_\xi(t_1, t_2)| \le c|t_2 - t_1|^{1+\varepsilon} , \tag{2.91}$$

then almost all realizations of process $\xi(t)$ are continuous over the interval $[a, b]$.
- All linear transformations of Gaussian processes yield Gaussian processes.
- A Gaussian process has only two first nonzero cumulants.

- If $x(t)$ is a zero-mean Gaussian process, then all its moments of odd order are equal to zero (for each finite $n \in \mathbf{N}$)

$$E[x(t_1)x(t_2)\ldots x(t_{2n+1})] = 0 , \quad n = 1, 2, \ldots , \tag{2.92}$$

and all even-order moments satisfy the following formula:

$$E[x(t_1)x(t_2)\ldots x(t_{2n})]$$
$$= \sum E[x(t_1)x(t_2)], \ldots , E[x(t_{2n-1})x(t_{2n})] , \quad n = 1, 2, \ldots , \tag{2.93}$$

where summation is over all possible ways of dividing $2n$ points into $n$ combinations of pairs $(i_1, i_2)$, $(i_3, i_4)$, $\ldots$, $(i_{2n-1}, i_{2n})$. The number of terms in summation is:

$$\frac{(2n)!}{2^n n!} = 1 \cdot 3 \cdot 5 \cdot \ldots \cdot (2n-1) = (2n-1)!!.$$

For example,

$$E[x^{2n}(t)] = (2n-1)!!(E[x^2(t)])^n . \tag{2.94}$$

- If $x(t)$ is a Gaussian process having the continuous covariance function $K(t_1, t_2)$ over a square $[a, b] \times [a, b] \subset R^2$, then the following relation holds

$$E\left[\exp\left\{\int_a^b x(t)dt\right\}\right] = \exp\left\{\frac{1}{2}\int_a^b \int_a^b K(t_1, t_2)dt_1 dt_2 + \int_a^b E[x(t)]dt\right\} . \tag{2.95}$$

- If $x = x(t)$, $y = y(t)$, $z = z(t)$ are zero-mean Gaussian processes and $f(x)$ is a nonlinear function, then

$$E[yf(x)] = \frac{E[xy]}{E[x^2]} E[xf(x)] , \tag{2.96}$$

$$E[yf(x)] = \frac{1}{E[x^2]} \left(E[x^2]E[yz] - E[xz]E[xy]\right) E[f(x)]$$
$$+ \frac{E[xy]E[xz]}{(E[x^2])^2} E[x^2 f(x)] . \tag{2.97}$$

In the case of vector continuous non-Gaussian processes it is possible to describe the joint probability density function of $r$-dimensional random process in a form similar to (2.87) in the form of Gram–Charlier series, i.e.,

$$g_N(\mathbf{x_1}, t_1, \ldots, \mathbf{x_n}, t_n) = g_G(\mathbf{x_1}, t_1, \ldots, \mathbf{x_n}, t_n)$$
$$\times \left[1 + \sum_{k=3}^{\infty} \sum_{\sigma(\nu)=k} \frac{c_{\nu_1, \ldots, \nu_n}(t_1, \ldots, t_n)}{\nu_{11}! \ldots \nu_{nr}!}\right.$$
$$\left. H_{\nu_1, \ldots, \nu_n}((\mathbf{x_1} - \mathbf{m_{x_1}}(t_1)), \ldots, (\mathbf{x_n} - \mathbf{m_{x_n}}(t_n)))\right], \tag{2.98}$$

where

$$g_G(\mathbf{x}_1, t_1, \ldots, \mathbf{x}_n, t_n) = [(2\pi)^{nr}|\mathbf{K}|]^{-1/2} \exp\left\{-\frac{1}{2}(\mathbf{u} - \mathbf{m_x})^T \mathbf{K}^{-1}(\mathbf{u} - \mathbf{m_x})\right\},$$

$$(2.99)$$

$$\mathbf{u}^T = [\mathbf{x_1}^T, \ldots, \mathbf{x_n}^T], \quad \mathbf{m_X}^T = [\mathbf{m}_{\mathbf{X_1}}^T(t_1), \ldots, \mathbf{m}_{\mathbf{X_n}}^T(t_n)],$$

$$c_{\nu_1, \ldots, \nu_n}(t_1, \ldots, t_n)$$

$$= \int_{-\infty}^{+\infty} \cdots \int_{-\infty}^{+\infty} g_G(\mathbf{x}_1, t_1, \ldots, \mathbf{x}_n, t_n)$$

$$G_{\nu_1, \ldots, \nu_n}((\mathbf{x_1} - \mathbf{m}_{\mathbf{x_1}}(t_1)), \ldots, (\mathbf{x_n} - \mathbf{m}_{\mathbf{x_n}}(t_n)))d\mathbf{x}_1, \ldots, d\mathbf{x}_n$$

$$= G_{\nu_1, \ldots, \nu_n}\left(\frac{\partial}{i\partial\Theta_1}, \ldots, \frac{\partial}{i\partial\Theta_n}\right)\Phi(\boldsymbol{\Theta})\Big|_{\boldsymbol{\Theta}=0} \qquad (2.100)$$

and

$$\nu_1 = [\nu_{11}, \ldots, \nu_{1r}], \ldots, \nu_n = [\nu_{n1}, \ldots, \nu_{nr}],$$

$$\sigma(\nu) = \nu_{11} + \ldots + \nu_{1r} + \ldots + \nu_{n1} + \ldots + \nu_{nr}.$$

The coefficients $c_{\nu_1, \ldots, \nu_n}(t_1, \ldots, t_n)$ have the following properties:

$$c_{\nu_1, \ldots, \nu_{n-1}, 0}(t_1, \ldots, t_n) = c_{\nu_1, \ldots, \nu_{n-1}}(t_1, \ldots, t_{n-1}),$$

$$c_{\nu_1, \ldots, \nu_{n-1}, \nu_n}(t_1, \ldots, t_{n-1}, t_{n-1}) = c_{\nu_1, \ldots, \nu_{n-1}+\nu_n}(t_1, \ldots, t_{n-1}),$$

for all $\nu_i$, $i = 1, \ldots, n$.

To obtain an approximate probability density function of a non-Gaussian process the first sum can be limited to $k = N$. Then a normalized constant must be introduced. This approximation can be used in both cases: when the process $x(t)$ has moments of higher order than $N$ or not. The expansion of a function with respect to an orthogonal basis has the property that the expansion of exact and approximate probability density functions gives the same coefficients of expansion $c_{\nu_1, \ldots, \nu_n}(t_1, \ldots, t_n)$. However, there are some problems with direct truncation of the approximation series. It may often appear meaningless negative values at the probability density function. To overcome this drawback one can approximate $g(\mathbf{x})$ by $|g(\mathbf{x})|$ and calculate the corresponding new normalized constant or can apply the idea of functional transformation of the Gaussian process proposed by Winterstein [54]. This method employs Hermite polynomials not to approximation of the probability density function but to the time history itself. This approach was used by Mohr and Ditlevsen also [39].

## 2.6 Markov Processes

We now discuss a wide class of stochastic processes having the property that the "future" does not depend on the "past" when the "present" is known. Such processes are defined as follows:

**Definition 2.9.** *A vector stochastic process* $\boldsymbol{\xi}(t)$, $t \in R^+$, *of r-dimension is called Markov process if for numbers* $n \in \mathbf{N}$ *and for any values of the parameter* $t_m \in R^+$, $m = 1, \ldots, n$, *where* $t_0 < t_1 < \ldots < t_n$ *and for any real vectors* $\mathbf{x}_1, \ldots, \mathbf{x}_n \in R^r$ *the following relation is satisfied:*

$$P\{\boldsymbol{\xi}(t_n) < \mathbf{x}_n | \boldsymbol{\xi}(t_{n-1}) = \mathbf{x}_{n-1}, \ldots, \boldsymbol{\xi}(t_1) = \mathbf{x}_1\}$$
$$= P\{\boldsymbol{\xi}(t_n) < \mathbf{x}_n | \boldsymbol{\xi}(t_{n-1}) = \mathbf{x}_{n-1}\}, \quad (2.101)$$

*i.e., conditional distribution* $\boldsymbol{\xi}(t_n)$ *for given values of* $\boldsymbol{\xi}(t_0)$, $\boldsymbol{\xi}(t_1), \ldots, \boldsymbol{\xi}(t_{n-1})$ *depends only on the most recent values of stochastic process* $\boldsymbol{\xi}(t_{n-1})$ *and does not depend on values* $\boldsymbol{\xi}(t_0)$, $\boldsymbol{\xi}(t_1), \ldots, \boldsymbol{\xi}(t_{n-2})$.

We introduce some notations

$$\mathcal{P}(s, \mathbf{x}; t, \mathbf{B}) = P\{\boldsymbol{\xi}(t) \in \mathbf{B} | \boldsymbol{\xi}(s) = \mathbf{x}\}, s \le t, \quad (2.102)$$

$$F(s, \mathbf{x}; t, \mathbf{y}) = P\{\boldsymbol{\xi}(t) < \mathbf{y} | \boldsymbol{\xi}(s) = \mathbf{x}\}, \quad (2.103)$$

where $\mathbf{B} \in B^r$, $B^r$ is $\sigma$-field of Borel sets in $R^r$.

The function $\mathcal{P}(s, x; t, B)$ called *transition probability function* or shortly *transition function* associated with Markov process $\boldsymbol{\xi}(t)$ has the following properties:

- For fixed $s, t, x$ the function $\mathcal{P}(s, \mathbf{x}; t, \cdot)$ is a probability measure on $R^r$; $\mathcal{P}(s, \mathbf{x}; t, R^r) = 1$.
- For fixed $s, t, \mathbf{B} \in B^r$ the function $\mathcal{P}(s, \cdot; t, B)$ is $B^r$-measurable.

-
$$\mathcal{P}(s, \mathbf{x}; t, B) = \begin{cases} 1 \ for \ \mathbf{y} \in \mathbf{B} \\ 0 \ for \ \mathbf{y} \notin \mathbf{B} \end{cases}.$$

- for $s < u < t$ such that $\boldsymbol{\xi}(u) = \mathbf{v}$ the following equality holds:

$$\mathcal{P}(s, \mathbf{x}; t, \mathbf{B}) = \int_{R^r} \mathcal{P}(u, \mathbf{v}; t, \mathbf{B}) d\mathcal{P}(s, \mathbf{x}; u, \mathbf{v}). \quad (2.104)$$

It is called *Chapman–Kolmogorov equation.*

If the function $\mathcal{P}(s, \mathbf{x}; t, \mathbf{B})$ has a density $g(s, \mathbf{x}; t, \mathbf{y})$ such that

$$\mathcal{P}(s, \mathbf{x}; t, \mathbf{B}) = \int_{R^r} g(u, \mathbf{x}; t, \mathbf{y}) d\mathbf{y}, \quad (2.105)$$

then (2.104) can be written in the form

$$g(s, \mathbf{x}; t, \mathbf{B}) = \int_{R^r} g(s, \mathbf{x}; u, \mathbf{v}) g(u, \mathbf{v}; t, \mathbf{y}) d\mathbf{v}. \quad (2.106)$$

In the analysis of Markov process we often use the property of homogeneity.

**Definition 2.10.** *A Markov process $\boldsymbol{\xi}(t), t \in R^+$, is called homogeneous (with respect to time) if for any $s, t \in R^+$, $s < t$, the transition probability depends only on the difference $t - s = \tau$, i.e.,*

$$\mathcal{P}(s, \mathbf{x}; t, \mathbf{B}) = \mathcal{P}(\mathbf{x}, \tau, \mathbf{B}) \,, \tag{2.107}$$

$$F(s, \mathbf{x}; t, \mathbf{y}) = F(\mathbf{x}, \tau, \mathbf{y}) \,. \tag{2.108}$$

**Remark.** A Markov process with the stationary transition probability is not a stationary stochastic process. An example of such process is a Wiener process that will be defined in Sect. 2.8.

## 2.7 Diffusion Processes and Kolmogorov Equations

Among Markov processes an important class are processes with continuous parameter $t$ and with continuous state space.

**Definition 2.11.** *A Markov vector process $\boldsymbol{\xi}(t), t \in R^+$, with values in $R^r$ is called r-dimensional diffusion process if its transition probability $F(s, \mathbf{x}; t, \mathbf{y})$ for any $t \in R^+$, $\mathbf{x}, \mathbf{y} \in R^r$ and any $\varepsilon > 0$ satisfies the following conditions*

*(i)*

$$\lim_{\Delta t \to 0} \frac{1}{\Delta t} \int_{|\mathbf{y} - \mathbf{x}| \geq \varepsilon} d_{\mathbf{y}} F(t, \mathbf{x}; t + \Delta t, \mathbf{y}) = 0 \,. \tag{2.109}$$

*(ii) There exists a vector function $\mathbf{A}(\mathbf{x}, t)$, such that*

$$\lim_{\Delta t \to 0} \frac{1}{\Delta t} \int_{|\mathbf{y} - \mathbf{x}| < \varepsilon} (\mathbf{y} - \mathbf{x}) d_{\mathbf{y}} F(t, \mathbf{x}; t + \Delta t, \mathbf{y}) = \mathbf{A}(\mathbf{x}, t) \,. \tag{2.110}$$

*(iii) There exists a vector function $\boldsymbol{\sigma}(\mathbf{x}, t)$, such that*

$$\lim_{\Delta t \to 0} \frac{1}{\Delta t} \int_{|\mathbf{y} - \mathbf{x}| < \varepsilon} (\mathbf{y} - \mathbf{x})(\mathbf{y} - \mathbf{x})^T d_{\mathbf{y}} F(t, \mathbf{x}; t + \Delta t, \mathbf{y})$$
$$= \boldsymbol{\sigma}(\mathbf{x}, t) \boldsymbol{\sigma}^T(\mathbf{x}, t) = \mathbf{B}(\mathbf{x}, t) > 0 \,, \tag{2.111}$$

*where $|.|$ is the Euclidean norm in $R^r$ and the convergence in conditions (2.110) and (2.111) is uniform with respect to $\mathbf{x}$. $d_{\mathbf{y}} F(t, \mathbf{x}; t + \Delta t, \mathbf{y})$ is the derivative of $F$ with respect to $\mathbf{y}$. The functions $\mathbf{A}(\mathbf{x}, t)$ and $\mathbf{B}(\mathbf{x}, t)$ are called drift vector and diffusion matrix, respectively.*

For a diffusion process there is a possibility of the determination of the density of transition probability on the basis of $\mathbf{A}(\mathbf{x}, t)$ and $\mathbf{B}(\mathbf{x}, t)$. It follows from the following theorem:

**Theorem 2.6.** *[46] Let $\boldsymbol{\xi}(t) = [\xi_1(t), \ldots, \xi_r(t)]^T$ be a r-dimensional diffusion process taking the values $\mathbf{x} = [x_1, \ldots, x_r]^T$. If the density corresponding to the transition of probability $F(s, \mathbf{x}; t, \mathbf{y})$ for every $s, \mathbf{x}, \mathbf{y}$, and $t > s$ is determined by*

$$g(s, \mathbf{x}; t, \mathbf{y}) = g(s, x_1, \ldots, x_r, t, y_1, \ldots, y_r) = \frac{\partial^r F(s, x_1, \ldots, x_r, t, y_1, \ldots, y_r)}{\partial y_1 \ldots \partial y_r}$$

*and there exist continuous derivatives,*

$$\frac{\partial g(s, \mathbf{x}; t, \mathbf{y})}{\partial t}, \quad \frac{\partial [A_i(\mathbf{y}, t) g(s, \mathbf{x}; t, \mathbf{y})]}{\partial y_i}, \quad \frac{\partial^2 [b_{ij}(\mathbf{y}, t) g(s, \mathbf{x}; t, \mathbf{y})]}{\partial y_i \partial y_j}, \quad (2.112)$$

*then $g(s, \mathbf{x}; t, \mathbf{y})$ as a function of t and $\mathbf{y}$ satisfies the equation*

$$\frac{\partial g(s, \mathbf{x}; t, \mathbf{y})}{\partial t} + \sum_{i=1}^{r} \frac{\partial [A_i(\mathbf{y}, t) g(s, \mathbf{x}; t, \mathbf{y})]}{\partial y_i}$$

$$- \frac{1}{2} \sum_{i=1}^{r} \sum_{j=1}^{r} \frac{\partial^2 [b_{ij}(\mathbf{y}, t) g(s, \mathbf{x}; t, \mathbf{y})]}{\partial y_i \partial y_j} = 0, \quad (2.113)$$

*where $A_i(\mathbf{x}, t)$ and $b_{ij}(\mathbf{x}, t)$ are elements of drift vector $\mathbf{A}(\mathbf{x}, t)$ and diffusion matrix $\mathbf{B}(\mathbf{x}, t)$, respectively.*

*$A_i(\mathbf{x}, t)$ and $b_{ij}(\mathbf{x}, t)$ can be also determined from the following relations:*

$$\lim_{\Delta t \to 0} \frac{1}{\Delta t} E\left[\{\xi_i(t + \Delta t) - \xi_i(t) | \boldsymbol{\xi}(t) = \mathbf{x}\}\right] = A_i(\mathbf{x}, t), \quad (2.114)$$

$$\lim_{\Delta t \to 0} \frac{1}{\Delta t} E[\{[\xi_i(t + \Delta t) - \xi_i(t)][\xi_j(t + \Delta t) - \xi_j(t)] | \boldsymbol{\xi}(t) = \mathbf{x}\}] = b_{ij}(\mathbf{x}, t). \quad (2.115)$$

Since the function $g(s, \mathbf{x}; t, \mathbf{y})$ appearing in (2.113) is an unknown probability density function, to solve the partial differential equation (2.113) one should take into account the bounds of the probabilistic nature, i.e.,

$$g(s, \mathbf{x}; t, \mathbf{y}) \geq 0, \quad \int_{R^r} g(s, \mathbf{x}; t, \mathbf{y}) d\mathbf{y} = 1,$$

$$\lim_{\Delta t \to 0} \frac{1}{\Delta t} \int_{|\mathbf{y} - \mathbf{x}| \geq \varepsilon} g(t, \mathbf{x}; t + \Delta t, \mathbf{y}) d\mathbf{y} = 0. \quad (2.116)$$

If the initial condition in (2.113) is random, then the corresponding probability density function is assumed in the form $g(t_0, \mathbf{x_0})$; if it is assumed to be a deterministic one, i.e., for $s = t_0$, $\mathbf{y} = \mathbf{x_0}$, then

$$g(t_0, \mathbf{x}_0; t, \mathbf{y}) = \prod_{j=1}^{r} \delta(y_j - x_{0j}) . \tag{2.117}$$

Equation (2.113) is known in the literature as *forward Kolmogorov equation* or *Fokker–Planck–Kolmogorov equation* (FPK. equation). Since it plays very important role in theory and applications we will discuss the methods of its solution in Sect. 2.13.

We also note that the probability density function $g(s, \mathbf{x}; t, \mathbf{y})$ because of its conditional property is often denoted in the literature as follows:

$$g(s, \mathbf{x}; t, \mathbf{y}) = g(\mathbf{y}, t | \mathbf{x}, s) . \tag{2.118}$$

## 2.8 Wiener Process

A particular important class of Markov processes are processes with independent increments defined by the following:

**Definition 2.12.** *A stochastic process $\xi(t)$, $t \in R^+$, is called process with independent increments if for any $t_i \in R^+$, $t_0 < t_1 < \ldots < t_n$, random variables being increments of process $\xi(t)$, i.e., $\xi(t_0)$, $\xi(t_1) - \xi(t_0)$, $\ldots$, $\xi(t_n) - \xi(t_{n-1})$, are independent.*

**Definition 2.13.** *A stochastic process $\xi(t)$, $t \in R^+$, with independent increments is called a process with stationary independent increments if its increments $\xi(t_1) - \xi(t_0)$, $\ldots$, $\xi(t_n) - \xi(t_{n-1})$ depend only on differences $t_1 - t_0, \ldots, t_n - t_{n-1}$, respectively.*

**Definition 2.14.** *A stochastic process $\xi(t)$, $t \in R^+$, is called martingale with respect to $\sigma$-field $\{\Im_t\}$, $t \in R^+$, if $E[|\xi(t)|] < \infty$, $t \in R^+$, and*

$$E[\xi(t)|\Im_s] = \xi(s) , \quad t \geq s . \tag{2.119}$$

The stochastic process with stationary independent increments called *Brownian motion* or *Wiener process* plays the fundamental role in stochastic analysis.

**Definition 2.15.** *A stochastic process $\xi(t)$, $t \in R^+$, is called Wiener process or Brownian motion, if*

*(i)* $P\{\xi(0) = 0\} = 1$.
*(ii)* $\xi(t)$ *is a process with independent increments.*
*(iii) The increments $\xi(t) - \xi(s)$ have a Gaussian distribution, such that*

$$E[\xi(t) - \xi(s)] = 0 , \tag{2.120}$$

$$E[(\xi(t) - \xi(s))^2] = \sigma^2 |t - s| , \quad \sigma^2 = \text{const} > 0 . \tag{2.121}$$

*(iv) Almost all realizations of process $\xi(t)$ are continuous.*
*In the case when $\sigma^2 = 1$, the process $\xi(t)$ is called standard Wiener process.*

The existence of such process follows from the construction given by Lipcer and Shiryayev [35]. Namely, let $\eta_1, \eta_2, \ldots$ be a sequence of zero-mean Gaussian random variables with variances equal to one and let $\phi_1(t), \phi_2(t), \ldots, t \in [0, T]$, be any complete and orthogonal sequence in $L^2[0, T]$. Then, the following theorem holds [35].

**Theorem 2.7.** *For every $t \in [0, T]$ series*

$$\xi(t) = \sum_{j=1}^{\infty} \eta_j \int_0^t \phi_j(s) ds \qquad (2.122)$$

*is convergent almost everywhere and defines a Wiener process over the interval $[0, T]$.*

From the definition of the standard Wiener process follow its properties

$$E[\xi(t)] = 0 , \qquad (2.123)$$

$$K(s, t) = E[\xi(s)\xi(t)] = \min(s, t) , \qquad (2.124)$$

$$P\{\xi(t) \le x\} = \frac{1}{\sqrt{2\pi t}} \int_{-\infty}^{x} \exp\left\{ -\frac{y^2}{2t} \right\} dy , \qquad (2.125)$$

$$E[|\xi(t)|] = \sqrt{\frac{2t}{\pi}} , \qquad (2.126)$$

$$E[(\xi(t+\Delta t)-\xi(t))^{2p}] = \frac{1}{\sqrt{2\pi\Delta t}} \int_{-\infty}^{+\infty} z^{2p} \exp\left\{ -\frac{z^2}{2\Delta t} \right\} dz = (2n-1)!!(\Delta t)^p . \qquad (2.127)$$

**Theorem 2.8.** *Let $a = t_0 < t_1 < \ldots < t_n = b$ be any partition of an interval $[a, b]$ and let $\Delta t = \max(t_{i+1} - t_i)$ and $\xi(t)$ be a Wiener process. Then*

$$\underset{\Delta t \to 0}{l.i.m.} \sum_{i=0}^{n-1} [\xi(t_{i+1}) - \xi(t_i)]^2 = b - a , \qquad (2.128)$$

*and with probability 1*

$$\lim_{\Delta t \to 0} \sum_{i=0}^{n-1} [\xi(t_{i+1}) - \xi(t_i)]^2 = b - a . \qquad (2.129)$$

Let $\Im_t^\xi$ be a family of growing $\sigma$-fields generated by the Wiener process $\xi(t)$, i.e., $\Im_t^\xi = \sigma(\xi(s), s \leq t)$. It is simple to verify that the Wiener process is a martingale with respect to $(\Im_t^\xi, t \in [0, T])$, i.e.,

$$E[\xi(t)|\Im_s^\xi] = \xi(s) , \quad for \quad s \leq t \tag{2.130}$$

and

$$E[(\xi(t) - \xi(s))^2|\Im_s^\xi] = t - s , \quad for\, s \leq t . \tag{2.131}$$

From Definition 2.15 it follows that almost all realizations of Wiener process are continuous. Additionally, one can prove a very important property of Wiener process [35].

**Theorem 2.9.** *Although almost all realizations of the Wiener process are continuous they are nondifferentiable for all $t \geq 0$ and over any finite interval they have unbounded variation (it can be interpreted as "very dense saw function").*

**Definition 2.16.** *A stochastic process $\boldsymbol{\xi}(t)$ is called r-dimensional Wiener process $\boldsymbol{\xi}(t) = [\xi_1(t), \ldots, \xi_r(t)]^T$ if each component $\xi_i(t)$, $i = 1, \ldots, r$, is a scalar Wiener process and if all components $\xi_i(t)$ are mutually independent processes.*

## 2.9 White and Colored Noise

A fundamental mathematical tool in stochastic analysis, particularly in applications to stochastic dynamic systems (similar to $i = \sqrt{-1}$ in electrical engineering), is an abstract stochastic process (physically nonrealizable) called *white noise*. In technical literature one can find different definitions of white noise. We quote the definition that uses the concept of generalized stochastic process, according to Sobczyk [46] independently introduced by Ito and Gelfand.

Let $D(T)$ be the space of test functions, i.e., all infinitely differentiable functions $\phi : T \to R^1$ vanishing identically outside a finite closed interval. The topology in $D(T)$ is defined in the same way as in the case of ordinary Schwartz generalized functions. The $D(T)$ is the topological vector space. Let $H$ be a Hilbert space of all $P$- equivalent random variables defined on $(\Omega, \Im, P)$ having finite moments.

**Definition 2.17.** *A continuous linear mapping $\Phi : D(T) \to H$ is called a generalized stochastic process. The value of the process $\Phi$ at $\phi$ will be denoted by $\{\phi, \Phi\}$ or $\Phi(\phi)$.*

An advantage of a generalized stochastic process is the existence of its derivative, which is also another generalized stochastic process. It follows from the following definition:

**Definition 2.18.** *The derivative $\dot{\Phi}$ with respect to $t$ of a generalized stochastic process $\Phi$ (generalized derivative) in $D(T)$ is determined by*

$$\{\phi, \dot{\Phi}\} = \{\frac{d\phi}{dt}, \Phi\}, \quad for\ all\ \phi \in D(T) . \tag{2.132}$$

Applying the definition of generalized derivative to a Wiener process one can obtain a new generalized stochastic process.

**Definition 2.19.** *A generalized derivative of a Wiener process $\xi(t)$ for $t \in [0, +\infty)$ denoted by $\eta(t) = \dot{\xi}(t)$, i.e.,*

$$\{\phi, \eta\} = \{\phi, \dot{\xi}\} = -\{\frac{d\phi}{dt}, \xi\}, \quad for\ all\ \phi \in D(T) , \tag{2.133}$$

*is called a Gaussian white noise.*

Equality (2.133) will also be denoted in the following form:

$$d\xi(t) = \eta(t)dt . \tag{2.134}$$

Not going into details one can show that for every $\phi \in D(T)$, $\{\phi, \eta\}$ is a Gaussian process and for every finite number of functions $\phi_1, \ldots, \phi_n \in D(T)$ the random variables $\{\phi_i, \eta\}$, $1 \leq i \leq n$, are jointly Gaussian. Additionally

$$E[\eta(t)] = 0 \tag{2.135}$$

and the covariance function is the Dirac generalized function,

$$K_{\eta\eta}(t_1, t_2) = c\delta(t_2 - t_1) = c\delta(\tau) , \quad c = \text{const} , \quad c > 0 , \tag{2.136}$$

(for details see, for instance, [46]). From last equality it follows directly that the variance of white noise is equal to infinity $K_{\eta\eta}(t, t) = \delta(0) = \infty$, and the power spectral density function is equal to a constant $c$

$$S_{\xi\xi}(\lambda) = c . \tag{2.137}$$

The property (2.136) confirms physical nonrealizability of such a process and property (2.137) indicates the genesis of the name "white noise" in analogy with the "white light" in which all frequency of electromagnetic waves (colors) are represented.

In real physical systems disturbances called *coloured noises* appear, i.e., stationary zero-mean second-order stochastic processes. They are characterized by correlation functions $R(\tau)$ or by the corresponding power spectral density functions $S(\lambda)$. As a criterion of the correlation degree (i.e., the departure from white noise) of the process $x(t)$ is proposed the value $t_c$ defined as follows:

$$t_c = \frac{\int_0^\infty \tau R_x(\tau)d\tau}{\int_0^\infty R_x(\tau)d\tau} , \qquad (2.138)$$

and called *correlation time*.

If $t_{ob} \in [0,T]$, called *observation time*, satisfies the inequality $t_c \ll t_{ob}$, then the correlation can be neglected and the considered process $x(t)$ can be treated as an approximation of white noise, otherwise as colored noise [41]. Another approximation of the white noise can be treated as a wideband process with the power spectral density function defined by

$$S(\lambda) = \begin{cases} S_0 & for \ \lambda_a \leq |\lambda| < \lambda_b \\ 0 & for \quad other \ \lambda , \end{cases} \qquad (2.139)$$

where $\lambda_a$ and $\lambda_b$ are fixed frequencies, in particular $\lambda_a = 0$.

We note that in the literature there are algorithms of numerical simulation of realizations of a process that approximate white noise. They are described, for instance, in [13, 50, 51].

## 2.10 Integration and Differentiation Formulas of Diffusion Processes

In the analysis of stochastic processes the constructions of integrals and differentiation formulas differ from the corresponding operations for deterministic functions. For stochastic processes they are more complicated, and the detailed derivations can be found, for instance, in [25, 35]. Here we quote only the basic definitions and theorems.

### 2.10.1 Stochastic Integrals

**Definition 2.20.** *Ito stochastic integral of the nonanticipating function* $f(x(t),t)$ *($\Im_t-$ measurable) over the interval* $[0,T]$ *with respect to a diffusion process* $x(t)$ *is called the mean-square limit of Riemann's sum, i.e.,*

$$\int_0^T f((x(t),t)dx(t) = \underset{\Delta t \to 0}{l.i.m.} \sum_{i=0}^N f(x(t_i),t_i)[x(t_{i+1}) - x(t_i)] , \qquad (2.140)$$

*where*

$$E\left[\int_0^T f^2((x(t),t)dt\right] < \infty , \quad 0 = t_1 < \ldots < t_{N+1} = T , \quad \Delta t = \max_i(t_{i+1} - t_i)$$

*and the limit does not depend on the selection of the points* $t_i$.

In contrast to the mean-square integral defined by (2.48) the value of stochastic integral of the nonanticipating function $f(x(t), t)$ over the interval $[0, T]$ with respect to a diffusion process $x(t)$ depends on the selection of the intermediate points $t_i'$ for which the value of the function $f(x(t), t)$ in Riemann sum is calculated, i.e., the interval $[t_i, t_{i+1}]$ is treated as a convex set and an intermediate point $t_i'$ can be treated as a convex combination of $t_i$ and $t_{i+1}$, i.e., $t_i' = \beta t_i + (1 - \beta) t_{i+1}$ and the value of the function $f(x(t), t)$ in this interval as $f(\beta x(t_i) + (1 - \beta)x(t_{i+1}), \beta t_i + (1 - \beta)t_{i+1})$, where $\beta$ is a real parameter $0 \leq \beta \leq 1$. Then the definition of stochastic integral of the nonanticipating function $f(x(t), t)$ ( $\Im_t-$ measurable) over the interval $[0, T]$ with respect to a diffusion process $x(t)$ has the form of the definition of Ito stochastic integral, where equality (2.140) is replaced by the following equation:

$$\int_0^T f(x(t), t) d_\beta x(t) = \underset{\Delta t \to 0}{l.i.m.} \sum_{i=0}^N f(\beta x(t_i)$$
$$+ (1 - \beta)x(t_{i+1}), \beta t_i + (1 - \beta)t_{i+1})[x(t_{i+1}) - x(t_i)] .$$
$$(2.141)$$

Such general definition of the stochastic integral includes definitions for two special cases $\beta = 1$ and $\beta = \frac{1}{2}$ called *Ito stochastic integral* and *Stratonovich stochastic integral*, respectively. The relationship between both integrals establishes the following theorem.

**Theorem 2.10.** *If $x(t)$, $t \in [0, T]$, is a diffusion process, $f(x(t), t)$ is a nonlinear nonanticipating function ( $\Im_t-$ measurable) over the interval $[0, T]$ having continuous derivatives with respect to both arguments and $E\left[\int_0^T f^2(x(t), t)dt\right] < \infty$, then the following equality holds*

$$\int_0^T f(x(t), t) d_\beta x(t) = \int_0^T f(x(t), t) dx(t) + (1 - \beta) \int_0^T \frac{\partial f}{\partial x}(x(t), t) B(x(t), t) dt ,$$
$$(2.142)$$

*where $B(x(t), t)$ is the diffusion coefficient defined by (2.111), $0 \leq \beta \leq 1$.*

The definition of stochastic integral and Theorem 2.10 can be extended to vector processes.

**Definition 2.21.** *Let $\mathbf{x}(t)$ be a $r$-dimensional diffusion process for $t \in [0, T]$ whose drift vector $\mathbf{A}(\mathbf{x}, t)$ and diffusion matrix $\mathbf{B}(\mathbf{x}, t)$ with first derivatives $\partial \mathbf{B}(\mathbf{x}, t)/\partial x_j$, $j = 1, \ldots, r$, are continuous with respect to both arguments. Let $\mathbf{f}(\mathbf{x}, t)$ be a nonlinear nonanticipating vector function with values in $R^r$, continuous with respect to $\mathbf{x}$, that for $t \in [0, T]$ satisfies the following conditions:*

*(i)   There exist partial derivatives $\frac{\partial \mathbf{f}(\mathbf{x}, t)}{\partial x_j}$ $j = 1, \ldots, r$.*

*(ii)  $\int_0^T E[|\mathbf{f}^T(\mathbf{x}(s), s)\mathbf{A}(\mathbf{x}(s), s)|]ds < \infty$.*

*(iii)* $\int_0^T E[|\mathbf{f}^T(\mathbf{x}(s), s)\mathbf{B}(\mathbf{x}(s), s)\mathbf{f}(\mathbf{x}(s), s)|]ds < \infty.$

*then vector stochastic integral is defined by*

$$\int_0^T \mathbf{f}^T(\mathbf{x}(s), s)d_\beta\mathbf{x}(t) = l.i.m. \sum_{i=0}^{N-1} \mathbf{f}^T(\beta\mathbf{x}(t_i)$$
$$+(1-\beta)\mathbf{x}(t_{i+1}), \beta t_i + (1-\beta)t_{i+1})[\mathbf{x}(t_{i+1}) - \mathbf{x}(t_i)] ,$$

(2.143)

*where* $\Delta t = \max[t_{i+1} - t_i], 0 = t_0 < \ldots < t_N = T.$

Like the scalar case the Ito and Stratonovich vector integrals are defined: for $\beta = 1$ – Ito vector integral and for $\beta = \frac{1}{2}$ Stratonovich vector integral.

Stratonovich [53] has shown that the relationship between Ito and Stratonovich vector integrals denoted, respectively, by

$$I_I = \int_0^T \mathbf{f}^T(\mathbf{x}(s), s)d_1\mathbf{x}(t) , \quad I_S = \int_0^T \mathbf{f}^T(\mathbf{x}(s), s)d_{\frac{1}{2}}\mathbf{x}(t) ,$$

(2.144)

is the following

$$I_S = I_I + \frac{1}{2}\sum_{j=1}^{r}\sum_{k=1}^{r}\int_0^T \frac{\partial f_j}{\partial x_k}(\mathbf{x}(t), t)b_{jk}(\mathbf{x}(t), t)dt ,$$

(2.145)

where $b_{jk}(\mathbf{x}(t), t)$ are the elements of the diffusion matrix of the diffusion process $\mathbf{x}(t)$.

## 2.10.2 Ito Formula

We now give the formula for differentiation of a vector function with respect to a vector diffusion process.

Let $\mathbf{x}(t)$ be an $n$-dimensional stochastic process for $t \in [0, T]$, i.e., $\mathbf{x}(t) = [x_1(t), \ldots, x_n(t)]^T$ having the stochastic differential

$$d\mathbf{x}(t) = \mathbf{a}(t, \omega)dt + \boldsymbol{\sigma}(t, \omega)d\boldsymbol{\xi}(t) ,$$

(2.146)

where $\boldsymbol{\xi}(t)$ is a $r$-dimensional Wiener process for $t \in [0, T]$, i.e., $\boldsymbol{\xi}(t) = [\xi_1(t), \ldots, \xi_r(t)]^T$. The vector $\mathbf{a}(t, \omega) = [a_1(t, \omega), \ldots, a_n(t, \omega)]^T$ and the matrix $\boldsymbol{\sigma}(t, \omega) = [\sigma_{ij}(t, \omega)]$, $i = 1, \ldots, n$, $j = 1, \ldots, r$, consist of nonlinear nonanticipating functions satisfying conditions

$$P\left\{\int_0^T |a_i(t, \omega)|dt < \infty\right\} = 1, \quad i = 1, \ldots, n ,$$

(2.147)

$$P\left\{\int_0^T |\sigma_{ij}^2(t, \omega)|dt < \infty\right\} = 1 , \quad i = 1, \ldots, n, \quad j = 1, \ldots, r .$$

(2.148)

Then the determination of differential of a nonlinear function $f(\mathbf{x}(t), t)$ is established by the following theorem:

**Theorem 2.11.** *Let $f(y_1, \ldots, y_n, t)$ be a continuous function having continuous derivatives $\partial f/\partial t$, $\partial f/\partial y_i$, $\partial^2 f/\partial y_i \partial y_j$, $i, j = 1, \ldots, n$. Then the stochastic process $f(x_1(t), \ldots, x_n(t), t)$ has the stochastic differential in the form*

$$
df(x_1(t), \ldots, x_n(t), t) = \left[ \frac{\partial f}{\partial t}(x_1(t), \ldots, x_n(t), t) \right.
$$

$$
+ \sum_{i=1}^{n} \frac{\partial f}{\partial y_i}(x_1(t), \ldots, x_n(t), t) + \frac{1}{2} \sum_{i=1}^{n} \sum_{j=1}^{n} \frac{\partial^2 f}{\partial y_i \partial y_j}
$$

$$
\left. (x_1(t), \ldots, x_n(t), t) \sum_{k=1}^{r} \sigma_{ik}(t, \omega) \sigma_{jk}(t, \omega) \right] dt
$$

$$
+ \sum_{i=1}^{n} \sum_{k=1}^{r} \frac{\partial f}{\partial y_i}(x_1(t), \ldots, x_n(t), t) \sigma_{ik}(t, \omega) d\xi_k.
$$

$$(2.149)$$

The proof of this theorem can be found in many monographs, for instance, in [25, 35].

In the particular case when $r = n$ and the function $f(\mathbf{x}(t), t)$ is a quadratic form, i.e.,

$$
f(\mathbf{x}(t), t) = \mathbf{x}^T(t) \mathbf{H}(t) \mathbf{x}(t) , \tag{2.150}
$$

where $\mathbf{H}(t)$ is a deterministic matrix, $\mathbf{x}(t)$ is a diffusion process with differential given by

$$
d\mathbf{x}(t) = \mathbf{a}(t)dt + \boldsymbol{\sigma}(t)d\boldsymbol{\xi}(t) , \tag{2.151}
$$

then

$$
d(\mathbf{x}^T(t) \mathbf{H}(t) \mathbf{x}(t)) = \left[ \mathbf{x}^T(t) \mathbf{H}(t) \mathbf{a}(t) + \mathbf{a}^T(t) \mathbf{H}(t) \mathbf{x}(t) \right.
$$

$$
\left. + \mathbf{x}^T(t) \frac{d\mathbf{H}(t)}{dt} \mathbf{x}(t) + tr(\boldsymbol{\sigma}(t) \boldsymbol{\sigma}^T(t) \mathbf{H}(t)) \right] dt
$$

$$
+ [\mathbf{x}^T(t) \mathbf{H}(t) \boldsymbol{\sigma}(t) + \boldsymbol{\sigma}^T(t) \mathbf{H}(t) \mathbf{x}(t)] d\boldsymbol{\xi}(t) , \tag{2.152}
$$

where $tr(\mathbf{A})$ denotes the trace of the matrix $\mathbf{A}$, i.e., if $\mathbf{A} = [a_{ij}]$, $i, j = 1, \ldots, n$, then

$$
tr(\mathbf{A}) = \sum_{i=1}^{n} a_{ii} . \tag{2.153}
$$

If we assume that $\mathbf{H}(t)$ is an identity matrix $\mathbf{H}(t) = \mathbf{I}$, then

$$
d(|\mathbf{x}(t)|^2) = \left[ \mathbf{x}^T(t) \mathbf{a}(t) + \mathbf{a}^T(t) \mathbf{x}(t) + \sum_{i=1}^{r} \sigma_{ii}^2(t) \right] dt
$$

$$
+ [\mathbf{x}^T(t) \boldsymbol{\sigma}(t) + \boldsymbol{\sigma}^T(t) \mathbf{x}(t)] d\boldsymbol{\xi}(t) . \tag{2.154}
$$

## 2.11 Stochastic Differential Equations

We consider the vector stochastic differential equation in the form

$$d\mathbf{x}(t) = \mathbf{F}(\mathbf{x}, t)dt + \sum_{k=1}^{M} \mathbf{G}_k(\mathbf{x}, t)d\xi_k(t) , \quad \mathbf{x}(t_0) = \mathbf{x}_0 , \qquad (2.155)$$

where $\mathbf{F}, \mathbf{G}_k : R^n \times [0, T] \to R^n$ are nonlinear vector deterministic functions, $\mathbf{F} = [F_1, \ldots, F_n]^T$, $\mathbf{G}_k = [\sigma_k^i]$, $i = 1, \ldots, n, k = 1, \ldots, M$, $\xi_k$ are standard independent Wiener processes measurable with respect to nondecreasing family of $\sigma$-fields $\Im_t$, $t \in [0, T]$.

We denote by $(\mathcal{C}_T, \mathcal{B}_T)$ measurable space of functions $(\mathbf{x}(t), t \in [0, T])$ continuous over $[0, T]$ with $\sigma$-field $\mathcal{B}_T = \sigma(\mathbf{x} : \mathbf{x}(s), s \leq T)$. Similarly, we denote $\mathcal{B}_t = \sigma(\mathbf{x} : \mathbf{x}(s), s \leq t)$. Let $F_i(\mathbf{x}, t)$ and $\sigma_k^i(\mathbf{x}, t)$ be nonanticipating functionals, i.e., $\mathcal{B}_t$ measurable for all $t \in [0, T]$.

**Definition 2.22.** *A continuous vector stochastic process* $\mathbf{x}(t)$, $t \in [0, T]$, *with probability 1 is called strong solution or shortly solution of stochastic differential equation (2.155) with* $\Im_0$-*measurable initial condition* $\mathbf{x}(t_0) = \mathbf{x}_0$ *if for all* $t \in [0, T]$ *vector random variables* $\mathbf{x}(t)$ *are* $\Im_t$ *measurable, where*

$$P\left\{ \int_0^T |\mathbf{F}(\mathbf{x}, t)|dt < \infty \right\} = 1 , \qquad (2.156)$$

$$P\left\{ \sum_{k=1}^{M} \int_0^T |\mathbf{G}_k(\mathbf{x}, t)|^2 dt < \infty \right\} = 1 \qquad (2.157)$$

*( $|.|$ is a norm in Euclidean space) and with probability 1 for* $t \in [0, T]$, *we have*

$$\mathbf{x}(t) = \mathbf{x}_0 + \int_0^t \mathbf{F}(\mathbf{x}, s)ds + \int_0^t \sum_{k=1}^{M} \mathbf{G}_k(\mathbf{x}, s)d\xi_k(s) . \qquad (2.158)$$

**Definition 2.23.** *A stochastic differential equation (2.155) has a unique strong solution, if for any two strong solutions* $\mathbf{x}(t)$, $\tilde{\mathbf{x}}(t)$, $t \in [0, T]$, *the following equality holds:*

$$P\left\{ \sup_{t \in [0, T]} |\mathbf{x}(t) - \tilde{\mathbf{x}}(t)| > 0 \right\} = 0 . \qquad (2.159)$$

We quote the theorem of the existence of unique solutions that proof can be found, for instance, in [35].

**Theorem 2.12.** *Let components of vectors* $\mathbf{F}(\mathbf{x}, t)$ *and* $\mathbf{G}_k(\mathbf{x}, t)$, $k = 1, \ldots, M$, *be nonanticipating functionals,* $\mathbf{x} \in C_T$, $t \in [0, 1]$, *and satisfy Lipschitz condition:*

$$|F_i(\mathbf{x}, t) - F_i(\mathbf{y}, t)|^2 + |\sigma_k^i(\mathbf{x}, t) - \sigma_k^i(\mathbf{y}, t)|^2$$

$$\leq L_1 \int_0^t |\mathbf{x}(s) - \mathbf{y}(s)|^2 dK(s) + L_2 |\mathbf{x}(t) - \mathbf{y}(t)|^2 \qquad (2.160)$$

*and*

$$F_i^2(\mathbf{x}, t) + (\sigma_k^i)^2(\mathbf{x}, t) \leq L_1 \int_0^t (1 + |\mathbf{x}(s)|^2) dK(s) + L_2(1 + (\mathbf{x}(t))^2 , \quad (2.161)$$

*for all* $i = 1, \ldots, n$, $k = 1, \ldots, M$, *where* $L_1 > 0$ *and* $L_2 > 0$ *are constants,* $K(s)$ *is the nondecreasing right continuous function,* $0 \leq K(s) \leq 1$, $\mathbf{x}, \mathbf{y} \in C_T$. *Let the initial condition* $\mathbf{x}_0 = \mathbf{x}_0(\omega)$ *be* $\Im_0$-*measurable vector random variable such that*

$$P\left\{ \sum_{i=1}^n |x_{0i}| < \infty \right\} = 1 . \qquad (2.162)$$

*Then (2.155) has unique strong solution* $\mathbf{x}(t)$ *measurable with respect to* $\Im_t$, $t \in [0, 1]$, *and additionally if* $E\left[\sum_{i=1}^n x_{0i}^{2m}\right] < \infty$, *then there exists a constant* $C_m$, *such that*

$$E\left[\sum_{i=1}^n x_i^{2m}(t)\right] \leq \left(1 + E\left[\sum_{i=1}^n x_{0i}^{2m}\right]\right) \exp\{C_m t\} - 1 \qquad (2.163)$$

From considerations given in monographs [22, 25] it follows that under assumptions of Theorem 2.12 the solution of (2.155) is a Markov process and can be extended for $t \in R^+$.

From inequality (2.163) it follows that the boundedness of conditional initial conditions implies the boundedness of even-order moments of solution for all $t \in [0, 1]$ (and generally for all $t$) if (2.155) is considered for $t \in R^+$.

Since linear dynamic systems are particularly important in modeling study and several methods of their analysis will be discussed in next chapters we quote a theorem of the existence of strong solution of vector linear stochastic differential equations (for proof, see [35]).

**Theorem 2.13.** *Let elements of the vector function* $\mathbf{A}_0(t) = [a_0^1(t), \ldots, a_0^n(t)]^T$ *and the matrices* $\mathbf{A} = [a_{ij}]$, $\mathbf{G}_k = [\sigma_{k0}^i]$, $i, j = 1, \ldots, n$, $k = 1, \ldots, M$, *be measurable functions (deterministic) of the variable* $t$, $t \in [0, 1]$, *satisfying conditions*

$$\int_0^1 |a_0^i(t)| dt < \infty , \quad \int_0^1 |a_{ij}(t)| dt < \infty , \quad \int_0^1 |(\sigma_{k0}^i)^2(t)| dt < \infty . \quad (2.164)$$

Then the vector stochastic differential equation

$$dx = [\mathbf{A}_0(t) + \mathbf{A}(t)\mathbf{x}(t)]dt + \sum_{k=1}^{M} \mathbf{G}_{k0}(t)d\xi_k(t) , \quad \mathbf{x}(t_0) = \mathbf{x}_0 , \quad (2.165)$$

where $\xi_k(t)$ are independent Wiener processes measurable with respect to $\Im_t$, $t \in [0,1]$, has the unique strong solution determined by

$$\mathbf{x}(t) = \mathbf{\Phi}(t,0)\left[\mathbf{x}(t_0) + \int_0^t \mathbf{\Phi}^{-1}(s,0)\mathbf{A}_0 ds + \int_0^t \sum_{k=1}^{M} \mathbf{\Phi}^{-1}(s,0)\mathbf{G}_{k0}(s)d\xi_k(s)\right] ,$$
$$(2.166)$$

where $\mathbf{\Phi}(t,0)$ is a fundamental matrix of $(n \times n)$ dimension

$$\mathbf{\Phi}(t,0) = \mathbf{I} + \int_0^t \mathbf{A}(s)\mathbf{\Phi}(s,0)ds . \quad (2.167)$$

$\mathbf{I}$ is the $n \times n$-dimensional identity matrix.

Solution (2.166) can be extended for any $t \in [0,T]$ under the assumption that the elements of the vector $\mathbf{A}_0$ and matrices $\mathbf{A}, \mathbf{G}_k$ are measurable and bounded functions for $t \in [0,T]$.

**Theorem 2.14.** *[25] Let the elements of vectors $\mathbf{F}(\mathbf{x},t)$ and $\mathbf{G}_k(\mathbf{x},t)$, $k = 1,\ldots,M$, satisfy the assumptions of Theorem 2.12, then the solution of (2.158) (process $\mathbf{x}(t)$) is a Markov process whose transfer function for $s < t$ is given by*

$$P(s,\mathbf{x};t,\mathbf{A}) = P\{\mathbf{x}(t) \in \mathbf{A}|\mathbf{x}(s) = \mathbf{x}\} = P\{\mathbf{x}_{s,\mathbf{x}}(t) \in \mathbf{A}\} , \quad (2.168)$$

*where*

$$\mathbf{x}_{s,\mathbf{x}}(t) = \mathbf{x} + \int_s^t \mathbf{F}(\mathbf{x}_{s,\mathbf{x}}(u),u)du + \int_s^t \sum_{k=1}^{M} \mathbf{G}_k(\mathbf{x}_{s,\mathbf{x}}(u),u)d\xi_k(u) . \quad (2.169)$$

Additionally, if elements of vectors $\mathbf{F}(\mathbf{x},t)$ and $\mathbf{G}_k(\mathbf{x},t)$, $k = 1,\ldots,M$, are continuous with respect to $t$, then the process $\mathbf{x}(t)$ is a diffusion process with drift vector $\mathbf{F}(\mathbf{x},t)$ and diffusion matrix $\mathbf{B}(\mathbf{x},t)$ satisfying the equality

$$\mathbf{z}^T \mathbf{B}(\mathbf{x},t)\mathbf{z} = \sum_{k=1}^{M} < \mathbf{G}_k(\mathbf{x},t), \mathbf{z} >^2 , \quad\quad\quad\quad \cdot (2.170)$$

for all $\mathbf{z} \in R^n$, where $< .,. >$ denotes the scalar product.

If we assume that there exists a function $V(\mathbf{x},t)$ having continuous and bounded derivatives of first order with respect to $t$ and second orders with respect to $\mathbf{x}$, for $t \in [0,T]$ and $\mathbf{x} \in R^n$ denoted by $V \in \mathbf{C}_2$, then from Theorems 2.11 and 2.12 it follows that

$$V(\mathbf{x}(t),t) - V(\mathbf{x}(s),s) = \int_s^t LV(\mathbf{x}(u),u)du + \int_s^t \sum_{k=1}^M \frac{\partial V}{\partial \mathbf{x}} \mathbf{G}_k(\mathbf{x}(u),u)d\xi_k(u) ,$$

(2.171)

where the operator $L(.)$ was determined by the definition of stochastic integral appearing on the right side of (2.158). It means that if the stochastic integral appearing in (2.155) is treated as the Ito integral or Stratonovich integral then the stochastic differential equation (2.155) is called *Ito stochastic differential equation or Stratonovich stochastic differential equation*, respectively. For simplicity we call these equations *Ito equation or Stratonovich equation*. Similarly, corresponding names are used for differentials $d_I\xi_k$ and $d_S\xi_k$ and also operators $L_I$ and $L_S$. From (2.146), (2.158), and (2.171) it follows that Ito and Stratonovich operators are defined as follows:

$$L_I(.) = \frac{\partial}{\partial t} + \mathbf{F}^T(\mathbf{x},t)\frac{\partial}{\partial \mathbf{x}} + \frac{1}{2}\sum_{k=1}^M \left\langle \frac{\partial}{\partial \mathbf{x}}, \mathbf{G}_k(\mathbf{x},t) \right\rangle^2 ,$$

(2.172)

$$L_S(.) = \frac{\partial}{\partial t} + \left( \mathbf{F}(\mathbf{x},t)\frac{\partial}{\partial \mathbf{x}} + \frac{1}{2}\sum_{k=1}^M \frac{\partial \mathbf{G}_k(\mathbf{x},t)}{\partial \mathbf{x}}\mathbf{G}_k(\mathbf{x},t) \right)^T \frac{\partial}{\partial \mathbf{x}}$$

$$+ \frac{1}{2}\sum_{k=1}^M \left\langle \frac{\partial}{\partial \mathbf{x}}, \mathbf{G}_k(\mathbf{x},t) \right\rangle^2 .$$

(2.173)

The stochastic differential equation (2.155) treated as Stratonovich stochastic differential equation is equivalent to the following Ito stochastic differential equation [53]:

$$d\mathbf{x} = \left[ \mathbf{F}(\mathbf{x},t) + \frac{1}{2}\sum_{k=1}^M \frac{\partial \mathbf{G}_k(\mathbf{x},t)}{\partial \mathbf{x}}\mathbf{G}_k(\mathbf{x},t) \right] dt + \sum_{k=1}^M \mathbf{G}_k(\mathbf{x},t)d\xi_k(t) . \quad (2.174)$$

It means that all obtained results for Ito equations can be used in the analysis of the corresponding Stratonovich equations. It should be stressed that in the case when functions $\mathbf{G}_k$ do not depend on vector state $\mathbf{x}$, then Ito equations and corresponding Stratonovich equations are the same. Since in further considerations in this book we will use Ito equations for simplicity of notation we will omit the subindex I in the operator $L_I$.

Using the properties of stochastic integrals and applying Fubini theorem to equality (2.171), we obtain

$$E[V(\mathbf{x}(t),t) - V(\mathbf{x}(s),s)] = \int_s^t LV(\mathbf{x}(u),u)du . \qquad (2.175)$$

We note that (2.175) is often used in the study of stochastic stability.

## Physical Interpretation of Stratonovich Equation

The differential equations with stochastic parameters are very often used in the modeling of real physical systems. An example of such equation is the scalar Langevin equation

$$\frac{dx(t)}{dt} = F(x(t), t) + G(x, t)\dot{\xi}(t) , \qquad (2.176)$$

where $F(x, t)$ and $G(x, t)$ are scalar nonlinear functions and $\dot{\xi}(t)$ is a Gaussian white noise. Since white noise is an abstraction while in real physical system can act only real processes, for instance, colored noises then arises the following problem. Let us consider a family of differential equations (2.176) where the process $\dot{\xi}(t)$ is replaced by a sequence of stationary Gaussian wideband processes $\{\eta^n(t)\}$, $n = 1, 2, \ldots$, and assume that $\{\eta^n(t)\}$ converges in some sense to a Gaussian white noise (for instance, with growing $n$ grows the width of the band of the corresponding power spectral density function) and for each $n$ processes $\eta^n(t)$ have regular realizations. Then the corresponding sequence $\{x_n(t)\}$ is the solution of this family of differential equations

$$\frac{dx_n(t)}{dt} = F(x_n(t), t) + G(x_n(t), t)\eta^n(t) . \qquad (2.177)$$

Suppose that the sequence $\{x_n(t)\}$ converges to a process $x(t)$. Then, two questions arise: what is the process $x(t)$ and what is the relation of the process $x(t)$ to the solution of the corresponding Ito equation

$$dx(t) = F(x(t), t)dt + G(x(t), t)d_I\xi(t) , \qquad (2.178)$$

Wong and Zakai [56, 57] have shown that the sequence of solutions $x_n(t)$ converges to the solution of the following corresponding Stratonovich equation:

$$dx(t) = F(x(t), t)dt + G(x(t), t)d_S\xi(t) , \qquad (2.179)$$

or equivalent Ito equation

$$dx(t) = \left[ F(x(t), t) + \frac{1}{2}\frac{\partial G}{\partial x}(x(t), t)G(x(t), t) \right] dt + G(x(t), t)d\xi(t) . \qquad (2.180)$$

This important result has been generalized by Papanicolau and Kohler for multidimensional case [42]. They considered a vector differential equation with stochastic coefficients characterized by a parameter $\varepsilon > 0$

$$\frac{d\mathbf{x}^\varepsilon(\tau)}{d\tau} = \mathbf{F}(\mathbf{x}^\varepsilon(\tau), \tau) + \sum_{k=1}^{M} \mathbf{G}_k(\mathbf{x}^\varepsilon(\tau), \tau)\frac{1}{\varepsilon}\eta_k\left(\frac{\tau}{\varepsilon^2}\right) , \quad \mathbf{x}^\varepsilon(0) = \mathbf{x}_0 \quad (2.181)$$

and the corresponding integral equation

$$\mathbf{x}^{\varepsilon}(\tau) = \mathbf{x}_0 + \int_0^{\tau} \mathbf{F}(\mathbf{x}^{\varepsilon}(s), s) ds + \int_0^{\tau} \sum_{k=1}^{M} \mathbf{G}_k(\mathbf{x}^{\varepsilon}(s), s) \frac{1}{\varepsilon} \eta_k \left( \frac{s}{\varepsilon^2} \right) ds , \quad (2.182)$$

where $\mathbf{F}(\mathbf{x}, t)$ and $\mathbf{G}_k(\mathbf{x}, t)$, $k = 1, \ldots, M$, are deterministic functions, $\mathbf{F}, \mathbf{G}_k :$ $R^n \times [0, T] \to R^n$ satisfying the following conditions:

- There exists a constant $C$ independent of $(\mathbf{x}, t)$ and initial conditions, such that

$$|\mathbf{G}_k(\mathbf{x}, t)| \leq C(1 + |\mathbf{x}|) , \quad (2.183)$$

$$\left| \frac{\partial \mathbf{G}_k(\mathbf{x}, t)}{\partial x_j} \right| \leq C , \quad j = 1, \ldots, n, \ k = 1, \ldots, M . \quad (2.184)$$

- Additionally, there exists a constant $q > 0$, such that

$$\left| \frac{\partial^2 \mathbf{G}_k(\mathbf{x}, t)}{\partial x_i \partial x_j} \right| \leq C(1 + |\mathbf{x}|^q) , \quad i, j = 1, \ldots, n , \ k = 1, \ldots, M , \quad (2.185)$$

$$\left| \frac{\partial^3 \mathbf{G}_k(\mathbf{x}, t)}{\partial x_i \partial x_j \partial x_l} \right| \leq C(1 + |\mathbf{x}|^q) , \quad i, j, l = 1, \ldots, n , \ k = 1, \ldots, M , \quad (2.186)$$

$$\left| \frac{\partial^4 \mathbf{G}_k(\mathbf{x}, t)}{\partial x_i \partial x_j \partial x_l \partial x_m} \right| \leq C(1 + |\mathbf{x}|^q) , \quad i, j, l, m = 1, \ldots, n , \ k = 1, \ldots, M . \quad (2.187)$$

Then the solution of (2.181) with $\varepsilon \to 0$ converges weakly to a diffusion process determined by differential operator (2.173), i.e., the probability distribution of the solution of the integral equation (2.182) converges to the probability distribution of the solution of the integral equation (2.158) treated in Stratonovich sense, where $\eta_k(t) = \eta_k(t, \omega)$ are stationary zero-mean second-order processes.

The approximation is chosen such that if $\varepsilon \to 0$ and $t = \frac{\tau}{\varepsilon^2} \to \infty$, then $\varepsilon^2 t = $ const. Then under additional assumptions about measurability of the processes $\eta_k^{\varepsilon}(\tau) = \frac{1}{\varepsilon} \eta_k \left( \frac{\tau}{\varepsilon^2} \right)$ one can show using central limit theorem that the sequence $\{\eta_k^{\varepsilon}(\tau)\}$ converges weakly for $k \to +\infty$ to a Gaussian white noise, i.e., the integrals of this sequence of processes converges to a corresponding Wiener process.

Papanicolau and Kohler [42] have also shown that moments of any order of the process $\mathbf{x}^{\varepsilon}(\tau)$ arising in (2.182) converge weakly to the moments of the solutions of the corresponding Stratonovich equation (2.155). We also note that other approximation techniques proposed in the literature [28] have been used in the analysis of solutions of stochastic differential equations.

We shortly come back to the problem of the existence of solution of the Ito stochastic differential equation considered in Theorem 2.12. The sufficient conditions formulated there and also in other references (see for instance

[20, 23]) use the assumptions that coefficients of the nonlinear equations satisfy the global Lipschitz conditions. These assumptions are of course not useful in the study of nonlinear vibrating systems, where usually the coefficients are polynomials of vector state variables. To omit this inconvenience, Khasminski [27] proposed to use the Lyapunov function approach for the determination of the sufficient conditions for the existence of strong solution of Ito stochastic differential equation. Following his approach we assume that the nonlinear functions $\mathbf{F}(\mathbf{x}, t)$ and $\mathbf{G}_k(\mathbf{x}, t)$ in (2.155) satisfy the conditions

$$|\mathbf{F}(\mathbf{x}, t) - \mathbf{F}(\mathbf{y}, t)| + \sum_{k=1}^{M} |\mathbf{G}_k(\mathbf{x}, t) - \mathbf{G}_k(\mathbf{y}, t)| \le B|\mathbf{x} - \mathbf{y}| \qquad (2.188)$$

and

$$|\mathbf{F}(\mathbf{x}, t)| + \sum_{k=1}^{M} |\mathbf{G}_k(\mathbf{x}, t)| \le B(1 + |\mathbf{x}|) , \qquad (2.189)$$

for a positive constant $B$, in all cylinders $R^+ \times U_\rho$ where $R^+ = [0, +\infty)$, $U_\rho = \{\mathbf{x} : |\mathbf{x}| < \rho\}$, where $\rho$ is a constant.

Then the sufficient conditions for the existence of the solution of (2.155) have the form.

**Theorem 2.15.** *[27] Let the vectors $\mathbf{F}(\mathbf{x}, t)$ and $\mathbf{G}_k(\mathbf{x}, t)$, $k = 1, \ldots, M$, be nonanticipating functionals satisfying Lipschitz conditions (2.188) and (2.189) in all cylinders $R^+ \times U_\rho$. Let there exist a nonnegative function $V \in \mathbf{C}_2$ in $R^n \times [0, T]$ satisfying for a constant $c > 0$ the following conditions*

$$LV \le cV , \qquad (2.190)$$

$$\lim_{\rho \to \infty} \inf_{|\mathbf{x}| > \rho} V(\mathbf{x}, t) = \infty . \qquad (2.191)$$

Let the initial condition $\mathbf{x}_0 = \mathbf{x}_0(\omega)$ be $\Im_0$-measurable vector random variable satisfying the condition (2.162).

Then (2.155) has unique strong solution $\mathbf{x}(t)$ measurable with respect to $\Im_t$, $t \in [0, 1]$ which is a Markov process.

Stationary solutions play a particular role in the analysis of random vibrations. To prove the sufficient conditions for the existence of stationary solution Khasminski [27] has used an assumption of regularity of solution.

**Definition 2.24.** *The stochastic process $\mathbf{x}(t)$ is called regular when*

$$P\{\tau = \infty\} = 1 , \qquad (2.192)$$

*where $\tau = \tau(\omega)$ is a Markov time with respect to family $\mathcal{B}_t$, $t \in [0, T]$, i.e., when for all $t \in [0, T]$*

$$\{\omega : \tau(\omega) \le t\} \in \mathcal{B}_t \qquad (2.193)$$

*is defined as* $\tau = \lim_{\rho \to \infty} \tau_\rho(\omega)$, *where*

$$\tau_\rho(\omega) = \inf_{t>0}\{|\mathbf{x}(t,\omega)| \geq \rho\} . \tag{2.194}$$

**Theorem 2.16.** *[27] Let the vectors* $\mathbf{F}(\mathbf{x},t) = \mathbf{F}(\mathbf{x})$ *and* $\mathbf{G}_k(\mathbf{x},t) = \mathbf{G}_k(\mathbf{x})$, $k = 1,\dots,M$, *be nonanticipating functionals satisfying Lipschitz conditions* *(2.188) and (2.189) in all sets* $U_R$. *Let there exist a nonnegative function* $V \in \mathbf{C}_2$ *in* $R^n$ *satisfying the following conditions*

$$V \geq 0 , \tag{2.195}$$

$$\lim_{\rho \to \infty} \sup_{|\mathbf{x}|>\rho} LV(\mathbf{x}) = \infty . \tag{2.196}$$

Let $\mathbf{x}_0 = \mathbf{x}_0(\omega)$ be a $\Im_0$-measurable vector random variable satisfying the condition (2.162). We assume that the solution $\mathbf{x}(t,\mathbf{x}_0)$ of (2.155) is regular for at least one initial condition $\mathbf{x}_0 \in R^n$. Then there exists the solution of (2.155), which is a stationary Markov process.

As an example of application of the Theorem 2.16 Khasminski [27] considered a wide class of nonlinear oscillators described by

$$dx_1 = x_2 dt ,$$
$$dx_2 = -[f(x_1)x_2 + g(x_1)]dt + \sigma d\xi(t) , \tag{2.197}$$

where $f(x_1)$ and $g(x_1)$ are nonlinear functions, $\sigma$ is a constant, and $\xi(t)$ is a standard Wiener process.

Then taking into account the Lyapunov function $V(x_1,x_2)$ in the form

$$V(x_1,x_2) = \frac{x_2}{2} + [\Phi(x_1) - p(x_1)]x_2 + G(x_1)$$
$$+ \int_0^{x_1} f(s)[\Phi(s) - p(s)]ds + \kappa , \tag{2.198}$$

where

$$\Phi(x_1) = \int_0^{x_1} f(s)ds , G(x_1) = \int_0^{x_1} g(s)ds , p(x_1) = \gamma \arctan x_1 , \tag{2.199}$$

with appropriate chosen parameters $\gamma$, $\kappa$ if for some $\delta > 0$ and $x_0 > 0$ the following conditions are satisfied

$$sign[g(x_1)] = sign(x_1) \quad for |x| > x_0 ,$$

$$\lim_{|x_1| \to \infty} [g(x_1)\Phi(x_1) - \delta|g(x_1)|] = \infty ,$$

$$\lim_{|x_1| \to \infty} \left[ G(x_1) + \delta \int_0^{x_1} \frac{\Phi(s)}{1 + s^2} ds \right] = \infty \; . \tag{2.200}$$

This result is very important for particular classes of nonlinear oscillators, namely for the Duffing oscillator ($f(x_1) = 2h = \text{const}$, $g(x_1) = \lambda_0^2 x_1 + \varepsilon x_1^3$) and the Van der Pol oscillator ($f(x_1) = \varepsilon(x_1^2 - 1)$, $g(x_1) = x_1$); $\lambda_0^2$, $\varepsilon$, and $h$ are constant positive parameters.

Another approach leading to sufficient conditions for the existence of the solution of (2.155) for $\mathbf{F}(\mathbf{x}, t) = \mathbf{F}(\mathbf{x})$ and $\mathbf{G}_k(\mathbf{x}, t) = \mathbf{G}_k(\mathbf{x})$, $k = 1, \ldots, M$, satisfying local Lipschitz conditions was proposed by Bernard and Fleury [5, 6] and developed by Fleury [18]. These authors considered different numerical (discrete time) schemes of the solution of stochastic differential equations such as explicit and implicit Euler, Milshtein, or Newmark schemes.

In the particular case for stochastic mechanics when the considered equations have the form

$$dX_1 = X_2 dt,$$
$$dX_2 = [\mathbf{M}^{-1} \mathbf{C} \mathbf{X}_2 + \mathbf{M}^{-1} \mathbf{K} \mathbf{X}_1] dt + \mathbf{M}^{-1} \mathbf{D} d\boldsymbol{\xi}(t) \; , \tag{2.201}$$

where $\mathbf{X}_1, \mathbf{X}_2 \in R^n$, $\mathbf{M} = \mathbf{M}(\mathbf{X}_1, \mathbf{X}_2)$, $\mathbf{C} = \mathbf{C}(\mathbf{X}_1, \mathbf{X}_2)$, $\mathbf{K} = \mathbf{K}(\mathbf{X}_1, \mathbf{X}_2)$, $\mathbf{D} = \mathbf{D}(\mathbf{X}_1, \mathbf{X}_2)$, $\mathbf{K}$ and $\mathbf{C}$ are symmetric nonnegative $n \times n$ matrices, $\mathbf{D}$ is a $n \times m$ matrix, $\mathbf{M}$ is an invertible symmetric $n \times n$ positive matrix, $\boldsymbol{\xi}$ is $m$-dimensional standard Wiener process, then the solutions are usually continuous and therefore more complicated Newmark scheme can be applied.

To obtain the sufficient conditions for the uniform convergence of the trajectories of the approximating processes to those of the solution process almost surely the authors have used the stopping time technique. According to Fleury's suggestion [18], this approach seems to be applicable also to other numerical schemes of solving stochastic differential equations presented, for instance, in Kloeden and Platen's book [28].

## 2.12 Stochastic Stability

The problem of stability of stochastic dynamic systems is one of the fundamental areas of research interest in control theory as well as in response analysis. It is closely related to the qualitative analysis of stochastic differential equations. Due to different definitions of convergence of stochastic processes several definitions of stochastic stability were introduced in the literature [27, 29, 37, 38]. In this section we quote only the basic definitions for the trivial solution of vector stochastic differential equation (2.155).

**Definition 2.25.** *The trivial solution* $\mathbf{x} \equiv 0$ *of (2.155) is said to be stochastically stable (or stable in probability) if*

$$\forall_{\varepsilon > 0} \forall_{r > 0} \exists_{\delta(\varepsilon) > 0} \quad |\mathbf{x}_0| < \delta \Rightarrow \forall_{t \geq t_0} P\{|\mathbf{x}(t; t_0, \mathbf{x}_0)| > r\} < \varepsilon \; . \tag{2.202}$$

**Definition 2.26.** *The trivial solution* $\mathbf{x} \equiv 0$ *of (2.155) is said to be asymptotically stochastically stable if it is stochastically stable and moreover if*

$$\forall_{r>0} \exists_{\delta>0} \ |\mathbf{x}_0| < \delta \Rightarrow \lim_{t \to \infty} P\{|\mathbf{x}(t; t_0, \mathbf{x}_0)| > r\} = 0 . \tag{2.203}$$

**Definition 2.27.** *The trivial solution* $\mathbf{x} \equiv 0$ *of (2.155) is said to be asymptotically stochastically stable in the large if it is stochastically stable and moreover if*

$$\forall_{\mathbf{x}_0} P\{\lim_{t \to \infty} |\mathbf{x}(t; t_0, \mathbf{x}_0)| = 0\} = 1 . \tag{2.204}$$

**Definition 2.28.** *The trivial solution* $\mathbf{x} \equiv 0$ *of (2.155) is said to be p-stable* $(p > 0)$ *if*

$$\forall_{\mathbf{x}_0} \lim_{\delta \to 0} \sup_{|\mathbf{x}_0| < \delta, \, t \geq t_0} E\left[|\mathbf{x}(t; t_0, \mathbf{x}_0)|^p\right] = 0 . \tag{2.205}$$

**Definition 2.29.** *The trivial solution* $\mathbf{x} \equiv 0$ *of (2.155) is said to be asymptotically p-stable* $(p > 0)$ *if it is p-stable and moreover if*

$$\lim_{t \to \infty} E\left[|\mathbf{x}(t; t_0, \mathbf{x}_0)|^p\right] = 0 . \tag{2.206}$$

**Definition 2.30.** *The trivial solution* $\mathbf{x} \equiv 0$ *of (2.155) is said to be pth moment exponentially stable* $(p > 0)$ *if there is a pair of positive constants c and* $\alpha$ *such that*

$$E\left[|\mathbf{x}(t; t_0, \mathbf{x}_0)|^p\right] \leq c|\mathbf{x}_0|^p \exp\{-\alpha(t - t_0)\} . \tag{2.207}$$

In the particular case for $p = 1$ and $p = 2$ the stability is called stability in mean and stability in mean-square sense, respectively.

**Definition 2.31.** *The trivial solution* $\mathbf{x} \equiv 0$ *of (2.155) is said to be stable with probability 1 in some sense if all trajectories except a set of measure zero are stable in appropriate sense.*

Now we quote a few criteria of stochastic stability.

**Theorem 2.17.** *[38] If there exists a positive-definite function* $V(\mathbf{x}, t) \in C_2$ *in* $U_\rho \times [t_0, \infty)$ *such that*
$$LV(\mathbf{x}, t) \leq 0 , \tag{2.208}$$

*for all* $(\mathbf{x}, t) \in U_\rho \times [t_0, \infty)$, *then the trivial solution* $\mathbf{x} \equiv 0$ *of (2.155) is stochastically stable.*

If the weak inequality (2.208) is replaced by the strong inequality $LV$ $(\mathbf{x}, t) < 0$, then the trivial solution $\mathbf{x} \equiv 0$ of (2.155) is stochastically asymptotically stable.

**Theorem 2.18.** *[38] If there exists a positive-definite function $V(\mathbf{x}, t) \in C_2$ in $R^n \times [t_0, \infty)$ such that*

$$\lim_{\rho \to \infty} \inf_{|\mathbf{x}| > \rho} V(\mathbf{x}, t) = \infty ,  \tag{2.209}$$

$$LV(\mathbf{x}, t) < 0 ,  \tag{2.210}$$

*for all $(\mathbf{x}, t) \in U_r \times [t_0, \infty)$, then the trivial solution $\mathbf{x} \equiv 0$ of (2.155) is stochastically asymptotically stable in large.*

**Theorem 2.19.** *[27] If there exists a positive-definite function $V(\mathbf{x}, t) \in C_2$ in $R^n \times [t_0, \infty)$ and positive constants $c_1, c_2, c_3$ such that*

$$c_1 |\mathbf{x}|^p \leq V(\mathbf{x}, t) \leq c_2 |\mathbf{x}|^p ,  \tag{2.211}$$

$$LV(\mathbf{x}, t) \leq -c_3 |\mathbf{x}|^p ,  \tag{2.212}$$

*for all $(\mathbf{x}, t) \in U_r \times [t_0, \infty)$, then the trivial solution $\mathbf{x} \equiv 0$ of (2.155) is pth moment exponentially stable.*

In this particular case, using Theorem 2.19 one can derive the following stability conditions.

**Theorem 2.20.** *[38] If there exists a symmetric positive-definite $n \times n$ matrix $\mathbf{H}$ and constants $\alpha_1, \alpha_2, \alpha_3$ such that for $(\mathbf{x}, t) \in R^n \times [t_0, \infty)$*

$$\mathbf{x}^T \mathbf{H} \mathbf{F}(\mathbf{x}, t) + \frac{1}{2} tr\{\mathbf{G}^T(\mathbf{x}, t) \mathbf{H} \mathbf{G}(\mathbf{x}, t)\} \leq \alpha_1 \mathbf{x}^T \mathbf{H} \mathbf{x}  \tag{2.213}$$

*and*

$$\alpha_2 \mathbf{x}^T \mathbf{H} \mathbf{x} \leq |\mathbf{x}^T \mathbf{H} \mathbf{G}(\mathbf{x}, t)| \leq \alpha_3 \mathbf{x}^T \mathbf{H} \mathbf{x} ,  \tag{2.214}$$

*where $\mathbf{G} = [\mathbf{G}_1 \mathbf{G}_2 \ldots \mathbf{G}_M]$.*

(i) *If $\alpha_1 < 0$ , then the trivial solution $\mathbf{x} \equiv 0$ of (2.155) is pth moment exponentially stable for $p < 2 + 2\frac{|\alpha_1|}{\alpha_3^2}$.*

(ii) *If $0 \leq \alpha_1 < \alpha_2^2$ , then the trivial solution $\mathbf{x} \equiv 0$ of (2.155) is pth moment exponentially stable for $p < 2 - 2\frac{\alpha_1}{\alpha_2^2}$.*

The proof of Theorem 2.20 [38] follows immediately from the calculation of $LV$ where the Lyapunov function $V(\mathbf{x}, t)$ is assumed in the form

$$V(\mathbf{x}, t) = (\mathbf{x}^T \mathbf{H} \mathbf{x})^{\frac{p}{2}} . \tag{2.215}$$

Then we obtain

$$\lambda_{\min}^{\frac{p}{2}}(\mathbf{H})|\mathbf{x}|^p \leq V(\mathbf{x}, t) \leq \lambda_{\max}^{\frac{p}{2}}(\mathbf{H})|\mathbf{x}|^p , \tag{2.216}$$

where $\lambda_{\min}(\mathbf{H})$ and $\lambda_{\max}(\mathbf{H})$ denote the smallest and the largest eigenvalue of $\mathbf{H}$, respectively;
and

$$LV(\mathbf{x}, t) = p(\mathbf{x}^T \mathbf{H} \mathbf{x})^{\frac{p}{2}-1} \left[ \mathbf{x}^T \mathbf{H} \mathbf{F}(\mathbf{x}, t) + \frac{1}{2} tr\{\mathbf{G}^T(\mathbf{x}, t)\mathbf{H}\mathbf{G}(\mathbf{x}, t)\} \right]$$
$$+ p \left( \frac{p}{2} - 1 \right) (\mathbf{x}^T \mathbf{H} \mathbf{x})^{\frac{p}{2}-2} |\mathbf{x}^T \mathbf{H} \mathbf{G}(\mathbf{x}, t)|^2 . \tag{2.217}$$

From equality (2.217) and assumptions (2.215, 2.216) and (i) it follows that

$$LV(\mathbf{x}, t) \leq -p \left[ |\alpha_1| - \left( \frac{p}{2} - 1 \right) \alpha_3^2 \right] V(\mathbf{x}, t) . \tag{2.218}$$

Similarly, from equality (2.217) and assumptions (2.215, 2.216) and (ii) we obtain

$$LV(\mathbf{x}, t) \leq -p \left[ -\alpha_1 + \left( \frac{p}{2} - 1 \right) \alpha_2^2 \right] V(\mathbf{x}, t) . \tag{2.219}$$

Hence the proof of Theorem 2.20 follows from Theorem 2.19.
We note that under assumptions of Theorem 2.19, one can prove [27] that

$$\exists_{\lambda>0} \forall_{\mathbf{x}_0} \forall_{t_0} \forall_{t \geq t_0} \ |x(t_0, \mathbf{x}_0; t)| \leq K(t_0, \mathbf{x}_0) \exp\{-\lambda t\} \tag{2.220}$$

where the random variable $K(\mathbf{x}_0, t_0)$ is almost sure (with probability 1) bounded.
We illustrate these criteria on two examples.

*Example 2.1.* Consider a scalar Ito stochastic differential equation

$$dx(t) = -\alpha x(t)dt + \sigma x(t)d\xi(t) , \tag{2.221}$$

where $\alpha$ and $\sigma$ are positive constants. The corresponding Ito operator has the form

$$L = L_I = \frac{\partial}{\partial t} - \alpha x(t) \frac{\partial}{\partial x} + \frac{1}{2} \sigma^2 x^2(t) \frac{\partial^2}{\partial x^2} , \tag{2.222}$$

and we assume that the Lyapunov function is

$$V(\mathbf{x}, t) = c|x|^p , \tag{2.223}$$

where $c, p$ are positive constants. Then the sufficient condition of exponential p-stability is

$$LV(x,t) = (-\alpha + \frac{p-1}{2}\sigma^2)pc|x|^p \leq -c_3|x|^p .\tag{2.224}$$

Hence it follows that for $p < 1$ the system can be stabilized by parametric noise. It means that even for the negative parameter $\alpha$ could be exponentially p-stable for sufficient big parameter $\sigma$ when the inequality $-\alpha + \frac{p-1}{2}\sigma^2 < 0$ holds. This statement contradicts our intuition. Therefore if we model a real physical system we describe it by the corresponding Stratonovich equation. In our example it will be

$$dx(t) = -\alpha x(t)dt + \sigma x(t)d_S\xi(t) .\tag{2.225}$$

Then the equivalent Ito equation has the form

$$dx(t) = \left(-\alpha + \frac{1}{2}\sigma^2\right)x(t)dt + \sigma x(t)d\xi(t) .\tag{2.226}$$

Assuming the Lyapunov function in the form (2.223) we calculate the sufficient condition of exponential p-stability

$$LV(\mathbf{x},t) = \left(-\alpha + \frac{1}{2}\sigma^2 + \frac{p-1}{2}\sigma^2\right)pc|x|^p = \left(-\alpha + \frac{p}{2}\sigma^2\right)pc|x|^p \leq -c_3|x|^p .\tag{2.227}$$

In condition (2.223) the mentioned contradiction disappears.

*Example 2.2.*

$$dx_1 = x_2 dt,$$
$$dx_2 = -[a_{21}x_1 + a_{22}x_2 + bg(x_1)]dt + \sigma x_1 d\xi(t) ,\tag{2.228}$$

where $a_{21} < 0$, $a_{22} < 0$, $b > 0$, $\sigma > 0$ are constant parameters, $g(x_1)$ is a nonlinear monotonically nondecreasing function satisfying the following inequalities

$$0 \leq g(x_1)x_1 \leq Mx_1^2 , \quad M > 0 .\tag{2.229}$$

We assume the Lyapunov function in the form

$$V(\mathbf{x},t) = \mathbf{x}^T\mathbf{W}\mathbf{x} = [x_1 x_2]\begin{bmatrix} w_1 & w_2 \\ w_2 & w_3 \end{bmatrix}\begin{bmatrix} x_1 \\ x_2 \end{bmatrix} ,\tag{2.230}$$

where $\mathbf{W}$ is a positive-definite matrix. Then condition (2.211) is satisfied for $p = 2$ if we denote the smallest and the greatest eigenvalue of the matrix $\mathbf{W}$ by $c_1$ and $c_2$, respectively, i.e.,

$$c_1 = \min\lambda(\mathbf{W}), \quad c_2 = \max\lambda(\mathbf{W}) .\tag{2.231}$$

Condition (2.212) for $p = 2$ is satisfied if the following inequalities hold

$$
\begin{aligned}
LV &= 2w_1x_1x_2 + 2w_2x_2^2 + 2w_2a_{21}x_1^2 + 2w_2a_{22}x_1x_2 + 2w_2bx_1g(x_1) \\
&\quad 2w_3a_{21}x_1x_2 + 2w_3a_{22}x_2^2 + 2w_3bx_2g(x_1) + w_3\sigma^2x_1^2 \\
&\leq (2w_2a_{21} + w_3\sigma^2)x_1^2 + 2(w_1 + w_2a_{22} + w_3a_{21})x_1x_2 + 2w_3a_{22}x_2^2 \\
&\quad +2w_2bMx_1^2 + 2w_3b|x_2||g(x_1)| \ \leq \ -2\mathbf{x}^T\mathbf{R}\mathbf{x} \ \leq \ -c_3\mathbf{x}^T\mathbf{x}\,, \quad (2.232)
\end{aligned}
$$

where

$$
\mathbf{R} = \begin{bmatrix} -w_2(a_{21} + bM - \frac{1}{2}w_3\sigma^2) - w_3bM & -\frac{1}{2}(w_1 + w_2a_{22} + w_3a_{21}) \\ -\frac{1}{2}(w_1 + w_2a_{22} + w_3a_{21}) & -(w_2 + w_3a_{22} + \frac{1}{2}w_3bM) \end{bmatrix}\,,
$$
$$(2.233)$$

$$
c_3 = \lambda_{\min}(\mathbf{R})\,. \tag{2.234}
$$

The first inequality in condition (2.232) follows from (2.229) and the second one from the property that the function $g(x_1)$ was assumed to be monotonically nondecreasing. For instance, if we assume that $a_{21} = -4$, $a_{22} = -2$, $b = 1$, $\sigma^2 = 1$, $M = 1$, then the matrix $\mathbf{R}$ takes the form

$$
\mathbf{R} = \begin{bmatrix} 3w_2 - w_3 & -\frac{1}{2}(w_1 - 2w_2 - 4w_3) \\ -\frac{1}{2}(w_1 - 2w_2 - 4w_3) & -w_2 + \frac{3}{2}w_3 \end{bmatrix}\,. \tag{2.235}
$$

Then one can find the sufficient conditions of mean-square exponential stability in terms of $w_i$, $i = 1, 2, 3$, from the following inequalities

$$
w_1 > 0\,, \quad w_1w_3 - (w_2)^2 > 0\,, \quad 3w_2 - w_3 > 0\,,
$$
$$
4(3w_2 - w_3)\left(\frac{3}{2}w_3 - w_2\right) - (w_1 - 2w_2 - 4w_3)^2 > 0\,. \tag{2.236}
$$

For instance, the elements $w_1 = 10$ and $w_2 = 2$ satisfy inequalities (2.236) and then the coefficients $c_i$, $i = 1, 2, 3$, determined by (2.231) and (2.234) are equal to

$$
c_1 = 6 - \sqrt{17}\,, \quad c_2 = 6 + \sqrt{17}\,, \quad c_3 = 1\,. \tag{2.237}
$$

## 2.13 The Method of Fokker–Planck–Kolmogorov Equations

To derive moment equations and to determine exact characteristics of the stationary solutions of nonlinear stochastic differential equations one can use the method of Fokker–Planck–Kolmogorov equations (FPK). We shortly discuss this approach.

Consider the nonlinear vector Ito stochastic differential equation

$$
d\mathbf{x}(t) = \mathbf{F}(\mathbf{x}(t), t)dt + \boldsymbol{\sigma}(\mathbf{x}(t), t)d\boldsymbol{\xi}(t)\,, \quad \mathbf{x}(t_0) = \mathbf{x}_0\,, \tag{2.238}
$$

where $\mathbf{x} = [x_1, \ldots, x_r]^T$ is the vector state, $\mathbf{x}_0 = [x_{10}, \ldots, x_{r0}]^T$ is the random vector of initial conditions independent of $\boldsymbol{\xi}(t)$, $\mathbf{F}(\mathbf{x}, t) = [a_1(\mathbf{x}, t), \ldots, a_r(\mathbf{x}, t)]^T$ is a nonlinear vector function, $\boldsymbol{\sigma}(\mathbf{x}, t) = [\sigma_{ik}(\mathbf{x}, t)]$, $i = 1, \ldots, r$, $k = 1, \ldots, M$, is a matrix whose elements are nonlinear functions; $\boldsymbol{\xi}(t) = [\xi_1(t), \ldots, \xi_M(t)]^T$ is an M-dimensional Wiener process whose differentials satisfy the condition

$$E[d\xi_k d\xi_l] = 2\pi K_{kl} dt , \quad k, l = 1, \ldots, M , \tag{2.239}$$

where $K_{kl}$ are constant positive parameters.

Then the FPK equation corresponding to (2.238) has the form (2.113), i.e.,

$$\frac{\partial g}{\partial t} + \sum_{i=1}^{r} \frac{\partial [A_i(\mathbf{x}, t) g]}{\partial x_i} - \frac{1}{2} \sum_{i=1}^{r} \sum_{j=1}^{r} \frac{\partial^2 [b_{ij}(\mathbf{x}, t) g]}{\partial x_i \partial x_j} = 0 , \tag{2.240}$$

where $g = g(t_0, \mathbf{x}_0; t, \mathbf{x})$ is the probability density function of the solution of (2.238), where for convenience the notation of arguments of function $g$ has been changed, i.e., $s = t_0$, $\mathbf{x} = \mathbf{x}_0$, $\mathbf{y} = \mathbf{x}$;

$$A_i(\mathbf{x}, t) = F_i(\mathbf{x}, t) \tag{2.241}$$

are components of drift vector and

$$b_{ij}(\mathbf{x}, t) = 2\pi \sum_{k=1}^{M} \sum_{l=1}^{M} K_{kl} \sigma_{ki} \sigma_{lj} \tag{2.242}$$

are elements of diffusion matrix $\mathbf{B}(\mathbf{x}, t) = [b_{ij}]$.

For simplicity we assume that the initial condition for (2.238) is deterministic. Hence, it follows the form of initial condition for (2.240)

$$g(t_0, \mathbf{x}_0; t, \mathbf{x}) = \prod_{j=1}^{n} \delta(x_j - x_{0j}) . \tag{2.243}$$

The details concerning the derivation of FPK equations as well as the determination of coefficients $A_i(\mathbf{x}, t)$ and $b_{ij}(\mathbf{x}, t)$ can be found in many monographs, for instance, [17, 21, 24, 25, 44, 47]. If (2.238) is treated in Stratonovich sense, then the elements of the diffusion matrix $\mathbf{B}(\mathbf{x}, t)$ are determined by (2.242), while the elements of drift vector have the form

$$A_i(\mathbf{x}, t) = F_i(\mathbf{x}, t) + \pi \sum_{j=1}^{r} \sum_{k=1}^{M} \sum_{l=1}^{M} K_{kl} \frac{\partial \sigma_{ki}}{\partial x_j} \sigma_{lj} . \tag{2.244}$$

FPK equation is a linear partial differential equation of second order (parabolic equation) with variable coefficients. It should be stressed that FPK equation can be used in the derivation of moment and cumulant equations of linear and nonlinear systems. Such derivations one can find, for instance, in

[16, 43]. Although, in the mathematical literature there are several methods of the solutions of different class of parabolic partial differential equations the general solution of (2.240) has not been obtained yet. An example of an exact analytical nonstationary solution was shown by Caughey and Dienes [11] for a particular class of scalar equation

$$dx(t) = -k\,\mathrm{sgn}(x)dt + d\xi(t), \quad x(t_0) = x_0, \quad k > 0 .$$ (2.245)

The solution of the corresponding FPK equation is the function depending on $t$ and $x$

$$g(0, x_0; t, x) = \frac{2k}{\sqrt{\pi}} \exp\{-2k|x|\} \int_{(x_0+|x|)/\sqrt{2t}}^{+\infty} \exp\{-u^2\}du$$
$$+ \frac{1}{\sqrt{2\pi t}} \exp\left\{ k(x - |x|) - \frac{(x - x_0 - kt)^2}{2t} \right\} .$$ (2.246)

In the case when the coefficients of the state equation (2.238) are constant, then under some additional assumptions one can find a stationary solution of the corresponding FPK equation, i.e., the probability density function depending only on vector state $g(t, \mathbf{x}) = g(\mathbf{x})$. The independence of the solution of (2.240) of variable $t$ implies that the derivative $\partial g/\partial t = 0$, i.e., the left side of (2.240), is equal to zero. Then the solution of (2.240) reduces to the solution of a boundary problem. The methods of the determination of this solution in an analytical form for different classes of nonlinear systems were given, for instance, by Andronov et al. [1], Fuller [19], Caughey [9], Stratonovich [52], Caughey and Ma [12], Dimentberg [14], and Soize [47]. These method are known in the literature as potential methods. They were generalized by Lin and his coworkers [31, 32, 58] who called their approach *dissipation energy balancing*.

We are going to present the method called *generalized potential method* [8], which is a further generalization of *dissipation energy balancing*. This approach is very useful in the determination of exact stationary solutions of a wide class of nonlinear stochastic dynamic systems.

We consider the simplified (reduced) FPK equation (2.240) in the following form

$$\sum_{i=1}^{r} \frac{\partial [A_i(\mathbf{x}, t)g_s(\mathbf{x})]}{\partial x_i} - \frac{1}{2} \sum_{i=1}^{r} \sum_{j=1}^{r} \frac{\partial^2 [b_{ij}(\mathbf{x}, t)g_s(\mathbf{x})]}{\partial x_i \partial x_j} = 0 ,$$ (2.247)

where $g_s(\mathbf{x}) = g(t_0, \mathbf{x}_0; t, \mathbf{x})$ is the sought stationary probability density function of the solution of (2.238). Equation (2.247) can be rewritten in the form

$$\sum_{i=1}^{r} \frac{\partial G_i(\mathbf{x})}{\partial x_i} = 0 , \quad i = 1, \ldots, r ,$$ (2.248)

where $G_i(\mathbf{x})$ is the *probability flow in ith direction* defined by

$$G_i(\mathbf{x}) = A_i(\mathbf{x}, t)g_s(\mathbf{x}) - \frac{1}{2}\sum_{j=1}^{r}\frac{\partial[b_{ij}(\mathbf{x}, t)g_s(\mathbf{x})]}{\partial x_j} , \quad i = 1,\ldots,r . \quad (2.249)$$

Usually $g_s(\mathbf{x})$ and hence also $G_i(\mathbf{x})$ satisfy the natural boundary conditions

$$G_i(\mathbf{x}) = 0, \quad for \quad x_i \in \partial D , \quad i = 1, \quad \ldots,r , \quad (2.250)$$

where $\partial D$ denotes the boundary of the state space of vector variable $\mathbf{x} \in D \subseteq R^r$. In this particular case the stationary potential is such a probability flow that vanish on the boundary. Therefore without loss of generality one can assume that $g_s(\mathbf{x})$ has the form

$$g_s(\mathbf{x}) = c\exp\{-\phi(\mathbf{x})\} , \quad (2.251)$$

where $\phi(\mathbf{x})$ is a nonlinear function and $c$ is a normalized constant

$$c^{-1} = \int_D g_s(\mathbf{x})d(\mathbf{x}) . \quad (2.252)$$

Substituting (2.251) in to (2.249) and equating to zero $G_i(\mathbf{x})$, $i = 1,\ldots,r$, we obtain

$$\sum_{j=1}^{r} b_{ij}\frac{\partial\phi(\mathbf{x})}{\partial x_j} = \sum_{j=1}^{r}\frac{\partial b_{ij}(\mathbf{x})}{\partial x_j} - 2A_i , \quad i = 1,\ldots,r . \quad (2.253)$$

If it is possible to find a solution of (2.253) in the form $\phi = \phi(\mathbf{x})$, then we say that the stochastic system (2.247) belongs to the class of stationary potential. Unfortunately, in many cases such a solution cannot be found. Particularly it concerns nonlinear systems with oscillating properties. To obtain a solution of (2.247) Cai and Lin [8] proposed a splitting of drift coefficients $A_i$ and diffusion coefficients $b_{ij}$, $i, j = 1,\ldots,r$, into the symmetric and asymmetric parts

$$A_i = A_i^{(1)} + A_i^{(2)} , \quad (2.254)$$

$$b_{ij} = b_{ij}^{(i)} + b_{ij}^{(j)} . \quad (2.255)$$

Equality (2.254) does retain the symmetric property of the diffusion matrix $b_{ij} = b_{ji}$. Substituting (2.251), (2.254), and (2.255) in to (2.247), we obtain

$$\sum_{j=1}^{r} b_{ij}^{(i)}\frac{\partial\phi(\mathbf{x})}{\partial x_j} = \sum_{j=1}^{r}\frac{\partial b_{ij}^{(i)}(\mathbf{x})}{\partial x_j} - A_i^{(i)} , \quad i = 1,\ldots,r , \quad (2.256)$$

$$\sum_{j=1}^{r} A_i^{(2)}\frac{\partial\phi(\mathbf{x})}{\partial x_j} = \sum_{j=1}^{r}\frac{\partial A_i^{(2)}(\mathbf{x})}{\partial x_j} , \quad i = 1,\ldots,r . \quad (2.257)$$

Condition (2.256) is connected with vanishing of the probability flow and condition (2.257) with his rotation. We show an application of the method of generalized potential to one degree-of-freedom (1–DOF) system.

Consider a nonlinear system described by the vector Stratonovich stochastic differential equation [8]

$$dx_1 = x_2 dt, \quad x_1(t_0) = x_{10} ,$$

$$dx_2 = -f(x_1, x_2)dt + \sum_{k=1}^{M} \sigma_k(x_1, x_2)d\xi_k , \quad x_2(t_0) = x_{20} , \quad (2.258)$$

where $f(.)$ and $\sigma(.)$ are nonlinear functions and $\xi_k(t)$, $k = 1, \ldots, M$, are mutually independent zero-mean Wiener processes with the variance of differentials given by (2.239), the initial conditions $x_{10}$ and $x_{20}$ are random variables independent of $\xi_k$, $k = 1, \ldots, M$.

By transforming (2.258) into the Ito equation and by using relations (2.174) one can obtain the corresponding FPK equation, which for stationary case has the form

$$x_2 \frac{\partial g_s}{\partial x_1} + \frac{\partial}{\partial x_2} \left\{ \left[ -f(x_1, x_2) + \pi \sum_{k=1}^{M} \sum_{l=1}^{M} K_{kl} \sigma_k(x_1, x_2) \frac{\partial \sigma_l(x_1, x_2)}{\partial x_2} \right] g_s \right\}$$

$$- \pi \sum_{k=1}^{M} \sum_{l=1}^{M} K_{kl} \frac{\partial^2}{\partial x_2^2} [\sigma_k(x_1, x_2)\sigma_l(x_1, x_2)g_s] = 0 , \quad (2.259)$$

i.e., the drift and diffusion coefficients are determined by relations

$$A_1 = x_2 ,$$

$$A_2 = -f(x_1, x_2) + \pi \sum_{k=1}^{M} \sum_{l=1}^{M} K_{kl} \sigma_k(x_1, x_2) \frac{\partial \sigma_l(x_1, x_2)}{\partial x_2} ,$$

$$b_{11} = b_{12} = b_{21} = 0 ,$$

$$b_{22} = 2\pi \sum_{k=1}^{M} \sum_{l=1}^{M} K_{kl} \sigma_k(x_1, x_2)\sigma_l(x_1, x_2) . \quad (2.260)$$

We split both groups of coefficients as follows:

$$A_1^{(1)} = 0 , \quad A_1^{(2)} = x_2 ,$$

$$A_2^{(1)} = -f(x_1, x_2) + \pi \sum_{k=1}^{M} \sum_{l=1}^{M} K_{kl} \sigma_k(x_1, x_2) \frac{\partial \sigma_l(x_1, x_2)}{\partial x_2} + \frac{\lambda_x}{\lambda_y} ,$$

$$A_2^{(2)} = -\frac{\lambda_x}{\lambda_y} , \quad b_{22}^{(1)} = \frac{1}{2}b_{22} , \quad b_{21}^{(2)} = -b_{12}^{(1)} , \quad (2.261)$$

where $x = x_1$, $y = \frac{1}{2}(x_2)^2$, $\lambda(x, y)$ is a proposed function twice differentiable with respect to $x$ and $y$, $\lambda_x = \partial\lambda/\partial x$, $\lambda_y = \partial\lambda/\partial y$, $\lambda_{yy} = \partial^2\lambda/\partial y^2$.

Taking into account the splitting (2.261) in (2.259) we obtain relations corresponding to (2.256) and (2.257)

$$b_{12}^{(1)} \frac{\partial \phi(\mathbf{x})}{\partial x_2} = \frac{\partial b_{12}^{(1)}(\mathbf{x})}{\partial x_2} \tag{2.262}$$

and

$$b_{21}^{(2)} \frac{\partial \phi}{\partial x_1} + \pi \sum_{k=1}^{M} \sum_{l=1}^{M} K_{kl} \sigma_k(x_1, x_2) \sigma_l(x_1, x_2) \frac{\partial \phi}{\partial x_2}$$

$$= \frac{\partial b_{21}^{(2)}}{\partial x_1} + f(x_1, x_2) - \pi \sum_{k=1}^{M} \sum_{l=1}^{M} K_{kl} \sigma_k(x_1, x_2) \frac{\partial \sigma_l(x_1, x_2)}{\partial x_2} - \frac{\lambda_x}{\lambda_y} , \tag{2.263}$$

$$-x_2 \frac{\partial \phi}{\partial x_1} - \frac{\partial}{\partial x_2} \left( \frac{\lambda_x}{\lambda_y} \right) + \frac{\lambda_x}{\lambda_y} \frac{\partial \phi}{\partial x_2} = 0 , \tag{2.264}$$

respectively.

The solution of (2.264) has the form

$$\phi(x_1, x_2) = -ln\lambda_y + \phi_0(\lambda) , \tag{2.265}$$

where $\phi_0(\lambda)$ is a function following from the integration of (2.264). Equation (2.262) can be presented in the form

$$\frac{\partial}{\partial x_2} \left( b_{12}^{(1)}(x_1, x_2) \exp\{-\phi(x_1, x_2)\} \right) = 0 . \tag{2.266}$$

Hence, we find that the term $b_{12}^{(1)} \exp\{-\phi(x_1, x_2)\}$ is only a function of $x_1$ and we denote it by $D_1(x_1)$, i.e.,

$$D_1(x_1) = b_{12}^{(1)} \exp\{-\phi(x_1, x_2)\} . \tag{2.267}$$

By using relations (2.261–2.267) Cai and Lin [8] have shown that nonlinear function $f(x_1, x_2)$ satisfies the following condition:

$$f(x_1, x_2) = \pi \sum_{k=1}^{M} \sum_{l=1}^{M} x_2 K_{kl} \sigma_k(x_1, x_2) \sigma_l(x_1, x_2) \left[ \lambda_y \frac{d\phi_0(\lambda)}{d\lambda} - \frac{\lambda_{yy}}{\lambda_y} \right]$$

$$- \pi \sum_{k=1}^{M} \sum_{l=1}^{M} K_{kl} \sigma_k(x_1, x_2) \frac{\partial \sigma_l(x_1, x_2)}{\partial x_2} + \frac{\lambda_x}{\lambda_y} + \frac{1}{\lambda_y} \frac{dD_1(x_1)}{dx_1} e^{\phi_0(\lambda)} . \tag{2.268}$$

Equality (2.268) establishes the relationship between the functions $f(x_1, x_2)$, $\sigma_k(x_1, x_2)$, $k = 1, \ldots, M$, those that define (2.258), their derivatives $\partial \sigma_k(x_1, x_2)/\partial x_2$, and the derivatives of a function $\lambda(x_1, x_2)$. Hence, one can find the stationary solution of the corresponding FPK equation (2.259) in the form (2.251). Cai and Lin [8] illustrated the discussed method in the following example.

*Example 2.3.* Consider a nonlinear oscillator described by the Ito equation
[10]

$$dx_1 = x_2 dt, \quad x_1(t_0) = x_{10},$$
$$dx_2 = [-h(\Lambda) - f(x_1)]dt + \sqrt{2\pi\rho}d\xi(t), x_2(t_0) = x_{20} , \qquad (2.269)$$

where $h(.)$ and $f(.)$ are nonlinear functions, $\rho$ is a positive constant, $\xi(t)$ is
the standard Wiener process. $\Lambda$ is determined by

$$\Lambda = \Lambda(x_1, x_2) = \frac{1}{2}(x_2)^2 + \int_0^{x_1} f(u)du . \qquad (2.270)$$

The simplified FPK equation corresponding to the system (2.269) has the
form

$$x_2 \frac{\partial g_s}{\partial x_1} - \frac{\partial}{\partial x_2} \{[h(\Lambda) + f(x_1)] g_s\} - \pi\rho \frac{\partial^2 g_s}{\partial x_2^2} = 0 . \qquad (2.271)$$

By splitting of drift and diffusion coefficients (2.261), we obtain

$$A_1^{(1)} = 0, \quad A_1^{(2)} = x_2 , \quad A_2^{(1)} = -h(\Lambda) - f(x_1) + \frac{\lambda_x}{\lambda_y} , \quad A_2^{(2)} = -\frac{\lambda_x}{\lambda_y} ,$$
$$b_{22}^{(1)} = \pi\rho , \quad b_{21}^{(2)} = b_{12}^{(1)} = 0 . \qquad (2.272)$$

Equation (2.262) is satisfied and (2.263) and (2.264) for $\lambda = \Lambda$, $y = \frac{1}{2}(x_2)^2$,
take the form

$$\pi\rho \frac{\partial\phi}{\partial x_2} = h(\Lambda)x_2 , \qquad (2.273)$$

$$-x_2 \frac{\partial\phi}{\partial x_1} + f(x_1)\frac{\partial\phi}{\partial x_2} = 0 . \qquad (2.274)$$

Solving (2.273), we obtain

$$\phi(x_1, x_2) = \frac{1}{\pi\rho} \int_0^{\Lambda} h(u)du + \alpha(x_1) , \qquad (2.275)$$

where $\alpha(x_1)$ is a function calculated from an integration of (2.273). Substituting the relation (2.275) into (2.274), we find

$$\alpha(x_1) = \text{const} \qquad (2.276)$$

and finally the probability density function of the stationary solution of
system (2.269) has the form (2.251), i.e.,

$$g_s(x_1, x_2) = g_s(\Lambda) = c_1 \exp\left\{-\frac{1}{\pi\rho}\int_0^{\Lambda} h(u)du\right\} , \qquad (2.277)$$

where $c_1$ is a normalized constant.

In the particular case if

$$h(\Lambda) \equiv h, \quad f(x_1) = \lambda_0^2 x_1 + \varepsilon x_1^3 \,, \tag{2.278}$$

(2.269) is called *Duffing equation or Duffing oscillator* and the corresponding probability density function of its stationary solution is found from the relation (2.277), i.e.,

$$g_s(x_1, x_2) = c_1 \exp\left\{ -\frac{h}{\pi\rho}\left( \frac{\lambda_0^2}{2}(x_1)^2 + \frac{\varepsilon}{4}(x_1)^4 + \frac{1}{2}(x_2)^2 \right) \right\} \,, \tag{2.279}$$

where

$$c_1^{-1} = \int_{-\infty}^{+\infty}\int_{-\infty}^{+\infty} \exp\left\{ -\frac{h}{\pi\rho}\left( \frac{\lambda_0^2}{2}(x_1)^2 + \frac{\varepsilon}{4}(x_1)^4 + \frac{1}{2}(x_2)^2 \right) \right\} dx_1 dx_2 \,. \tag{2.280}$$

Since the exact solution of FPK equation could be found only in limited number of cases and only if the random excitations are Gaussian white noises, approximation methods of solving FPK equation were developed parallely. It includes *the method of eigenfunctions (variational approach or perturbation approach), the two-dimensional Hermite polynomials*, and *difference numerical methods*. A broad discussion of these methods is in monographs , for instance, Feller [17], Fuller [19], Gardiner [21], Risken [44], Krasowski [30], and Soize [47]. Also the generalized potential method was used by Lin and his coworkers [34] in new proposed approximation methods.

## Bibliography Notes

A wider study of stochastic calculus including mean-square analysis of stochastic processes can be found in many text books, for instance, [24, 25, 43, 46, 48]. The systematic study of Markov processes and stochastic differential equations can be found, for instance, in [20, 23, 35].

A wider treatment of the material collected in this subsection in the context of random vibration can be found in many books, for instance, [34, 41, 43, 45, 46, 49]. In particular, the problem of stochastic stability was considered by the Lyapunov function approach [7, 26, 33, 40], as well as by the Lyapunov exponent approach [2, 3, 4, 15].

A broad discussion of the methods of solution of the Fokker–Planck–Kolmogorov equations are in monographs , for instance, Feller [17], Fuller [19], Gardiner [21], Risken [44], Krasowski [30], and Soize [47]. Also the generalized potential method was used by Lin and his coworkers [34] in the new proposed approximation methods.

# References

1. Andronov, A., Wit, A., and Pontriagin, L.: On statistical analysis of dynamic systems. Zur. Vithisl. Math. Math. Phys. **3**, 165–180, 1933 (in Russian).
2. Ariaratnam, S. and Xie, W.: Lyapunov exponents and stochastic stability of two-dimensional parametrically excited random systems. ASME J. Appl. Mech. **60**, 677–682, 1993.
3. Ariaratnam, S. and Xie, W.: Almost sure stochastic stability of coupled non-linear oscillators. Int. J. Non-Linear Mech. **29**, 197–204, 1994.
4. Ariaratnam, S., Tam, D., and Xie, W.: Lyapunov exponents and stochastic stability of coupled linear systems under white noise excitations. Probab. Eng. Mech. **6**, 51–56, 1991.
5. Bernard, P. and Fleury, G.: Convergence of numerical schemes for stochastic differential equations. Monte Carlo Methods Appl. **7**, 35–533, 2001.
6. Bernard, P. and Fleury, G.: Stochastic newmark scheme. Probab. Eng. Mech. **17**, 45–61, 2002.
7. Brouwers, J.: Stability of non-linearly damped second-order system with randomly fluctuating restoring coefficient. Int. J. Non-Linear Mech. **21**, 1–13, 1986.
8. Cai, G. and Lin, Y.: On exact stationary solutions of equivalent nonlinear stochastic systems. Int. J. Non-Linear Mech. **23**, 315–325, 1988.
9. Caughey, T.: Derivation and application of the Fokker–Planck equation to discrete nonlinear dynamic systems subjected to white random excitation. J. Acoust. Soc. Am. **35**, 1683–1692, 1963.
10. Caughey, T.: Nonlinear theory of random vibrations. Adv. Appl. Mech. **11**, 209–253, 1971.
11. Caughey, T. and Dienes, J.: Analysis of a nonlinear first order system with a white noise input. J. Appl. Phys. **23**, 2476–2479, 1961.
12. Caughey, T. and Ma, F.: The exact steady-state solution of a class of nonlinear stochastic systems. Int. J. Non-Linear Mech. **18**, 137–142, 1983.
13. Clough, R. and Penzien, J.: Dynamics of Structures. IV Random Vibrations. McGraw-Hill, New York, 1972.
14. Dimentberg, M.: An exact solution to a certain nonlinear random vibration problem. Int. J. Non-Linear Mech. **17**, 231–236, 1982.
15. Doyle, D., Namachchivaya, N., and Van Roessel, H.: Asymptotic stability of structural systems based on Lyapunov exponents and moment Lyapunov exponents. Int. J. Non-Linear Mech. **32**, 681–692, 1997.
16. Evlanov, L. and Konstantinov, V.: Systems with Random Parameters. Nauka, Moskwa, 1976 (in Russian).
17. Feller, W.: An Introduction to Probability Theory and Its Applications, Vol. II. Wiley Eastern Ltd., New Delhi, 1966.
18. Fleury, G.: Convergence of schemes for stochastic differential equations. Probab. Eng. Mech. **21**, 35–43, 2006.
19. Fuller, A.: Analysis of nonlinear stochastic systems by means of the Fokker–Planck equation. Int. J. Control **9**, 603–655, 1969.
20. Gard, T.: Introduction to Stochastic Differential Equations. Marcel Dekker, New York, 1988.
21. Gardiner, C.: Handbook of Stochastic Methods for Physics, Chemistry and Natural Sciences. Springer, New York, 1983.
22. Gihman, I. and Skorochod, A.: Stochastic Differential Equations and Their Applications. Naukovaja Dumka, Kiew, 1982 (in Russian).

23. Gihman, I. and Skorochod, A.: Stochastic Differential Equations. Springer, Berlin, 1968.
24. Gihman, I. and Skorochod, A.: The Theory of Stochastic Processes, Vol. II. Springer, Berlin, 1974.
25. Gihman, I. and Skorochod, A.: The Theory of Stochastic Processes, Vol. I. Springer, Berlin, 1974.
26. Katafygiotis, I., Papadimitriou, C., and Tsarkov, Y.: Mean-square stability of linear stochastic dynamical systems under parametric wide-band excitations. Probab. Eng. Mech. **12**, 137–147, 1997.
27. Khasminski, R.: Stochastic Stability of Differential Equations. Sijthoff and Noordhoff, Alpen aan den Rijn, the Netherlands, 1980.
28. Kloeden, P. and Platen, E.: The Numerical Solution of Stochastic Differential Equations, 2nd edn. Springer, Berlin, 1995.
29. Kozin, F.: A survey of stability of stochastic systems. Automatica **5**, 95–112, 1969.
30. Krasowski, A.: Phase Space and Statistical Theory of Dynamic Systems. Nauka, Moskwa, 1974 (in Russian).
31. Lin, Y. and Cai, G.: Equivalent stochastic systems. Trans. ASME J. Appl. Mech. **55**, 918–922, 1988.
32. Lin, Y. and Cai, G.: Exact stationary response for second order nonlinear systems under parametric and external white-noise excitations II. Trans. ASME J. Appl. Mech. **55**, 702–705, 1988.
33. Lin, Y. and Cai, G.: Stochastic stability of non-linear systems. Int. J. Non-Linear Mech. **29**, 539–553, 1994.
34. Lin, Y. and Cai, G.: Probabilistic Structural Dynamics. McGraw Hill, New York, 1995.
35. Lipcer, R. and Shiryayev, A.: Statistics of Random Processes I : General Theory. Springer, New York, 1977.
36. Loeve, M.: Probability Theory. Van Nostrand, New York, 1955.
37. Mao, X.: Exponential Stability of Stochastic Differential Equations. Marcel Dekker, New York, 1994.
38. Mao, X.: Stochastic Differential Equations and Their Applications. Horwood Publishing Limited, Westergate, Chichester, West Sussex, 1997.
39. Mohr, G. and Ditlevsen, O.: Partial summations of stationary sequences of non-Gaussian random variables. Probab. Eng. Mech. **11**, 25–30, 1996.
40. Naprstek, J.: Stochastic exponential and asymptotic stability of simple nonlinear systems. Int. J. Non-Linear Mech. **31**, 693–705, 1996.
41. Nigam, N.: Introduction to Random Vibrations. M.I.T. Press, Cambridge, Mass., 1983.
42. Papanicolau, G. and Kohler, W.: Asymptotic theory of mixing stochastic ordinary differential equation. Commun. Pure Appl. Math. **27**, 641–668, 1974.
43. Pugachev, W. and Sinitsyn, I.: Stochastic Differential Systems. Wiley, Chichester, 1987.
44. Risken, H.: Fokker-Planck Equation: Methods of Solution and Applications. Springer, Berlin, 1984.
45. Roberts, J. and Spanos, P.: Random Vibration and Statistical Linearization. John Wiley and Sons, Chichester, 1990.
46. Sobczyk, K.: Stochastic Differential Equations for Applications. Kluwer, The Netherlands, 1991.

47. Soize, C.: The Fokker-Planck Equation for Stochastic Dynamical Systems and Its Explicit Steady State Solution. World Scientific, Singapore, 1994.
48. Soong, T.: Random Differential Equations in Science and Engineering. Academic Press, New York, 1973.
49. Soong, T. and Grigoriu, M.: Random Vibration of Mechanical and Structural Systems. PTR Prentice Hall, Englewood Cliffs, NJ, USA, 1993.
50. Spanos, P.: Numerical simulations of Van der Pol oscillator. Comput. Math. Appl. **6**, 135–145, 1980.
51. Spanos, P.: Monte Carlo simulations of responses of nonsymmetric dynamic system to random excitations. Comput. Struct. **13**, 371–376, 1980.
52. Stratonovich, R.: Topics in the Theory of Random Noise, Vol. 1. Gordon and Breach, New York, 1963.
53. Stratonovich, R.: A new form of representation of stochastic integrals and equations. SIAM J. Control **4**, 362–371, 1966.
54. Winterstein, S.: Nonlinear vibration models for extremes and fatigue. ASCE J. Eng. Mech. **114**, 1172–1790, 1988.
55. Wong, E.: Stochastic Processes in Information Theory and Dynamical Systems. McGraw-Hill, New York, 1971.
56. Wong, E. and Zakai, M.: On the convergence of ordinary integrals to stochastic integrals. Ann. Math. Stat. **36**, 1560–1564, 1965.
57. Wong, E. and Zakai, M.: On the relation between ordinary and stochastic differential equations. Int. J. Eng. Sci. **3**, 213–229, 1965.
58. Yong, Y. and Lin, Y.: Exact stationary response for second order nonlinear systems under parametric and external white-noise excitations. Trans. ASME J. Appl. Mech. **54**, 414–418, 1987.

# 3

# Moment Equations for Linear Stochastic Dynamic Systems (LSDS)

The determination of response moments in stochastic dynamic systems is one of the fundamental problems in stochastic analysis. The response moments are particularly important in the study of linearization methods. In this chapter we discuss the basic methods of solving linear stochastic differential equations. We start our considerations with two simple classes of scalar linear stochastic differential equations.

### Homogeneous Equation

Consider a scalar linear homogeneous Ito stochastic equation

$$dx(t) = a(t)x(t)dt + \sigma(t)x(t)d\xi(t), \quad x(t_0) = x_0 , \tag{3.1}$$

where $a(t)$ and $\sigma(t)$ are nonlinear time-dependent functions, the initial condition $x_0$ is a random variable independent of standard Wiener process $\xi(t)$. Using Ito formula one can show that the solution of (3.1) is the following stochastic process:

$$x(t) = \psi(t, t_0)x_0 , \tag{3.2}$$

where

$$\psi(t, t_0) = \exp\left\{ \int_{t_0}^{t} \left[ a(s) - \frac{\sigma^2(s)}{2} \right] ds + \int_{t_0}^{t} \sigma(s)d\xi(s) \right\} , \tag{3.3}$$

and his $p$th order moment has the form

$$E[x^p(t)] = E[x_0^p] \exp\left\{ p \int_{t_0}^{t} \left[ a(s) - \frac{\sigma^2(s)}{2} \right] ds + \frac{p^2}{2} \int_{t_0}^{t} \sigma^2(s)ds \right\} . \tag{3.4}$$

### Nonhomogeneous Equation

Reference [28]. Consider a scalar linear nonhomogeneous Ito stochastic equation

$$dx(t) = [a(t)x(t) + b(t)]dt + [\sigma(t)x(t) + q(t)]d\xi(t), \quad x(t_0) = x_0 , \tag{3.5}$$

L. Socha: *Moment Equations for Linear Stochastic Dynamic Systems (LSDS)*, Lect. Notes Phys. **730**, 59–84 (2008)
DOI 10.1007/978-3-540-72997-6_3       © Springer-Verlag Berlin Heidelberg 2008

where $b(t)$ and $q(t)$ are nonlinear time-dependent functions, and all other notations are the same as in (3.1).

Introducing a new variable

$$z(t) = x(t)[\psi(t, t_0)]^{-1} \,, \tag{3.6}$$

where $\psi(t, t_0)$ was defined by (3.3), and using Ito formula one can obtain the stochastic differential equation for the process $z(t)$

$$dz(t) = \{[b(t) - q(t)\sigma(t)]dt + q(t)d\xi(t)\}[\psi(t, t_0)]^{-1} \,. \tag{3.7}$$

Integrating (3.7) and introducing the inverse transformation to (3.6) one can find that

$$x(t) = \psi(t, t_0)\{x(t_0) + \int_{t_0}^{t} [\psi(s, t_0)]^{-1}[b(s) - q(s)\sigma(s)]ds$$

$$+ \int_{t_0}^{t} [\psi(s, t_0)]^{-1}q(s)d\xi(s)\} \,. \tag{3.8}$$

In contrast to a homogeneous equation it is more convenient to find the differential equation for the $p$th order moment than to average the $p$th power of the quantity (3.8) ($p > 0$). Therefore by applying again the Ito formula to the function $x^p$ for $p > 0$ with (3.5) and next averaging the obtained equation we find

$$\frac{dE[x^p(t)]}{dt} = E[x^p(t)]\left[pa(t) + \frac{p(p-1)}{2}\sigma^2(t)\right]$$

$$+ E[x^{p-1}(t)][pb(t) + p(p-1)q(t)\sigma(t)]$$

$$+ E[x^{p-2}(t)]\frac{p(p-1)}{2}q^2(t) \,. \tag{3.9}$$

Unfortunately, in a general case of a linear vector stochastic differential equation with parametric excitation, it is impossible to find an analytical solution. Such opportunity exists only for a linear vector stochastic differential equation with external excitation [58].

## 3.1 Gaussian White Noise External Excitation

Consider a linear Ito vector stochastic differential equation with external excitation

$$d\mathbf{x}(t) = [\mathbf{A}_0(t) + \mathbf{A}(t)\mathbf{x}(t)]dt + \sum_{k=1}^{M} \mathbf{G}_{k0}(t)d\xi_k(t), \quad \mathbf{x}(t_0) = \mathbf{x}_0 \,, \tag{3.10}$$

where $\mathbf{x}(t) = [x_1(t), \ldots, x_n(t)]^T$, $\mathbf{A}_0(t) = [a_0^1(t), \ldots, a_0^n(t)]^T$, $\mathbf{G}_{k0}(t) = [\sigma_{k0}^1(t), \ldots, \sigma_{k0}^n(t)]^T$, are $n$-dimensional vectors, $\mathbf{A}(t) = [a_{ij}(t)]$, $\xi_k(t)$ are independent standard Wiener processes, $i, j = 1, \ldots, n$, $k = 1, \ldots, M$, the initial condition $\mathbf{x}_0$ is a vector random variable independent of $\xi_k(t)$, $k = 1, \ldots, M$; $a_0^i$, $a_{ij}$, and $\sigma_{k0}^i$ are bounded measurable deterministic functions of the variable $t \in R^+$. Then the solution (strong) is determined by

$$\mathbf{x}(t) = \mathbf{\Psi}(t, t_0)\mathbf{x}(t_0) + \int_{t_0}^t \mathbf{\Psi}(t, s)\mathbf{A}_0(s)ds + \int_{t_0}^t \mathbf{\Psi}(t, s) \sum_{k=1}^M \mathbf{G}_{k0}(s)d\xi_k(s) ,$$

(3.11)

where $\mathbf{\Psi}(t, t_0)$ is an $n \times n$-dimensional fundamental matrix of homogeneous equation

$$\frac{d\mathbf{x}(t)}{dt} = \mathbf{A}(t)\mathbf{x}(t), \quad \mathbf{x}(t_0) = \mathbf{x}_0 .$$

(3.12)

In particular, when $\mathbf{A}_0, \mathbf{A}, \mathbf{G}_0$ are constant matrices, then

$$\mathbf{\Psi}(t, t_0) = \mathbf{\Psi}(t - t_0) = \exp\{\mathbf{A}(t - t_0)\} = \sum_{l=0}^{\infty} \frac{1}{l!}\mathbf{A}^l(t - t_0)^l$$

(3.13)

and the solution (3.11) simplifies to the form

$$\mathbf{x}(t) = \exp\{\mathbf{A}(t - t_0)\}\mathbf{x}_0 + \int_{t_0}^t \exp\{\mathbf{A}(t - s)\}\mathbf{A}_0(s)ds$$

$$+ \int_{t_0}^t \exp\{\mathbf{A}(t - s)\} \sum_{k=1}^M \mathbf{G}_{k0}(s)d\mathbf{\xi}(s) .$$

(3.14)

## 3.2 Gaussian White Noise External and Parametric Excitation

Consider a linear Ito vector stochastic differential equation with external and parametric excitation

$$d\mathbf{x}(t) = [\mathbf{A}_0(t) + \mathbf{A}(t)\mathbf{x}(t)]dt + \sum_{k=1}^M [\mathbf{G}_{k0}(t) + \mathbf{G}_k(t)\mathbf{x}(t)]d\xi_k(t), \quad \mathbf{x}(t_0) = \mathbf{x}_0 ,$$

(3.15)

where $\mathbf{x}(t) = [x_1(t), \ldots, x_n(t)]^T$, $\mathbf{A}_0(t) = [a_0^1(t), \ldots, a_0^n(t)]^T$, $\mathbf{G}_{k0}(t) = [\sigma_{k0}^1(t), \ldots, \sigma_{k0}^n(t)]^T$, are $n$-dimensional vectors, $\mathbf{A}(t) = [a_{ij}(t)]$, $\mathbf{G}_k(t) = [\sigma_{kj}^i(t)]$ are $n \times n$-dimensional matrices, $\xi_k(t)$ are independent standard Wiener processes, $i, j = 1, \ldots, n$, $k = 1, \ldots, M$, the initial condition $\mathbf{x}_0$ is a vector random variable independent of $\xi_k(t)$, $k = 1, \ldots, M$; $a_0^i$, $a_{ij}$, and $\sigma_{k0}^i$ are bounded measurable deterministic functions of the variable $t \in R^+$.

For simplicity we assume that the initial condition $\mathbf{x}_0 = [x_{01}, \ldots, x_{0n}]^T$ is a vector random variable independent of $\xi_k(t)$. Using the Ito formula and the averaging operation one can find differential equations for first- and second-order moments

$$\frac{d\mathbf{m}(t)}{dt} = \mathbf{A}_0(t) + \mathbf{A}(t)\mathbf{m}(t), \quad \mathbf{m}(t_0) = \mathbf{m}_0 , \tag{3.16}$$

$$\frac{d\mathbf{\Gamma}(t)}{dt} = \mathbf{m}(t)\mathbf{A}_0^T(t) + \mathbf{A}_0(t)\mathbf{m}^T(t) + \mathbf{\Gamma}(t)\mathbf{A}^T(t) + \mathbf{A}(t)\mathbf{\Gamma}(t)$$

$$+ \sum_{k=1}^{M}[\mathbf{G}_{k0}(t)\mathbf{G}_{k0}^T(t) + \mathbf{G}_k(t)\mathbf{m}(t)\mathbf{G}_{k0}(t) + \mathbf{G}_{k0}(t)\mathbf{m}^T(t)\mathbf{G}_k^T(t)$$

$$+ \mathbf{G}_k(t)\mathbf{\Gamma}(t)\mathbf{G}_k^T(t)] , \quad \mathbf{\Gamma}(t_0) = \mathbf{\Gamma}_0 , \tag{3.17}$$

where $\mathbf{m}(t) = E[\mathbf{x}(t)]$, $\mathbf{\Gamma}(t) = E[\mathbf{x}(t)\mathbf{x}^T(t)]$, $\mathbf{m}_0 = E[\mathbf{x}(t_0)]$, $\mathbf{\Gamma}_0 = E[\mathbf{x}(t_0)\mathbf{x}^T(t_0)]$.

Equations (3.16) and (3.17) for coordinates have the form

$$\frac{dm_i(t)}{dt} = a_0^i(t) + \sum_{j=1}^{n} a_{ij}(t)m_j(t), \quad m_i(t_0) = m_{i0} , \tag{3.18}$$

$$\frac{d\Gamma_{ij}(t)}{dt} = a_0^i(t)m_j(t) + a_0^j(t)m_i(t)$$

$$+ \sum_{l=1}^{n}[a_{il}(t)\Gamma_{lj}(t) + a_{jl}(t)\Gamma_{li}(t)] + \sum_{k=1}^{M} \sigma_{k0}^i(t)\sigma_{k0}^j(t)$$

$$+ \sum_{k=1}^{M}\sum_{\alpha=1}^{n}[\sigma_{k\alpha}^i(t)\sigma_{k0}^j(t)m_\alpha(t) + \sigma_{k\alpha}^j(t)\sigma_{k0}^i(t)m_\alpha(t)] \quad (3.19)$$

$$+ \sum_{k=1}^{M}\sum_{\alpha=1}^{n}\sum_{\beta=1}^{n} \sigma_{k\alpha}^i(t)\sigma_{k\alpha}^j(t)\Gamma_{\alpha\beta}(t)$$

$$\Gamma_{ij}(t_0) = \Gamma_{ij0}, \quad i, j = 1, \ldots, n ,$$

where $m_i(t) = E[x_i(t)]$, $\Gamma_{ij}(t) = E[x_i(t)x_j(t)]$, $m_{i0} = E[x_i(t_0)]$, $\Gamma_{ij0} = E[x_i(t_0)x_j(t_0)]$.

We note that the obtained moment equations have the closed form, i.e., on the right-hand side there are not higher-order moments than on the left-hand side and we also note that the second-order moments depend only on one variable $t$.

*Example 3.1.* Consider a linear oscillator with deterministic and stochastic coefficients described by a linear Ito vector stochastic differential equation with external and parametric excitation and deterministic initial conditions

$$\frac{d\mathbf{x}(t)}{dt} = \mathbf{A}_0 + \mathbf{A}\mathbf{x}dt + \sum_{k=0}^{2}[\mathbf{G}_{k0} + \mathbf{G}_k\mathbf{x}]d\xi_k(t), \quad \mathbf{x}(t_0) = \mathbf{x}_0 , \qquad (3.20)$$

where

$$\mathbf{x} = \begin{bmatrix} x_1 \\ x_2 \end{bmatrix}, \quad \mathbf{x}_0 = \begin{bmatrix} x_{10} \\ x_{20} \end{bmatrix}, \quad \mathbf{A}_0 = \begin{bmatrix} 0 \\ -a_0 \end{bmatrix}, \quad \mathbf{A} = \begin{bmatrix} 0 & 1 \\ -\lambda_0^2 & -2\zeta\lambda_0 \end{bmatrix},$$

$$\mathbf{G}_{00} = \begin{bmatrix} 0 \\ \sigma_0 \end{bmatrix}, \quad \mathbf{G}_{01} = \mathbf{G}_{02} = 0, \quad \mathbf{G}_0 = 0, \quad \mathbf{G}_1 = \begin{bmatrix} 0 & 0 \\ \sigma_1 & 0 \end{bmatrix}, \quad \mathbf{G}_2 = \begin{bmatrix} 0 & 0 \\ 0 & \sigma_2 \end{bmatrix}.$$
$$(3.21)$$

$\lambda_0, \zeta, a_0, \sigma_0, \sigma_1$, and $\sigma_2$ are constant positive parameters, $\xi_1(t)$ and $\xi_2(t)$ are independent Wiener processes, the initial conditions $x_{10}$ and $x_{20}$ are random variables independent of $\xi_k(t)$, $k = 1, \ldots, M$.

The first- and second-order moment equations for coordinates have the form

$$\begin{aligned}
\frac{dm_1}{dt} &= m_2, & m_1(t_0) &= E[x_{10}] , \\
\frac{dm_2}{dt} &= -\lambda_0^2 m_1 - 2\zeta\lambda_0 m_2 - a_0, & m_2(t_0) &= E[x_{20}] , \\
\frac{d\Gamma_{11}}{dt} &= 2\Gamma_{12}, & \Gamma_{11}(t_0) &= E[(x_{10})^2] , \\
\frac{d\Gamma_{12}}{dt} &= \Gamma_{22} - a_0 m_1 - 2\zeta\lambda_0\Gamma_{12} - \lambda_0^2\Gamma_{11}, & \Gamma_{12}(t_0) &= E[x_{10}x_{20}] , \\
\frac{d\Gamma_{22}}{dt} &= -2a_0 m_2 - 4\zeta\lambda_0\Gamma_{22} - \lambda_0^2\Gamma_{12} + \sigma_0^2 + \sigma_1^2\Gamma_{11}
\end{aligned}$$

$$+\sigma_2^2\Gamma_{22} + 2\sigma_0\sigma_1 m_1 + 2\sigma_0\sigma_2 m_2 + 2\sigma_1\sigma_2\Gamma_{12}, \quad \Gamma_{22}(t_0) = E[(x_{20})^2] ,$$
$$(3.22)$$

where $m_i = E[x_i]$, $\Gamma_{ij} = E[x_i x_j]$, $i, j = 1, 2$.

## 3.3 Gaussian Colored Noise External and Parametric Excitation

Consider a linear system with Gaussian colored noise external and parametric excitation described by a class of nonlinear Ito vector stochastic differential equation

$$d\mathbf{x}(t) = [\mathbf{A}_0(t) + \mathbf{A}(t)\mathbf{x}(t)]dt + \sum_{k=1}^{M}[\mathbf{G}_{k0}(t) + \mathbf{G}_k(t)\mathbf{x}(t)]\eta_k(t)dt, \quad \mathbf{x}(t_0) = \mathbf{x}_0 ,$$
$$(3.23)$$

$$d\boldsymbol{\eta}(t) = \mathbf{D}\boldsymbol{\eta}(t)dt + \boldsymbol{\Theta}d\boldsymbol{\xi}(t), \quad \boldsymbol{\eta}(t_0) = \boldsymbol{\eta}_0 , \qquad (3.24)$$

where $\mathbf{A}_0(t), \mathbf{A}(t), \mathbf{G}_{k0}(t)$, and $\mathbf{G}_k(t)$ are defined as in (3.15), $\boldsymbol{\eta}(t) = [\eta_1(t), \ldots, \eta_M(t)]^T$ is a vector stationary Gaussian colored noise treated as an output of a linear filter subjected to Gaussian white noise; $\mathbf{D} = [d_{ik}]$ and $\boldsymbol{\Theta} = [\Theta_{ik}]$, $i, k = 1, \ldots, M$, are $M \times M$-dimensional constant matrices, $\mathbf{D}$ is the negative definite matrix, i.e., real parts of all its eigenvalues are negative $Re\lambda_j(\mathbf{D}) < 0$, $j = 1, \ldots, M$; the initial conditions $\mathbf{x}_0$ and $\boldsymbol{\eta}_0$ are independent random variables; $\boldsymbol{\xi}(t)$ is an $M$-dimensional standard Wiener process independent of $\mathbf{x}_0$ and $\boldsymbol{\eta}_0$.

Equations (3.23) and (3.24) can be transformed to the vector form as

$$d\begin{bmatrix} \mathbf{x}(t) \\ \boldsymbol{\eta}(t) \end{bmatrix} = \left[ \begin{bmatrix} \mathbf{A}_0(t) \\ \mathbf{0} \end{bmatrix} + \begin{bmatrix} \mathbf{A}(t) + \sum_{k=1}^{M} \mathbf{G}_k(t)\eta_k(t) & \mathbf{0} \\ \mathbf{0} & \mathbf{D} \end{bmatrix} \begin{bmatrix} \mathbf{x}(t) \\ \boldsymbol{\eta}(t) \end{bmatrix} \right.$$

$$\left. + \sum_{k=1}^{M} \begin{bmatrix} \mathbf{G}_{k0}(t)\eta_k(t) \\ \mathbf{0} \end{bmatrix} \right] dt + \begin{bmatrix} \mathbf{0} \\ \boldsymbol{\Theta}(t) \end{bmatrix} d\boldsymbol{\xi}(t) . \tag{3.25}$$

We note that the vector equation (3.25) is a nonlinear one and the corresponding moment differential equations for vector $[\mathbf{x}^T \boldsymbol{\eta}^T]^T$ are not closed. For instance, for first- and second-order moments (for coordinates) they have the form

$$\frac{dm_i(t)}{dt} = a_0^i(t) + \sum_{j=1}^{n} a_{ij}(t)m_j(t)$$

$$+ \sum_{k=1}^{M} \sum_{\alpha=1}^{n} \sigma_{k\alpha}^i(t)E[x_\alpha(t)\eta_k(t)], \quad m_i(t_0) = E[x_{i0}] , \tag{3.26}$$

$$m_i(t) = E[x_i(t)], \quad i = 1, \ldots, n ,$$

$$\frac{dm_i(t)}{dt} = \sum_{j=n+1}^{n+M} d_{ij}m_j(t), \quad m_i(t_0) = E[\eta_0] ,$$

$$m_i(t) = E[\eta_{i-n}(t)], \quad i = n+1, \ldots, n+M , \tag{3.27}$$

$$\frac{d\Gamma_{ij}(t)}{dt} = a_0^i(t)m_j(t) + a_0^j(t)m_i(t) + \sum_{\alpha=1}^{n} [a_{i\alpha}(t)\Gamma_{\alpha j}(t) + \Gamma_{i\alpha}(t)a_{\alpha j}(t)]$$

$$+ \sum_{k=1}^{M} [\sigma_{k0}^i(t)E[x_j(t)\eta_k(t)] + \sigma_{k0}^j(t)E[x_i(t)\eta_k(t)]]$$

$$+ \sum_{k=1}^{M} \sum_{\alpha=1}^{n} [\sigma_{k\alpha}^i(t)E[x_\alpha(t)x_j(t)\eta_k(t)] + \sigma_{k\alpha}^j(t)E[x_\alpha(t)x_i(t)\eta_k(t)]] ,$$

$$\Gamma_{ij}(t_0) = E[x_{i0}x_{j0}], \quad \Gamma_{ij}(t) = E[x_i(t)x_j(t)], \quad i,j = 1, \ldots, n ,$$
$$\tag{3.28}$$

$$\frac{d\Gamma_{ij}(t)}{dt} = \sum_{\alpha=1}^{n} a_{i\alpha}(t)\Gamma_{\alpha j} + \sum_{\alpha=1}^{n} \sum_{k=1}^{M} [\sigma_{k\alpha}^i(t)E[x_\alpha(t)\eta_k(t)\eta_{j-n}(t)]$$

$$+ \sum_{k=1}^{M} \sigma_{k0}^i(t)\Gamma_{k+nj}(t)] + \sum_{k=1}^{M} d_{j-nk}\Gamma_{ik+n}(t) ,$$

$$\Gamma_{ij}(t_0) = E[x_{i0}\eta_{j-n0}], \Gamma_{ij}(t)$$
$$= E[x_i(t)\eta_{j-n}(t)], i = 1, \ldots, n, j = n+1, \ldots, n+M , \quad (3.29)$$

$$\frac{d\Gamma_{ij}(t)}{dt} = \sum_{k=1}^{M} \Theta_{i-nk}(t)\Theta_{j-nk}(t)$$
$$+ \sum_{k=1}^{M}[d_{i-nk}\Gamma_{k+nj}(t) + d_{j-nk}\Gamma_{k+ni}(t)] ,$$
$$\Gamma_{ij}(t_0) = E[\eta_{i-n0}\eta_{j-n0}], \Gamma_{ij}(t)$$
$$= E[\eta_{i-n}(t)\eta_{j-n}(t)], \quad n+1 \le i, \quad j \le n+M . \quad (3.30)$$

As was mentioned earlier, in the differential equations for first-order moments, second-order moments appear on the right-hand side, while in equations for second-order moments, third-order moments appear. To solve such a system of equations one has to apply one of the closure techniques that are discussed in Chap. 4.

## 3.4 Nonstationary Gaussian External Excitation

The problem of the determination of the exact covariance function of the response of a linear system subjected to special classes of nonstationary excitation is well known in the literature. In the first works in the literature about differential equations with stochastic parameters and excitations usually for linear systems the operator methods were used. The generalization of Green function methods used in the response analysis of linear systems under deterministic nonstationary excitations to stochastic excitations belongs to Zadeh.

The first paper in the field of random vibration was written by Caughey and Stumpf [14]. They considered the transient mean-square response of a linear oscillator subjected to a unit step-modulated white noise.

In this section we present only the classical literature method called *Green function method for stochastic systems*. We consider a particular case of linear equation (3.25) where only external excitations are second-order stochastic processes and the coefficients are deterministic function of $t$. We illustrate this approach for the following system:

$$\frac{d\mathbf{x}(t)}{dt} = \mathbf{A}_0(t) + \mathbf{A}(t)\mathbf{x}(t) + \mathbf{G}_0(t)\boldsymbol{\eta}(t), \quad \mathbf{x}(t_0) = \mathbf{x}_0 , \quad (3.31)$$

where $\mathbf{x}(t) = [x_1(t), \ldots, x_n(t)]^T$ is an $n$-dimensional vector state, $\mathbf{A}_0(t) = [a_0^1(t), \ldots, a_0^n(t)]^T$ is an $n$-dimensional vector of deterministic external excitation, $\mathbf{A}(t) = [a_{ij}(t)]$ is an $n \times n$-dimensional matrix, $\mathbf{G}_0(t) = [\sigma_{0k}^i(t)]$ is an $n \times M$-dimensional matrix of intensity of external stochastic excitation,

$\boldsymbol{\eta}(t) = [\eta_1(t), \ldots, \eta_M(t)]^T$ is an $M$-dimensional second-order stochastic process with the matrix correlation function $\mathbf{K}_\eta(s, t)$, the initial condition $\mathbf{x}_0$ is a vector random variable independent of $\eta_k(t)$, $k = 1, \ldots, M$; $a_0^i$, $a_{ij}$, and $\sigma_{0l}^i$ are bounded measurable deterministic functions of variable $t \in R^+$; $i, j = 1, \ldots, n$, $l = 1, \ldots, M$.

The solution of (3.31) has the form

$$\mathbf{x}(t) = \boldsymbol{\Psi}(t, t_0)\mathbf{x}_0 + \int_{t_0}^t \boldsymbol{\Psi}(t, s)\mathbf{A}_0(s)ds + \int_{t_0}^t \boldsymbol{\Psi}(t, s)\mathbf{G}_0(s)\boldsymbol{\eta}(s)ds, \quad \mathbf{x}(t_0) = \mathbf{x}_0 .$$
$$(3.32)$$

where $\boldsymbol{\Psi}(t, t_0)$ is an $n \times n$-dimensional fundamental matrix of homogeneous equation (3.12).

Using equality (3.32), one can calculate the mean value and the matrix covariance function for solution $\mathbf{x}(t)$, i.e.,

$$\mathbf{m_x}(t) = \boldsymbol{\Psi}(t, t_0)\mathbf{m_{x0}} + \int_{t_0}^t \boldsymbol{\Psi}(t, s)\mathbf{A}_0(s)ds$$

$$+ \int_{t_0}^t \boldsymbol{\Psi}(t, s)\mathbf{G}_0(s)E[\boldsymbol{\eta}(s)]ds, \quad \mathbf{m}_x(t_0) = E[\mathbf{x}_0] , \qquad (3.33)$$

$$\mathbf{K_x}(t, s) = \boldsymbol{\Psi}(t, t_0)\mathbf{K_{x0}}\boldsymbol{\Psi}(s, t_0)^T$$

$$+ \int_{t_0}^t \int_{t_0}^s \boldsymbol{\Psi}(t, \tau_1)\mathbf{G}_0(\tau_1)\mathbf{K}_\eta(\tau_1, \tau_2)\mathbf{G}_0^T(\tau_2)\boldsymbol{\Psi}(s, \tau_2)^T d\tau_1 d\tau_2 ,$$
$$(3.34)$$

where

$$\mathbf{m_x}(t) = E[\mathbf{x}(t)], \quad \mathbf{K_x}(t, s) = E[(\mathbf{x}(t) - \mathbf{m}_x(t))(\mathbf{x}(s) - \mathbf{m}_x(s))^T], \quad t > s$$

$$\mathbf{m_{x0}} = E[\mathbf{x}_0], \quad \mathbf{K_{x0}} = E[(\mathbf{x}(t_0) - \mathbf{m}(t_0))(\mathbf{x}(t_0) - \mathbf{m}(t_0))^T]. \quad (3.35)$$

The matrix covariance function for $t < s$ is equal to $\mathbf{K_x}^T(s, t)$. In the particular case when $t = s$, the matrix covariance function reduces to the matrix variance function $\mathbf{K_x}(t, s) = \mathbf{K_x}(t, t) = \mathbf{K_x}(t)$ determined by

$$\mathbf{K_x}(t) = \boldsymbol{\Psi}(t, t_0)\mathbf{K_{x0}}\boldsymbol{\Psi}(t, t_0)^T$$

$$+ \int_{t_0}^t \int_{t_0}^t \boldsymbol{\Psi}(t, \tau_1)\mathbf{G}_0(\tau_1)\mathbf{K}_\eta(\tau_1, \tau_2)\mathbf{G}_0^T(\tau_2)\boldsymbol{\Psi}(t, \tau_2)^T d\tau_1 d\tau_2 . (3.36)$$

The integral form (3.36) is often not convenient for calculations and therefore the differential equation for the variance matrix function is mainly used. It has the following form:

$$\frac{d\mathbf{K_x}(t)}{dt} = \mathbf{A}(t)\mathbf{K_x}(t) + \mathbf{K_x}(t)\mathbf{A}^T(t) + \mathbf{P}(t) + \mathbf{P}^T(t), \quad \mathbf{K}(t_0) = \mathbf{K}_0 , \quad (3.37)$$

where

$$\mathbf{P}(t) = \mathbf{G}_0(t) \int_{t_0}^{t} \mathbf{K}_\eta(t, \tau) \mathbf{G}_0^T(\tau) \mathbf{\Psi}(t, \tau)^T d\tau \ . \tag{3.38}$$

The derivation of (3.37) can be found, for instance, in [49, 77].

We note that the process $\boldsymbol{\eta}(t)$ in (3.37) is a second-order stochastic process that can be stationary or nonstationary. If we assume that $\boldsymbol{\eta}(t)$ is a zero-mean Gaussian white noise with the matrix correlation function

$$\mathbf{K}_\eta(s, t) = \mathbf{Q}(s)\delta(t - s) \ , \tag{3.39}$$

where $\mathbf{Q}(s)$ is the matrix intensity function, then (3.37) simplifies to the form

$$\frac{d\mathbf{K_x}(t)}{dt} = \mathbf{A}(t)\mathbf{K_x}(t) + \mathbf{K_x}(t)\mathbf{A}^T(t) + \mathbf{G}_0(t)\mathbf{Q}(t)\mathbf{G}_0^T(t), \quad \mathbf{K_x}(t_0) = \mathbf{K_{x0}} \ . \tag{3.40}$$

We also note that (3.40) can be derived directly from (3.23) and (3.24) by substituting $\mathbf{G}_k(t) = 0$ and taking into account the matrix correlation function of Gaussian white noise and the equality

$$\mathbf{K_x}(t) = E[\mathbf{x}(t)\mathbf{x}^T(t)] - \mathbf{m}(t)\mathbf{m}^T(t) = \mathbf{\Gamma}(t) - \mathbf{m}(t)\mathbf{m}^T(t) \ . \tag{3.41}$$

In a similar way one can derive partial differential equations for $\mathbf{K_x}(t, s)$, when $t > s$ and $t < s$. The details are given, for instance, in [49, 77]. In the case of linear systems with constant coefficients, relations (3.31–3.41) simplify because the fundamental matrix $\mathbf{\Phi}(t, s)$ takes the form (3.13). One can show [49] that although system (3.31) is a linear one with constant coefficients if the excitation is a nonstationary second-order process, the solution is also a nonstationary process. In particular case, when the system (3.31) is a linear one with constant coefficients and the excitation is stationary process, then the solution can be a stationary or a nonstationary process. It depends on the matrix $\mathbf{A}$ determining a linear system and on the choice of the initial condition $\mathbf{x}(t_0)$. If $Re\lambda_i(\mathbf{A}) < 0$, for $1 \leq i \leq n$, and $\mathbf{x}(t_0)$ is the zero-mean vector random variable with the matrix variance function equal to the variance matrix of the stationary solution $\mathbf{x}(\infty) = \lim_{t \to \infty} x(t)$, then $\mathbf{x}(t)$ is a second-order stationary stochastic process. We illustrate these statements on an example.

*Example 3.2.* Consider the scalar Ito equation

$$d\eta(t) = -a\eta(t)dt + bd\xi(t), \quad \eta(0) = \eta_0 \ , \tag{3.42}$$

where $a$ and $b$ are positive constants, $\xi(t)$ is a standard Wiener process, $\eta_0$ is a zero-mean random variable with the variance equal to $\sigma_0^2$. If we denote

$$\sigma^2(t) = E[\eta^2(t)] \ , \tag{3.43}$$

then $\sigma^2(t)$ satisfies the differential equation

$$\frac{d\sigma^2(t)}{dt} = -2a\sigma^2(t) + b^2, \quad \sigma^2(0) = \sigma_0^2 . \tag{3.44}$$

The solution of (3.44) has the form

$$\sigma^2(t) = \frac{b^2}{2a} + \left(\sigma_0^2 - \frac{b^2}{2a}\right) \exp\{-2at\} \tag{3.45}$$

and moreover

$$\lim_{t\to\infty} \sigma^2(t) = \frac{b^2}{2a} . \tag{3.46}$$

Hence it follows that if we assume $\sigma_0^2 = \frac{b^2}{2a}$, then

$$\sigma^2(t) \equiv \frac{b^2}{2a} . \tag{3.47}$$

For other values $\sigma_0^2$ the variance $\sigma^2(t)$ depends on $t$, i.e., the process $\eta(t)$ is a nonstationary one.

Similar analysis can be done for an $n$-dimensional system

$$d\mathbf{x}(t) = \mathbf{A}\mathbf{x}(t)dt + \mathbf{G}_0 d\boldsymbol{\xi}(t), \quad \mathbf{x}(t_0) = \mathbf{x}_0 , \tag{3.48}$$

where $\mathbf{A}$ and $\mathbf{G}_0$ are constant matrices of $n \times n$ and $n \times M$ dimension, respectively; $Re\lambda_i(\mathbf{A}) < 0$ for $1 \leq i \leq n$, $\boldsymbol{\xi}(t)$ is a vector standard Wiener process $(\mathbf{Q} = \mathbf{I})$. Then, according to (3.40) the matrix variance function satisfies the following equation:

$$\frac{d\mathbf{K_x}(t)}{dt} = \mathbf{A}\mathbf{K_x}(t) + \mathbf{K_x}(t)\mathbf{A}^T + \mathbf{G}_0\mathbf{G}_0^T, \quad \mathbf{K_x}(t_0) = \mathbf{K_{x0}} . \tag{3.49}$$

If we assume that $\mathbf{K_{x0}}$ is a solution of an algebraic matrix Lyapunov equation

$$\mathbf{A}\mathbf{K} + \mathbf{K}\mathbf{A}^T + \mathbf{G}_0\mathbf{G}_0^T = \mathbf{0} , \tag{3.50}$$

then $\mathbf{K_x}(t) \equiv \mathbf{K_{x0}}$, i.e., the process $\mathbf{x}(t)$ that is the solution of (3.48) is a stationary Gaussian stochastic process.

Here, an important fact should be stressed that the stationary solution of a stochastic differential equation does not depend on initial conditions, see, for instance, formulas (3.46) and (3.50) in scalar and vector case, respectively. Therefore in some further considerations regarding stationary response characteristics we will omit the description of initial conditions.

## 3.5 Spectral Method

By the discussion of the Green function method we have noted that the moment equations of linear systems with constant coefficients and stationary excitation simplify. If additionally we reduce our consideration to stationary solutions, then one can obtain simple algebraic formulas for moments of solutions using properties of Fourier transforms.

Consider a differential equation

$$\frac{d\mathbf{x}(t)}{dt} = \mathbf{A}\mathbf{x}(t) + \mathbf{G}_0\boldsymbol{\eta}(t), \quad \mathbf{x}(t_0) = \mathbf{x}_0 , \qquad (3.51)$$

where $\mathbf{A}$ is an $n \times n$-dimensional constant matrix, $Re\lambda_i(\mathbf{A}) < 0$ for $1 \leq i \leq n$, $\mathbf{G}_0 = [G_{0\alpha\beta}]$ is an $n \times M$-dimensional constant matrix, $\boldsymbol{\eta}(t) = [\eta_1(t), \ldots, \eta_M(t)]^T$ is a zero-mean second-order vector stationary process, the initial condition $\mathbf{x}_0$ is a vector random variable independent of $\eta_k(t)$, $k = 1, \ldots, M$.

The stationary processes have been introduced in previous sections by the discussion of linear filter equations and additionally it has been assumed that the initial condition is a random variable with the mean value and the covariance matrix equal to the corresponding ones for stationary solution. There is another opportunity making allowance for the stationarity of the process $\boldsymbol{\eta}(t)$ by taking into account the assumption that $\boldsymbol{\eta}(t)$ is acting from $-\infty$, i.e., $t_0 = -\infty$. From this assumption and negative-definite constant matrix $\mathbf{A}$ it follows additionally that the fundamental matrix for homogeneous system (3.12) in the form (3.13) satisfies the condition

$$\lim_{t_0 \to -\infty} \boldsymbol{\Psi}(t - t_0) = \lim_{t_0 \to -\infty} \exp\{\mathbf{A}(t - t_0)\} = 0 . \qquad (3.52)$$

Using (3.32) and corollaries following from the assumption of stationarity of $\boldsymbol{\eta}(t)$ the solution $\mathbf{x}(t)$ has the form

$$\mathbf{x}(t) = \int_{-\infty}^{t} \boldsymbol{\Psi}(t - \tau)\mathbf{G}_0\boldsymbol{\eta}(\tau)d\tau \qquad (3.53)$$

or

$$\mathbf{x}(t) = \int_{-\infty}^{+\infty} \boldsymbol{\Psi}(t - \tau)\mathbf{G}_0\boldsymbol{\eta}(\tau)d\tau , \qquad (3.54)$$

because $\boldsymbol{\Psi}(u) \equiv 0$ for $u < 0$ .

Using equality (3.54) we find the mean value and the correlation matrix of the solution $\mathbf{x}(t)$

$$\mathbf{m}_x(t) = \int_{-\infty}^{+\infty} \boldsymbol{\Psi}(t - \tau)\mathbf{G}_0(s)\mathbf{m}_\eta(\tau)d\tau , \qquad (3.55)$$

$$\mathbf{R}_x(t_1, t_2) = \int_{-\infty}^{+\infty} \int_{-\infty}^{+\infty} \boldsymbol{\Psi}(t_1 - \tau_1)\mathbf{G}_0\mathbf{R}_\eta(\tau_1, \tau_2)\mathbf{G}_0^T \boldsymbol{\Psi}^T(t_2 - \tau_2)d\tau_1 d\tau_2 .$$

$$(3.56)$$

From the assumption of stationarity of process $\boldsymbol{\eta}(t)$ it follows

$$\mathbf{R}_\eta(\tau_1, \tau_2) = \mathbf{R}_\eta(\tau_1 - \tau_2) \ . \tag{3.57}$$

Introducing notations

$$t_1 - t_2 = \tau, \quad t_1 - \tau_1 = \tau_1' \quad t_1 - \tau_2 = \tau_2' \tag{3.58}$$

and changing variables in equality (3.56), we find

$$\mathbf{R}_x(t_1, t_2) = \mathbf{R}_x(t_1 - t_2) = \mathbf{R}_x(\tau)$$

$$= \int_{-\infty}^{+\infty} \int_{-\infty}^{+\infty} \boldsymbol{\Psi}(\tau_1')$$

$$\times \mathbf{G}_0 \mathbf{R}_\eta(-\tau_1' + \tau_2' + \tau) \mathbf{G}_0^T \boldsymbol{\Psi}^T(\tau_2') d\tau_1' d\tau_2' \ . \tag{3.59}$$

For further analysis we use the Fourier transform of correlation matrix of the stochastic process $\mathbf{x}(t)$ discussed in previous chapter satysfying the condition

$$\int_{-\infty}^{+\infty} |\mathbf{R}_x(\tau)| d\tau < \infty \ . \tag{3.60}$$

Then the Fourier transform of correlation matrix called *power spectral density matrix of process* $\mathbf{x}(t)$ has the form

$$\mathbf{S}_x(\lambda) = \frac{1}{2\pi} \int_{-\infty}^{+\infty} \mathbf{R}_x(\tau) e^{-i\lambda\tau} d\tau, \quad \lambda \in R \ . \tag{3.61}$$

The knowledge of the power spectral density matrix of the process $\mathbf{x}(t)$ allows to obtain the corresponding correlation matrix by the inverse Fourier transform

$$\mathbf{R}_x(\tau) = \int_{-\infty}^{+\infty} \mathbf{S}_x(\lambda) e^{i\lambda\tau} d\lambda \ . \tag{3.62}$$

Substituting (3.59) in to (3.60), we find

$$\mathbf{S}_x(\lambda) = \frac{1}{2\pi} \int_{-\infty}^{+\infty} \int_{-\infty}^{+\infty} \int_{-\infty}^{+\infty} \boldsymbol{\Psi}(\tau_1') \mathbf{G}_0 \mathbf{R}_\eta(-\tau_1' + \tau_2' + \tau)$$

$$\times \mathbf{G}_0^T \boldsymbol{\Psi}^T(\tau_2') e^{-i\lambda\tau} d\tau_1' d\tau_2' d\tau$$

$$= \int_{-\infty}^{+\infty} \boldsymbol{\Psi}(\tau_1') e^{-i\lambda\tau_1'} d\tau_1' \mathbf{G}_0 \frac{1}{2\pi} \int_{-\infty}^{+\infty} \mathbf{R}_\eta(z) e^{-i\lambda z} dz$$

$$\times \mathbf{G}_0^T \int_{-\infty}^{+\infty} \boldsymbol{\Psi}(\tau_2') e^{-i\lambda\tau_2'} d\tau_2'$$

$$= \mathbf{G}_1(i\lambda) \mathbf{G}_0 \mathbf{S}_\eta(\lambda) \mathbf{G}_0^T \mathbf{G}_1^T(-i\lambda), \tag{3.63}$$

where $z$ is a variable defined by $z = -\tau_1' + \tau_2' + \tau$. $\mathbf{G}(s) = \mathbf{G}_1(s)\mathbf{G}_0$ is called *operational transfer function* of the linear system (3.51) defined by

$$\mathbf{G}(s) = (s\mathbf{I} - \mathbf{A})^{-1}\mathbf{G}_0 \ . \tag{3.64}$$

It is obtained directly from the Laplace transform (with zero initial conditions) of the linear system (3.51).

The last equality in (3.63) described by elements of matrix $\mathbf{S}_x(\lambda)$ has the form

$$\mathbf{S}_{xpq}(\lambda) = \sum_{\alpha=1}^{n}\sum_{\beta=1}^{n} G_{1p\alpha}(i\lambda)\left(\sum_{k=1}^{M}\sum_{l=1}^{M} G_{0\alpha l}S_{\eta lk}(\lambda)G_{0\beta k}\right)$$
$$G_{1q\beta}(-i\lambda), p, \quad q = 1, \ldots, n \ . \tag{3.65}$$

If the stationary solution $\mathbf{x}(t)$ is treated as an output of a linear filter defined by an operational transfer function $\mathbf{G}(s)$ when on the input of this filter is acting the zero-mean stationary process with the power spectral density matrix $\mathbf{S}_\eta(\lambda)$, then the relation (3.63) reduces to the following one:

$$\mathbf{S}_x(\lambda) = \mathbf{G}(i\lambda)\mathbf{S}_\eta(\lambda)\mathbf{G}^*(-i\lambda) \ , \tag{3.66}$$

where $\mathbf{G}^*(-i\lambda)$ is the transpose and complex-conjugated matrix of the matrix $\mathbf{G}(i\lambda)$.

Based on the power spectral density matrix of the process $\mathbf{x}(t)$ one can calculate the corresponding correlation matrix and variance matrix

$$\mathbf{R}_x(\tau) = \int_{-\infty}^{+\infty} \mathbf{G}(i\lambda)\mathbf{S}_\eta(\lambda)\mathbf{G}^*(-i\lambda)e^{i\lambda\tau}d\lambda \ , \tag{3.67}$$

$$\boldsymbol{\sigma}_x^2(\tau) = \int_{-\infty}^{+\infty} \mathbf{G}(i\lambda)\mathbf{S}_\eta(\lambda)\mathbf{G}^*(-i\lambda)d\lambda - \mathbf{m}_x\mathbf{m}_x^* \ . \tag{3.68}$$

*Example 3.2.* (*continued*) We calculate the correlation function and the power spectral density function for the stationary solution of the system (3.42) with initial condition equal to zero, $\eta(0) = 0$. First we find the operational transfer function using Laplace transforms of input and output variables under the assumption that the initial condition is equal to zero

$$G(s) = \frac{b}{a+s} \ . \tag{3.69}$$

Remembering that the power spectral density function of the input process $\dot{\xi}$ (white noise) is according to (2.137) equal to the constant $c$ one can calculate the power spectral density function of the solution of (3.42) by the formula (3.66)

$$S_\eta(\lambda) = \frac{c}{|a+i\lambda|^2} = \frac{c}{a^2 + \lambda^2} \ . \tag{3.70}$$

If we consider a zero-mean stationary process $x(t)$ with the correlation function $K_x(\tau)$ defined by

$$K_x(\tau) = \sigma_0^2 \exp\{-a|\tau|\} \ . \tag{3.71}$$

then it is simple to show by direct integration that the power spectral density function $S_x(\lambda)$ corresponding to the correlation function $K_x(\tau)$ is

$$S_\eta(\lambda) = \frac{\sigma_0^2}{\pi} \frac{a}{a^2 + \lambda^2} \ . \tag{3.72}$$

Comparing equalities (3.70) and (3.72) we find that

$$\sigma_0^2 = \frac{\pi c}{a} = \frac{b^2}{2a} \ . \tag{3.73}$$

Hence

$$c = \frac{b^2}{2\pi} \ . \tag{3.74}$$

This simple example shows that the zero-mean stationary process $x(t)$ with the correlation function defined by (3.71) can be treated as an output process from the linear filter defined by (3.42) and excited by the white noise with the constant power spectral density function equal to $S_{\dot\xi}(\lambda) = \frac{b^2}{2\pi}$.

*Example 3.3.* Consider a linear oscillator subjected to stationary colored noise described by

$$\frac{d\mathbf{x}(t)}{dt} = \mathbf{A}\mathbf{x} + \mathbf{G}_0\eta(t), \mathbf{x}(t_0) = \mathbf{0} \ , \tag{3.75}$$

where

$$\mathbf{x} = \begin{bmatrix} x_1 \\ x_2 \end{bmatrix}, \quad \mathbf{A} = \begin{bmatrix} 0 & 1 \\ -\lambda_0^2 & -2\zeta\lambda_0 \end{bmatrix}, \quad \mathbf{G}_0 = \begin{bmatrix} 0 \\ \delta \end{bmatrix}, \tag{3.76}$$

$\lambda_0, \zeta$, and $\delta$ are constant positive parameters; $\eta(t)$ is a scalar zero-mean stationary second-order stochastic process with the correlation function

$$\mathbf{R}_\eta(\tau) = C \exp\{-\alpha|\tau|\} \left( \cos\beta\tau + \frac{\alpha}{\beta}\sin\beta|\tau| \right) \ , \tag{3.77}$$

where $\alpha, \beta$, and $C$ are constant positive parameters.

The corresponding power spectral density function of the process $\eta(t)$ is determined from (3.61)

$$\mathbf{S}_\eta(\lambda) = \frac{1}{2\pi} \int_{-\infty}^{+\infty} \mathbf{R}_\eta(\tau)e^{-i\lambda\tau} d\tau = \frac{2}{\pi} \frac{C\alpha(\alpha^2 + \beta^2)}{(\lambda^2 - \alpha^2 - \beta^2)^2 + 4\alpha^2\lambda^2} \ . \tag{3.78}$$

The operational transfer function of system (3.75) has the form

$$\mathbf{G}(s) = (s\mathbf{I} - \mathbf{A})^{-1}\mathbf{G}_0 = \begin{bmatrix} G_1(s) \\ G_2(s) \end{bmatrix}, \tag{3.79}$$

where

$$G_1(s) = \frac{\delta}{s^2 + 2\zeta\lambda_0 s + \lambda_0^2} \ , \tag{3.80}$$

$$G_2(s) = \frac{\delta s}{s^2 + 2\zeta\lambda_0 s + \lambda_0^2} \ . \tag{3.81}$$

Using the relation (3.65) one can calculate the power spectral density matrix $\mathbf{S_x}(\lambda)$ of the solution $\mathbf{x}(t)$

$$\mathbf{S}_x(\lambda) = \begin{bmatrix} S_{x11}(\lambda) & S_{x12}(\lambda) \\ S_{x21}(\lambda) & S_{x22}(\lambda) \end{bmatrix} , \tag{3.82}$$

where

$$S_{x11}(\lambda) = \frac{2}{\pi} \frac{\delta^2}{(\lambda_0^2 - \lambda^2)^2 + 4\zeta^2\lambda_0^2\lambda^2} \frac{C\alpha(\alpha^2 + \beta^2)}{(\lambda^2 - \alpha^2 - \beta^2)^2 + 4\alpha^2\lambda^2} , \tag{3.83}$$

$$S_{x12}(\lambda) = S_{x21}(\lambda) = \frac{2}{\pi} \frac{\delta^2\lambda}{(\lambda_0^2 - \lambda^2)^2 + 4\zeta^2\lambda_0^2\lambda^2} \frac{C\alpha(\alpha^2 + \beta^2)}{(\lambda^2 - \alpha^2 - \beta^2)^2 + 4\alpha^2\lambda^2} , \tag{3.84}$$

$$S_{x22}(\lambda) = \frac{2}{\pi} \frac{\delta^2\lambda^2}{(\lambda_0^2 - \lambda^2)^2 + 4\zeta^2\lambda_0^2\lambda^2} \frac{C\alpha(\alpha^2 + \beta^2)}{(\lambda^2 - \alpha^2 - \beta^2)^2 + 4\alpha^2\lambda^2} . \tag{3.85}$$

Comparing two factors in the power spectral density function (3.83) one can present the first factor as the Fourier transform of a zero-mean stationary process $z(t)$ with the correlation function $R_z(\tau)$

$$S_{x11}(\lambda) = \frac{\delta^2}{(\lambda_0^2 - \lambda^2)^2 + 4\zeta^2\lambda_0^2\lambda^2} = \frac{1}{2\pi} \int_{-\infty}^{+\infty} R_z(\tau)e^{-i\lambda\tau} d\tau , \tag{3.86}$$

where the process $z(t)$ can be treated as an output process from the two-dimensional linear filter defined by (3.75) and excited by the white noise with the constant power spectral density function equal to $S_\eta(\lambda) = c > 0$. The correlation function $R_z(\tau)$ can be presented in the form

$$R_z(\tau) = C_z \exp\{-\alpha_z|\tau|\} \left( \cos\beta_z\tau + \frac{\alpha_z}{\beta_z} \sin\beta_z|\tau| \right) , \tag{3.87}$$

where

$$\alpha_z = \zeta\lambda_0, \quad \beta_z = \lambda_0\sqrt{1 - \zeta^2}, \quad C_z = \frac{\pi c}{2\zeta\lambda_0^3} . \tag{3.88}$$

Also the zero-mean process $\eta(t)$ with the correlation function defined by (3.77) can be treated as an output process from the two-dimensional linear filter defined by the equation

$$\frac{d\mathbf{x}_\eta(t)}{dt} = \mathbf{A}_\eta\mathbf{x}_\eta + \mathbf{G}_{0\eta}\dot{\xi}, \quad \mathbf{x}_\eta(t_0) = \mathbf{0} , \tag{3.89}$$

where

$$\mathbf{x}_\eta = \begin{bmatrix} \eta \\ \dot{\eta} \end{bmatrix}, \quad \mathbf{A}_\eta = \begin{bmatrix} 0 & 1 \\ -\lambda_{0\eta}^2 & -2\zeta_\eta\lambda_{0\eta} \end{bmatrix}, \quad \mathbf{G}_{0\eta} = \begin{bmatrix} 0 \\ \delta_\eta \end{bmatrix} . \tag{3.90}$$

$\dot{\xi}$ is a standard white noise with the constant power spectral density function equal to $S_{\dot{\xi}}(\lambda) = c$. $\lambda_{0\eta}, \zeta_\eta$, and $\delta_\eta$ are constant positive parameters defined by parameters of the correlation function $R_\eta(\tau)$, i.e.,

$$\alpha = \zeta_\eta \lambda_{0\eta}, \quad \beta = \lambda_{0\eta} \sqrt{1 - \zeta_\eta^2}, \quad C = \frac{\pi c}{2\zeta_\eta \lambda_{0\eta}^3} . \tag{3.91}$$

## 3.6 Non-Gaussian External Excitation

There are two general groups of modeling of non-Gaussian excitations, namely by *"memoryless transformation of Gaussian processes"* and by *"transformation with memory of the Brownian and Levy processes"* [73, 33, 36]. The first example was proposed by Grigoriu [38] and Kotulski and Sobczyk [51] in the form of quadratic Gaussian excitation. It was used to describe the wind pressure proportional to the square of the wind velocity, which is a Gaussian process. Further development of these results in the response analysis of linear systems was done by Grigoriu and Ariaratnam [37], Krenk and Gluver [52] who considered non-Gaussian excitations in the form of a polynomial of Gaussian processes. We discuss a few basic representations of non-Gaussian processes defined by memory transformation of Gaussian processes externally exciting a linear oscillator.

### Quadratic Gaussian Excitation

First we consider a linear oscillator with non-Gaussian colored noise excitation proposed by Roberts and Vasta [71]

$$dx_1(t) = x_2(t)dt ,$$
$$dx_2(t) = [-\lambda_0^2 x_1(t) - 2hx_2(t) + \eta(t)]dt, \tag{3.92}$$

where $\lambda_0, h$ are positive constant parameters, $\eta(t)$ is a continuous stationary non-Gaussian process defined by

$$\eta(t) = k_1 v(t) + k_2 v(t)|v(t)| , \tag{3.93}$$

$v(t)$ is a stationary colored Gaussian process, $k_1$ and $k_2$ are constant parameters.

Roberts and Vasta [71] considered, for instance, $v(t)$ as follows:

$$v(t) = \rho[\zeta_1(t) \cos \lambda_p t + \zeta_2(t) \sin \lambda_p t] , \tag{3.94}$$

where $\rho$ is a scaling constant, $\lambda_p$ is a "frequency shift" parameter, and $\zeta_1, \zeta_2$ are two independent processes treated as outputs of first-order linear filters excited by a Gaussian white noise, i.e.,

$$d\zeta_i(t) = -\alpha\zeta_i(t)dt + \alpha d\xi(t), \quad i = 1, 2 , \tag{3.95}$$

$\alpha$ is a positive parameter, $\xi(t)$ is a standard Wiener process.

Treating $\zeta_i(t)$ as new state variables, i.e., $x_3(t) = \zeta_1(t)$, $x_4(t) = \zeta_2(t)$, (3.92–3.95) can be presented in a vector form

$$dx_1(t) = x_2(t)dt \;,$$
$$dx_2(t) = \left[-\lambda_0^2 x_1(t) - 2hx_2(t) + k_1\rho[x_3(t)\cos\lambda_p t + x_4(t)\sin\lambda_p t]\right.$$
$$\left.+k_2\rho[x_3(t)\cos\lambda_p t + x_4(t)\sin\lambda_p t]|\rho[x_3(t)\cos\lambda_p t + x_4(t)\sin\lambda_p t]|\right] dt \;,$$
$$dx_3(t) = -\alpha x_3(t)dt + \alpha d\xi(t),$$
$$dx_4(t) = -\alpha x_4(t)dt + \alpha d\xi(t). \qquad (3.96)$$

System (3.96) represents the nonlinear dynamic system under Gaussian excitation and the corresponding methods of the determination of moment equations will be discussed in next chapters.

### Polynomial Gaussian Excitation

Now we consider a linear oscillator with non-Gaussian excitation in the form of a polynomial of filtered coloured noises.

$$\eta(t) = \sum_{k=1}^{M} \alpha_k y^k(t) \qquad (3.97)$$

and

$$dy(t) = -\alpha y(t)dt + qd\xi(t) \;, \qquad (3.98)$$

where $\alpha, \alpha_k (k = 1,\ldots,M)$, and $q$ are positive constant parameters, $y(t)$ is a one-dimensional colored Gaussian process, and $\xi(t)$ is a standard Wiener process.

Treating the powers of $y(t)$ denoted by $y^k(t)$, $k = 1,\ldots,M$, as new state variables $x_{2+k}(t) = y^k(t)$, (3.92), (3.97), and (3.98) can be presented in a vector form

$$dx_1(t) = x_2(t)dt \;,$$
$$dx_2(t) = \left[-\lambda_0^2 x_1(t) - 2hx_2(t) + \sum_{k=1}^{M}\alpha_k x_{2+k}(t)\right] dt \;,$$
$$\vdots \qquad\qquad \vdots$$
$$dx_{2+k}(t) = \left[-\alpha k x_{2+k}(t) + \frac{k(k-1)}{2}q^2 x_k\right] dt + kx_{k+1}qd\xi(t) \;. \qquad (3.99)$$

In this case the moment equations can be solved without applying any closure technique. It will be illustrated in an example in Sect. 5.6.

## Trigonometric Function Representation
## of Non-Gaussian Processes

The following trigonometric representation of continuous stationary non-Gaussian processes was proposed by Iyengar and Jaiswal [48]

$$\eta(t) = \sum_{k=0}^{M} a_k \psi_k(t) , \tag{3.100}$$

where $a_i$ are constant parameters, $\psi_0(t) = 1$, and $\psi_1(t) = \xi(t)$ is a Gaussian process with zero mean and unit standard deviation. The remaining functions $\psi_k(t)$ are selected as normalized Hermitian polynomials in $\zeta(t)$:

$$\psi_k = \frac{(-1)^k}{\sqrt{k!}} \exp\left\{\frac{\xi^2}{2}\right\} \frac{d^k}{d\xi^k} \exp\left\{-\frac{\xi^2}{2}\right\}, \quad k = 2, 3, \dots . \tag{3.101}$$

The probabilistic characteristics of the process $\eta(t)$ are given in [48]. As an example of the process $\xi(t)$, Iyengar and Jaiswal have proposed

$$\zeta(t) = R\cos(\lambda t - \Theta) , \tag{3.102}$$

where $\lambda$ is a constant positive parameter (frequency), $R$ and $\Theta$ are Rayleigh and uniformly distributed random variables, respectively, i.e., the probability density function $g_\xi(r, \theta)$ has the form

$$g_\xi(r, \theta) = g_1(r)g_2(\theta) , \tag{3.103}$$

where

$$g_1(r) = \begin{cases} r\exp\{-\frac{r^2}{2}\}, & r > 0 , \\ 0, & r \leq 0, \end{cases} \tag{3.104}$$

and

$$g_2(\theta) = \begin{cases} \frac{1}{2\pi}, & \text{for } \theta \in [0, 2\pi] , \\ 0, & \text{for } \theta \notin [0, 2\pi] . \end{cases} \tag{3.105}$$

In this case the stationary solution of (3.92) can be found in the analytical form, for instance, for $M = 3$ and $a_0 = 0$ the process $\eta(t)$ has the form

$$\eta(t) = a_1 R\cos(\lambda t - \Theta) + \frac{a_2}{\sqrt{2}}[R^2\cos^2(\lambda t - \Theta) - 1]$$

$$+\frac{a_3}{\sqrt{6}}[R^3\cos^3(\lambda t - \Theta) - 3R\cos(\lambda t - \Theta)] = \alpha_0 + \sum_{k=1}^{3} \alpha_k \cos(k(\lambda t - \Theta)) , \tag{3.106}$$

where $\alpha_0 = \frac{1}{\sqrt{2}}a_2(R^2 - 1)$, $\alpha_1 = a_1 R + \frac{3a_3 R}{\sqrt{6}}(\frac{R^2}{4} - 1)$, $\alpha_2 = \frac{1}{\sqrt{2}}a_2 R^2$, $\alpha_3 = \frac{1}{4\sqrt{6}}a_3 R^3$, and the corresponding stationary solution of (3.92)

$$x_1(t) = \frac{1}{\lambda_0^2}[c_0 + \sum_{k=1}^{3} c_k \cos(k(\lambda t - \Theta - \delta_k))] \,, \tag{3.107}$$

where $c_0 = \alpha_0$, $c_k = H(\lambda_e, k\lambda)^{-1}\alpha_k$,

$$\delta_k = \frac{1}{k} \arctan \frac{2\zeta i \lambda}{\omega_0^2 - (i\lambda)^2} \,, \tag{3.108}$$

for $k = 1, 2, 3$ and

$$H(\lambda_e, \lambda) = \left[\left(1 - \frac{\lambda^2}{\lambda_0^2}\right)^2 + \left(\frac{2\zeta\lambda}{\lambda_0^2}\right)^2\right]^{-\frac{1}{2}} \,. \tag{3.109}$$

Hence one can calculate the response statistics numerically.

Another trigonometric representation of non-Gaussian excitation was proposed by Dimentberg [20]. He considered the non-Gaussian excitation $\eta(t)$ as a harmonic excitation with random phase modulation defined by

$$\eta(t) = \lambda \cos \nu \,, \tag{3.110}$$

$$\dot{\nu} = \mu + 1(t)w(t) \,, \tag{3.111}$$

where $\lambda$ and $\nu$ are constant parameters, $1(t)$ is a unit step envelope function

$$1(t) = \begin{cases} 1 & t \geq 0 \,, \\ 0 & t < 0, \end{cases} \tag{3.112}$$

$w(t)$ is a Gaussian stationary white noise with zero mean and the power spectral density constant $K > 0$.

Introducing new state variables $x_3(t) = \cos \nu$, $x_4(t) = \sin \nu$, (3.92), (3.110–3.112) can be presented in a vector form

$$\begin{aligned}
dx_1(t) &= x_2(t)dt \,, \\
dx_2(t) &= [-\lambda_0^2 x_1(t) - 2hx_2(t) + \lambda x_3(t)]dt \,, \\
dx_3(t) &= -\mu x_4(t)dt - \sqrt{2\pi K}1(t)x_4(t)d\xi(t) \,, \\
dx_4(t) &= \mu x_3(t)dt + \sqrt{2\pi K}1(t)x_3(t)d\xi(t) \,,
\end{aligned} \tag{3.113}$$

where $\xi(t)$ is the standard Wiener process and a constraint equation $x_3^2(t) + x_4^2(t) = 1$ must be satisfied. Hence one can derive the moment equations for linear system with parametric excitations (3.113).

A modification of the above representation of non-Gaussian excitation was proposed by Hou et al. [41] who considered the non-Gaussian excitation $\eta(t)$ in the form

$$\eta(t) = [\lambda + w_1(t)] \cos \nu \,, \tag{3.114}$$

$$\dot{\nu} = \mu + 1(t)w_2(t) \,, \tag{3.115}$$

where $\lambda$ and $\nu$ are constant parameters, $w_1(t)$ and $w_2(t)$ are two independent Gaussian white noises with spectral densities $K_1 > 0$ and $K_2 > 0$, respectively.

Introducing new state variables $x_3(t) = \lambda \cos \nu$, $x_4(t) = \lambda \sin \nu$, (3.92, 3.114–3.115) can be presented in a vector form

$$
\begin{aligned}
dx_1(t) &= x_2(t)dt \ , \\
dx_2(t) &= [-\lambda_0^2 x_1(t) - 2hx_2(t) + \lambda x_3(t)]dt + \sqrt{2\pi K_1} x_3(t)d\xi_1(t) \ , \\
dx_3(t) &= (\pi K_2 x_3(t) - \mu x_4(t))dt - \sqrt{2\pi K_2} x_4(t)d\xi_2(t) \ , \\
dx_4(t) &= (\mu x_3(t) - \pi K_2 x_4(t))dt + \sqrt{2\pi K_2} x_3(t)d\xi_2(t) \ , \qquad (3.116)
\end{aligned}
$$

where $\xi_1(t)$ and $\xi_2(t)$ are independent Wiener processes and a constraint equation $x_3^2(t) + x_4^2(t) = 1$ must be satisfied. Hence one can derive the moment equations for linear system with parametric excitations (3.116).

### Transformation with Memory of the Brownian and Levy Processes

An illustration of the second group of generation of non-Gaussian processes, namely, *transformation with memory of the Brownian and Levy processes* is the treatment of the stationary process $\eta(t)$ as a stationary solution of a nonlinear system under Gaussian excitation. We illustrate the idea of this approach using the Duffing oscillator, i.e., we consider the following equation for stochastic process $\eta(t) = \eta_1(t)$ appearing in (3.92)

$$
\begin{aligned}
d\eta_1(t) &= \eta_2(t)dt \ , \\
d\eta_2(t) &= [-a_1\eta_1(t) - 2a_2\eta_2(t) + \varepsilon\eta_1^3(t)]dt + qd\xi \ . \qquad (3.117)
\end{aligned}
$$

By introducing new state variables $x_3(t) = \eta_1(t)$, $x_4(t) = \eta_2(t)$, (3.92, 3.117) can be presented in a vector form

$$
\begin{aligned}
dx_1(t) &= x_2(t)dt \ , \\
dx_2(t) &= [-\lambda_0^2 x_1(t) - 2hx_2(t) + x_3(t)]dt \ , \\
dx_3(t) &= x_4(t)dt \ , \\
dx_4(t) &= [-a_1 x_3(t) - 2a_2 x_4(t) + \varepsilon x_3^3(t)]dt + qd\xi(t) \ , \qquad (3.118)
\end{aligned}
$$

where $a_1$, $a_2$, $\varepsilon$, and $q$ are constant parameters.

Further examples of the generation of non-Gaussian stationary processes by solutions of stochastic differential equations are given, for instance, by Lyandres and Shahaf [60] who studied narrow band non-Gaussian stationary processes, Grigoriu [35] who considered non-Gaussian processes with both finite and infinite variance.

To obtain moments of solution we have to use one of the approximate methods. This problem will be discussed in Chap. 4.

# Bibliography Notes

The problem of the determination of response moments in linear dynamic systems under external stochastic excitation is a basic problem and was considered in many monographs, for instance, [28, 49, 56, 57, 75, 77, 78].

A particular class of researchers' interest are linear systems with time-dependent coefficients under stationary stochastic excitations, see for instance [18, 66, 67, 85]. The application of Voltera second-order integral equations for the determination of response probabilistic characteristics of these systems was proposed by Szopa [81]. Also an explicit solutions for multi-degree-of-freedom (MDOF) systems with coefficients being random variables uncorrelated with external Gaussian white noise excitation was given in [43].

The case of stochastic parametric excitations in linear systems that is more complicated than external excitation was discussed in [1, 2, 3, 4, 6, 7, 8, 9, 21, 25, 26, 31, 42, 50, 53, 65, 86].

A wide group of methods for linear systems with parametric excitations and special structures defined by some Lie algebra properties are algebraic methods presented, for instance, in [10, 11, 87]. These method are very useful for such group of linear systems with parametric excitations that can be transformed by suitable choice of linear transformation to a linear system with triangular matrices corresponding to deterministic and stochastic parts of the considered system. This approach is valid to white noise as well as to colored noise excitation.

The problem of the determination of response moments in linear dynamic systems under external stochastic nonstationary excitations was discussed very intensively in the literature. For instance, Barnoski and Maurer [5] studied the mean-square response of a linear oscillator excited by both white noise and noise with an exponentially decaying harmonic correlation function for both the unit step and boxcar time-modulating function. An exact second-order moment of response of a linear oscillator subjected to a white noise modulated with the unit step, boxcar and gamma envelope functions was obtained in [44]. The evolutionary response covariance matrices of a multi–degree–of–freedom system excited by a piecewise linear-modulated white noise were derived by Gasparini and DebChaudhury [19, 30]. Also evolutionary (time dependent) response spectrum was obtained for special classes of nonstationary excitation by Corotis and Vanmarcke [16], Corotis and Marshall [17].

Some other results can be found, for instance, in [12, 29, 40, 45, 54, 55, 61, 70, 72, 74, 76, 79, 82, 83, 84].

A general explicit, closed-form solutions for the correlation matrix and an evolutionary power spectral density matrix of the response of a linear multi–degree–of–freedom system subjected to a uniformly modulated stochastic process with the gamma envelope function were presented by Conte and Peng [15, 68].

The problem of the determination of response moments in linear dynamic systems under external stationary non-Gaussian excitations was discussed theoretically [13, 59] as well as in the application of modeling physical random excitations [27, 62, 64, 80]. Poisson-driven pulses were shown, for instance in [22, 34, 46, 47, 69] and polynomial forms of non-Gaussian processes in [23, 24, 63].

The problem of the construction and the simulation of non-Gaussian processes was discussed in [32, 39, 80].

# References

1. Adomian, G.: Stochastic Systems. Academic Press, New York, 1983.
2. Ariaratnam, S. and Graefe, P.: Linear systems with stochastic coefficients.I. Int. J. Control **1**, 239–250, 1965.
3. Ariaratnam, S. and Graefe, P.: Linear systems with stochastic coefficients.II. Int. J. Control **2**, 161–169, 1965.
4. Ariaratnam, S. and Graefe, P.: Linear systems with stochastic coefficients.III. Int. J. Control **2**, 205–210, 1965.
5. Barnoski, R. and Maurer, J.: Mean-square response of simple mechanical systems to nonstationary random excitation. Trans. ASME J. Appl. Mech. **36**, 221–227, 1969.
6. Benaroya, H. and Rehak, M.: Parametric random excitation. I. Exponentially correlated parameters. ASCE J. Eng. Mech. **113**, 861–874, 1986.
7. Benaroya, H. and Rehak, M.: Parametric random excitation. II. White-noise parameters. ASCE J. Eng. Mech. **113**, 875–884, 1986.
8. Benaroya, H. and Rehak, M.: Response and stability of random differential equation. I. Moment equation method. ASME J. Appl. Mech. **56**, 192–195, 1989.
9. Benaroya, H. and Rehak, M.: Response and stability of random differential equation. II. Expansion method. ASME J. Appl. Mech. **56**, 196–201, 1989.
10. Brockett, R.: System theory on group manifolds and coset spaces. SIAM J. Control **10**, 265–284, 1972.
11. Brockett, R.: Lie algebras and Lie groups in control theory. In: R. Brockett (ed.), Geometric Methods in System Theory, 49–62. Reidel, The Netherlands, 1973.
12. Bucher, C.: Approximate nonstationary random vibration analysis for MDOF systems. Trans. ASME J. Appl. Mech. **55**, 197–200, 1988.
13. Bucher, C. and Schuëller, G.: Non-Gaussian response of nonlinear systems. In: G. Schuëller (ed.), Structural Dynamics-Recent Advances, 103–127. Springer-Verlag, Berlin, 1991.
14. Caughey, T. and Stumpf, H.: Transient response of a dynamic system under random excitation. Trans. ASME J. Appl. Mech. **28**, 563–566, 1961.
15. Conte, J. and Peng, B.: An explicit closed form solution for linear systems subjected to nonstationary random excitations. Probab. Eng. Mech. **11**, 37–50, 1996.
16. Corotis, R. and Vanmarcke, E.: Time-dependent spectral content of system response. ASCE J. Eng. Mech. Div. **101**, 623–637, 1975.
17. Corotis, R. and Marshall, T.: Oscillator response to modulated random excitation. ASCE J. Eng. Mech. Div. **103**, 501–513, 1977.
18. Czerny, L. and Popp, K.: Analyse instationärer Zufallschwingungen. Z. Angew. Math. Mech. **64**, 40–42, 1984.

19. DebChaudhury, A. and Gasparini, D.: Response of MDOFS systems to vector random excitation. ASCE J. Eng. Mech. Div. **108**, 367–385, 1982.

20. Dimentberg, M.: Statistical Dynamics of Nonlinear and Time-Varying systems. Research Studies Press, London, 1988.

21. Dimentberg, M. and Sidorenko, A.: On joint interaction between vibrations generated in linear system under external and parametric excitations. Iz. AN SSSR Mech. Tverd. Tela **3**, 3–7, 1978.

22. Di Paola, M.: Stochastic dynamics of MDOF structural systems under non-normal filtered inputs. Probab. Eng. Mech. **9**, 265–272, 1994.

23. Di Paola, M.: Linear systems to polynomials of nonnormal processes and quasi-linear systems. Trans. ASME J. Appl. Mech. **64**, pp. in press, 1997.

24. Di Paola, M. and Falsone, G.: Higher order statistics of the response of MDOF linear systems under polynomials of filtered normal white noises. Probab. Eng. Mech. **12**, 189–196, 1997.

25. Evlanov, L. and Konstantinov, V.: Systems with Random Parameters. Nauka, Moskwa, 1976 (in Russian).

26. Fang, T. and Dowell, E.: Transient mean square response of randomly damped linear systems. J. Sound Vib. **113**, 71–79, 1987.

27. Floris, C.: Equivalent Gaussian process in stochastic dynamics with application to along-wind response of structures. Int. J. Non-Linear Mech. **31**, 779–794, 1996.

28. Gard, T.: Introduction to Stochastic Differential Equations. Marcel Dekker, New York, 1988.

29. Gasparini, D.: Response of MDOF systems to nonstationary random excitation. ASCE J. Eng. Mech. Div. **105**, 13–27, 1979.

30. Gasparini, D. and DebChaudhury, A.: Dynamic response to nonstationary non-white excitation. ASCE J. Eng. Mech. Div. **106**, 1233–1248, 1980.

31. Gopalsamy, K.: On a class of linear systems with random coefficients. Z. Angew. Math. Mech. **56**, 453–459, 1976.

32. Grigoriu, M.: Applied non-Gaussian Processes: Examples, Theory, Simulation, Linear Random Vibration, and MATLAB Solutions. Prentice Hall, Englewood Cliffs, 1995.

33. Grigoriu, M.: Equivalent linearization for Poisson white noise input. Probab. Eng. Mech. **10**, 45–51, 1995.

34. Grigoriu, M.: Linear and nonlinear systems with non-Gaussian white noise input. Probab. Eng. Mech. **10**, 171–180, 1995.

35. Grigoriu, M.: Equivalent linearization for systems driven by Levy white noise. Probab. Eng. Mech. **15**, 185–190, 2000.

36. Grigoriu, M.: Non-Gaussian models for stochastic mechanics. Probab. Eng. Mech. **15**, 15–23, 2000.

37. Grigoriu, M. and Ariaratnam, S.: Response of linear systems to polynomials of Gaussian process. Trans. ASME J. Appl. Mech. **55**, 905–910, 1988.

38. Grigoriu, M.: Response of linear systems to quadratic Gaussian excitations. ASCE J. Eng. Mech. **110**, 523–535, 1984.

39. Gurley, K., Kareem, A., and Tognarelli, M.: Simulation of a class of non-normal random process. Int. J. Non-Linear Mech. **31**, 601–617, 1996.

40. Hammond, J.: On the response of single and multidegree of freedom systems to nonstationary random excitations. J. Sound Vib. **7**, 393–416, 1968.

41. Hou, Z., Zhou, Y., Dimentberg, M., and Noori, M.: A stationary model for periodic excitation with uncorrelated random disturbances. Probab. Eng. Mech. **11**, 191–203, 1996.
42. Ibrahim, R.: Parametric Random Vibration. Research Studies Press, Letchworth United Kingdom, 1985.
43. Impollonia, N. and Riccardi, G.: Explicit solutions in the stochastic dynamics of structural systems. Probab. Eng. Mech. **21**, 171–181, 2006.
44. Iwan, W. and Hou, Z.: Explicit solutions for the response of simple systems subjected to nonstationary random excitation. Struct. Safety **6**, 77–86, 1989.
45. Iwan, W. and Smith, K.: On the nonstationary response of stochastically excited secondary systems. Trans. ASME J. Appl. Mech. **54**, 688–694, 1987.
46. Iwankiewicz, R.: Response of linear vibrating systems driven by renewal point processes. Probab. Eng. Mech. **5**, 111–121, 1990.
47. Iwankiewicz, R. and Sobczyk, K.: Dynamic response of linear structures to correlated random impulses. J. Sound Vib. **86**, 303–317, 1983.
48. Iyengar, R. N. and Jaiswal, O. R.: A new model for non-Gaussian random excitations. Probab. Eng. Mech. **8**, 281–287, 1993.
49. Kazakov, I. and Malczikow, S.: Analysis of Stochastic Systems in State Space. Nauka, Moskwa, 1975 (in Russian).
50. Khasminski, R.: Stochastic Stability of Differential Equations. Sijthoff and Noordhoff, Alpen aan den Rijn, the Netherlands, 1980.
51. Kotulski, Z. and Sobczyk, K.: Linear systems and normality. J. Stat. Phys. **24**, 359–373, 1981.
52. Krenk, S. and Gluver, H.: An algorithm for moments of response from nonnormal excitation of linear systems. In: S. Ariaratnam, G. Schueller, and I. Elishakoff (eds.), Stochastic Structural Dynamics, 181–195. Elsevier, London, 1988.
53. Lambert, L.: Berechnung der Momente stochastisch parametererregter Systeme. Z. Angew. Math. Mech. **59**, 397–398, 1979.
54. Langley, R.: On quasi-stationary approximations to nonstationary random vibration. J. Sound Vib. **113**, 365–375, 1987.
55. Leung, A.: A simple method for exponentially modulated random excitation. J. Sound Vib. **112**, 273–282, 1987.
56. Lin, Y.: Probabilistic Theory of Structural Dynamics. McGraw Hill, New York, 1967.
57. Lin, Y. and Cai, G.: Probabilistic Structural Dynamics. McGraw Hill, New York, 1995.
58. Lipcer, R. and Shiryayev, A.: Statistics of Random Processes I : General Theory. Springer, New York, 1977.
59. Lutes, L. and Hu, S.: Non-normal stochastic response of linear systems. ASCE J. Eng. Mech. **112**, 127–141, 1986.
60. Lyandres, V. and Shahaf, M.: Envelope correlation function of the narrow-band non-Gaussian process. Int. J. Non-Linear Mech. **30**, 359–369, 1995.
61. Masri, S.: Response of multidegree of freedom system to nonstationary random excitation. Trans. ASME J. Appl. Mech. **45**, 649–656, 1978.
62. Moshchuk, N. and Ibrahim, R.: Response statistics of ocean structures to nonlinear hydrodynamic loading: Part 2. Non-Gaussian ocean waves. J. Sound Vib. **191**, 107–128, 1996.
63. Muscolino, G.: Linear systems excited by polynomial forms of non-Gaussian filtered process. Probab. Eng. Mech. **10**, 35–44, 1995.

64. Ochi, M.: Non-Gaussian random processes in ocean engineering. Probab. Eng. Mech. **1**, 28–31, 1986.

65. Parajev, U.: Introduction to Stochastic Dynamics of Control and Filtering Processes. Sovetzkie Radio, Moskwa, 1976 (in Russian).

66. Pekala, W. and Szopa, J.: The application of Green's function to the investigation of the response of vehicle accelerating over random profile. Rev. Roum. Acad. des Sci. Tech. Mechanique Appliquee **28**, 295–310, 1983.

67. Pekala, W. and Szopa, J.: The application of Green's multidimensional function to the investigation of the stochastic vibrations of dynamical systems. Ing. Arch. **54**, 91–97, 1984.

68. Peng, B. and Conte, J.: Closed form solutions for the response of linear systems to fully nonstationary earthquake excitations. ASCE J. Eng. Mech. **124**, 684–694, 1998.

69. Roberts, J.: The response of linear vibratory systems to random impulses. J. Sound Vib. **2**, 375–390, 1965.

70. Roberts, J. and Spanos, P.: Random Vibration and Statistical Linearization. John Wiley and Sons, Chichester, 1990.

71. Roberts, J. and Vasta, M.: Parametric identification of systems with non-Gaussian excitation using measured response spectra. Probab. Eng. Mech. **15**, 59–71, 2001.

72. Sakata, M. and Kimura, K.: The use of moment equation for calculating the mean square response of a linear system to nonstationary random excitation. J. Sound Vib. **67**, 383–393, 1979.

73. Samorodnitsky, G. and Taqqu, M.: Stable non-Gaussian Random Processes, Stochastic Models with Infinite Variance. Chapman and Hall, London, 1994.

74. Shihab, S. and Preumont, A.: Nonstationary random vibrations of linear multidegree of freedom systems. J. Sound Vib. **132**, 457–471, 1989.

75. Sobczyk, K.: Stochastic Differential Equations for Applications. Kluwer, The Netherlands, 1991.

76. Solomos, G. and Spanos, P.: Oscillator response to nonstationary excitation. Trans. ASME J. Appl. Mech. **51**, 907–912, 1984.

77. Soong, T.: Random Differential Equations in Science and Engineering. Academic Press, New York, 1973.

78. Soong, T. and Grigoriu, M.: Random Vibration of Mechanical and Structural Systems. PTR Prentice Hall, Englewood Cliffs, NJ, USA, 1993.

79. Spanos, P.: Nonstationary random vibration of linear a structure. Int. J. Solid Struct. **14**, 861–867, 1978.

80. Steinwolf, A., Ferguson, N., and White, R.: Variations in steepness of probability density function of beam random vibration. Eur. J. Mech. A/Solids **19**, 319–341, 2000.

81. Szopa, J.: Application of volterra stochastic integral equation of the II-nd kind to the analysis of dynamic systems of variable inertia. J. Tech. Phys. **17**, 423–433, 1976.

82. To, C.: Non-stationary random responses of a multidegree of freedom system by the theory of evolutionary spectra. J. Sound Vib. **83**, 273–291, 1982.

83. To, C.: Time-dependent variance and covariance of responses of structures to nonstationary random excitations. J. Sound Vib. **93**, 135–156, 1984.

84. To, C.: Response statistics of discretized structures to nonstationary random excitation. J. Sound Vib. **105**, 217–231, 1986.

85. Wan, F.: Nonstationary response of linear time varying dynamical systems to random excitation. ASME J. Appl. Mech. **40**, 422–428, 1973.

86. Wedig, W.: Stochastische Schwingungen-Simulation, Schätzung und Stabilität. Z. Angew. Math. Mech. **67**, 34–42, 1987.

87. Willems, J.: Moment stability of linear white noise and coloured noise systems. In: B. Clarkson (ed.), Stochastic Problems in Dynamics, 67–89. Pitman Press, London, 1976.

# 4

# Moment Equations for Nonlinear Stochastic Dynamic Systems (NSDS)

The determination of characteristics of solutions of nonlinear stochastic differential equations is much more complicated than for linear equations and in general case it is the unsolved problem. The main reason is the fact that if we derive the differential moment equations then on the right-hand sides appear complicated functions of moments or/and moments of higher order than on the left-hand sides. This basic difficulty stimulated as well the study of the widest class of nonlinear stochastic differential equations having exact solutions in analytical forms or corresponding moment equations in a closed form as the search of approximate methods of the determination of characteristics of solutions of nonlinear stochastic differential equations. These analytical approximate methods were developed intensively in past 50 years. They can be divided in several groups:

- simple and complex closure techniques
- Gaussian and non-Gaussian closure techniques
- method of Fokker–Planck–Kolmogorov equation
- stochastic averaging
- perturbation methods
- spectral methods
- operator methods

These methods are described in many monographs and books, for instance, [1, 12, 13, 27, 30, 32, 33, 34].

Except the main groups there are many special methods that can be used only for particular classes of nonlinear systems. The main fault of the approximate methods (except stochastic linearization) is the fact that even for a few dimensional nonlinear systems of stochastic differential equations the number of moment differential equations is extremely large and these moments are only approximate ones. Therefore, in this chapter, we consider only stochastic dynamic systems (SDS) with nonlinearities in the form of polynomials under parametric and external Gaussian excitation and two groups of closure techniques, namely simple and non-Gaussian ones.

L. Socha: *Moment Equations for Nonlinear Stochastic Dynamic Systems (NSDS)*, Lect. Notes Phys. **730**, 85–102 (2008)
DOI 10.1007/978-3-540-72997-6_4

## 4.1 Moment Equations for Polynomial SDS Under Parametric and External Gaussian Excitation

Consider the vector nonlinear Ito stochastic differential equation

$$dx(t) = \mathbf{a}(\mathbf{x}(t), t)dt + \boldsymbol{\sigma}(\mathbf{x}(t), t)d\boldsymbol{\xi}(t), \quad \mathbf{x}(t_0) = \mathbf{x}_0 , \qquad (4.1)$$

where $\mathbf{x} = [x_1, \ldots, x_n]^T$ is a vector state, $\mathbf{a}(\mathbf{x}, t) = [a_1(\mathbf{x}, t), \ldots, a_n(\mathbf{x}, t)]^T$ is a vector nonlinear function, $\boldsymbol{\sigma}(\mathbf{x}, t) = [\sigma_{ij}(\mathbf{x}, t)]$, $i = 1, \ldots, n$, $j = 1, \ldots, M$ is a matrix, that elements are nonlinear functions; $\boldsymbol{\xi}(t) = [\xi_1(t), \ldots, \xi_M(t)]^T$ is a vector standard Wiener process, $\mathbf{x}_0 = [x_{10}, \ldots, x_{n0}]^T$ is a vector random initial condition independent of $\boldsymbol{\xi}(t)$ (elements of one vector are independent of elements of second vector).

Denote the $p$th moment of vector $\mathbf{x}$ by $\alpha_p$, the $p$th central moment by $\mu_p$, and the $p$th correlation moment with respect to the coordinate $x_r$ by $\rho_p^r(\tau)$, i.e.,

$$\alpha_p = \alpha_{p_1 \ldots p_n} = E[x_1^{p_1} \ldots x_n^{p_n}],$$

$$\mu_p = \mu_{p_1 \ldots p_n} = E[(x_1 - m_1)^{p_1} \ldots (x_n - m_n)^{p_n}],$$

$$\rho_p^r(\tau) = \rho_{p_1 \ldots p_n}^r(\tau) = E[x_1^{p_1}(t) \ldots x_n^{p_n}(t)x_r(t - \tau)] , \qquad (4.2)$$

where $p = [p_1, \ldots, p_n]$ is the multiindex, $\sigma(p) = \sum_{i=1}^p p_i$, $m_i = E[x_i]$.

Using Ito formula one can derive the Ito differentials of $x_1^{p_1} \ldots x_n^{p_n}$ and $(x_1 - m_1)^{p_1} \ldots (x_n - m_n)^{p_n}$ [27] as follows:

$$d(x_1^{p_1} \ldots x_n^{p_n}) = \left[ \sum_{i=1}^n p_i a_i(\mathbf{x}, t) x_1^{p_1} \ldots x_i^{p_i - 1} \ldots x_n^{p_n} \right.$$

$$+ \frac{1}{2} \sum_{k=1}^n p_k(p_k - 1) b_{kk}(\mathbf{x}, t) x_1^{p_1} \ldots x_k^{p_k - 2} \ldots x_n^{p_n}$$

$$\left. + \sum_{l=2}^n \sum_{k=1}^{l-1} p_l p_k b_{kl}(\mathbf{x}, t) x_1^{p_1} \ldots x_k^{p_k - 1} \ldots x_l^{p_l - 1} \ldots x_n^{p_n} \right] dt$$

$$+ \sum_{i=1}^n \sum_{j=1}^M p_i a_i(\mathbf{x}, t) x_1^{p_1} \ldots x_i^{p_i - 1} \ldots x_n^{p_n} \sigma_{ij}(\mathbf{x}, t) d\xi_j, \qquad (4.3)$$

$$d\left((x_1 - m_1)^{p_1} \ldots (x_n - m_n)^{p_n}\right)$$

$$= \left[ \sum_{i=1}^n p_i a_i(\mathbf{x}, t)(x_1 - m_1)^{p_1} \ldots (x_i - m_i)^{p_i - 1} \ldots (x_n - m_n)^{p_n} \right.$$

$$+ \frac{1}{2} \sum_{k=1}^n p_k(p_k - 1) b_{kk}(\mathbf{x}, t)(x_1 - m_1)^{p_1}$$

$$\ldots (x_k - m_k)^{p_k - 2} \ldots (x_n - m_n)^{p_n}$$

$$+ \sum_{l=2}^{n} \sum_{k=1}^{l-1} p_l p_k b_{kl}(\mathbf{x},t)(x_1 - m_1)^{p_1} \ldots (x_k - m_k)^{p_k-1} \ldots$$

$$\times (x_l - m_l)^{p_l-1} \ldots (x_n - m_n)^{p_n} \Big] dt$$

$$+ \sum_{i=1}^{n} \sum_{j=1}^{M} p_i a_i(\mathbf{x},t)(x_1 - m_1)^{p_1} \ldots (x_i - m_i)^{p_i-1} \ldots$$

$$\times (x_n - m_n)^{p_n} \sigma_{ij}(\mathbf{x},t) d\xi_j, \tag{4.4}$$

where

$$b_{ij}(\mathbf{x},t) = \sum_{k=1}^{M} \sigma_{ik}(\mathbf{x},t)\sigma_{jk}(\mathbf{x},t), \quad i,j = 1,\ldots n . \tag{4.5}$$

The direct averaging of (4.3) and (4.4) gives differential equations for moments and central moments

$$\frac{d\alpha_p}{dt} = \sum_{i=1}^{n} p_i E[a_i(\mathbf{x},t)x_1^{p_1} \ldots x_i^{p_i-1} \ldots x_n^{p_n}]$$

$$+ \frac{1}{2} \sum_{k=1}^{n} p_k(p_k - 1) E[b_{kk}(\mathbf{x},t)x_1^{p_1} \ldots x_k^{p_k-2} \ldots x_n^{p_n}]$$

$$+ \sum_{l=2}^{n} \sum_{k=1}^{l-1} p_l p_k E[b_{kl}(\mathbf{x},t)x_1^{p_1} \ldots x_k^{p_k-1} \ldots x_l^{p_l-1} \ldots x_n^{p_n}], \tag{4.6}$$

$$\frac{d\mu_p}{dt} = \sum_{i=1}^{n} p_i E[a_i(\mathbf{x},t)(x_1 - m_1)^{p_1} \ldots (x_i - m_i)^{p_i-1} \ldots (x_n - m_n)^{p_n}]$$

$$+ \frac{1}{2} \sum_{k=1}^{n} p_k(p_k - 1) E[b_{kk}(\mathbf{x},t)(x_1 - m_1)^{p_1} \ldots$$

$$\times (x_k - m_k)^{p_k-2} \ldots (x_n - m_n)^{p_n}]$$

$$+ \sum_{l=2}^{n} \sum_{k=1}^{l-1} p_l p_k E[b_{kl}(\mathbf{x},t)(x_1 - m_1)^{p_1} \ldots (x_k - m_k)^{p_k-1} \ldots$$

$$\times (x_l - m_l)^{p_l-1} \ldots (x_n - m_n)^{p_n}]. \tag{4.7}$$

If we multiply each equation in the equation set (4.3) by $x_r(t_0)$, and next average, then we obtain the differential equations for coordinates of correlation moments $x_r$, $r = 1, \ldots, n$

$$\frac{d\rho_p^r}{dt} = \sum_{i=1}^{n} p_i E[a_i(\mathbf{x},t)x_1^{p_1}(t) \ldots x_i^{p_i-1}(t) \ldots x_n^{p_n}(t)x_r(t - \tau)]$$

$$+ \frac{1}{2} \sum_{k=1}^{n} p_k(p_k - 1) E[b_{kk}(\mathbf{x},t)x_1^{p_1}(t) \ldots$$

$$\times x_k^{p_k-2}(t)\dots x_n^{p_n}(t)x_r(t-\tau)] \tag{4.8}$$

$$+\sum_{l=2}^{n}\sum_{k=1}^{l-1}p_lp_kE[b_{kl}(\mathbf{x},t)x_1^{p_1}(t)\dots x_k^{p_k-1}(t)$$

$$\times\dots x_l^{p_l-1}(t)\dots x_n^{p_n}(t)x_r(t-\tau)]\,,$$

$$r=1,\dots,n\,,$$

where $\rho_p^r(t)=E[x_1^{p_1}(t)\dots x_i^{p_i-1}(t)\dots x_n^{p_n}(t)x_r(t-\tau)]$, $\tau=t-t_0$.

The initial conditions in moment differential equations can be determined by using the joint probability distribution of coordinates of $\mathbf{x}_0$, then

$$\alpha_p(t_0)=E[x_{10}^{p_1}\dots x_{n0}^{p_n}]\,. \tag{4.9}$$

We note that the quantities:

$$E[a_i(\mathbf{x},t)x_1^{p_1}\dots x_i^{p_i-1}\dots x_n^{p_n}], \quad E[b_{kk}(\mathbf{x},t)x_1^{p_1}\dots x_k^{p_k-2}\dots x_n^{p_n}],$$

$$E[b_{kl}(\mathbf{x},t)x_1^{p_1}\dots x_k^{p_k-1}\dots x_l^{p_l-1}\dots x_n^{p_n}],$$

$$E[a_i(\mathbf{x},t)(x_1-m_1)^{p_1}\dots(x_i-m_i)^{p_i-1}\dots(x_n-m_n)^{p_n}],$$

$$E[b_{kk}(\mathbf{x},t)(x_1-m_1)^{p_1}\dots(x_k-m_k)^{p_k-2}\dots(x_n-m_n)^{p_n}],$$

$$E[b_{kl}(\mathbf{x},t)(x_1-m_1)^{p_1}\dots(x_k-m_k)^{p_k-1}\dots(x_l-m_l)^{p_l-1}\dots(x_n-m_n)^{p_n}],$$

$$E[a_i(\mathbf{x},t)x_1^{p_1}(t)\dots x_i^{p_i-1}(t)\dots x_n^{p_n}(t)x_r(t-\tau)], \quad E[b_{kk}(\mathbf{x},t)x_1(t)^{p_1}\dots$$

$$x_k(t)^{p_k-2}\dots x_n^{p_n}(t)x_r(t-\tau)],$$

$$E[b_{kl}(\mathbf{x},t)x_1^{p_1}(t)\dots x_k^{p_k-1}(t)\dots x_l^{p_l-1}(t)\dots x_n^{p_n}(t)x_r(t-\tau)],$$

$$i,k=1,\dots,n,\ l=2,\dots,n$$

appearing in (4.6–4.8) are nonlinear functions of moments $\alpha_p$, $\mu_p$, $\rho_p^r$, $r=1,2\dots,n$, $p=1,2\dots$, that in general case one can determine only in approximate way applying, for instance, an approximation of probability density function by Gramm–Charlier or Edgeworth series [27].

Consider the particular case when the elements of vector function $a(\mathbf{x},t)$ and $b(\mathbf{x},t)$ are polynomials of coordinates of vector $\mathbf{x}$, i.e.,

$$a_i(\mathbf{x},t)=\sum_{\sigma(p)=1}^{N_i}a_{ip_1\dots p_n}(t)x_1^{p_1}\dots x_n^{p_n}\,, \tag{4.10}$$

where $i=1,\dots,n$, $\sigma(p)=\sum_{i=1}^{n}p_i$, $p_i=0,1,2,\dots$, $N_i=0,1,\dots$,

$$b_{ij}(\mathbf{x},t)=\sum_{\sigma(r)=1}^{M_{ij}}b_{ijr_1\dots r_n}(t)x_1^{r_1}\dots x_n^{r_n}\,, \tag{4.11}$$

where $i,j=1,\dots,n$, $\sigma(r)=\sum_{i=1}^{n}r_i$, $r_i=0,1,2,\dots$, $M_{ij}=0,1,\dots$; $a_{ip_1\dots p_n}(t)$ and $b_{ijr_1\dots r_n}(t)$ are functions of variable $t$.

Summation with respect to $\sigma(p)$ in (4.10) and with respect to $\sigma(r)$ in (4.11) should be treated as the summation with respect to the corresponding multiindices.

The moment equations for system (4.1), (4.10), and (4.11) were given by Dashevski [10]. They have the following form:

$$\frac{dE[x_1^{k_1}\ldots x_n^{k_n}]}{dt} = \sum_{i=1}^{n}\sum_{\sigma(p)=1}^{N_i} a_{ip_1\ldots p_n}(t)$$

$$\times E[x_1^{p_1+k_1}\ldots x_{i-1}^{p_{i-1}+k_{i-1}}x_i^{p_i+k_i-1}x_{i+1}^{p_{i+1}+k_{i+1}}\ldots x_n^{p_n+k_n}]$$

$$+\frac{1}{2}\sum_{i=1}^{n}k_i(k_i-1)\sum_{\sigma(r)=1}^{M_{ii}}b_{iir_1\ldots r_n}(t)$$

$$\times E[x_1^{r_1+k_1}\ldots x_{i-1}^{r_{i-1}+k_{i-1}}x_i^{r_i+k_i-2}x_{i+1}^{r_{i+1}+k_{i+1}}\ldots x_n^{r_n+k_n}]$$

$$+\frac{1}{2}\sum_{\substack{i,j=1\\i\neq j}}^{n}k_ik_j\sum_{k=1}^{M_{ij}}b_{ijr_1\ldots r_n}(t)$$

$$\times E[x_1^{r_1+k_1}\ldots x_{i-1}^{r_{i-1}+k_{i-1}}x_i^{r_i+k_i-1}x_{i+1}^{r_{i+1}+k_{i+1}}\ldots$$
$$x_{j-1}^{r_{j-1}+k_{j-1}}x_j^{r_j+k_j-1}x_{j+1}^{r_{j+1}+k_{j+1}}\ldots x_n^{r_n+k_n}], \qquad (4.12)$$

where

$$\sum_{i=1}^{n}k_i>0,\ \sigma(p)=\sum_{i=1}^{n}p_i,\ \sigma(r)=\sum_{i=1}^{n}r_i,\ p_i,r_i=0,1,2,\ldots,N_i,\ M_{ij}=0,1,\ldots;$$

$$\sum_{\sigma(r)=1}^{M_{ii}}b_{iir_1\ldots r_n}(t)E[.]=0 \text{ for } k_i<2, \sum_{k=1}^{M_{ij}}b_{ijr_1\ldots r_n}(t)E[.]=0 \text{ for } k_i=0,k_j=0.$$

In a similar way, one can derive the differential equations for central moments and cumulants. We note that the moments appearing on the right-hand side of (4.12) are higher order than the corresponding moments on the left-hand side. To solve this system of equations, one has to introduce some algebraic relations that describe higher-order moments as nonlinear functions of lower-order moments. This problem is discussed in the next sections of this chapter.

Now, we present two examples of creation of moment differential equations.

*Example 4.1.* Determine moment equations for scalar Ito stochastic differential equation

$$dx(t) = [x(t)+a(x(t))]dt+(b_1x(t)+b_2)d\xi(t), \quad x(t_0)=x_0, \qquad (4.13)$$

where $a(x)$ is a nonlinear scalar function, and $b_1$ and $b_2$ are constants. The moment equations have the form

$$\frac{dE[x^k]}{dt} = kE[x^k] + kE[a(x)x^{k-1}] + \frac{1}{2}k(k-1)E[(b_1x + b_2)^2 x^{k-2}] ,$$

$$E[x^k(t_0)] = E[x_0^k], \quad k = 1, 2, \dots . \tag{4.14}$$

In the particular case when $a(x)$ is a polynomial described by

$$a(x) = \sum_{i=2}^{N} a_i x^i , \tag{4.15}$$

Equation (4.14) takes the form

$$\frac{dE[x^k]}{dt} = kE[x^k] + k\sum_{i=2}^{N} E[a_i x^{i+k-1}] + \frac{1}{2}k(k-1)E[(b_1x + b_2)^2 x^{k-2}] ,$$

$$E[x^k(t_0)] = E[x_0^k], \quad k = 1, 2, \dots . \tag{4.16}$$

*Example 4.2.* Determine moment equations for the Duffing oscillator

$$d\mathbf{x}(t) = [\mathbf{A}\mathbf{x}(t) + \mathbf{B}(\mathbf{x}(t))]dt + \mathbf{G}d\xi(t), \quad \mathbf{x}(t_0) = \mathbf{x}_0 , \tag{4.17}$$

where

$$\mathbf{x} = \begin{bmatrix} x_1 \\ x_2 \end{bmatrix}, \ \mathbf{A} = \begin{bmatrix} 0 & 1 \\ -\lambda_0^2 & -2h \end{bmatrix}, \ \mathbf{B}(\mathbf{x}) = \begin{bmatrix} 0 \\ -\varepsilon x_1^3 \end{bmatrix}, \ \mathbf{G} = \begin{bmatrix} 0 \\ \delta \end{bmatrix}, \ \mathbf{x}_0 = \begin{bmatrix} x_{10} \\ x_{20} \end{bmatrix}. \tag{4.18}$$

where $\lambda_0, h, \varepsilon$, and $\delta$ are constant parameters; $\xi(t)$ is a standard Wiener process; the initial conditions $x_{10}$ and $x_{20}$ are independent variables of $\xi(t)$.

Using relations (4.12), we find

$$\frac{dE[x_1]}{dt} = E[x_2],$$

$$\frac{dE[x_2]}{dt} = -\lambda_0^2 E[x_1] - 2hE[x_2] - \varepsilon E[(x_1)^3],$$

$$\frac{dE[(x_1)^2]}{dt} = 2E[x_1 x_2],$$

$$\frac{dE[x_1 x_2]}{dt} = E[(x_2)^2] - \lambda_0^2 E[(x_1)^2] - 2hE[x_1 x_2] - \varepsilon E[(x_1)^4],$$

$$\frac{dE[(x_2)^2]}{dt} = -2\lambda_0^2 E[x_1 x_2] - 2hE[(x_2)^2] - 2\varepsilon E[(x_1)^3 x_2] + \delta^2,$$

$$\frac{dE[(x_1)^3]}{dt} = 3E[(x_1)^2 x_2],$$

$$\frac{dE[(x_1)^2 x_2]}{dt} = 2E[x_1(x_2)^2] - \lambda_0^2 E[(x_1)^3] - 2hE[(x_1)^2 x_2] - \varepsilon E[(x_1)^5],$$

$$\frac{dE[x_1(x_2)^2]}{dt} = E[(x_2)^3] - 2\lambda_0^2 E[(x_1)^2 x_2] - 4hE[x_1(x_2)^2]$$
$$-2\varepsilon E[(x_1)^4 x_2] + \delta^2 E[x_1],$$

$$\frac{dE[(x_2)^3]}{dt} = -3\lambda_0^2 E[x_1(x_2)^2] - 6hE[(x_2)^3] - 3\varepsilon E[(x_1)^3 (x_2)^2]$$
$$+3\delta^2 E[x_2],$$

$$\frac{dE[(x_1)^4]}{dt} = 4E[(x_1)^3 x_2],$$

$$\frac{dE[(x_1)^3 x_2]}{dt} = 3E[(x_1)^2 (x_2)^2] - \lambda_0^2 E[(x_1)^4] - 2hE[(x_1)^3 x_2] - \varepsilon E[(x_1)^6],$$

$$\frac{dE[(x_1)^2 (x_2)^2]}{dt} = 2E[x_1(x_2)^3] - 2\lambda_0^2 E[(x_1)^3 x_2] - 4hE[(x_1)^2 (x_2)^2]$$
$$-2\varepsilon E[(x_1)^5 x_2] + \delta^2 E[(x_1)^2],$$

$$\frac{dE[x_1(x_2)^3]}{dt} = E[(x_2)^4] - 3\lambda_0^2 E[(x_1)^2 (x_2)^2] - 6hE[x_1(x_2)^3]$$
$$-3\varepsilon E[(x_1)^4 (x_2)^2] + 3\delta^2 E[x_1 x_2],$$

$$\frac{dE[(x_2)^4]}{dt} = -4\lambda_0^2 E[x_1(x_2)^3] - 8hE[(x_2)^4] - 4\varepsilon E[(x_1)^3 (x_2)^3]$$
$$+6\delta^2 E[(x_2)^2],$$

$$\frac{dE[(x_1)^5]}{dt} = 5E[(x_1)^4 x_2],$$

$$\frac{dE[(x_1)^4 x_2]}{dt} = 4E[(x_1)^3 (x_2)^2] - \lambda_0^2 E[(x_1)^5] - 2hE[(x_1)^4 x_2] - \varepsilon E[(x_1)^7],$$

$$\frac{dE[(x_1)^3 (x_2)^2]}{dt} = 3E[(x_1)^2 (x_2)^3] - 2\lambda_0^2 E[(x_1)^4 x_2] - 4hE[(x_1)^3 (x_2)^2]$$
$$-2\varepsilon E[(x_1)^6 x_2] + \delta^2 E[(x_1)^3],$$

$$\frac{dE[(x_1)^2 (x_2)^3]}{dt} = 2E[x_1(x_2)^4] - 3\lambda_0^2 E[(x_1)^3 (x_2)^2] - 6hE[(x_1)^2 (x_2)^3]$$
$$-3\varepsilon E[(x_1)^5 (x_2)^2] + 3\delta^2 E[(x_1)^2 x_2],$$

$$\frac{dE[x_1(x_2)^4]}{dt} = E[(x_2)^5] - 4\lambda_0^2 E[(x_1)^2 (x_2)^3] - 8hE[x_1(x_2)^4]$$
$$-4\varepsilon E[(x_1)^4 (x_2)^3] + 6\delta^2 E[x_1(x_2)^2],$$

$$\frac{dE[(x_2)^5]}{dt} = -5\lambda_0^2 E[x_1(x_2)^4] - 10hE[(x_2)^5]$$
$$-5\varepsilon E[(x_1)^3 (x_2)^4] + 10\delta^2 E[(x_2)^3],$$

$$\frac{dE[(x_1)^6]}{dt} = 6E[(x_1)^5 x_2],$$

$$\frac{dE[(x_1)^5 x_2]}{dt} = 5E[(x_1)^4 (x_2)^2] - \lambda_0^2 E[(x_1)^6] - 2hE[(x_1)^5 x_2] - \varepsilon E[(x_1)^8],$$

$$\frac{dE[(x_1)^4 (x_2)^2]}{dt} = 4E[(x_1)^2 (x_2)^3] - 2\lambda_0^2 E[(x_1)^5 x_2] - 4hE[(x_1)^4 (x_2)^2]$$
$$- 2\varepsilon E[(x_1)^7 x_2] + \delta^2 E[(x_1)^4],$$

$$\frac{dE[(x_1)^3 (x_2)^3]}{dt} = 3E[(x_1)^2 (x_2)^4] - 3\lambda_0^2 E[(x_1)^4 (x_2)^2] - 6hE[(x_1)^3 (x_2)^3]$$
$$- 3\varepsilon E[(x_1)^6 (x_2)^2] + 3\delta^2 E[(x_1)^3 x_2],$$

$$\frac{dE[(x_1)^2 (x_2)^4]}{dt} = 2E[x_1 (x_2)^5] - 4\lambda_0^2 E[(x_1)^3 (x_2)^3] - 8hE[(x_1)^2 (x_2)^4]$$
$$- 4\varepsilon E[(x_1)^5 (x_2)^3] + 6\delta^2 E[(x_1)^2 (x_2)^2],$$

$$\frac{dE[x_1 (x_2)^5]}{dt} = E[(x_2)^6] - 5\lambda_0^2 E[(x_1)^2 (x_2)^4] - 10hE[x_1 (x_2)^5]$$
$$- 5\varepsilon E[(x_1)^4 (x_2)^4] + 10\delta^2 E[x_1 (x_2)^3],$$

$$\frac{dE[(x_2)^6]}{dt} = -6\lambda_0^2 E[x_1 (x_2)^5] - 12hE[(x_2)^6]$$
$$- 6\varepsilon E[(x_1)^3 (x_2)^5] + 15\delta^2 E[(x_2)^4] . \tag{4.19}$$

The initial conditions are

$$E[x_1^p x_2^q](t_0) = E[(x_{10})^p (x_{20})^q], \quad p, q = 0, 1, \ldots 6, \quad p + q \le 6 . \tag{4.20}$$

One can derive in a similar way differential equations for correlation moments defined by $\rho_{ij}(\tau) = E[x_1^i(t) x_2^j(t) x_1(t - \tau)]$, $i, j = 0, \ldots 5$ with respect to the variable $\tau = t - t_0 \in R^+$

$$\frac{d\rho_{10}}{d\tau} = \rho_{01},$$

$$\frac{d\rho_{01}}{d\tau} = -\lambda_0^2 \rho_{10} - 2h\rho_{01} - \varepsilon \rho_{30},$$

$$\frac{d\rho_{20}}{d\tau} = 2\rho_{11},$$

$$\frac{d\rho_{11}}{d\tau} = \rho_{02} - \lambda_0^2 \rho_{20} - 2h\rho_{11} - \varepsilon \rho_{40},$$

$$\frac{d\rho_{02}}{d\tau} = -2\lambda_0^2 \rho_{11} - 2h\rho_{02} - 2\varepsilon \rho_{31} + \delta^2 \rho_{00},$$

$$\frac{d\rho_{30}}{d\tau} = 3\rho_{21},$$

$$\frac{d\rho_{21}}{d\tau} = 2\rho_{12} - \lambda_0^2 \rho_{30} - 2h\rho_{21} - \varepsilon \rho_{50},$$

$$\frac{d\rho_{12}}{d\tau} = \rho_{03} - 2\lambda_0^2\rho_{21} - 4h\rho_{12} - 2\varepsilon\rho_{41} + \delta^2\rho_{10},$$

$$\frac{d\rho_{03}}{d\tau} = -3\lambda_0^2\rho_{12} - 6h\rho_{03} - 3\varepsilon\rho_{32} + 3\delta^2\rho_{01},$$

$$\frac{d\rho_{40}}{d\tau} = 4\rho_{31},$$

$$\frac{d\rho_{31}}{d\tau} = 3\rho_{22} - \lambda_0^2\rho_{40} - 2h\rho_{31} - \varepsilon\rho_{60},$$

$$\frac{d\rho_{22}}{d\tau} = 2\rho_{13} - 2\lambda_0^2\rho_{31} - 4h\rho_{22} - 2\varepsilon\rho_{51} + \delta^2\rho_{20},$$

$$\frac{d\rho_{13}}{d\tau} = \rho_{04} - 3\lambda_0^2\rho_{22} - 6h\rho_{13} - 3\varepsilon\rho_{42} + 3\delta^2\rho_{11},$$

$$\frac{d\rho_{04}}{d\tau} = -4\lambda_0^2\rho_{13} - 8h\rho_{04} - 4\varepsilon\rho_{33} + 6\delta^2\rho_{02},$$

$$\frac{d\rho_{50}}{d\tau} = 5\rho_{41},$$

$$\frac{d\rho_{41}}{d\tau} = 4\rho_{32} - \lambda_0^2\rho_{50} - 2h\rho_{41} - \varepsilon\rho_{70},$$

$$\frac{d\rho_{32}}{d\tau} = 3\rho_{23} - 2\lambda_0^2\rho_{41} - 4h\rho_{32} - 2\varepsilon\rho_{61} + \delta^2\rho_{30},$$

$$\frac{d\rho_{23}}{d\tau} = 2\rho_{14} - 3\lambda_0^2\rho_{32} - 6h\rho_{23} - 3\varepsilon\rho_{52} + 3\delta^2\rho_{21},$$

$$\frac{d\rho_{14}}{d\tau} = \rho_{05} - 4\lambda_0^2\rho_{23} - 8h\rho_{14} - 4\varepsilon\rho_{43} + 6\delta^2\rho_{12},$$

$$\frac{d\rho_{05}}{d\tau} = -5\lambda_0^2\rho_{14} - 10h\rho_{05} - 5\varepsilon\rho_{34} + 10\delta^2\rho_{03} . \tag{4.21}$$

The initial conditions are

$$\rho_{ij}(0) = E[x_1^{i+1}x_2^j], \quad i,j = 0,\dots 5 . \tag{4.22}$$

They can be calculated from the moment equations (4.19). We note also that the correlation function of $x_1(t)$ is defined by the correlation moment $\rho_{10}$, namely,

$$R_{11}(\tau) = E[x_1(t)x_1(t-\tau)] = \rho_{10}(\tau) . \tag{4.23}$$

Similar derivation is also possible for other differential equations for correlation moments, for instance, $p_{ij}(\tau) = E[x_1^i(t)x_2^j(t)x_2(t-\tau)]$ with initial conditions $p_{ij}(0) = E[x_1^i x_2^{j+1}]$ and next correlation functions

$$R_{12}(\tau) = E[x_1(t)x_2(t-\tau)] = p_{10}(\tau), R_{22}(\tau) = E[x_2(t)x_2(t-\tau)] = p_{01}(\tau) . \tag{4.24}$$

For details, see for instance, [6].

## 4.2 Simple Closure Techniques

In Sect. 4.1 we have derived differential equations for moments in the case when the nonlinear functions are polynomials of components of vector state. We have noted that on right-hand sides there are higher-order moments than on the left-hand sides. To solve these equations one has to close them, i.e., to replace higher-order moments by functions of lower-order moments in such a way that on the right-hand side of the whole system of moment equations will be moments of the less or equal order than on the left-hand side. It can be done by simple or complex closure techniques. They have been described in many monographs and survey papers, for instance, [13, 21, 27, 31, 33]. It should be stressed that not every individual moment differential equation has to be closed. It is important that the whole system of moment differential equations has to be closed.

Application of simple closure techniques usually leads to incorrect or inaccurate results, while application of complex techniques gives complicated formulas that are not useful in practical calculations for multidimensional systems, except the situation when computer symbolic transformation programs can be used. In what follows, we discuss in this subsection only the basic simple techniques.

### The Method of Neglecting Higher-Order Moments or Cumulants

In this closure technique, if the moments (or central moments or cumulants or quasi-moments) on the right-hand side are higher order than the corresponding ones on the left-hand side of the system then they are neglected. We illustrate this technique with an example.

*Example 4.3.* Consider a particular case of (4.13) from Example 4.1

$$dx(t) = [x(t) + ax^3(t)]dt + [b_1 x(t) + b_2]d\xi(t), \quad x(t_0) = x_0 . \qquad (4.25)$$

The moment equations have the form

$$\frac{dE[x^k]}{dt} = kE[x^k] + kaE[x^{k+2}] + \frac{1}{2}k(k-1)E[(b_1 x + b_2)^2 x^{k-2}],$$

$$E[x^k](t_0) = E[x_0^k], \quad k = 1, 2, \ldots . \qquad (4.26)$$

After application of the method of neglecting higher-order moments, for instance, for $k \geq N - 2$, the moment equations have the form (4.26) for $k = 1, 2, \ldots, N - 2$ and

$$\frac{dE[x^k]}{dt} = kE[x^k] + \frac{1}{2}k(k-1)E[(b_1 x + b_2)^2 x^{k-2}],$$

$$E[x^k](t_0) = E[x_0^k], \quad for \quad k = N - 1, \quad k = N , \qquad (4.27)$$

i.e., the two last equations do not depend on nonlinearity $ax^3$. The same method can be applied to the derivation of differential equations for central moments or cumulants or quasi-moments.

### The Method of Steady-State Values of Moments or Cumulants

Consider again moment differential equations (4.12) to the $N$th order. Then on the right-hand side appear moments until $N+L$ order, where $L$ is a number depending on orders of polynomials $N_i$ and $M_{ij}$ (we use notation introduced by the derivation of (4.12)). To close the considered system of $N$ moment equations, the system is extended to $N+L$ equations using general formula (4.12). Next, the right-hand sides of equations from order $N+1$ to $N+L$ and also all moments of order higher than $N+L$ are equated to zero. From the so obtained algebraic equations we find moments of order from $N+1$ to $N+L$ as functions of moments till $N$ order. The same procedure can be applied to the central moments or cumulants.

*Example 4.4.* If we apply the method of steady-state values of moments to system (4.25), then for the moment of order $N+2$ we obtain the relation

$$0 = (N+2)E[x^{N+2}] + \frac{1}{2}(N+2)(N+1)E[(b_1 x + b_2)^2 x^N] , \qquad (4.28)$$

hence

$$E[x^{N+2}] = -\frac{2(N+1)b_1 b_2}{2 + (N+1)b_1^2} E[x^{N+1}] - \frac{(N+1)b_2^2}{2 + (N+1)b_1^2} E[x^N] . \qquad (4.29)$$

Similarly, for $(N+1)$-order moment we find

$$E[x^{N+1}] = -\frac{2N}{2 + Nb_1^2} E[x^N] - \frac{Nb_2^2}{2 + Nb_1^2} E[x^{N-1}] . \qquad (4.30)$$

Substituting equalities (4.29) and (4.30) into moment equations (4.26), we obtain equations for $k = N-1$ and $N-2$

$$\frac{dE[x^{N-1}]}{dt} = (N-1)E[x^{N-1}] + (N-1)$$

$$\times a\left[-\frac{2N}{2 + Nb_1^2} E[x^N] - \frac{Nb_2^2}{2 + Nb_1^2} E[x^{N-1}]\right] + \frac{1}{2}(N-1)(N-2)$$

$$\times E[(b_1 x + b_2)^2 x^{N-3}], \quad E[x^{N-1}](t_0) = E[x_0^{N-1}], \qquad (4.31)$$

$$\frac{dE[x^N]}{dt} = NE[x^N]$$

$$+ Na\left[-\frac{2(N+1)b_1 b_2}{2 + (N+1)b_1^2}\left[-\frac{2N}{2 + Nb_1^2} E[x^N] - \frac{Nb_2^2}{2 + Nb_1^2} E[x^{N-1}]\right]\right.$$

$$\left. -\frac{(N+1)b_2^2}{2 + (N+1)b_1^2} E[x^N]\right] + \frac{1}{2}N(N-1)E[(b_1 x + b_2)^2 x^{N-2}],$$

$$\times E[x^N](t_0) = E[x_0^N] . \qquad (4.32)$$

The same procedures for both simple closure techniques can be applied to the central moments or cumulants or quasi-moments.

## 4.3 Non-Gaussian Closure Techniques

In this subsection we present two basic approaches, namely central moment and cumulant closure techniques.

### Method of Central Moments

In the method of central moments, it is assumed that starting from an order $N$ the central moments of a vector random variable $\mathbf{x}$ are equated to zero, i.e.,

$$E[(x_1 - m_1)(x_2 - m_2)\ldots(x_s - m_s)] = 0, \quad s > N , \tag{4.33}$$

where $m_i = E[x_i]$, $i = 1\ldots, s$. From relation (4.33) one can determine moments of variables $x_i$ as functions of lower-order moments

$$E\left[\prod_{k=1}^{n} x_k\right] = \sum_{i=1}^{n-2} \binom{n}{i}(-1)^{i+1}\left\{\prod_{k=1}^{i} E[x_k]E\left[\prod_{k=i+1}^{n} x_k\right]\right\}_s$$

$$+(-1)^n(n-1)\prod_{k=1}^{n} E[x_k] , \tag{4.34}$$

where $n = 2, 3, \ldots$, $\{.\}_s$ denotes a symmetrizing operation with respect to all its arguments, i.e., the arithmetic mean of different permuted terms similar to the one within the braces. For example,

$$\{E[x_i]E[x_kx_l]\}_s = \frac{1}{3}\{E[x_i]E[x_kx_l] + E[x_k]E[x_ix_l] + E[x_l]E[x_kx_i]\} . \tag{4.35}$$

In particular, for the first six-order moments, if for different $k$ we assume different indices, equality (4.34) takes the forms

$$E[x_i] = m_i ,$$

$$E[x_ix_j] = E[x_i]E[x_j] ,$$

$$E[x_ix_jx_k] = 3\{E[x_i]E[x_kx_l]\}_s - 2E[x_i]E[x_j]E[x_k],$$

$$E[x_ix_jx_kx_l] = 4\{E[x_i]E[x_jx_kx_l]\}_s - 6\{E[x_i]E[x_j]E[x_kx_l]\}_s$$
$$+3E[x_i]E[x_j]E[x_k]E[x_l],$$

$$E[x_ix_jx_kx_lx_m] = 5\{E[x_i]E[x_jx_kx_lx_m]\}_s - 10\{E[x_i]E[x_j]E[x_kx_lx_m]\}_s$$
$$+10\{E[x_i]E[x_j]E[x_k]E[x_lx_m]\}_s$$
$$-4E[x_i]E[x_j]E[x_k]E[x_l]E[x_m],$$

$$E[x_ix_jx_kx_lx_mx_n] = 6\{E[x_i]E[x_jx_kx_lx_mx_n]\}_s$$
$$-15\{E[x_i]E[x_j]E[x_kx_lx_mx_n]\}_s$$
$$+20\{E[x_i]E[x_j]E[x_k]E[x_lx_mx_n]\}_s$$
$$-15\{E[x_i]E[x_j]E[x_k]E[x_l]E[x_mx_n]\}_s$$
$$+5E[x_i]E[x_j]E[x_k]E[x_l]E[x_m]E[x_n]. \tag{4.36}$$

## Cumulant Method

The cumulant method was proposed by Cramer [7], and next by Stratonovich [37]. It was applied by many authors to the approximation of the characteristics of the solutions of stochastic differential equations, for instance, by Dashevski [10], Wu and Lin [42], Sun and Hsu [38], and Pugachev and Sinitsyn [27]. To our knowledge based on several numerical examples discussed in the literature, one can recognize the cumulant method as one of the best approximation approaches.

The definition of cumulants and the method of their calculation were presented in Sect. 2.2. We recall that for the determination of formulas for the cumulants, the joint characteristic function defined by

$$\Phi(\Theta) = \Phi(\Theta_1, \dots, \Theta_n) = E\left[\exp\left\{i\sum_{j=1}^{n}\Theta_j x_j\right\}\right] \tag{4.37}$$

is used. When the number of random variables is equal to $m$ for $m < n$, we substitute

$$\Theta_{m+1} = \Theta_{m+2} = \dots = \Theta_n = 0. \tag{4.38}$$

$\Phi(\Theta)$ can be presented in the form of a series

$$\Phi(\Theta) = \Phi_1(\Theta) = 1 + i\sum_{j=1}^{n}\Theta_j E[X_j] + \frac{i^2}{2!}\sum_{j=1}^{n}\sum_{k=1}^{n}\Theta_j\Theta_k E[X_j X_k] + \dots \tag{4.39}$$

or in an equivalent form

$$\Phi(\Theta) = \Phi_2(\Theta) = \exp\left\{i\sum_{j=1}^{n}\Theta_j\kappa_1[X_j] + \frac{i^2}{2!}\sum_{j=1}^{n}\sum_{k=1}^{n}\Theta_j\Theta_k\kappa_2[X_j X_k] + \dots\right\}, \tag{4.40}$$

where $E[X_1 \dots X_n]$ and $\kappa_n[X_1 \dots X_n]$ are the moments and cumulants of the random variables $X_1 \dots X_n$, respectively.

We then find the moments and cumulants from the relations

$$E[x_1^{k_1} \dots x_m^{k_m}] = \frac{1}{i^k}\frac{\partial^k\Phi_1(\Theta)}{\partial\Theta_1^{k_1}\dots\partial\Theta_m^{k_m}}\Big|_{\Theta=0}, \quad \sum_{j=1}^{m}k_j = k, k \leq n, \tag{4.41}$$

$$\kappa_m[x_1^{q_1} \dots x_m^{q_m}] = \frac{1}{i^q}\frac{\partial^q ln(\Phi_2(\Theta))}{\partial\Theta_1^{q_1}\dots\partial\Theta_m^{q_m}}\Big|_{\Theta=0}, \quad \sum_{j=1}^{m}q_j = q, q \leq n. \tag{4.42}$$

After the $k$th and $q$th differentiation of quantities $\Phi_2$ and $ln(\Phi_1)$ in equalities (4.41) and (4.42), respectively, we obtain

$$E[x_1^{k_1} \dots x_m^{k_m}] = \sum_{q_1=1}^{k_1}\binom{k_1-1}{q_1-1}\sum_{q_2=0}^{k_2}\binom{k_2}{q_2}\dots$$

$$\times \sum_{q_m=0}^{k_m} \binom{k_m}{q_m} \kappa_m [x_1^{q_1} \dots x_m^{q_m}] E[x_1^{k_1-q_1} \dots x_m^{k_m-q_m}], (4.43)$$

$$\kappa_m [x_1^{q_1} \dots x_m^{q_m}]$$
$$= E[x_1^{q_1} \dots x_m^{q_m}] - \sum_{l_1=1}^{q_1} \binom{q_1-1}{l_1-1} \sum_{l_2=0}^{q_2} \binom{q_2}{l_2} \dots$$
$$\times \sum_{l_m=0}^{q_m} \binom{q_m}{l_m} \kappa_m [x_1^{l_1} \dots x_m^{l_m}] E[x_1^{q_1-l_1} \dots x_m^{q_m-l_m}]. \qquad (4.44)$$

For instance, from equality (4.43) one can determine the relations between the moments and the cumulants for the first six-order moments, see, for instance, [13]

$$E[x_j] = \kappa_1[x_j]$$
$$E[x_ix_j] = \kappa_2[x_j, x_k] + \kappa_1[x_j]\kappa_1[x_k]$$
$$E[x_ix_jx_k] = \kappa_3[x_j, x_k, x_l] + 3\{\kappa_1[x_i]\kappa_2[x_j, x_k]\}_s + \kappa_1[x_i]\kappa_1[x_j]\kappa_1[x_k]$$
$$E[x_ix_jx_kx_l] = \kappa_4[x_ix_jx_kx_l] + 4\{\kappa_1[x_i]\kappa_3[x_j, x_k, x_l]\}_s$$
$$+3\{\kappa_2[x_i, x_j]\kappa_2[x_k, x_l]\}_s$$
$$+6\{\kappa_1[x_i]\kappa_1[x_j]\kappa_2[x_kx_l]\}_s + \kappa_1[x_i]\kappa_1[x_j]\kappa_1[x_k]\kappa_1[x_l]$$

$$E[x_ix_jx_kx_lx_m] = \kappa_5[x_i, x_j, x_k, x_l, x_m] + 5\{\kappa_1[x_i]\kappa_4[x_j, x_k, x_l, x_m]\}_s$$
$$+10\{\kappa_2[x_i, x_j]\kappa_3[x_k, x_l, x_m]\}_s$$
$$+10\{\kappa_1[x_i]\kappa_1[x_j]\kappa_3[x_k, x_l, x_m]\}_s$$
$$+10\{\kappa_1[x_i]\kappa_1[x_j]\kappa_1[x_k]\kappa_2[x_l, x_m]\}_s$$
$$+15\{\kappa_2[x_i, x_j]\kappa_2[x_k, x_l]\kappa_1[x_m]\}_s$$
$$+\kappa_1[x_i]\kappa_1[x_j]\kappa_1[x_k]\kappa_1[x_l]\kappa_1[x_m]$$

$$E[x_ix_jx_kx_lx_mx_n] = \kappa_6[x_i, x_j, x_k, x_l, x_m, x_n] + 6\{\kappa_1[x_i]\kappa_5[x_j, x_k, x_l, x_m, x_n]\}_s$$
$$+10\{\kappa_3[x_i, x_j, x_k]\kappa_3[x_l, x_m, x_n]\}_s$$
$$+15\{\kappa_1[x_i]\kappa_1[x_j]\kappa_4[x_k, x_l, x_m, x_n]\}_s$$
$$+15\{\kappa_1[x_i]\kappa_1[x_j]\kappa_1[x_k]\kappa_1[x_l]\kappa_2[x_m, x_n]\}_s$$
$$+20\{\kappa_1[x_i]\kappa_1[x_j]\kappa_1[x_k]\kappa_3[x_l, x_m, x_n]\}_s$$
$$+45\{\kappa_1[x_i]\kappa_1[x_j]\kappa_2[x_k, x_l]\kappa_2[x_m, x_n]\}_s$$
$$+60\{\kappa_1[x_i]\kappa_2[x_j, x_k]\kappa_3[x_l, x_m, x_n]\}_s$$

$$+15\{\kappa_2[x_i, x_j]\kappa_2[x_k, x_l]\kappa_2[x_m, x_n]\}_s$$
$$+15\{\kappa_2[x_i, x_j]\kappa_4[x_k, x_l, x_m, x_n]\}_s$$
$$+\kappa_1[x_i]\kappa_1[x_j]\kappa_1[x_k]\kappa_1[x_l]\kappa_1[x_m]\kappa[x_n]. \tag{4.45}$$

Using relations (4.43) and (4.44), one can establish relations between moments of higher order and lower orders, i.e., one can find moments of higher order than $N$ as nonlinear functions of moments till order $N$. This can be done by equating to zero all cumulants of higher order than $N$ and from obtained nonlinear algebraic equations one can calculate higher order than $N$ moments. This procedure is called cumulant closure technique. For instance, for moments till six orders, we obtain the following relations;

$$E[x_i] = m_i,$$
$$E[x_i x_j] = E[x_i]E[x_j],$$
$$E[x_i x_j x_k] = 3\{E[x_i](E[x_j x_k] - E[x_j]E[x_k])\}_s + E[x_i]E[x_j]E[x_k],$$
$$E[x_i x_j x_k x_l] = 4\{E[x_i]E[x_j x_k x_l]\}_s + 3\{E[x_i x_j]E[x_k x_l]\}_s$$
$$-2 \cdot 6\{E[x_i]E[x_j]E[x_k x_l]\}_s$$
$$+6E[x_i]E[x_j]E[x_k]E[x_l], \tag{4.46}$$

$$E[x_i x_j x_k x_l x_m] = 5\{E[x_i]E[x_j x_k x_l x_m]\}_s - 2 \cdot 10\{E[x_i]E[x_j]E[x_k x_l x_m]\}_s$$
$$+6 \cdot 10\{E[x_i]E[x_j]E[x_k]E[x_l x_m]\}_s$$
$$-2 \cdot 15\{E[x_i]E[x_j x_k]E[x_l x_m]\}_s$$
$$+10\{E[x_i x_j]E[x_k x_l x_m]\}_s - E[x_i]E[x_j]E[x_k]E[x_l]E[x_m],$$

$$E[x_i x_j x_k x_l x_m x_n] = 6\{E[x_i]E[x_j x_k x_l x_m x_n]\}_s$$
$$-2 \cdot 15\{E[x_i]E[x_j]E[x_k x_l x_m x_n]\}_s$$
$$+6 \cdot 20\{E[x_i]E[x_j]E[x_k]E[x_l x_m x_n]\}_s$$
$$-24 \cdot 15\{E[x_i]E[x_j]E[x_k]E[x_l]E[x_m x_n]\}_s$$
$$+6 \cdot 45\{E[x_i]E[x_j]E[x_k x_l]E[x_m x_n]\}_s$$
$$+15\{E[x_i x_j]E[x_k x_l x_m x_n]\}_s$$
$$+10\{E[x_i x_j x_k]E[x_l x_m x_n]\}_s$$
$$-2 \cdot 15\{E[x_i x_j]E[x_k x_l]E[x_m x_n]\}_s$$
$$-2 \cdot 60\{E[x_i]E[x_j x_k]E[x_l x_m x_n]\}_s$$
$$+120E[x_i]E[x_j]E[x_k]E[x_l]E[x_m]E[x_n]. \tag{4.47}$$

If before the symmetrization operation there are two factors, then the second one determines the number of different terms between braces of symmetrization operation, for instance,

$$2 \cdot 3 \{ E[x_i] E[x_k x_l] \}_s = 2 ( E[x_i] E[x_k x_l] + E[x_k] E[x_i x_l] + E[x_l] E[x_k x_i] ) \ . \quad (4.48)$$

## Bibliography Notes

The problem of the determination of moment equations in nonlinear dynamic systems was presented and discussed in many books and papers. In this chapter we have used the presentation proposed partially in [27] and [13]. The other recommended books with applications of non-Gaussian closure techniques are [4, 19, 22, 23].

The problem of the determination of differential equations for other characteristics such as central moments, cumulants, and quasi-moments for nonlinear dynamic systems was discussed, for instance, in [27]. We note that the method of quasi-moments was proposed by Kuznecov et al. [20] and developed first by Stratonovich [37] and next by Sperling [35, 36]. A list of joint relations between moments and quasi-moments one can find, for instance, in [13].

The most popular non-Gaussian closure technique called cumulant neglect closure was studied and applied by many authors, mainly to nonlinear parametric excited systems with polynomial nonlinearities, for instance, [2, 3, 5, 8, 9, 14, 15, 16, 24, 25, 26, 38, 39, 40, 41]. It was also modified by (Kasharova and Shin) [17, 18]. A non-Gaussian closure technique was also applied to a linear oscillator with hysteresis [11].

Since the number of differential equations for higher-order moments is very large, for instance, in the case of $p$-dimensional system the number of equations $Q(p, N)$ is equal to

$$Q(p, N) = \frac{(p + N)!}{p! N!} - 1 \ , \quad (4.49)$$

where $N$ is the highest order of moment used in the approximation, a few software including symbolic transformation programs were proposed. They are described in [13, 28, 29].

## References

1. Adomian, G.: Stochastic Systems. Academic Press, New York, 1983.
2. Assaf, S. A. and Zirkle, L.: Approximate analysis of nonlinear stochastic systems. Int. J. Control **23**, 477–492, 1976.
3. Boguslavski, I.: Statistical analysis of multidimensional dynamical system by application of hermite polynomials of multivariables. Avtomatika i Telemekhanika **30**, 36–51, 1969.
4. Bolotin, V.: Random Vibration of Elastic Systems. Martinus Nijhoff Publishers, Boston, 1984.
5. Bover, D.: Moment equation methods for nonlinear stochastic systems. J. Math. Anal. Appl. **65**, 306–320, 1978.

6. Cai, G. Q. and Lin, Y.: Response spectral densities of strongly nonlinear systems under random excitation. Probab. Eng. Mech. **12**, 41–47, 1997.

7. Cramer, H.: Mathematical Methods of Statistics. Princeton University Press, New York, 1946.

8. Crandall, S.: Non-Gaussian closure for random vibration of nonlinear oscillators. Int. J. Non-Linear Mech. **15**, 303–313, 1980.

9. Crandall, S.: Non-Gaussian closure for stationary random vibration. Int. J. Non-Linear Mech. **20**, 1–8, 1985.

10. Dashevski, M.: Technical realization of moment-semiinvariant method of analysis of random processes. Avtomatika i Telemekhanika **37**, 23–26, 1976.

11. Davoodi, H. and Noori, M.: Extension of an ito-based general approximation technique for random vibration of a BBW general hysteresis model Part II: Non-Gaussian analysis. J. Sound Vib. **140**, 319–339, 1990.

12. Evlanov, L. and Konstantinov, V.: Systems with Random Parameters. Nauka, Moskwa, 1976 (in Russian).

13. Ibrahim, R.: Parametric Random Vibration. Research Studies Press, Letchworth United Kingdom, 1985.

14. Ibrahim, R. and Li, W.: Structural modal interaction with combination internal resonance under wide-band random excitation. J. Sound Vib. **123**, 473–495, 1988.

15. Ibrahim, R. and Soundararajan, A.: An improved approach for random parametric response of dynamic systems with nonlinear inertia. Int. J. Non-Linear Mech. **20**, 309–323, 1985.

16. Ibrahim, R., Soundararajan, A., and Heo, H.: Stochastic response of nonlinear dynamic systems based on a nonGaussian closure. Trans. ASME J. Appl. Mech. **52**, 965–970, 1985.

17. Kasharova, A. and Shin, V.: Modified cumulant method for the analysis of nonlinear stochastic systems. Avtomatika i Telemekhanika **47**, 69–80, 1986.

18. Kasharova, A. and Shin, V.: Approximate method for determining the moments of phase coordinates of multidimensional stochastic systems. Avtomatika i Telemekhanika **51**, 43–52, 1990.

19. Klackin, W.: Statistical Description of Dynamic Systems with Fluctuated Parameters. Nauka, Moskwa, 1975 (in Russian).

20. Kuznecov, P., Stratonovich, R., and Tichonov, W.: Quasi-moment functions in theory of random processes. Dok. Acad. Nau. SSSR **94**, 1615–1618, 1954.

21. Lin, Y.: Probabilistic Theory of Structural Dynamics. McGraw Hill, New York, 1967.

22. Lin, Y. and Cai, G.: Probabilistic Structural Dynamics. McGraw Hill, New York, 1995.

23. Malachov, A.: Cumulant analysis of random processes and their transformations. Sovetzkie Radio, Moskwa, 1978 (in Russian).

24. Papadimitriou, C. and Lutes, L.: Approximate analysis of higher cumulants for multi-degree-of-freedom random vibration. Probab. Eng. Mech. **9**, 71–82, 1994.

25. Papadimitriou, C. and Lutes, L.: Stochastic cumulant analysis of MDOF systems with polynomial type nonlinearities. Probab. Eng. Mech. **11**, 1–13, 1994.

26. Pawleta, M. and Socha, L.: Cumulant-neglect closure of nonstationary solutions of stochastic systems. Trans. ASME J. Appl. Mech. **57**, 776–779, 1990.

27. Pugachev, V. and Sinitsyn, I.: Stochastic Differential Systems: Analysis and Filtering. John Wiley and Sons, Chichester, 1987.

28. Pugachev, V., et al.: Mathematical software for analysis of multidimensional nonlinear stochastic systems. Avtomatika i Telemekhanika **52**, 87–97, 1991.

29. Rehak, M., et al.: Random vibration with Macsyma. Comput. Meth. Appl. Mech. Eng. **61**, 61–70, 1987.

30. Risken, H.: Fokker-Planck Equation: Methods of Solution and Applications. Springer, Berlin, 1984.

31. Roberts, J.: Techniques for nonlinear random vibration problems. Shock Vib. Dig. **16**, 3–14, 1984.

32. Roberts, J. and Spanos, P.: Random Vibration and Statistical Linearization. John Wiley and Sons, Chichester, 1990.

33. Socha, L.: Moment Equations in Stochastic Dynamic Systems. PWN, Warszawa, 1993 (in Polish).

34. Soize, C.: The Fokker-Planck Equation for Stochastic Dynamical Systems and Its Explicit Steady State Solution. World Scientific, Singapore, 1994.

35. Sperling, L.: Analyse stochastisch erregter nichtlinearer systeme mittels linearer differentialgleichungen für verallgemeinerte quasi-momentenfunktionen. Z. Angew. Math. Mech. **59**, 169–173, 1979.

36. Sperling, L.: Approximate analysis of nonlinear stochastic differential equations using certain generalized quasi-momen functions. Acta Mechanica **59**, 183–200, 1986.

37. Stratonovich, R.: Topics in the Theory of Random Noise, Vol. 1. Gordon and Breach, New York, 1963.

38. Sun, J. and Hsu, C.: Cumulant-neglect closure method for nonlinear systems under random excitations. Trans. ASME J. Appl. Mech. **54**, 649–655, 1987.

39. Sun, J. and Hsu, C.: Cumulant-neglect of closure method for asymmetric nonlinear systems driven by Gaussian white noise. J. Sound Vib. **135**, 338–345, 1989.

40. Van Kampen, N.: A cumulant expansion for stochastic linear differential equations II. Physica **74**, 239–247, 1974.

41. Van Kampen, N.: A cumulant expansion for stochastic linear differential equations I. Physica **74**, 215–238, 1974.

42. Wu, W. F. and Lin, Y.: Cumulant-neglect closure for nonlinear oscillators under random parametric and external excitations. Int. J. Non-Linear Mech. **19**, 349–362, 1984.

# Statistical Linearization of Stochastic Dynamic Systems Under External Excitations

As was mentioned in Chap. 1, the first work on the theory of statistical linearization (SL) was carried out virtually simultaneously by Booton [5] and Kazakov [44, 45]. The objective of this method is to replace the nonlinear elements in a model by linear forms where the coefficients of linearization can be found based on a specified criterion of linearization. A systematic treatment of methods of statistical linearization and their application were first given by Pugachev [66]. This approach has been extended by Kazakov [46] to multidimensional inertialess nonlinearities and to nonstationary models by Alexandrov et al. [2]. These results were presented in a more general form by Kazakov [48] and we will follow this historical presentation in the next section using the stochastic differential equation description. We note that statistical linearization can be applied to inertialess as well as time-dependent nonlinearities. In next subsections of this chapter we discuss the most important new approaches that appeared recently in the literature, in particular energy criteria, criteria in probability density functions space, and application of statistical linearization to response analysis of nonlinear dynamic systems subjected to nonstationary or non-Gaussian excitation.

## 5.1 Moment Criteria

We start our consideration with the presentation of the standard *classical* results about statistical linearization presented by Kazakov [48, 45].

Consider a dynamic system with external excitation described by the nonlinear vector Ito stochastic differential equation

$$dx(t) = \Phi(x, t)dt + \sum_{k=1}^{M} G_{k0}(t)d\xi_k(t), \quad x(t_0) = x_0, \tag{5.1}$$

where $x = [x_1, \ldots, x_n]^T$ is the vector state, $\Phi = [\Phi_1, \ldots, \Phi_n]^T$ is a nonlinear vector function, $G_{k0} = [\sigma_{k0}^1, \ldots, \sigma_{k0}^n]^T$, $k = 1, \ldots, M$, are deterministic vectors of intensities of noise, $\xi_k$, $k = 1, \ldots, M$ are independent standard Wiener

L. Socha: *Statistical Linearization of Stochastic Dynamic Systems Under External Excitations*, Lect. Notes Phys. **730**, 103–146 (2008)
DOI 10.1007/978-3-540-72997-6_5 © Springer-Verlag Berlin Heidelberg 2008

process; the initial condition $\mathbf{x}_0$ is a vector random variable independent of $\xi_k(t)$, $k = 1, \ldots, M$. We assume that the solution of (5.1) exists.

The objective of statistical linearization is to find for nonlinear vector $\boldsymbol{\Phi}(\mathbf{x}, t)$ an equivalent one "in some sense" but in a linear form, i.e., replacing

$$\mathbf{y} = \boldsymbol{\Phi}(\mathbf{x}, t) \tag{5.2}$$

(5.1) by a linearized form

$$\mathbf{y} = \boldsymbol{\Phi}_0(\mathbf{m}_x, \boldsymbol{\Theta}_{\mathbf{x}}, t) + \mathbf{K}(\mathbf{m}_x, \boldsymbol{\Theta}_{\mathbf{x}}, t)\mathbf{x}^0 , \tag{5.3}$$

where $\boldsymbol{\Phi}_0 = [\Phi_0^1, \ldots, \Phi_0^n]^T$ is a nonlinear vector function of the moments of variables $x_j$, $\mathbf{x}^0 = [x_1^0, \ldots, x_n^0]^T$, $\mathbf{K} = [k_{ij}]$ is an $n \times n$ matrix of statistical linearization coefficients.

$\mathbf{m}_{\mathbf{x}} = E[\mathbf{x}] = [m_{x_1}, \ldots, m_{x_n}]^T$, $m_{x_i} = E[x_i]$, $\boldsymbol{\Theta}_{\mathbf{x}} = [\Theta_{ij}] = [E[x_i^0 x_j^0]]$, $i, j = 1, \ldots, n$, and $x_i^0$ is the centralized $i$th coordinate of the vector state $\mathbf{x}$

$$x_i^0 = x_i - m_{x_i} . \tag{5.4}$$

Consider first the case of one-dimensional nonlinearity. Then $\boldsymbol{\Phi}$ and $\boldsymbol{\Phi}_0$ are scalars and $\mathbf{K}$ is the transpose of the vector, i.e., $\mathbf{K} = [k_1, \ldots, k_n]^T$, where $k_i$ are scalars. Their determination depends upon the choice of an equivalence criterion. In what follows, a few standard equivalence criteria are presented. First we quote two basic criteria introduced by Kazakov [45].

**Criterion 1-$SL$.** Equality of the first and second moments of nonlinear and linearized variables

$$E[y] = m_y = \Phi_0 , \tag{5.5}$$

$$E[(y - E[y])^2] = E[(\Phi - \Phi_0)^2] = \sum_{i=1}^{n} \sum_{j=1}^{n} k_i k_j \Theta_{ij} . \tag{5.6}$$

Except for the one-dimensional case, (5.5) and (5.6) do not determine coefficients $k_i$ uniquely. Hence, additional relations are needed. This can be done by introducing, for example, the equality of cross second-order moments, i.e.,

$$E[(y - E[y])(x_j - m_{x_j})] = E\left[\sum_{i=1}^{n} k_i x_i^0 x_j^0\right] , \quad j = 1, \ldots, n . \tag{5.7}$$

One then obtains a system of linear equations in the form

$$\Theta_{\phi_j} = \sum_{i=1}^{n} k_i \Theta_{ij} , \quad j = 1, \ldots, n , \tag{5.8}$$

where

$$\Theta_{\phi_j} = E[y x_j^0] = E[\Phi x_j^0] . \tag{5.9}$$

Equations (5.8) and (5.9) combined with (5.6) produce a system of $(n+1)$ equations for $n$ unknowns $k_i$, $i = 1, \ldots, n$. One can now either eliminate one of the equations or use another approximate means to determine $k_i$. One of the procedures proposed by Kazakov [45] suggests that the coefficients $k_i$ are sought in the form

$$k_i = \left[ \frac{E[\Phi^2] - \Phi_0^2}{\Theta_{ii}} \mu \right]^{1/2} sgn \left( \frac{\partial \Phi_0}{\partial m_x} \right) , \quad i = 1, \ldots, n , \qquad (5.10)$$

where $sgn(.)$ denotes the signum function and $\mu$ is a parameter that is found by substituting (5.10) into (5.6). With the requirement that the signs for $k_i$ should agree with the signs of the corresponding terms of $\partial \Phi_0 / \partial m_x$ for normally distributed variables $x_i$, one obtains

$$k_i = \left\{ \frac{E[\Phi^2] - \Phi_0^2}{\Theta_{ii}} \left[ \sum_{r=1}^{n} \sum_{l=1}^{n} \frac{\Theta_{rl}}{\sqrt{\Theta_{rr}\Theta_{ll}}} sgn \left( \frac{\partial \Phi_0}{\partial m_{x_r}} \right) sgn \left( \frac{\partial \Phi_0}{\partial m_{x_l}} \right) \right]^{-1} \right\}^{1/2}$$

$$sgn \left( \frac{\partial \Phi_0}{\partial m_x} \right) , \quad i = 1, \ldots, n . \qquad (5.11)$$

In the particular case of one-dimensional nonlinearity with $\Phi(\mathbf{x}, t) = \Phi(x_1, t)$ in (5.1) one finds that $n = 1$ and

$$k_1 = \sqrt{\frac{D\Phi}{\Theta_{11}}} sgn \left( \frac{\partial \Phi_0}{\partial m_{x_1}} \right) , \qquad (5.12)$$

where

$$\Phi_0 = \int_{-\infty}^{+\infty} \Phi(x_1, t) g(x_1, t) dx_1 , \qquad (5.13)$$

$$D\Phi = \int_{-\infty}^{+\infty} \Phi^2(x_1, t) g(x_1, t) dx_1 - \Phi_0^2 , \qquad (5.14)$$

$$\Theta_{11} = \int_{-\infty}^{+\infty} x_1^2 g(x_1, t) dx_1 - m_{x_1}^2 , \qquad (5.15)$$

$$g(x_1, t) = \frac{1}{\sqrt{2\pi\Theta_{11}}} \exp \left\{ -\frac{(x_1 - m_{x_1})^2}{2\Theta_{11}} \right\} . \qquad (5.16)$$

The function $\Phi_0$ in (5.3) in this case usually takes the form

$$\Phi_0 = k_0 m_x , \qquad (5.17)$$

where $k_0$ is the linearization coefficient with respect to the mean value $m_x$.

**Criterion 2-$SL$.** Mean-square error of approximation

Consider the mean-square error of the linear approximation of function $y = \Phi(\mathbf{x}, t)$ by the linearized form (5.3), i.e.,

$$\delta = E\left[\left(\Phi(x_1, \ldots, x_n, t) - \Phi_0 - \sum_{i=1}^{n} k_i x_i^0\right)^2\right].$$  (5.18)

The necessary conditions for the minimum of this criterion have the form

$$\frac{\partial \delta}{\partial \Phi_0} = 0, \quad \frac{\partial \delta}{\partial k_i} = 0, \quad i = 1, \ldots, n.$$  (5.19)

From (5.19) we calculate the following relations

$$\Phi_0 = E[\Phi],$$  (5.20)

$$\Theta_{\Phi_j} = \sum_{i=1}^{n} k_i \Theta_{ij}, \quad j = 1, \ldots, n,$$  (5.21)

where $\Theta_{ij}$ and $\Theta_{\Phi_j}$ are defined by (5.3) and (5.9), respectively, i.e.,

$$\Theta_{ij} = E[x_i^0 x_j^0], \qquad \Theta_{\Phi_j} = E[\Phi x_j^0], \qquad i, j = 1, \ldots, n.$$  (5.22)

Solving the system of equations (5.21) we find the solutions

$$k_i = \sum_{j=1}^{n} (-1)^{i+j} \frac{\Delta_{ij}}{\Delta} \Theta_{\Phi_j}, \quad i = 1, \ldots, n,$$  (5.23)

where $\Delta = det[\Theta_{ij}]$, $\Delta_{ij}$ is the cofactor of the $i$th column and $j$th row of the determinant $\Delta$.

In the case of uncorrelated variables, (5.23) leads

$$k_i = \frac{1}{D_i} \Theta_{\Phi_i}, \quad i = 1, \ldots, n,$$  (5.24)

where $D_i = \Theta_{ii}$ is the variance of $x_i$.

It is simple to show that conditions (5.20) and (5.21) give the minimum of the criterion $\delta$ in (5.18). To do it we calculate the Hessian of $\delta$ with respect to $\Phi_0$ and $k_i$ and we substitute the quantities $\Phi_0$ and $k_i$ given by (5.20) and (5.24). Then, we find that the obtained Hessian is positive defined. It has been done in a little different way, for instance, as in [49].

It should be stressed that the averaging operations appearing in both criteria are with respect to the distribution of the output variables $Y_j$. In general case, Kazakov [45] assumed that the averaging operations are defined by non-Gaussian processes distribution with the corresponding probability density functions described in the form of Gram–Charlier series. It means that the expectation values appearing in equalities (5.5), (5.6) and (5.18) for a nonlinear element $Y_j = \phi_j(x_j)$ are defined by

$$E[.] = \int_{-\infty}^{+\infty} [.] g_{Y_j}(x_j) dx_j,$$  (5.25)

where

$$g_{Y_j}(x_j) = \frac{1}{\sqrt{2\pi c_j}\sigma_{x_j}} exp\left\{-\frac{(x_j - m_{x_j})^2}{2\sigma_{x_j}^2}\right\}$$

$$\times \left[1 + \sum_{\nu=3}^{N} \frac{c_{\nu j}}{\nu!} H_\nu\left(\frac{x_j - m_{x_j}}{\sigma_{x_j}}\right)\right], \qquad (5.26)$$

$m_{x_j} = E[X_j], \ \sigma_{x_j}^2 = E[(X_j - m_{x_j})^2]; \ c_{\nu j} = E[G_\nu(x_j - m_{x_j})], \ j = 1, \ldots, n,$
$\nu = 3, 4, \ldots, N$, are quasi-moments, $c_j$ are normalized constants, $H_\nu(x)$ and $G_\nu(x)$ are Hermite polynomials of one variable

$$H_\nu(x) = (-1)^\nu \exp\left\{\frac{x^2}{2\sigma^2}\right\} \frac{d^\nu}{dx^\nu} \exp\left\{-\frac{x^2}{2\sigma^2}\right\}, \qquad (5.27)$$

$$G_\nu(x) = (-1)^\nu \exp\left\{\frac{x^2}{2\sigma^2}\right\} \left[\frac{d^\nu}{dy^\nu} \exp\left\{-\frac{y^2\sigma^2}{2}\right\}\right]_{y=(x/\sigma)^2}, \qquad (5.28)$$

where $\sigma^2 = \sigma_{x_j}^2$ for $x = x_j$.

The calculations are greatly simplified when $x_i$ are Gaussian; for instance, Kazakov [45] has shown that in this case the coefficients $k_i$ for Criterion $2 - SL$ are

$$k_i = \frac{\partial \Phi_0}{\partial m_{x_i}}, \qquad i = 1, \ldots, n. \qquad (5.29)$$

Equality (5.29) is of course a particular case of (5.23). Statistical linearization in the case when the probability density function in Criterion $2 - SL$ is Gaussian and independent of linearization coefficients is also called *Gaussian statistical linearization* or *Gaussian linearization* or *standard statistical linearization*

We note that although the linearization coefficients have been determined separately, they are nonlinear functions of moments of solution of system (5.1). Since the moments arising in linearization coefficients formulas are unknown, they are approximated by the corresponding moments obtained for linearized system, i.e., for system (5.1), where the nonlinear vector (5.2) is replaced by its linearized form (5.3)

$$d\mathbf{x}(t) = [\mathbf{A}_0(t) + \mathbf{A}(t)\mathbf{x}(t)]dt + \sum_{k=1}^{M} \mathbf{G}_{k0}(t)d\xi_k(t), \quad \mathbf{x}(t_0) = \mathbf{x}_0, \qquad (5.30)$$

where $\mathbf{A}_0 = \Phi_0 - \mathbf{Km_x} = [A_{01}, \ldots, A_{0n}]^T$, $\mathbf{A} = \mathbf{K} = [a_{ij}], i, j = 1, \ldots, n$ are a vector and a matrix of linearization coefficients, respectively.

First, by using Ito formula and next by averaging one can obtain the differential equations for first- and second-order moments of the vector state $\mathbf{x}(t)$

$$\frac{d\mathbf{m_x}(t)}{dt} = \mathbf{A}_0(t) + \mathbf{A}(t)\mathbf{m_x}(t) , \quad \mathbf{m_x}(t_0) = E[\mathbf{x}_0] , \tag{5.31}$$

$$\frac{d\mathbf{\Gamma}_L(t)}{dt} = \mathbf{m_x}(t)\mathbf{A}_0^T(t) + \mathbf{A}_0(t)\mathbf{m_x}^T(t)$$

$$+ \mathbf{\Gamma}_L(t)\mathbf{A}^T(t) + \mathbf{A}(t)\mathbf{\Gamma}_L(t) + \sum_{k=1}^{M} \mathbf{G}_{k0}(t)\mathbf{G}_{k0}^T(t),$$

$$\mathbf{\Gamma}_L(t_0) = E[\mathbf{x}_0\mathbf{x}_0^T], \tag{5.32}$$

where $\mathbf{m_x} = E[\mathbf{x}]$, $\mathbf{\Gamma}_L = E[\mathbf{xx}^T]$
or in an equivalent form

$$\frac{d\mathbf{m}_{\mathbf{x}i}(t)}{dt} = \Phi_0^i(\mathbf{m_x}, \mathbf{\Theta_x}, t) , \quad \mathbf{m}_{\mathbf{x}i}(t_0) = E[\mathbf{x}_{i0}] , \quad i = 1, \dots, n, \tag{5.33}$$

$$\frac{d\Theta_{ij}(t)}{dt} = \sum_{l=1}^{n}(k_{il}\Theta_{lj}(t) + k_{jl}\Theta_{il}(t)) + \sum_{k=1}^{M} \sigma_{k0}^i(t)\sigma_{k0}^j(t),$$

$$\Theta_{ij}(t_0) = E[x_{i0}^0 x_{j0}^0] = \Theta_{ij0} \quad i, j = 1, \dots, n , \tag{5.34}$$

where $m_{xi}$ are elements (components) of vector $\mathbf{m_x}$, i.e., $\mathbf{m_x} = [m_{x1}, \dots, m_{xn}]^T$, $\Theta_{ij}$ are elements of the matrix $\mathbf{\Theta_x}$, i.e., $\mathbf{\Theta_x} = [\Theta_{ij}]$.

Since the elements $\Phi_0^i$ and $k_{ij}$ are functions depending only on $\mathbf{m_x}$ and $\mathbf{\Theta_x}$, (5.33) and (5.34) are nonlinear differential equations and can be solved by standard numerical methods, for instance, by Runge–Kutta method. In the case of the determination of stationary solutions (for $t \to \infty$) right-hand sides of (5.33) and (5.34) are equated to zero. Then we obtain a nonlinear system of algebraic equations

$$\lim_{t \to \infty} \Phi_0^i(\mathbf{m_x}, \mathbf{\Theta_x}, t) = 0 , \quad i = 1, \dots, n , \tag{5.35}$$

$$\lim_{t \to \infty} \left\{ \sum_{l=1}^{n}[k_{il}(\mathbf{m_x}, \mathbf{\Theta_x}, t)\Theta_{lj}(t) + k_{jl}(\mathbf{m_x}, \mathbf{\Theta_x}, t)\Theta_{il}(t)] \right.$$

$$\left. + \sum_{k=1}^{M} \sigma_{k0}^i(t)\sigma_{k0}^j(t) \right\} = 0 ,$$

$$i, j = 1, \dots, n. \tag{5.36}$$

One can solve systems (5.10) and (5.36) by optimization methods or by the following iterative procedure

$$m_i^1 = m_{i0} , \quad \lim_{t \to \infty} \Phi_0^i(\mathbf{m_x}^h, \mathbf{\Theta_x}^h, t)\frac{1}{m_i^h}m_i^{h+1} = 0 , \ i = 1, \dots, n, \ h = 0, 1, 2, \dots \tag{5.37}$$

$$\Theta_{ij}^1 = \Theta_{ij0}$$

$$\lim_{t\to\infty} \left\{ \sum_{l=1}^n [k_{il}(\mathbf{m_x}^h, \mathbf{\Theta_x}^h, t)\Theta_{lj}^{h+1} + k_{jl}(\mathbf{m_x}^h, \mathbf{\Theta_x}^h, t)\Theta_{il}^{h+1}] \right.$$

$$\left. + \sum_{k=1}^M \sigma_{k0}^i(t)\sigma_{k0}^j(t) \right\} = 0 ,$$

$$i, j = 1, \ldots, n, \quad h = 0, 1, 2, \ldots \tag{5.38}$$

where $m_i^h$ and $\Theta_{ij}^h$ are the mean values of $i$th element and the variance $ij$th element of the $h$th iteration of the approximation of the stationary solution, respectively.

One can transform system of equations (5.37) and (5.38) to the form

$$\mathbf{Y}^{h+1} = \mathbf{F}(\mathbf{Y}^h) , \tag{5.39}$$

where $\mathbf{Y}$ is a vector constructed from elements $m_i$ and $\Theta_{ij}$, i.e.,

$$\mathbf{Y} = [m_1, \ldots, m_n, \Theta_{11}, \ldots, \Theta_{1n}, \Theta_{21}, \ldots, \Theta_{nn}]^T , \tag{5.40}$$

$\mathbf{F}(.)$ is a nonlinear vector function defined by (5.37) and (5.38).

We note that usually in the stationary case when the mean values of excitations are equal to zero, then also

$$\mathbf{m}_x^\infty = \lim_{t\to\infty} \mathbf{m}_x(t) = 0 . \tag{5.41}$$

Functioning of the iterative procedure (5.37) and (5.38) will be shown after the presentation of other moment criteria.

Coming back to the relation (5.39) one can write the $h$th iteration of the vector $\mathbf{Y}$ in the form $\mathbf{Y}^h = [Y_1^h, \ldots, Y_n^h]^T$ and if we assume that the following inequalities are satisfied

$$|\mathbf{F}(Y_1, \ldots, Y_n) - \mathbf{F}(Y_1', \ldots, Y_n')| \leq \sum_{j=1}^n B_{ij}|Y_j - Y_j'| , \quad i = 1, \ldots, n , \tag{5.42}$$

where $\mathbf{Y} = [Y_1, \ldots, Y_n]^T$, $\mathbf{Y}' = [Y_1', \ldots, Y_n']^T$ belong to the domain of the operator $\mathbf{F}$, and $B_{ij}$, $i, j = 1, \ldots, n$ are constants, then inequality (5.42) can be replaced by

$$|\mathbf{F}(\mathbf{Y}) - \mathbf{F}(\mathbf{Y}')| \leq |\mathbf{B}||\mathbf{Y} - \mathbf{Y}'| , \tag{5.43}$$

where the norms of the vector $\mathbf{Y}$ and the matrix $\mathbf{B}$ are defined as follows:

$$|\mathbf{Y}| = \max_i |\mathbf{Y_i}| , \quad |\mathbf{B}| = \max_i \left( \sum_{j=1}^n |B_{ij}| \right) . \tag{5.44}$$

One can show, for instance [52], that

$$|\mathbf{B}| < 1 \Rightarrow \lim_{h \to \infty} \mathbf{Y}^h = \mathbf{Y}^\infty = F(\mathbf{Y}^\infty) . \qquad (5.45)$$

Another method of the determination of stationary values $\mathbf{m_x}$ and $\mathbf{\Theta_x}$ is applying an iterative procedure to (5.35) and (5.36) involving algorithms of solving Lyapunov equation. A wide review of the literature and a comparison of such algorithms can be found, for instance, in [43].

We note that the discussed iterative procedure for the determination of stationary solutions can also be applied to the determination of nonstationary solutions. Then instead of solving the algebraic equations (5.35) and (5.36) the following differential equations are solved

$$\frac{dm_i^{h+1}(t)}{dt} = \Phi_0^i(\mathbf{m_x}^h, \mathbf{\Theta_x}^h, t) \frac{1}{m_i^h} m_i^{h+1}(t) ,$$

$$m_i^0(t_0) = E[x_{i0}] , \quad i = 1, \dots, n , \quad h = 0, 1, \dots , \quad (5.46)$$

$$\frac{d\Theta_{ij}^{h+1}(t)}{dt} = \sum_{l=1}^{n} (k_{il}(\mathbf{m_x}^h, \mathbf{\Theta_x}^h, t) \Theta_{lj}^{h+1}(t)$$

$$+ k_{jl}(\mathbf{m_x}^h, \mathbf{\Theta_x}^h, t) \Theta_{il}^{h+1}(t)) + \sum_{k=1}^{M} \sigma_{k0}^i(t) \sigma_{k0}^j(t) ,$$

$$\Theta_{ij}^0(t_0) = E[x_{i0}^0 x_{j0}^0] = \Theta_{ij0} , \quad i, j = 1, \dots, n, h = 0, 1, \dots . \ (5.47)$$

As in the stationary case, an iterative procedure involving the solution of Lyapunov differential equation can be used to obtain the linearization coefficients, i.e., vector $\mathbf{\Phi_0}$ and matrix $\mathbf{K}$. We note that the limits of solutions of (5.46) and (5.47), i.e., $\lim_{\substack{t \to +\infty \\ h \to +\infty}} m_i^h(t)$ and $\lim_{\substack{t \to +\infty \\ h \to +\infty}} \Theta_{ij}^h(t)$, satisfy algebraic equations (5.35) and (5.36).

The first two criteria discussed are the most popular linearization criteria of stochastic dynamic systems and tables of linearization coefficients for both criteria can be found in many books, for instance, Csaki [11], Naumov [58], and Kazakov [48].

At the end of our discussion about iterative procedures we write the standard procedure that follows from previous considerations and is very often used in the literature. We formulate this procedure for stationary solutions in the case of the mean-square criterion.

**Procedure** *SL–MC*

(1) Substitute initial values for linearization coefficients $\Phi_0$, $k_i$, $i = 1, \dots, n$, and calculate response moments defined by stationary solutions of moment equations (5.31) and (5.32) for linearized system.

(2) Calculate new linearization coefficients $\Phi_0$, $k_i$, $i = 1, \dots, n$, using, for instance, conditions (5.20) and (5.29).

(3) Substitute the coefficient $\Phi_0$, $k_i$, $i = 1, \ldots, n$, into moment equations (5.31 and 5.32) and solve them.

(4) Repeat steps (2) and (3) until convergence.

The name of this procedure is from the first letters of Statistical Linearization–Moment Criteria. We will introduce similar abbreviation for other procedures also.

Approximation errors of the characteristics of stationary response of stochastic dynamic systems obtained by application of the first two criteria of linearization were not satisfying for researchers. It caused the strong development of statistical linearization method. Many new criteria of linearization have appeared. We discuss the basic three groups of them in detail.

### 5.1.1 Higher-Order Moment Criteria

A natural generalization of the mean-square criterion is a higher-order moment criterion. Naess et al. [57] proposed two classes of criteria for equivalent linearization. For one-dimensional function $f(x, \dot{x})$ the first class has the form

$$\varepsilon_{(n)} = E[|f(x, \dot{x}) - c_{(n)}\dot{x} - k_{(n)}x - d(k_{(n)})|^n] , \quad \text{for} \quad n = 3, 4, \ldots , \quad (5.48)$$

where $c_{(n)}, k_{(n)}$, and $d(k_{(n)})$ are linearization coefficients and

$$d(k_{(n)}) = E[f(x, \dot{x})] - c_{(n)}E[\dot{x}] - k_{(n)}E[x] , \quad \text{for} \quad n = 3, 4, \ldots . \quad (5.49)$$

If the variables $x$ and $\dot{x}$ in the nonlinear function $f(x, \dot{x})$ can be separated, i.e., if $f(x, \dot{x}) = g(\dot{x}) + h(x)$, then the second class of criteria is defined by

$$\epsilon_{(m,n)} = E[|g(\dot{x}) - c_{(m)}\dot{x}|^m] + E[|h(x) - k_{(n)}x - d(k_{(n)})|^n], \quad \text{for} \ m, n = 3, 4, \ldots , \quad (5.50)$$

where

$$d(k_{(n)}) = E[h(x)] - k_{(n)}E[x] , \quad \text{for} \quad n = 3, 4, \ldots \quad (5.51)$$

It is clear that the first class is not a particular case of the second class, i.e., $\varepsilon_{(n)} \neq \epsilon_{(n,n)}$. The authors have extended the second class of criteria to the multidimensional case. A special case of the first class of criteria proposed by Naess et al. [57] was considered by Anh and Schiehlen [3] who studied a special class of nonlinear functions $f(x, \dot{x})$ such that $f(0, 0) = 0$ and the error are defined by

$$\epsilon = f(x, \dot{x}) - c\dot{x} - kx . \quad (5.52)$$

The proposed criterion has the form

$$E[a^2(\epsilon, \alpha_2, \ldots, \alpha_n)] \longrightarrow \min , \quad (5.53)$$

where $a$ is an arbitrary function of the error $\epsilon$ and parameters $\alpha_2, \ldots, \alpha_n$ are called the error sample function.

The necessary conditions for optimality are obtained in the form

$$\frac{\partial}{\partial \alpha_i} E[a^2(\epsilon, \alpha_2, \ldots, \alpha_n)] = 0 , \quad \frac{\partial}{\partial s} E[a^2(\epsilon, \alpha_2, \ldots, \alpha_n)] = 0 , \quad (5.54)$$

for $i = 2, \ldots, n$ and $s = c, k$.

As an example of such function Anh and Schiehlen proposed a polynomial form of error sample functions defined by

$$a_1 = \epsilon , \quad a_k = \epsilon - \sum_{i=2}^{k} \alpha_i \epsilon^{2i-1} , \quad \text{for} \quad k = 2, 3, \ldots \quad (5.55)$$

Substituting (5.55) into (5.54) yields

$$\sum_{i=1}^{n} E[\epsilon^{2i+2j-2}] \alpha_i = E[\epsilon^{2j}] , \quad \text{for} \quad j = 2, 3, \ldots, n . \quad (5.56)$$

Hence the parameters $\alpha_j$, $j = 2, 3, \ldots, n$, are the solutions of system (5.56), i.e.,

$$\alpha_j = \frac{\Delta_j}{\Delta} , \quad \text{for} \quad j = 2, 3, \ldots, n , \quad (5.57)$$

where $\Delta_j$ and $\Delta$ are the corresponding determinants depending on the error of even moments $E[\epsilon^2]$, $E[\epsilon^4]$, $\ldots$, $E[\epsilon^{4n-2}]$.

Anh and Schiehlen [3] discussed the three approximate solutions in detail

$$k = 1, \quad \alpha_1 = \epsilon , \quad (5.58)$$

$$k = 2, \quad \alpha_2 = \frac{E[\epsilon^4]}{E[\epsilon^8]} , \quad (5.59)$$

$$k = 3, \quad \alpha_2 = \frac{E[\epsilon^4]E[\epsilon^{10}] - E[\epsilon^6]E[\epsilon^8]}{E[\epsilon^6]E[\epsilon^{10}] - (E[\epsilon^8])^2} , \quad \alpha_3 = \frac{(E[\epsilon^6])^2 - E[\epsilon^4]E[\epsilon^8]}{E[\epsilon^6]E[\epsilon^{10}] - (E[\epsilon^8])^2} . \quad (5.60)$$

### 5.1.2 Energy Criteria

A new class of linearization criteria for nonlinear stochastic dynamic systems was developed by Elishakoff and his coworkers. The idea of these criteria was the replacement of the displacements in Kazakov's criteria by the corresponding energies of displacements. Then the corresponding criteria of the equality of the first two order moments of potential energies of nonlinear and linearized displacement and mean-square error of the potential energies have the following form:

**Criterion 3-SL.** Criterion of equality of first two order moments of potential energies of nonlinear and linearized displacement [21]

$$E[(U(x))^2] = E\left[\left(\frac{1}{2}k_{eq}x^2\right)^2\right] , \tag{5.61}$$

where $U(x)$ is the potential energy of the nonlinear displacement (static element) $f(x)$,

$$U(x) = \int_0^x f(s)ds . \tag{5.62}$$

Hence, we find

$$k_{eq} = 2\sqrt{\frac{E[U^2(x)]}{E[x^4]}} . \tag{5.63}$$

**Criterion 4-SL.** Mean-square criterion for potential energies [94]

$$E\left[\left(U(x) - \frac{1}{2}k_{eq}x^2\right)^2\right] \longrightarrow \min . \tag{5.64}$$

Assuming that the averaging operation arising in criterion (5.64) does not depend on the linearization coefficient $k_{eq}$ one can find the necessary condition for minimum of criterion (5.64) and then calculate

$$k_{eq} = 2\frac{E[U(x)x^2]}{E[x^4]} . \tag{5.65}$$

Further generalized energy linearization criteria proposed by Elishakoff and Zhang [22] are

$$E\left[(w(x)[f(x) - k_{eq}^w x])^2\right] \longrightarrow \min , \tag{5.66}$$

$$E[(w(x)f(x))^2] = E[(w(x)k_{eq}^w x)^2] , \tag{5.67}$$

$$E\left[\left(w(x)[U(x) - \frac{1}{2}k_{eq}^w x^2]\right)^2\right] \longrightarrow \min , \tag{5.68}$$

$$E[(w(x)U(x))^2] = E[(w(x)\frac{1}{2}k_{eq}^w x^2)^2] , \tag{5.69}$$

where $w(x)$ is a weighting function defined by

$$w(x) = \sqrt{1 + \alpha U(x) + \beta U^2(x)} , \tag{5.70}$$

$\alpha > 0$ and $\beta > 0$ are weighting (undetermined) parameters, $k_{eq}^w$ is the corresponding linearization coefficient.

As for potential energies, Elishakoff and Colombi [18] introduced linearization criteria based on the energy dissipation function $R(\dot{x})$ of the original system and its counterpart possessed by the linear system, i.e., $\frac{1}{2}c_{eq}^v\dot{x}^2$. The criterion has the form

$$E\left[\left(v(\dot{x})[R(\dot{x}) - \frac{1}{2}c_{eq}^v \dot{x}^2]\right)^2\right] \longrightarrow \min, \qquad (5.71)$$

where $v(\dot{x})$ is also a weighting function, $c_{eq}^v$ is the corresponding linearization coefficient.

The corresponding criterion of the equality of second-order moments of the energy dissipation functions has the form

$$E[(v(\dot{x})R(\dot{x}))^2] = E\left[\left(v(\dot{x})\frac{1}{2}c_{eq}^v \dot{x}^2\right)^2\right]. \qquad (5.72)$$

Elishakoff and Colombi [18] proposed to consider generalized joint energy criteria for many nonlinear elements by introducing the generalized forms of weighting functions $w(x)$ and $v(\dot{x})$, i.e.,

$$w(x) = \sqrt{1 + \sum_{i=1}^{n} \alpha_i U^i(x)}, \qquad v(\dot{x}) = \sqrt{1 + \sum_{i=1}^{n} \beta_i R^i(\dot{x})}, \quad (5.73)$$

where $\alpha_i > 0$ and $\beta_i > 0$ are weighting (undetermined) parameters. They should be determined from the numerical experiments by the Monte Carlo method; $n$ signifies the number of series of Monte Carlo simulations.

Like Criteria $1 - SL$ and $2 - SL$ the linearization coefficients obtained by energy criteria also depend on moments of the solution of nonlinear dynamic system (higher-order moments than in the case of criteria $1 - SL$ and $2 - SL$). To obtain the required higher-order moments one can solve a system of nonlinear algebraic equations or nonlinear differential equations in the case of stationary or nonstationary moments, respectively. Another possibility is the replacement of the moments of a nonlinear dynamic system by the corresponding moments of the linearized system in an iterative procedure in a similar way as for criteria $1 - SL$ and $2 - SL$, for instance, by the application of Procedure $SL - MC$, with linearization coefficients determined by a proper energy criterion. Then the situation simplifies because the considered processes are assumed to be Gaussian and all higher-order moments can be expressed as polynomials of the first two order moments. Now, we show the functioning of the iterative procedure (5.37) and (5.38) for criteria $1 - SL$, $2 - SL$, $3 - SL$, and $4 - SL$ on the following example.

*Example 5.1.* Consider the nonlinear scalar dynamic system

$$dx = -(x + Ax^3)dt + \sqrt{q}d\xi, \qquad x(t_0) = x_0, \qquad (5.74)$$

where $A > 0$ and $q > 0$ are constant coefficients and $\xi$ is a standard Wiener process, the initial condition $x_0$ is a random variable independent of $\xi$.

First we calculate linearization coefficients for the nonlinear function $Y = Ax^3$. The statistically linearized function has the form

$$Y = k_0 m_x + k_1(x - m_x) . \tag{5.75}$$

We assume that the solution $x(t)$ is approximately a Gaussian process $N(m_x, \sigma_x)$, and the probability distribution function is given by (5.16) for $\sigma_x^2 = \theta_{11}$. The coefficient $k_0$ for all the considered criteria is the same and is calculated as follows:

$$k_0 = \frac{1}{\sqrt{2\pi}\sigma_x} \int_{-\infty}^{+\infty} A(\sigma_x z + m_x)^3 \exp\left\{-\frac{1}{2}z^2\right\} dz = A(m_x^2 + 3\sigma_x^2) . \tag{5.76}$$

By calculating the integrals of polynomials with respect to Gaussian measure we use the property that the integral $\int_{-\infty}^{+\infty} \phi(.) dx$ of an odd function is equal to zero and from the following properties

$$\int_0^{+\infty} z^{2k+1} \exp\{-az^2\} dz = \int_{-\infty}^0 z^{2k+1} \exp\{-az^2\} dz = \frac{k!}{2a^{k+1}} , \tag{5.77}$$

$$\int_0^{+\infty} z^{2k} \exp\{-az^2\} dz = \int_{-\infty}^0 z^{2k} \exp\{-az^2\} dz = \sqrt{\frac{\pi}{a}} \frac{(2k-1)!!}{2(2a)^k} . \tag{5.78}$$

where $a > 0$, $k = 1, 2, \ldots$

The linearization coefficients $k_1^i$ are different for all criteria and are calculated from the corresponding relations.

**Criterion 1-SL.** We use the relations (5.12–5.16)

$$\Phi_0 = k_0 m_x = A(m_x^3 + 3\sigma_x^2 m_x) , \tag{5.79}$$

$$k_1^{(1)} = \sqrt{\frac{D\Phi}{\sigma_x^2}} = A\sqrt{9m_x^4 + 36m_x^2\sigma_x^2 + 15\sigma_x^4} . \tag{5.80}$$

**Criterion 2-SL.** We use the relation (5.29)

$$k_1^{(2)} = \frac{\partial \Phi_0}{\partial m_x} = 3A(m_x^2 + \sigma_x^2) . \tag{5.81}$$

**Criterion 3-SL.** We use the relation (5.63)

$$k_1^{(3)} = 2\sqrt{\frac{E[U^2(x)]}{E[x^4]}} = \frac{\sqrt{35}}{2}A(m_x^2 + \sigma_x^2) . \tag{5.82}$$

**Criterion 4-SL.** We use the relation (5.65)

$$k_1^{(4)} = 2\frac{E[U(x)x^2]}{E[x^4]} = \frac{5}{2}A(m_x^2 + \sigma_x^2) . \tag{5.83}$$

Now, we calculate the differential equations for first- and second-order moments (5.74). They have the form

$$\frac{dm_x}{dt} = \Phi_0 , \quad m_x(t_0) = E[x_0] , \qquad (5.84)$$

$$\frac{dE[x^2]}{dt} = -2k_1^{(i)} E[x^2] + q, \, E[x^2](t_0) = E[x_0^2] , \quad i = 1,\ldots,4 . \qquad (5.85)$$

Hence, we find that the mean value of the stationary solution is equal to zero $m_x = 0$ and holds for the stationary second-order moment

$$E[x^2]^{h+1} = \kappa^{h+1} = \frac{q}{2k_1^{(i)}(m_x^h, \sigma_x^h)} = \frac{q}{2[1 + \alpha_i A \kappa^h]} , \quad h = 0, 1, \ldots \qquad (5.86)$$

where

$$\alpha_1 = \sqrt{15}, \quad \alpha_2 = 3 , \quad \alpha_3 = \frac{\sqrt{35}}{2} , \quad \alpha_4 = 2.5 . \qquad (5.87)$$

One can rewrite each of the four procedures corresponding to four criteria in the form

$$\kappa^{h+1} = F(\kappa^h) , \qquad (5.88)$$

where

$$\kappa = E[x^2], \quad F(y) = \frac{q}{2[1 + \alpha_i A y]} , \quad \alpha_i > 0, y > 0 . \qquad (5.89)$$

If $|F| = \sup\limits_{|y|=1} |F(y)|$ is smaller than 1 for sufficiently small $q$, iterative procedures (5.86) for $i = 1, 2, 3, 4$ converge for $h \to \infty$. In the stationary case, the second-order moment can be determined from the following algebraic equation:

$$2\alpha_i(\kappa_i^\infty)^2 + 2(\kappa_i^\infty) + q = 0 , \quad i = 1, 2, 3, 4 , \qquad (5.90)$$

where $\kappa_i^\infty = \lim\limits_{h \to \infty} \kappa_i^h$, then

$$(\kappa_i^\infty) = \frac{-1 + \sqrt{1 + 2A\alpha_i q}}{2A\alpha_i} , \quad i = 1, 2, 3, 4 . \qquad (5.91)$$

A comparison of the characteristics of stationary responses obtained by statistical linearization with cited energy criteria and standard mean-square criterion for displacements (criterion $2 - SL$) for Gaussian stationary excitations was done on simple examples in [19, 22, 14, 15]. The energy criterion (5.61) was also compared with statistical linearization with mean-square criterion on more complex examples such as $n$-storey structure with installed tuned liquid dampers with crossed tube-like containers [96], nonlinear sliding structure [93], and simply supported or clamped beam on elastic foundation [20, 27]. These linearization methods were compared with the exact solutions available in these cases and with simulations. The comparison study shows that statistical linearization with energy criteria yields the results (response characteristics) in closer vicinity with the exact or simulation results than the standard statistical linearization (criteria $1^0$ and $2^0$).

### 5.1.3 Moment Criteria with Non-Gaussian Distributions

As was mentioned in Sect. 5.1 Kazakov [45] proposed to use a non-Gaussian distribution in averaging operations in linearization criteria $1 - SL$ and $2 - SL$. He discussed the application of a non-Gaussian distribution defined by Gram–Charlier series. It means the expectation values appearing in equalities (5.5),(5.6), and (5.18) for a nonlinear element $Y_j = \phi_j(x_j)$ are defined by

$$E[.] = \int_{-\infty}^{+\infty} [.]g_{Y_j}(y_j)dy_j , \qquad (5.92)$$

where $g_{Y_j}(y_j)$ is defined by (5.26–5.28). We illustrate this approach for Criterion $2-SL$ on a simple example of one-dimensional nonlinearity $y = f(x)$ and non-Gaussian probability density function $g(x)$ represented by a truncated Gram–Charlier series as follows:

$$g(x) \cong \frac{1}{c_4\sigma\sqrt{2\pi}} \exp\left\{ -\frac{(x - \kappa_1)^2}{2\sigma^2} \right\}$$
$$\times \left[ 1 + \frac{\kappa_3}{3!\sigma^3}H_3\left( \frac{(x - \kappa_1)}{\sigma} \right) + \frac{\kappa_4}{4!\sigma^4}H_4\left( \frac{(x - \kappa_1)}{\sigma} \right) \right] , \quad (5.93)$$

where $c_4$ is a normalized constant,

$$\sigma^2 = \kappa_2 , \quad H_3(z) = z^3 - 3z , \quad H_4(z) = z^4 - 6z^2 + 3 ,$$
$$\kappa_1 = E[x], \quad \kappa_2 = E[x^2] - (E[x])^2 , \quad \kappa_3 = E[x^3] - 3E[x^2]E[x] + 2(E[x])^3 ,$$
$$\kappa_4 = E[x^4] - 4E[x^3]E[x] + 12E[x^2](E[x])^2 - 3(E[x^2])^2 - 6(E[x])^4 .$$
$$(5.94)$$

Then the linearization coefficient $k_1$ defined by (5.18) has the form

$$k_1 = \frac{E[f(x)x]}{E[x^2]} , \qquad (5.95)$$

where the averaged values $E[f(x)x]$ and $E[x^2]$ are determined by probability density function (5.93–5.94), i.e.,

$$E[f(x)x] = \int_{-\infty}^{+\infty} f(x)x \left( 1 + \frac{\kappa_3}{3!\sigma^3}\left[ \left( \frac{(x - \kappa_1)}{\sigma} \right)^3 - 3\left( \frac{(x - \kappa_1)}{\sigma} \right) \right] \right.$$
$$\left. + \frac{\kappa_4}{4!\sigma^4}\left[ \left( \frac{(x - \kappa_1)}{\sigma} \right)^4 - 6\left( \frac{(x - \kappa_1)}{\sigma} \right)^2 + 3 \right] \right) g_G(x)dx , \quad (5.96)$$

$$E[x^2] = \int_{-\infty}^{+\infty} x^2 \left( 1 + \frac{\kappa_3}{3!\sigma^3}\left[ \left( \frac{(x - \kappa_1)}{\sigma} \right)^3 - 3\left( \frac{(x - \kappa_1)}{\sigma} \right) \right] \right.$$
$$\left. + \frac{\kappa_4}{4!\sigma^4}\left[ \left( \frac{(x - \kappa_1)}{\sigma} \right)^4 - 6\left( \frac{(x - \kappa_1)}{\sigma} \right)^2 + 3 \right] \right) g_G(x)dx, \quad (5.97)$$

where

$$g_G(x) = \frac{1}{\sigma\sqrt{2\pi}} \exp\left\{-\frac{(x - \kappa_1)^2}{2\sigma^2}\right\} . \tag{5.98}$$

If the nonlinear function $f(x)$ is a polynomial, then the term $E[f(x)x]$ can be calculated as a function of higher moments of the nonlinear element. This is the fault in this approach of the determination of the linearization coefficient in a dynamic system. For instance, if we consider a scalar dynamic system

$$dx(t) = -f(x)dt + qd\xi(t) , \tag{5.99}$$

where the nonlinear function $f(x)$ has the form

$$f(x) = \sum_{p=1}^{N} a_{2p-1}x^{2p-1} , \tag{5.100}$$

$q$ and $a_p$ are positive constants, and if we try to obtain the linearization coefficient by applying an iterative procedure, then we have to replace the higher-order moments of nonlinear system (5.99) by the corresponding moments of linearized system. In this case higher-order moments for linearized system are equal to zero for odd order and $E[x^{2p}] = (2p - 1)!!(E[x^2])^p$ for even order. The cumulants of higher order than two are equal to zero.

Therefore, this procedure will give the same results as the Gaussian statistical linearization.

To omit this difficulty Beaman and Hedrick [4] have used an important property of the Gaussian distribution and its connection with Hermite polynomials. It was then possible to convert the calculation of the higher moments appearing in $E[f(x)x]$ into the evaluation of derivatives of the generating function

$$< f(x) >_0 = \int_{-\infty}^{+\infty} f(x)g(x)dx . \tag{5.101}$$

where $g(x)$ is a probability density function.

However, this procedure leads to a large number of differential equations. For first-order systems, this requires solving two extra equations beyond that required for Gaussian statistical linearization. The situation becomes even worse for higher-order systems. Therefore, Beaman and Hedrick [4] proposed a reduced expansion method that reduces the number of equations but still gives better results than the Gaussian statistical linearization. It is, however, noteworthy that this approach is in fact no longer statistical linearization but a modified version of cumulant discard closure technique.

## 5.2 Criteria in Probability Density Functions Space

As was mentioned in Chap. 1, in the case of moment criteria some information is lost, because not all order moments of the response are taken into account.

Since the complete information about a continuous random variable is contained in its probability density function, the present author proposed new criteria of linearization depending on the difference between probability densities of responses of nonlinear and linearized systems and two approximate approaches (see, for instance, [82]). One of these approaches is a method called *statistical linearization with probability density criteria* [81]. The objective of this method is to replace the nonlinear elements in a model by the corresponding linear forms, where the coefficients of linearization can be found separately for every element based on the criterion of linearization, which is a probabilistic metric in probability density functions space. The elements of this space are found as probability density functions of random variables obtained by linear and nonlinear transformation of one-dimensional Gaussian variable. In this section we present the basic linearization criteria and procedures of the determination of linearization coefficients.

We consider nonlinear static elements in the form

$$Y_j = \phi_j(x_j), \quad j = 1, \ldots, n_y \tag{5.102}$$

and the corresponding linearized elements

$$Y_j = k_j x_j, \quad j = 1, \ldots, n_y . \tag{5.103}$$

To determine the linearization coefficients $k_j$ it was proposed in [81] the following two criteria for scalar nonlinear functions $\phi_j(x_j)$ for $j = 1, \ldots, n_y$

**Criterion 1-*SLPD*.** Probabilistic square metric

$$I_{1_j} = \int_{-\infty}^{+\infty} (g_N(y_j) - g_L(y_j))^2 dy_j, j = 1, \ldots, n_y , \tag{5.104}$$

where $g_N(y_j)$ and $g_L(y_j)$ are probability density functions of stationary solutions of nonlinear elements (5.102) and linearized elements (5.103), respectively.

**Criterion 2-*SLPD*.** Pseudomoment metric

$$I_{1_j} = \int_{-\infty}^{+\infty} |y_j|^{2l} |g_N(y_j) - g_L(y_j)| dy_j, j = 1, \ldots, n_y . \tag{5.105}$$

If we assume that the input processes acting on static elements are Gaussian processes with mean values $m_{x_j}(t) = 0$, for $j = 1, \ldots, n$, and probability density functions

$$g_I(x_j(t)) = \frac{1}{\sqrt{2\pi}\sigma_{x_j}(t)} \exp\left\{-\frac{x_j^2}{2\sigma_{x_j}^2(t)}\right\}, \tag{5.106}$$

where $\sigma_{x_j}^2(t) = E[x_j^2(t)]$, then the output processes $Y_j$, for $j = 1, \ldots, n_y$, from the static linear elements defined by equality (5.103) are also zero-mean Gaussian processes with the corresponding probability density functions

$$g_{L_j}(x_j(t), k) = \frac{1}{\sqrt{2\pi}k_j\sigma_{x_j}(t)} \exp\left\{-\frac{x_j^2}{2k_j^2\sigma_{x_j}^2(t)}\right\}. \qquad (5.107)$$

To apply the proposed criteria $1 - SLPD$ and $2 - SLPD$ we have to find probability density functions $g_N(y_j)(t)$, $j = 1, \ldots, n_y$. Unfortunately, except for some special cases it is impossible to find them in analytical forms [65]. It is well known that one of these special cases is for a scalar strictly monotonically increasing or decreasing function

$$Y_j = \psi_j(x_j), \quad j = 1, \ldots, n_y, \qquad (5.108)$$

with continuous derivatives $\psi_j'(x_j)$ for all $x_j \in R$. Then the probability density functions of the output variables (5.108) are given by

$$g_N(y_j) = g_{Y_j}(y_j) = g_I(h(y_j))|h'(y_j)|, \quad j = 1, \ldots, n_y, \qquad (5.109)$$

where $g_I(x_j)$, $j = 1, \ldots, n$, are the probability density functions of the input variables and $h_j$ are the inverse functions to $\psi_j(x_j)$, i.e.,

$$x_j = h_j(Y_j) = \psi_j^{-1}(Y_j), \quad j = 1, \ldots, n_y. \qquad (5.110)$$

Both Criteria $1 - SLPD$ and $2 - SLPD$ can be also adjusted to energies of displacement, for instance, Criterion $1 - SLPD$ for energies of displacements can be considered in the following form.

**Criterion 3-*SLPD*.** Probabilistic square metric

$$I_{3_j} = \int_0^{+\infty} (g_{EN}(y_j) - g_{EL}(y_j))^2 dy_j, j = 1, \ldots, n_y, \qquad (5.111)$$

where $g_{EN}(y_j)$ and $g_{EL}(y_j)$ are probability density functions of stationary solutions of energies of nonlinear elements (5.102) and linearized elements (5.103), respectively.

*Example 5.2.* Consider the static nonlinear element defined by

$$Y = \lambda_0^2 x + \varepsilon x^3, \qquad (5.112)$$

where $\lambda_0$ and $\varepsilon > 0$ are constant parameters.

For simplicity we limit our considerations to the stationary case and we assume that the input process $x(t)$ is a stationary zero- mean Gaussian process described by the corresponding probability density function (5.106), i.e.,

$$g_I(x) = \frac{1}{\sqrt{2\pi}\sigma_x} \exp\left\{-\frac{x^2}{2\sigma_x^2}\right\}, \qquad (5.113)$$

where $\sigma_x^2 = E[x^2]$ is the variance of the process $x$.

Then the probability density function of the stationary output process $Y(t)$ for the nonlinear element defined by (5.112) has the form

$$g_Y(y) = \frac{1}{\sqrt{2\pi}\sigma_x} \exp\left\{-\frac{(v_1 + v_2)^2}{2\sigma_x^2}\right\} \frac{1}{6a\varepsilon}\left[\frac{a+y}{v_1^2} + \frac{a-y}{v_2^2}\right], \tag{5.114}$$

where

$$v_1 = \left[\frac{1}{2\varepsilon}\left(y + \sqrt{y^2 + 4\lambda_0^6/27\varepsilon}\right)\right]^{\frac{1}{3}}, \quad v_2 = \left[\frac{1}{2\varepsilon}\left(y - \sqrt{y^2 + 4\lambda_0^6/27\varepsilon}\right)\right]^{\frac{1}{3}},$$

$$a = \sqrt{y^2 + 4\lambda_0^6/27\varepsilon}. \tag{5.115}$$

The probability density function of the linearized variable (stationary process $Y(t)$) defined by

$$Y = kx \tag{5.116}$$

has the form

$$g_L(y) = g_L(y,k) = \frac{1}{\sqrt{2\pi}k\sigma_x}\exp\left\{-\frac{x^2}{2k^2\sigma_x^2}\right\}, \tag{5.117}$$

where $k$ is the linearization coefficient.

In the case of Criterion $3 - SLPD$ the energies of nonlinear and linearized elements are given by

$$E_N = \frac{\lambda_0^2}{2}x^2 + \frac{\varepsilon}{4}x^4, \tag{5.118}$$

$$E_L = \frac{k}{2}x^2, \tag{5.119}$$

and the corresponding probability density functions $g_{E_N}(y_j)$ and $g_{E_L}(y_j)$ of energies of nonlinear and linearized elements have the form

$$g_{E_N}(E_{N_j}) = \frac{2}{\sqrt{2\pi}\sigma_{x_j}|\lambda_0^2 z_j + \varepsilon z_j^3|}\exp\left\{-\frac{z_j^2}{2\sigma_{x_j}^2}\right\}, \tag{5.120}$$

$$g_{E_L}(E_{L_j}) = \frac{2}{\sqrt{2\pi}\sigma_{x_j}|\lambda_0^2 E_{L_j}|}\exp\left\{-\frac{E_{L_j}}{2\sigma_{x_j}^2}\right\}, \tag{5.121}$$

respectively, where

$$z_j = \sqrt{\frac{-\lambda_0^2 + \sqrt{\lambda_0^4 + 4\varepsilon E_{N_j}}}{\varepsilon}}, \quad j = 1,\dots,n_y. \tag{5.122}$$

In general case when the nonlinear functions $\phi_j(x)$, $j = 1,\dots,n_y$, are not strongly monotonically increasing or decreasing or not differentiable everywhere the approximation methods have to be used.

To obtain approximate probability density functions of nonlinear random variables (5.108) one can use, for instance, the Gram–Charlier expansion [65]. In the particular case for a scalar function $Y_j = \phi_j(x_j)$, $j = 1, \ldots, n_y$, of a scalar random variable $x_j$ the nonlinear variable has the probability density function

$$g_{Y_j}(y_j) = \frac{1}{\sqrt{2\pi}c_j\sigma_{Y_j}} exp\left\{-\frac{(y_j - m_{Y_j})^2}{2\sigma_{Y_j}^2}\right\}$$

$$\times \left[1 + \sum_{\nu=3}^{N} \frac{c_{\nu j}}{\nu!} H_\nu\left(\frac{y_j - m_{Y_j}}{\sigma_{Y_j}}\right)\right], \qquad (5.123)$$

where $m_{Y_j} = E[Y_j]$, $\sigma_{Y_j}^2 = E[(Y_j - m_{Y_j})^2]$; $c_{\nu j} = E[G_\nu(y_j - m_{y_j})]$, $j = 1, \ldots, n$, $\nu = 3, 4, \ldots, N$ are quasi-moments, $c_j$ are normalized constants, $H_\nu(x)$ and $G_\nu(x)$ are Hermite polynomials of one variable

$$H_\nu(x) = (-1)^\nu exp\left\{\frac{x^2}{2\sigma^2}\right\} \frac{d^\nu}{dx^\nu} exp\left\{-\frac{x^2}{2\sigma^2}\right\}, \qquad (5.124)$$

$$G_\nu(x) = (-1)^\nu exp\left\{\frac{x^2}{2\sigma^2}\right\} \left[\frac{d^\nu}{dy^\nu} exp\left\{-\frac{y^2\sigma^2}{2}\right\}\right]_{y=(x/\sigma)^2}, \qquad (5.125)$$

where $\sigma^2 = \sigma_{Y_j}^2$ for $x = Y_j$.

In contrast to standard statistical linearization with moment criteria in state space, one cannot find expressions for linearization coefficients in an analytical form. However, in some particular cases some analytical considerations can be done. For instance, for criterion $1 - SLPD$ defined by (5.104) and for an input zero-mean Gaussian process, the necessary condition for minimum can be derived in the following form:

$$\frac{\partial I_{1_j}}{\partial k_j} = \int_{-\infty}^{+\infty} (g_N(y_j, t) - g_L(y_j, t)) \frac{1}{k_j}\left(1 - \frac{y_j^2}{k_j^2\sigma_x^2}\right)$$

$$\times g_L(y_j, t)dy_j = 0 \ , j = 1, \ldots, n_y \ . \qquad (5.126)$$

To apply the proposed linearization method to the determination of the linearization coefficients $k_j$, $j = 1, \ldots, n_y$, and approximate characteristics of the stationary response of system (5.1) one can use moment equations for the linearized systems (5.31) and (5.32), i.e.,

$$\frac{d\mathbf{m_x}(t)}{dt} = \mathbf{A_0}(t) + \mathbf{A}(t)\mathbf{m_x}(t), \quad \mathbf{m_x}(t_0) = E[\mathbf{x_0}] \ ,$$

$$\frac{d\mathbf{\Gamma}_L(t)}{dt} = \mathbf{m_x}(t)\mathbf{A}_0^T(t) + \mathbf{A_0}(t)\mathbf{m_x}^T(t) + \mathbf{\Gamma}_L(t)\mathbf{A}^T(t) + \mathbf{A}(t)\mathbf{\Gamma}_L(t)$$

$$+ \sum_{k=1}^{M} \mathbf{G}_k(t)\mathbf{G}_k^T(t) \ ,$$

$$\mathbf{\Gamma}_L(t_0) = E[\mathbf{x}_0\mathbf{x}_0^T] \, ,$$

where $\mathbf{m_x} = E[\mathbf{x}]$, $\mathbf{\Gamma}_L = E[\mathbf{xx}^T]$.

These equations are solved in the following iterative procedure [82].

**Procedure** $SL$–$PD$

*Step 1*: Substitute initial values of linearization coefficients $k_j$, $j = 1, \ldots, n_y$, and calculate moments of the stationary response of linearized system by solving systems (5.31) and (5.32).

*Step 2*: For all nonlinear and linearized elements (one dimensional) defined by (5.102) and (5.103), calculate probability density functions using $m_j$, $\sigma_{x_j}$, $j = 1, \ldots, n_y$, obtained from elements of response moments $\mathbf{m}_x$ and $\mathbf{\Gamma}_L$ and relations between moments of the stationary response obtained in Step 1 and Hermite polynomials defined by (5.124) and (5.125).

*Step 3*: Consider a criterion, for instance, $I_1$, and apply it to all nonlinear functions separately, i.e.,

$$I_{1_j} = \int_{-\infty}^{+\infty} (g_{Y_j}(y_j) - g_{L_j}(y_j, k_j))^2 dy_j, \quad j = 1, \ldots, n_y \, , \tag{5.127}$$

where $g_{Y_j}(y_j, k_1)$ and $g_{L_j}(y_j, k_j)$, $j = 1, \ldots, n_y$, are probability density functions defined by (5.123) and (5.107), respectively, and find the coefficients $k_{j_{\min}}$, $j = 1, \ldots, n_y$, which minimize criterion (5.127) separately for each non-linear function. Next, substitute $k_j = k_{j_{\min}}$, $j = 1, \ldots, n_y$.

*Step 4*: Calculate moments of the stationary response of the linearized system by solving systems (5.31) and (5.32).

*Step 5*: For all nonlinear and linearized (one–dimensional) elements defined by (5.102) and (5.103) redefine probability density functions using new $m_j$, $\sigma_{x_j}$, $j = 1, \ldots, n_y$, obtained from elements of response moments $\mathbf{m}_x$ and $\mathbf{\Gamma}_L$ in Step 4 and using the relations between moments and Hermite polynomials defined by (5.124) and (5.125).

*Step 6*. Repeat Steps 3–5 until $m_j$, $\sigma_{x_j}$, $k_j$ converge (with a given accuracy).

## 5.3 Stationary Gaussian Excitations

In the considered models of stochastic dynamic systems it was assumed that the external excitations are Gaussian white noises. The derived relations for linearization coefficients and iterative procedures are also correct for the case when the excitations are assumed to be Gaussian stationary processes with a given correlation matrix or power spectral density matrix, i.e., for colored noise excitations. The only one difference is in the calculation of stationary response moments. The following considerations will illustrate this fact.

Consider the nonlinear vector stochastic differential equation

$$\frac{d\mathbf{x}(t)}{dt} = \mathbf{\Phi}(\mathbf{x}, t) + \sum_{k=1}^{M} \mathbf{G}_{\eta k} \eta_k(t), \quad \mathbf{x}(t_0) = \mathbf{x}_0 , \qquad (5.128)$$

where $\mathbf{x} = [x_1, \ldots, x_n]^T$ is the vector state, $\mathbf{\Phi} = [\Phi_1, \ldots, \Phi_n]^T$ is a nonlinear vector function, $\mathbf{G}_{\eta k} = [\sigma_{\eta k}^1, \ldots, \sigma_{\eta k}^n]^T$, $k = 1, \ldots, M$, are constant deterministic vectors of noise intensity, $\eta_k(t)$, $k = 1, \ldots, M$, are independent zero-mean Gaussian processes; the initial condition $\mathbf{x}_0$ is a vector random variable independent of $\eta_k(t)$, $k = 1, \ldots, M$ . We assume that $\eta_k(t)$, $k = 1, \ldots, M$, can be treated as outputs from linear filters where inputs are excited by Gaussian white noises, i.e., the filters satisfy the following algebraic equation and linear Ito stochastic differential equations

$$\eta_k(t) = \mathbf{C}_k^T \zeta_k(t) , \qquad (5.129)$$

$$d\zeta_k(t) = \mathbf{A}_{\zeta k} \zeta_k dt + \mathbf{G}_{\zeta k} d\xi_k, \quad k = 1, \ldots, M , \qquad (5.130)$$

where $\mathbf{A}_{\zeta k}$ are $n_k \times n_k$-dimensional negative definite matrices, $Re\lambda_{ki} (\mathbf{A}_{\zeta k}) < 0$, $\mathbf{C}_k, \mathbf{G}_{\zeta k}$, are $n_k$ dimensional vectors of the real coefficients modeling the linear filter with single input and single output (SISO); $\xi_k$ are independent standard Wiener processes.

As in the case of Gaussian white noise excitation we find for system (5.128) the corresponding equations for linearized system

$$\frac{d\mathbf{x}(t)}{dt} = [\mathbf{A}_0(t) + \mathbf{A}(t)\mathbf{x}(t)] + \sum_{k=1}^{M} \mathbf{G}_{\eta k} \eta_k(t) , \quad \mathbf{x}(t_0) = \mathbf{x}_0 , \qquad (5.131)$$

where $\mathbf{A}_0 = \mathbf{\Phi}_0 - \mathbf{K}\mathbf{m_x} = [A_{01}, \ldots, A_{0n}]^T$, $\mathbf{A} = \mathbf{K} = [a_{ij}]$, $i, j = 1, \ldots, n$, are the vector and the matrix of linearization coefficients, respectively.

If we join system (5.131) with filter equations (5.129 and 5.130), then we obtain a new extended linear vector Ito stochastic differential equation

$$d\mathbf{X}(t) = [\bar{\mathbf{A}}_0(t) + \bar{\mathbf{A}}(t)\mathbf{X}(t)]dt + \sum_{k=1}^{M} \bar{\mathbf{G}}_k d\xi_k(t) , \quad \mathbf{X}(t_0) = \mathbf{X}_0 , \qquad (5.132)$$

where

$$\mathbf{X} = \begin{bmatrix} \mathbf{x} \\ \zeta_1 \\ \zeta_2 \\ \vdots \\ \zeta_M \end{bmatrix}, \bar{\mathbf{A}} = \begin{bmatrix} \mathbf{A} & \mathbf{G}_{\eta 1} \mathbf{C}_1^T & \mathbf{G}_{\eta 2} \mathbf{C}_2^T & \cdots & \mathbf{G}_{\eta M} \mathbf{C}_M^T \\ \mathbf{0} & \mathbf{A}_{\zeta 1} & \mathbf{0} & \cdots & \mathbf{0} \\ \mathbf{0} & \mathbf{0} & \mathbf{A}_{\zeta 2} & \cdots & \mathbf{0} \\ \vdots & \vdots & \vdots & \ddots & \vdots \\ \mathbf{0} & \mathbf{0} & \mathbf{0} & \cdots & \mathbf{A}_{\zeta M} \end{bmatrix}, \bar{\mathbf{A}}_0 = \begin{bmatrix} \mathbf{A}_0 \\ \mathbf{0} \\ \mathbf{0} \\ \vdots \\ \mathbf{0} \end{bmatrix} , \qquad (5.133)$$

$$\bar{\mathbf{G}}_k = \left[ \mathbf{0}^T, \ldots, \mathbf{0}^T \mathbf{G}_{\zeta k}^T \mathbf{0}^T, \ldots, \mathbf{0}^T \right]^T , k = 1, \ldots, M, \mathbf{X}_0 = [\mathbf{x}_0^T \mathbf{0}^T, \ldots, \mathbf{0}^T]^T.$$

Hence, we find moment equations that have the form similar to (5.31) and (5.32), i.e.,

$$\frac{d\bar{\mathbf{m}}_{\mathbf{x}}(t)}{dt} = \bar{\mathbf{A}}_0(t) + \bar{\mathbf{A}}(t)\bar{\mathbf{m}}_{\mathbf{x}}(t) \;, \quad \bar{\mathbf{m}}_{\mathbf{x}}(t_0) = E[\mathbf{X}_0] = E[[\mathbf{x}_0^T \mathbf{0}^T, \dots, \mathbf{0}^T]^T] \;,$$

(5.134)

$$\frac{d\bar{\mathbf{\Gamma}}_L(t)}{dt} = \bar{\mathbf{m}}_{\mathbf{x}}(t)\bar{\mathbf{A}}_0^T(t) + \bar{\mathbf{A}}_0(t)\bar{\mathbf{m}}_{\mathbf{x}}^T(t) + \bar{\mathbf{\Gamma}}_L(t)\bar{\mathbf{A}}^T(t)$$

$$+\bar{\mathbf{A}}(t)\bar{\mathbf{\Gamma}}_L(t) + \sum_{k=1}^{M} \bar{\mathbf{G}}_k \bar{\mathbf{G}}_k^T \;,$$

(5.135)

$$\bar{\mathbf{\Gamma}}_L(t_0) = E[\mathbf{X}_0 \mathbf{X}_0^T] \;,$$

where $\bar{\mathbf{m}}_{\mathbf{x}} = E[\mathbf{X}]$, $\bar{\mathbf{\Gamma}}_L = E[\mathbf{X}\mathbf{X}^T]$. To determine linearization coefficients we use only a part of moments connected with the vector $\mathbf{x}$, i.e., $\mathbf{m}_{\mathbf{x}} = E[\mathbf{x}]$, $\mathbf{\Gamma}_L = E[\mathbf{x}\mathbf{x}^T]$.

At the end of this section we note that some applications of this approach have been presented by Lyon et al. [53] and by Dimentberg [13].

## 5.4 Nonstationary Gaussian Excitations

An application of statistical linearization to the response analysis of stochastic dynamic systems under external nonstationary excitations is more complicated for some classes of excitations than for stationary excitations. Historically, the first idea of the approximation of nonlinear systems under combined deterministic and Gaussian excitation by statistical and harmonic linearization was presented and investigated by Kazakov and Dostupov [49], Pervozwanskii [62], and by Sawargai et al. [72]. Following Kazakov and Dostupov [49] we consider the case of nonlinear function

$$y(t_1, t_2) = \phi(x(t_1, t_2)) \;,$$

(5.136)

where $\phi(.)$ is a nonlinear function and the input variable $x$ can be written as

$$x(t_1, t_2) = m_x + A_0(1 + \delta \cos(\lambda_1 t_2)) \cos(\lambda_0 t_1) + x^0(t_1) \;,$$

(5.137)

where $m_x$ is the constant component of the signal $x(t_1, t_2)$, $A_0(1 + \delta \cos(\lambda_1 t_2))$ $\cos(\lambda_0 t_1)$ represents a deterministic amplitude-modulated signal, $A_0$, $\delta$, $\lambda_0$, and $\lambda_1$ are constant parameters, $x^0(t_1) = x(t_1) - E[x(t_1)]$ is the zero-mean random component. If the distribution of the harmonic part deviates substantially from normal it is convenient to approximate the nonlinear variable $y$ defined by (5.136) by the function

$$u(t_1, t_2) = \phi_0 + \varepsilon\delta \cos(\lambda_1 t_2) + \kappa A_0(1 + \delta \cos(\lambda_1 t_2)) \cos(\lambda_0 t_1) + \kappa_1 x^0(t_1) \;,$$

(5.138)

where

$$\phi_0 = \frac{1}{T_1}\frac{1}{T_0}\int_0^{T_1} dt_2 \int_0^{T_0} E[\phi(x(t_1,t_2))]dt_1 \ , \quad T_1 = \frac{2\pi}{\lambda_1} \ , \quad T_0 = \frac{2\pi}{\lambda_0} \ , \tag{5.139}$$

$\phi_0$, $\kappa$, and $\kappa_1$ are linearization coefficients, $\varepsilon$ is the detection coefficient. In a special case it is assumed that $\phi_0 = \kappa_0 m_x$, where $\kappa_0$ is a linearization coefficient.

Similar to statistical linearization Kazakov and Dostupov [49] considered the following two criteria:

**Criterion 1-*SLNONST*.** Equality of the first and second moments of nonlinear and linearized variables

$$E[y] = E[u] \ , \tag{5.140}$$

$$K_y(0) = K_u(0) \ , \tag{5.141}$$

where

$$E[y] = \frac{1}{T_1}\frac{1}{T_0}\int_0^{T_1} dt_2 \int_0^{T_0} E[y(t_1,t_2)]dt_1 \ , \tag{5.142}$$

$$E[u] = \frac{1}{T_1}\frac{1}{T_0}\int_0^{T_1} dt_2 \int_0^{T_0} E[u(t_1,t_2)]dt_1 \ , \tag{5.143}$$

$$K_y(\tau) = \frac{1}{T_1}\frac{1}{T_0}\int_0^{T_1} dt_2 \int_0^{T_0} E[y(t_1,t_2)y(t_1+\tau,t_2+\tau)]dt_1$$
$$-\frac{1}{T_1}\int_0^{T_1} E[y(t_2)]^* E[y(t_2+\tau)]^* dt_2 \ , \tag{5.144}$$

$$K_u(\tau) = \frac{1}{T_1}\frac{1}{T_0}\int_0^{T_1} dt_2 \int_0^{T_0} E[u(t_1,t_2)u(t_1+\tau,t_2+\tau)]dt_1$$
$$-\frac{1}{T_1}\int_0^{T_1} E[u(t_2)]^* E[u(t_2+\tau)]^* dt_2 \ , \tag{5.145}$$

$$E[y(t_2)]^* = \frac{1}{T_1}\int_0^{T_1} E[y(t_1,t_2)]dt_1 \ , \tag{5.146}$$

$$E[u(t_2)]^* = \frac{1}{T_1}\int_0^{T_1} E[u(t_1,t_2)]dt_1 \ . \tag{5.147}$$

Additionally if we want to determine the coefficient $\varepsilon$ we use the following condition:

$$\frac{1}{T_1}\int_0^{T_1} (E[y(t_2)]^*)^2 dt_2 = \frac{1}{T_1}\int_0^{T_1} (E[u(t_2)]^*)^2 dt_2 \ . \tag{5.148}$$

**Criterion 2-*SLNONST*.** Mean-square error of approximation

$$I = \frac{1}{T_0 T_1} \int_0^{T_1} \int_0^{T_0} E[(y(t_1, t_2) - u(t_1, t_2))^2] dt_1 dt_2 . \qquad (5.149)$$

The calculation of linearization coefficients in this case is much more complicated than for statistical linearization. The details can be found, for instance, in [49].

Historically, the first work of application of linearization method to dynamic systems under nonstationary excitation in the context of random vibration was presented by Iwan and Mason [39], who considered a nonlinear stochastic system in the special form

$$\mathbf{M}\ddot{\mathbf{x}} + \mathbf{f}(\mathbf{x}, \dot{\mathbf{x}}) = \mathbf{\Theta}(t)\eta(t) , \qquad (5.150)$$

where $\mathbf{x} \in R^n$ is the vector of displacements, $\mathbf{M}$ is the mass matrix, $\mathbf{f}(\mathbf{x}, \dot{\mathbf{x}})$ is a slightly vector nonlinear function of $\mathbf{x}$ and $\dot{\mathbf{x}}$, $\eta(t)$ is a stationary zero-mean Gaussian random process (white or colored) with the given power spectral density function $S_\eta(\lambda)$, and $\mathbf{\Theta}(t)$ is a deterministic modulated vector function of time.

By transforming the corresponding linearized equation into a vector form, Iwan and Mason [39] obtained the nonstationary equation for the covariance matrix (5.32) ($\mathbf{m_x} = \mathbf{0}, \mathbf{A_0} = \mathbf{0}$). However, Iwan and Mason [39] first applied this approach to the Duffing oscillator when the deterministic modulated function has the form

$$\Theta(t) = te^{-\beta t} , \qquad (5.151)$$

where $\beta > 0$ is a parameter and the process $\eta(t)$ is a Gaussian white noise.

Similar approach for the Duffing oscillator was proposed by Sakata and Kimura [70] for different $\Theta(t)$ functions and for the case when $\eta(t)$ is a zero-mean Gaussian colored noise. Modifications were made by Spanos [83] and Kimura and Sakata [50] to cases of nonsymmetric nonlinearities and multi-degree-of-freedom systems [50].

The most common type of nonstationary inputs for which analytical solutions (exact or approximate) have been proposed is the uniformly (amplitude) modulated or separable random process, which is defined as the product of a stationary process and a deterministic envelope function, also called *the time-modulating function*. For the application of statistical linearization it is convenient to assume that the component stationary process is a Gaussian process. There are two basic approaches of application of statistical linearization to the response analysis of stochastic dynamic systems under external nonstationary excitations in the form of the product of a Gaussian colored noise and a deterministic envelope function called by Roberts and Spanos [68] *pre-filters method* and *decomposition method*.

To present both approaches we consider the nonlinear vector stochastic differential equation

$$\frac{d\mathbf{x}(t)}{dt} = \mathbf{\Phi}(\mathbf{x}, t) + \sum_{k=1}^{M} \mathbf{G}_{\eta k}(t)\eta_k(t), \quad \mathbf{x}(t_0) = \mathbf{x_0} , \qquad (5.152)$$

where $\mathbf{x} = [x_1, \ldots, x_n]^T$ is the vector state, $\boldsymbol{\Phi} = [\Phi_1, \ldots, \Phi_n]^T$ is a nonlinear vector function, $\mathbf{G}_{\eta k}(t) = [\sigma_{\eta k}^1(t), \ldots, \sigma_{\eta k}^n(t)]^T$, $k = 1, \ldots, M$, are time-dependent deterministic vectors of noise intensity, $\eta_k(t)$, $k = 1, \ldots, M$, are independent zero-mean Gaussian stationary processes with a given correlation functions $K_{\eta_k}(\tau)$ or power spectral density functions $S_{\eta_k}(\lambda)$; $\mathbf{x}_0$ is a vector of initial conditions independent of $\eta_k(t)$, $k = 1, \ldots, M$.

In the pre–filter method it is assumed that $\eta_k(t)$ can be treated as outputs from linear filters whose inputs are excited by Gaussian white noises, i.e., the filters satisfy the following linear algebraic equation and Ito stochastic differential equations

$$\eta_k(t) = \mathbf{C}_k^T \zeta_k(t) , \tag{5.153}$$

$$d\zeta_k(t) = \mathbf{A}_{\zeta \mathbf{k}} \zeta_k dt + \mathbf{G}_{\zeta \mathbf{k}} d\xi_k, \quad k = 1, \ldots, M , \tag{5.154}$$

where $\mathbf{A}_{\zeta k}$ are $n_k \times n_k$-dimensional negative- definite matrices, $Re\lambda_{ki}(\mathbf{A}_{\zeta k}) < 0$, $\mathbf{C}_k, \mathbf{G}_{\zeta k}$ are $n_k$-dimensional vectors of the real coefficients modeling the linear filter with single input and single output (SISO); $\xi_k$ are independent standard Wiener processes.

As in the case of Gaussian white or colored noise excitation we find for system (5.152) the corresponding equations for linearized system

$$\frac{d\mathbf{x}(t)}{dt} = [\mathbf{A}_0(t) + \mathbf{A}(t)\mathbf{x}(t)] + \sum_{k=1}^{M} \mathbf{G}_{\eta k}(t)\eta_k(t) , \quad \mathbf{x}(t_0) = \mathbf{x}_0 , \tag{5.155}$$

where $\mathbf{A}_0 = \boldsymbol{\Phi}_0 - \mathbf{K}\mathbf{m}_{\mathbf{x}} = [A_{01}, \ldots, A_{0n}]^T$, $\mathbf{A} = \mathbf{K} = [a_{ij}], i, j = 1, \ldots, n$, are the vector and the matrix of linearization coefficients, respectively; $\eta_k$ are defined by (5.153) and (5.154).

Next, repeating considerations for stationary excitations case one can join system equations (5.155) with filter equations (5.153–5.154) and rewrite them in the form of the linear vector Ito stochastic differential equation (5.132–5.133) for $\mathbf{G}_k = \mathbf{G}_k(t)$. Then the corresponding moment equations have the form (5.134–5.135).

We note that the problem of the determination of the coefficients of filter equations for a zero-mean stationary process with a given correlation function or a power spectral density function was presented in many monographs, for instance [64].

To determine linearization coefficients we have to solve moment equations (5.134–5.135) for $\mathbf{G}_k = \mathbf{G}_k(t)$ and for linearization matrices $\mathbf{A}_0(t)$ and $\mathbf{A}(t)$ depending on the moments $\mathbf{m}_{\mathbf{x}}(t) = E[\mathbf{x}(t)]$ and $\boldsymbol{\Gamma}_L(t) = E[\mathbf{x}(t)\mathbf{x}^T(t)]$ with a given set of initial conditions for $\mathbf{m}_{\mathbf{x}}(t)$, $\boldsymbol{\Gamma}_L(t)$ and separately calculate initial conditions for the filter part of moment equations. These initial conditions for the filter part have to ensure the existence of a stationary solution for all $t \geq 0$ for second-order moments of filter variables (for details see, for instance, [68]).

In the second approach of the application of statistical linearization to stochastic dynamic systems under external nonstationary excitations in the form of the product of a Gaussian colored noise and a deterministic envelope function called *decomposition method* by Roberts and Spanos [68], first we use the notation of Green function method presented in Chap. 2.

We consider the simplified version of nonlinear system (5.152) for a scalar stochastic process $\eta(t)$, i.e.,

$$\frac{d\mathbf{x}(t)}{dt} = \mathbf{\Phi}(\mathbf{x}, t) + \mathbf{G}_0(t)\eta(t), \quad \mathbf{x}(t_0) = \mathbf{x}_0 , \qquad (5.156)$$

where $\mathbf{x}(t) = [x_1(t), \dots, x_n(t)]^T$ is an $n$-dimensional vector state, $\mathbf{x}_0$ is a vector of initial conditions, $\mathbf{\Phi} = [\Phi_1, \dots, \Phi_n]^T$ is a nonlinear vector function, $\mathbf{G}_0(t) = \mathbf{v}a(t)$, $\mathbf{v} = [0, \dots, 0, 1]^T$ is an $n$- dimensional vector, $a(t)$ is a scalar nonlinear function that can be decomposed into a sum of exponential terms, i.e.,

$$a(t) = \begin{cases} \sum_{i=1}^{M} a_i \exp\{-\lambda_i t\} & \text{for } t \geq 0 \\ 0 & \text{for } t < 0 \end{cases} , \qquad (5.157)$$

where $a_i$, and $\lambda_i$ are positive constants. Similarly, it is assumed that $\eta(t)$ is a zero-mean Gaussian colored noise with correlation function $K_\eta(t_1, t_2)$ that can be also decomposed into a sum of exponential terms, i.e.,

$$K_\eta(t_1, t_2) = K_\eta(t_2 - t_1) = \sum_{j=1}^{N} \alpha_j \exp\{-\beta_j |t_2 - t_1|\} , \qquad (5.158)$$

where $\alpha_j$ and $\beta_j$ are positive constants, $N$ is a given natural number.

The corresponding vector equation for the linearized system is

$$\frac{d\mathbf{x}(t)}{dt} = \mathbf{A}(t)\mathbf{x}(t) + \mathbf{G}_0(t)\eta(t) , \quad \mathbf{x}(t_0) = \mathbf{x}_0 , \qquad (5.159)$$

where $\mathbf{A}(t) = [a_{ij}(t)], i, j = 1, \dots, n$, is a matrix of linearization coefficients.

The solution of (5.159) has the form

$$\mathbf{x}(t) = \mathbf{\Psi}(t, t_0)\mathbf{x}_0 + \int_{t_0}^{t} \mathbf{\Psi}(t, s)\mathbf{G}_0(s)\eta(s)ds , \quad \mathbf{x}(t_0) = \mathbf{x}_0 , \qquad (5.160)$$

where $\mathbf{\Psi}(t, t_0)$ is an $n \times n$-dimensional fundamental matrix of the homogeneous equation

$$\frac{d\mathbf{x}(t)}{dt} = \mathbf{A}(t)\mathbf{x}(t), \mathbf{x}(t_0) = \mathbf{x}_0 . \qquad (5.161)$$

In particular, if $\mathbf{A}$ and $\mathbf{G}_0$ are constant matrix and vector, respectively, then

$$\mathbf{\Psi}(t, t_0) = \mathbf{\Psi}(t - t_0) = \exp\{\mathbf{A}(t - t_0)\} = \sum_{i=0}^{\infty} \frac{1}{i!}\mathbf{A}^i(t - t_0)^i . \qquad (5.162)$$

First, using condition (3.32) one can find that the mean value in (5.159) $\mathbf{m}(t) = E[\mathbf{x}(t)] = \mathbf{0}$. Next, following the considerations for the Green function method one can show that the variance matrix $\mathbf{K}_x(t) = E[\mathbf{x}(t)\mathbf{x}^T(t)]$ satisfies the differential equation (3.38), i.e.,

$$\frac{d\mathbf{K}_x(t)}{dt} = \mathbf{A}(t)\mathbf{K}_x(t) + \mathbf{K}_x(t)\mathbf{A}^T(t) + \mathbf{P}(t) + \mathbf{P}^T(t) , \quad \mathbf{K}_x(t_0) = \mathbf{K}_0 , \quad (5.163)$$

where

$$\mathbf{P}(t) = \mathbf{G}_0(t) \int_{t_0}^{t} \mathbf{K}_\eta(t,\tau)\mathbf{G}_0^T(\tau)\mathbf{\Psi}(t,\tau)^T \, d\tau = \int_{t_0}^{t} \mathbf{K}_G(t,\tau)\mathbf{\Psi}(t,\tau)^T \, d\tau ,$$
$$(5.164)$$

where $\mathbf{K}_G(t,\tau)$ is the correlation matrix of the vector $\mathbf{G}_0(t) = \mathbf{v}a(t)$, defined by

$$\mathbf{K}_G(t_1,t_2) = \mathbf{R}a(t_1)a(t_2)K_\eta(t_2 - t_1) , \quad (5.165)$$

where $\mathbf{R} = \mathbf{v}\mathbf{v}^T$.

Substituting equalities (5.165),(5.157), and (5.158) into (5.164) we obtain

$$\mathbf{P}(t) = \sum_{k=1}^{M}\sum_{l=1}^{M}\sum_{j=1}^{N} a_k a_l \alpha_j \mathbf{P}^{(klj)}(t) , \quad (5.166)$$

where

$$\mathbf{P}^{(klj)}(t) = \exp\{-(\lambda_k + \beta_j)t\} \int_{t_0}^{t} \mathbf{\Psi}(t,\tau)^T \mathbf{R} \exp\{-(\lambda_l - \beta_j)\tau\} d\tau . \quad (5.167)$$

Differentiating the right-hand side of (5.167) with respect to $t$ we obtain

$$\frac{d\mathbf{P}^{(klj)}(t)}{dt} = -(\lambda_k + \beta_j)\mathbf{P}^{(klj)}(t) + \mathbf{A}\mathbf{P}^{(klj)}(t) + \exp\{-(\lambda_k + \lambda_l)t\}\mathbf{R} ,$$
$$k,l = 1,\ldots,M, \quad j = 1,\ldots,N . \quad (5.168)$$

To determine linearization coefficients we have to solve (5.166–5.168) and (5.163) for linearization matrix $\mathbf{A}(t)$ depending on the second-order moments $\mathbf{K}_x(t) = E[\mathbf{x}(t)\mathbf{x}^T(t)]$ with a given set of initial conditions for $\mathbf{K}_x(t_0)$ (for details see, for instance, [68]).

The *decomposition method* was proposed by Sakata and Kimura [70] and Roberts and Spanos [68] and was developed by Smyth and Masri [74] who used a method based on the spectral decomposition of the random process by the orthogonal Karhunen–Loeve expansion with an application of least-squares approaches to develop an approximate analytical fit for the eigenvectors of the underlying random process [86]. In this case the approximate covariance matrix $\mathbf{K}_\eta(t_1,t_2)$ can be constructed with the least-squares estimated eigenvectors and presented as the Chebyshev polynomial series in the form [54]

$$\mathbf{K}_\eta(t_1,t_2) = \sum_{j=1}^{N}\lambda_j \sum_{k=1}^{M_j-1}\sum_{l=1}^{M_j-1} H_{jk}H_{jl} T_k(t_1') T_l(t_2') , \quad (5.169)$$

where $\lambda_j$ are the truncated series of eigenvalues, $M_j > 1$ is a given natural number, $T_k$'s are Chebyshev polynomials defined as

$$T_k(t') = \cos(k \cos^{-1} t'), \quad for -1 \le t' \le 1, \quad k = 1, \ldots M_j - 1 . \quad (5.170)$$

$H_{jk}$ are weighting coefficients for the Chebyshev polynomials of the order $j$ in the least–square fitting of the eigenvectors, $0 \le t_i \le t_{max}$, $i = 1, 2$,

$$t'_i = \frac{2t_i}{t_{max}} - 1 . \quad (5.171)$$

To obtain the second-order moments in the generalized decomposition method [74] one has to solve matrix equation (5.163) where the matrix $\mathbf{P}$ is defined by

$$\mathbf{P}(t) = \sum_{j=1}^{N} \lambda_j \sum_{k=1}^{M_j-1} \sum_{l=1}^{M_j-1} H_{jk} H_{jl} T_k(t'_1) \mathbf{P}^{(l)}(t) , \quad (5.172)$$

where

$$\mathbf{P}^{(l)}(t) = \int_{t_0}^{t} \mathbf{\Psi}(t, \tau)^T \mathbf{R} T_l(\tau') d\tau . \quad (5.173)$$

$$\tau' = \frac{2\tau}{t_{max}} - 1 . \quad (5.174)$$

The matrices $\mathbf{P}^{(l)}(t)$ satisfy the following differential equations:

$$\frac{d\mathbf{P}^{(l)}(t)}{dt} = \mathbf{A}(t)\mathbf{P}^{(l)}(t) + \mathbf{R} T_l(t') , \quad \mathbf{P}^{(l)}(t_0) = 0, \quad l = 1, \ldots M_j - 1 . \quad (5.175)$$

To determine linearization coefficients we have to solve (5.172–5.175) and (5.163) for linearization matrix $\mathbf{A}(t)$ depending on the second-order moments $\mathbf{K}_x(t) = E[\mathbf{x}(t)\mathbf{x}^T(t)]$ with a given set of initial conditions for $\mathbf{K}_x(t_0)$ (for details see, for instance, [74]).

The problem of approximation of any covariance function or covariance matrix of a nonstationary process has been discussed by many authors, for instance, in [32, 68, 74].

At the end of this section we propose an alternative approach that use the exact or approximate covariance function of the response of linear dynamic system subjected to nonstationary stochastic excitation. The main idea of the proposed approach is the following.

Consider again the nonlinear system (5.152) and the corresponding linearized equation (5.155), i.e.,

$$\frac{d\mathbf{x}(t)}{dt} = \mathbf{\Phi}(\mathbf{x}, t) + \sum_{k=1}^{M} \mathbf{G}_{\eta k}(t) \eta_k(t) , \quad \mathbf{x}(t_0) = \mathbf{x}_0 , \quad (5.176)$$

where $\mathbf{x} = [x_1, \dots, x_n]^T$ is the vector state, $\mathbf{x}_0$ is a vector of initial conditions, $\boldsymbol{\Phi} = [\Phi_1, \dots, \Phi_n]^T$ is a nonlinear vector function, $\mathbf{G}_{\eta k}(t) = [\sigma_{\eta k}^1(t), \dots, \sigma_{\eta k}^n(t)]^T$, $k = 1, \dots, M$, are time-dependent deterministic vectors of noise intensity, $\eta_k(t)$ are independent zero-mean Gaussian stationary processes with a given correlation functions $K_{\eta_k}(\tau)$ or power spectral density functions $S_{\eta_k}(\lambda)$; $\mathbf{x}_0$ is a vector of initial conditions independent of $\eta_k(t)$, $k = 1, \dots, M$,

$$\frac{d\mathbf{x}(t)}{dt} = [\mathbf{A}_0(t) + \mathbf{A}(t)\mathbf{x}(t)] + \sum_{k=1}^{M} \mathbf{G}_{\eta k}(t)\eta_k(t) , \quad \mathbf{x}(t_0) = \mathbf{x}_0 , \qquad (5.177)$$

where $\mathbf{A}_0 = \boldsymbol{\Phi}_0 - \mathbf{K}\mathbf{m_x} = [A_{01}, \dots, A_{0n}]^T$, $\mathbf{A} = \mathbf{K} = [a_{ij}], i, j = 1, \dots, n$, are the vector and the matrix of linearization coefficients, respectively.

Assume that $\mathbf{K}_x(t_1, t_2)$ is the correlation matrix obtained by the procedure proposed by Conte and Peng [8] or by another one. Then the variance matrix equal to $\mathbf{K}_x(t) = \mathbf{K}_x(t, t)$ depends on linearization coefficients (separately derived and collected in the matrix $\mathbf{A}(t)$ and vector $\mathbf{A}_0(t)$). Since the linearization coefficients also depend on first-and second-order moments, then an iterative procedure should be used to calculate them, i.e., for all $t$, two steps are repeated until convergence is established

$1^0$ Let $i = 0$. For a given initial value of the correlation matrix $\mathbf{K}_x^i(t)$, calculate the matrix $\mathbf{A}^i(t)$ and the vector $\mathbf{A}^i(t)$ of linearization coefficients.

$2^0$ For the calculated $\mathbf{A}^i(t)$ and $\mathbf{A}^i(t)$ in step $1^0$ calculate $\mathbf{K}_x^{i+1}(t)$ and go to step $1^0$ of linearization coefficients

$3^0$ Repeat procedure until converge.

## 5.5 Non-Gaussian Excitations

The statistical linearization method was also applied to nonlinear systems with non-Gaussian excitations. First remarks about linearization for dynamic systems with external Poisson process excitation were made by Sinitsyn [73] and by Romanov and Khuberyan [69]. First systematic study was proposed by Tylikowski and Marowski [87] and Grigoriu [30, 31] for excitation in the form of stationary Poisson processes. This approach was generalized for stationary continuous non-Gaussian processes and compound Poisson processes by Sobiechowski and the present author [77, 78]. The fundamental difference between statistical linearization methods for Gaussian and non-Gaussian excitations is the method of calculation of moments (stationary or nonstationary) for linearized systems. Since this method is more complex in the case of non-Gaussian excitations we illustrate it on an example of the Duffing oscillator described by the following equation

$$\ddot{x}(t) + 2\zeta\dot{x}(t) + x(t) + \epsilon x^3(t) = \eta(t) , \qquad (5.178)$$

where $\zeta$ and $\epsilon$ are constant parameters, $\eta(t)$ is a non-Gaussian stochastic process.

The objective of statistical linearization is to replace the nonlinear element $\phi = \epsilon x^3(t)$ by the linear form $\phi_L = k_0 + k_e(x(t) - E[x(t)])$, where $k_0$ and $k_e$ are linearization coefficients, such that a certain equivalence criterion is satisfied. Then, the linearized system has the form

$$\ddot{x}(t) + 2h\dot{x}(t) + \lambda_e^2 x(t) + \epsilon_e = \eta(t) , \tag{5.179}$$

where $\lambda_e^2 = 1 + k_e$ and $\epsilon_e = k_0 - k_e E[x(t)]$.

We consider the following four equivalence criteria:

1. Criterion of equality of the first and second moments of nonlinear and linearized variables [45],

$$\epsilon^k E[x^{3k}(t)] = E[(\phi_L(t))^k] , \quad \text{for} \quad k = 1, 2 . \tag{5.180}$$

2. Minimization of the mean-square error of approximation [5, 45],

$$E\left[(\phi_L(t) - \epsilon x^3(t))^2\right] . \tag{5.181}$$

3. Criterion of equality of the first moments of variables and the corresponding potential energies [77, 78]

$$\epsilon E[x^3(t)] = E[(\phi_L(t))] ,$$
$$\epsilon E\left[\int_0^{x(t)} \xi^3 d\xi\right] = E\left[\int_0^{x(t)} \phi_L d\xi\right] . \tag{5.182}$$

4. Minimization of the mean-square difference of the potential energies [94]

$$E\left[\left(\int_0^{x(t)} (\phi_L(t) - \epsilon\xi^3)d\xi\right)^2\right] . \tag{5.183}$$

If the stochastic process $\eta(t)$ has vanishing odd moments, the corresponding odd moments of the response are equal to zero. In this case the linearization coefficients are

$$k_{0i} = 0 , \quad \text{for} \quad i = 1, 2, 3, 4 , \tag{5.184}$$

$$k_{e1} = \epsilon\sqrt{\frac{E[x^6(t)]}{E[x^2(t)]}}, \quad k_{e2} = \epsilon\frac{E[x^4(t)]}{E[x^2(t)]}, \quad k_{e3} = \epsilon\frac{E[x^4(t)]}{2E[x^2(t)]}, \quad k_{e4} = \epsilon\frac{E[x^6(t)]}{2E[x^4(t)]} . \tag{5.185}$$

We note that the linearization coefficients for moment criteria as well as for criteria in probability density functions space depend in a nonlinear way on moments of solutions of (5.178). Therefore, to calculate the linearization coefficients one has to calculate the moments of solutions of (5.178). However, such a method is too complicated and as in Gaussian excitations case we replace the moments of solution of (5.178) by the corresponding moments of solutions of linearized system (5.179). For convenience we rewrite (5.179) in the form of two-dimensional vector equation

$$
\begin{aligned}
dx_1(t) &= x_2(t)dt \,, \\
dx_2(t) &= [-\lambda_e^2 x_1(t) - 2hx_2(t) - \epsilon_1 + \eta(t)]dt \,.
\end{aligned} \tag{5.186}
$$

In this case, a modification of Procedure $SL - MC$ for the determination of the linearized system can be formulated.

1. Guess initial values for $k_{ei}$, $i = 1, 2, 3, 4$; for instance, $k_{ei} = 0$, $i = 1, 2, 3, 4$.
2. Calculate $E[x^k]$ for $k = 1, 2, 3, 4$ for the linear system (5.186).
3. Calculate coefficients $k_{ei}$ from $i = 1, 2, 3, 4$ from corresponding relations (5.185).
4. Go back to step 2 and iterate until convergence.

Thus, the whole problem consists in the determination of the moments $E[x^k]$, $k = 1, \ldots, 4$, for the linear system (5.186). This is described below. As we only want to compare predictions for the variance of the stationary state, we limit ourselves to the stationary moments.

## Calculation of the Stationary Moments of Solutions of Linearized System

We assume that the stochastic process $\eta(t)$ is non-Gaussian and can be represented by a polynomial form of a normal filtered process described by

$$
\eta(t) = \sum_{i=1}^{M} \alpha_i y^i(t) \tag{5.187}
$$

and

$$
dy(t) = -\alpha y(t)dt + qd\xi(t) \,, \tag{5.188}
$$

where $\alpha, \alpha_i (i = 1, \ldots, M)$, and $q$ are constant positive parameters, $y(t)$ is a one-dimensional colored Gaussian process, and $\xi(t)$ is a standard Wiener process.

The moment equations corresponding to the linearized system (5.186–5.188) have the form

$$
\begin{aligned}
\frac{dE[x_1^{p_1} x_2^{p_2}]}{dt} =\ & p_1 E[x_1^{p_1-1} x_2^{p_2+1}] - \lambda_e^2 p_2 E[x_1^{p_1+1} x_2^{p_2-1}] \\
& -2hp_2 E[x_1^{p_1} x_2^{p_2}] - \epsilon_1 p_2 E[x_1^{p_1} x_2^{p_2-1}] \\
& + \sum_{i=1}^{M} \alpha_i E[x_1^{p_1} x_2^{p_2-1} y^i],
\end{aligned} \tag{5.189}
$$

for $p_1, p_2 = 0, 1, \ldots, p$, $p_1 + p_2 = p$, $p = 1, 2, \ldots, N_p$ ,

$$\frac{dE[x_1^{p_1} x_2^{p_2-1} y^i]}{dt} = p_1 E[x_1^{p_1-1} x_2^{p_2} y^i] - \lambda_e^2 (p_2 - 1) E[x_1^{p_1+1} x_2^{p_2-2} y^i]$$

$$-2h(p_2 - 1) E[x_1^{p_1} x_2^{p_2-1} y^i] - \epsilon_1 (p_2 - 1) E[x_1^{p_1} x_2^{p_2-2} y^i]$$

$$+(p_2 - 1) \sum_{j=1}^{M} \alpha_j E[x_1^{p_1} x_2^{p_2-2} y^{i+j}]$$

$$-\alpha i E[x_1^{p_1} x_2^{p_2-1} y^i] + \frac{1}{2} i(i-1) q^2 E[x_1^{p_1} x_2^{p_2-1} y^{i-2}], \quad (5.190)$$

for $p_1, p_2 - 1 = 0, 1, \ldots, p - 1$, $p_1 + p_2 = p$, $p = 1, \ldots, N_p - 1$ and

$$\frac{dE[y^i]}{dt} = -\alpha i E[y^i] + \frac{1}{2} i(i-1) q^2 E[y^{i-2}], \quad (5.191)$$

for $i = 1, \ldots, MN_p$.

The number of equations is equal to $\frac{N_p(N_p+3)}{2} + M[N_p + 2(N_p - 1) + \ldots + (N_p - 1)2 + N_p]$. For example, if $M = 3$ and $N_p = 6$, we obtain 195 equations. We note that although system (5.186–5.188) is nonlinear, the moment equations (5.189–5.191) are in exact closed form and no closure technique has to be applied.

To illustrate the obtained results, a comparison of mean-square displacements $E[x_1^2]$ for four criteria of statistical linearization for systems (5.186–5.188) is shown in Figs. 5.1 and 5.2. The notation SL–$i$, $i = 1, \ldots, 4$, is the key of the figures, which refers to the four linearization criteria.

In the simulation of the response of the Duffing oscillator with excitations represented by a polynomial form of a normal filtered process, the sample functions of Gaussian white noise are modeled by piecewise constant functions with a sample interval $\Delta t = 0.005$. The equations of motion are solved using fourth-order Runge–Kutta scheme with time step equal to 0.00002 in the interval $[0, 200]$. The transient solutions of (5.178) are discarded to ensure that the sample functions of solution is from a stationary process. For each set of parameters, 1000 sample functions of the response are obtained, and for each sample function, only the last 50 points from a total of 40,000 retained as the stationary response. Next, to calculate the estimation of the mean-square displacement, they were divided into 50 batches of 1000 random points.

Figures 5.1 and 5.2 show that there are no significant differences between response characteristics obtained by the considered criteria and simulation results. However, with increasing values for $\varepsilon$ the approximation error increases and the mean-square criteria for variables and their potential energies yield better approximations than the criterion of equality of the first and second moments of nonlinear and linearized variables and the criterion of equality of the first moments of variables and potential energies.

Another representation of continuous stationary non-Gaussian processes was proposed by [41], and similar derivation of moment equations can be done. For details, see [41] and [78].

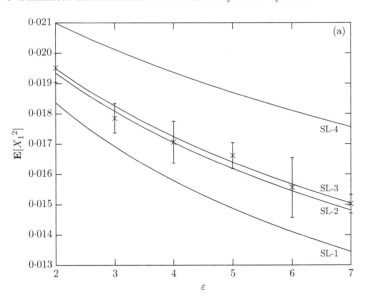

**Fig. 5.1.** Prediction of $E[X_1^2]$ for the Duffing oscillator under continuous non-Gaussian external excitation. Equations (5.187) and (5.188) with $\alpha_1 = 0.25$, $\alpha_2 = 0.25$, $\alpha_3 = 0.25$, $\alpha = 1$, $q^2 = 0.1$, $\zeta = 0.05$,   ×,– Simulation; reproduced from [78] with permission

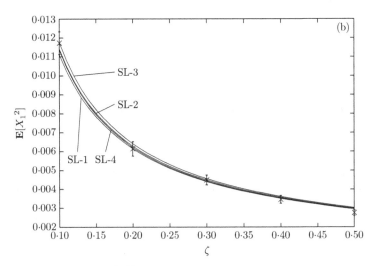

**Fig. 5.2.** Prediction of $E[X_1^2]$ for the Duffing oscillator under continuous non-Gaussian external excitation. Equations (5.187) and (5.188) with $\alpha_1 = 0.25$, $\alpha_2 = 0.25$, $\alpha_3 = 0.25$, $\alpha = 1$, $q^2 = 0.1$, $\epsilon = 1$   ×– Simulation; reproduced from [78] with permission

# Bibliography Notes

### Ref.5.1.

The moment criteria were the most popular criteria at the beginning of the research study of statistical linearization. The development of these criteria includes new moment criteria as well as new iterative procedures for the determination of linearization coefficients for special types of nonlinear functions or for nonlinear dynamic systems under non-Gaussian excitations.

An extension of statistical linearization based on the theory of vibrational correctness of solutions of differential equations was proposed by Kazakov [47]. He used for a given nonlinear vector function $\phi(x)$ and for an input vector Gaussian process $N(x, \theta_0)$ the following expectation

$$\phi_0(x, \theta_0) = \frac{1}{\sqrt{(2\pi)^n \theta_0}} \int_{R^n} \phi(y) \exp\left\{-\frac{1}{2}(y-x)^T \theta_0^{-1}(y-x)\right\} dy \quad (5.192)$$

for the determination of a linear approximation of the function $\phi(x)$ around $x_0$, i.e.,

$$\phi_0(x, \theta_0) \cong \phi_0(x_0, \theta_0) + K_\phi(x_0, \theta_0)(x - x_0), \quad (5.193)$$

where

$$K_\phi(x_0, \theta_0) = \left(\frac{\partial \phi_0^T(x, \theta_0)}{\partial x}\right)^T \bigg|_{x=x_0}. \quad (5.194)$$

Kazakov used this approach in application to the problems of analysis, filtering, and synthesis of the controls of nonlinear statistical control systems.

Friedrich [29] has derived linearization coefficients using Isserlis formula [38] and mean-square criterion for the following nonlinear function of two variables

$$f(x_1, x_2) = \sum_{k=0}^{m} \sum_{l=0}^{m} c_{kl} x_1^k x_2^l, \quad k+l \geq 2. \quad (5.195)$$

This class of linearized functions was extended by Cheded [7] to infinite sums.

Zaydenberg [91] proposed a criterion of equality of approximation error in the intervals $x < m_x$ and $x > m_x$. This criterion was investigated by Wicher [90] and extended by Socha [80].

The linearization coefficients for discontinuous and hysteretic elements were presented by Morosanov [56], and exponential, trigonometric, and $\delta$ functions were obtained by Melt's, Pukhova, and Uskov [55]. Also other types of excitations were considered in the literature. Evlanov and Kazakov [23, 24] showed that if the distribution of the harmonic part of the input variable $x$ is nearly normal, the calculation can be simplified by applying a single statistical linearization for the harmonic and nonharmonic parts. It was shown for stationary response [24] as well as for the nonstationary response [23].

Davis and Spanos [12] used statistical linearization in conjunction with the frequency shift method in the modal and residual mass (MRM) 2-DOF system to derive force-limiting specification for a nonlinear load mass modeled as the Duffing-type (cubic stiffness) and the Rayleigh-type (cubic damping) oscillators.

Also further results were obtained in the field of energy criteria. Elishakoff and Bert [17] and Elishakoff [16] generalized energy criteria using functions $U(x)$ and $R(\dot{x})$. Zhang [95, 92] proposed the following energy criterion:

$$E\left[\left(U^{\mu}(x) - (\frac{1}{2}k_{eq}x^2)^{\mu}\right)^2\right] \longrightarrow \min , \qquad (5.196)$$

where $\mu$ is an unknown parameter that has to be determined in an optimization process.

In this method, first an approximation of the covariance matrix of the response has to be found and next the application of Karhunen–Loeve expansion leads to the reduced number of new introduced variables. In this case in the transformed space, variables with smaller variance are approximated by Gaussian variables.

For a given value of the parameter $\mu$ criterion (5.196) was applied by Wang and Dai [89] to the response analysis of MDOF systems under Gaussian excitations.

To improve the standard equivalent linearization, Lee [51] approximated the ratio

$E[x^4]/E[x^2]$ appearing in the linearization coefficients by application of non-Gaussian closure by Edgeworth series expansions. He considered approximations $2p$ order $p = 1, 2, \ldots, 6$ corresponding to the equating to zero $2(p+1)$-order cumulants.

The application of non-Gaussian probability density function in averaging operation appearing in mean-square criterion for a special class of nonlinear functions called *soft nonlinearities* was proposed by some authors, for instance, Valsakumar et al. [88], Indira et al. [37], and Crandall [9, 10, 67]. They considered the following function:

$$f(x) = -\lambda_0^2 x + \varepsilon x^3 , \qquad (5.197)$$

where $\lambda_0, \varepsilon$ are positive parameters in a scalar dynamic system [88, 37] and in an oscillator [9, 10]. In the case of a scalar dynamic system, the authors proposed a bimodal Gaussian distribution, while in the case of a nonlinear oscillator a form of non-Gaussian probability density function. Also Ricciardi [67] considered a non-Gaussian probability density function in the form of a modified A-type of Gram–Charlier series expansion.

### Ref.5.2.

Another approach called local statistical linearization, similar to statistical linearization in probability density space, was proposed by Pradlwarter [63].

The author used the square metric for approximation of a given probability density of zero-mean Gaussian variable by a linear combination of probability density functions of nonzero-mean Gaussian variables, i.e.,

$$p_i(y_i) \simeq \sum_{k=-m}^{+m} B_{ik} p_{ki}(y_i) \,, \tag{5.198}$$

where

$$p_i(y_i) = \frac{1}{\sigma_i \sqrt{2\pi}} \exp\left\{ -\frac{1}{2} \left( \frac{y_i}{\sigma_i} \right)^2 \right\} \,, \tag{5.199}$$

$$p_{ik}(y_i) = \frac{1}{\sigma_0 \sqrt{2\pi}} \exp\left\{ -\frac{1}{2} \left( k - \frac{y_i}{\sigma_i} \right)^2 \right\} \,, \tag{5.200}$$

$\sigma_0 < \sigma_i$, $i = 1, \ldots, N$ is considerably smaller than $\sigma_i$. The coefficients $B_{ik}$ in (5.198) are determined from the square criterion

$$\int_{-\infty}^{+\infty} \left[ p_i(y_i) - \sum_{k=-m}^{+m} B_{ik} p_{ki}(y_i) \right]^2 dy_i \;\rightarrow\; \min . \tag{5.201}$$

Pradlwarter [63] assumed that the global probability density function $p(x,t)$ can be represented as a convex combination of other probability density functions

$$p(x,t) = \sum_{i=1}^{N} A_i p_i(x,t), \sum_{i=1}^{N} A_i = 1 \,, \tag{5.202}$$

where the local densities $p_i(x,t)$ have a considerable smaller standard deviation $\sigma_i < \sigma$ than the standard deviation $\sigma$ of the original probability density $p(x,t)$. The word "local" is used here in the sense that global probability density function is not approximated. Since the Gaussian probability density function can always be represented as a product of one- dimensional probability density functions by introducing suitable coordinates, it is sufficient to consider the decomposition of one- dimensional Gaussian probability density functions $p_i(y_i)$.

### Ref.5.3.

An application of statistical linearization to the response analysis of the Duffing oscillator under colored noise excitation was presented, for instance, in [26, 79] and by Iyengar [40] to the Duffing oscillator excited by narrowband colored noise. In this case multivalued stationary response moments were obtained.

### Ref.5.4.

A generalization of statistical linearization in the response analysis of a nonlinear oscillator under stationary excitations modulated by an envelope function was proposed by Ahmadi [1] also. He has used an assumption that linearization coefficients are smoothly varying functions of time. In further generalization of this approach for a nonlinear oscillator, Jahedi and Ahmadi [42], Orabi and Ahmadi [59, 60, 61] have used iterative procedures with application of Wiener–Hermite expansion to the determination of response characteristics and including mean-square response and correlation function.

The further generalization of Smyth and Masri approach for nonstationary and non-Gaussian excitation was proposed in their next paper [75].

**Ref.5.5.**

The problem of the generation of non-Gaussian processes was studied by many authors. Analytical and simulation methods of approximation of non-Gaussian probability density functions were discussed, for instance, in [28, 33, 35, 71, 85, 84].

The derivation of statistical linearization for nonlinear dynamic systems under other continuous non-Gaussian excitation was proposed by Sobiechowski and Socha [77, 78]. They have used another representation of continuous stationary non-Gaussian processes proposed by Iyengar and Jaiswal [41] and presented in Sect. 3.6.

Also Hou et al. [36] and Cai and Suzuki [6] applied statistical linearization to a nonlinear oscillator under non-Gaussian excitation represented by trigonometric functions (3.110–3.112) and (3.114–3.115), respectively. Cai and Suzuki called their approach *"statistical quasilinearization"*.

Hou et al. and Cai and Suzuki calculated the stationary response moments for linearized system appearing in linearization coefficients from a new linear system with parametric excitations.

The derivation of statistical linearization for nonlinear dynamic systems under other than continuous non-Gaussian excitation was proposed by several authors. In the case of impulse external excitations it was done by Sobiechowski and Socha [77, 78]. These authors assumed that the impulse occurrence time is a Poisson process and a renewal process, respectively. For the renewal-driven impulse process, the interarrival times have the probability density function

$$p(x) = \begin{cases} \lambda^2 x \exp(-\lambda x) & x > 0 , \\ 0 & x \le 0. \end{cases} \tag{5.203}$$

In all cases the moment equations were found in a closed form.

Grigoriu [34] derived also equivalent linearization for nonlinear systems with Levy white noise and mean-square criterion described by the stochastic differential equation

$$dx(t) = g(x(t))dt + GdL(t) , \tag{5.204}$$

where $x(t) \in R^n, t \ge 0, g : R^n \to R^n$ is a nonlinear vector function, $G$ denotes a constant $n \times m$ matrix and $L = [L_1, \ldots, L_m]^T$ is a vector of $m$ independent standard Levy processes. $L_k, k = 1, \ldots, m$ are real-valued, start at zero, and have stationary independent increments.

The characteristic function of an arbitrary increment $L_k(t) - L_k(s), t > s$ of $L_k$ is

$$\phi_{L_k(t)-L_k(s)} = E[\exp\{iu(L_k(t) - L_k(s))\}] = \exp\{-(t-s)|u|^\alpha\}, \quad k = 1, \ldots, m$$
(5.205)

where $\alpha$ is a parameter taking values in the range $(0, 2]$

We note that the standard Levy process is a generalization of the Gaussian and the compound Poisson processes and in the case when $\alpha \in (0, 2)$ it does not have finite variance and higher-order moments.

Therefore, the $n \times n$ matrix $A_{eq}$ of linearization coefficients appearing in the linearized system

$$dy(t) = A_{eq}y(t)dt + GdL(t)$$
(5.206)

with the same input noise as the nonlinear system(5.205) cannot be found as in the cases of nonlinear systems with mean-square criterion under excitations in the form of Gaussian or compound Poisson processes. To omit this difficulty, Grigoriu [34] proposed a new criterion of equivalent linearization in the space of characteristics functions,

$$I_{Levi} = \int_{R^n} |\phi_{Ay}(u) - \phi_{g(y)}(u)|w(u)(du) ,$$
(5.207)

where $w : R^n \to [0, 1]$ is an arbitrary weighting function. In this case the characteristic function $\phi_{Ay}(u)$ corresponding to linear system (5.206) can be found analytically in contrast to the characteristic function $\phi_{g(y)}(u)$ corresponding to nonlinear system (5.205). To determine $\phi_{g(y)}(u)$ Grigoriu [34] used Monte Carlo simulations.

A comparative study of three statistical linearization methods with simulation results for the nonlinear oscillator with non-Gaussian parametric delta-excitation with different criteria was given by Sobiechowski [76]. He considered the following model described by the Ito equation:

$$\begin{aligned}
dx_1(t) &= x_2(t)dt , \\
dx_2(t) &= f(x_1(t), x_2(t))dt + g(x_1(t))dC_1(t) \\
&\quad + x_2(t)(\exp\{b_2dC_2(t)\} - 1) + b_3dC_3(t),
\end{aligned}$$
(5.208)

where $C_1(t), C_2(t), C_3(t)$ are independent compound Poisson processes and $f(x_1, x_2), g(x_1)$ are nonlinear functions. The intensities of the corresponding Poisson counting processes are $\lambda_i, i = 1, 2, 3$. An example of application of equivalent linearization with mean-square criterion to the nonlinear system with parametric delta excitation was also given in [76].

Falsone [25] considered mean-square criterion of linearization with non–Gaussian averaging operation defined by Gram–Charlier series in application to the determination of linearization coefficients and response moments of one- and two-dimensional nonlinear dynamic systems under external non-Gaussian ($\delta$-correlated) process.

# References

1. Ahmadi, G.: Mean square response of a Duffing oscillator to a modulated white noise excitation by the generalized method of equivalent linearization. J. Sound Vib. **71**, 9–15, 1980.
2. Alexandrov, V., et al.: Study of the accuracy of nonlinear nonstationary systems by the method of statistical linearization. Avtomatika i Telemekhanika **24**, 488–495, 1965.
3. Anh, N. and Schiehlen, W.: A technique for obtaining approximate solutions in Gaussian equivalent linearization. Comput. Methods Appl. Mech. Eng. **168**, 113–119, 1999.
4. Beaman, J. and Hedrick, J.: Improved statistical linearization for analysis of control of nonlinear stochastic systems: Part 1: An extended statistical linearization method. Trans. ASME J. Dyn. Syst. Meas. Control **101**, 14–21, 1981.
5. Booton, R. C.: The analysis of nonlinear central systems with random inputs. IRE Trans. Circuit Theory **1**, 32–34, 1954.
6. Cai, G. and Suzuki, Y.: On statistical quasi-linearization. Int. J. Non-Linear Mech. **40**, 1139–1147, 2005.
7. Cheded, L.: Invariance property: Higher-order extension and application to Gaussian random variables. Signal Process. **83**, 1545–1551, 2003.
8. Conte, J. and Peng, B.: An explicit closed form solution for linear systems subjected to nonstationary random excitations. Probab. Eng. Mech. **11**, 37–50, 1996.
9. Crandall, S.: On using non-Gaussian distributions to perform statistical linearization. In: W. Zhu, G. Cai, and R. Zhang (eds.), Advances in Stochastic Structural Dynamics, pp. 49–62. CRC Press, Boca Raton, 2003.
10. Crandall, S.: On using non-Gaussian distributions to perform statistical linearization. Int. J. Non-Linear Mech. **39**, 1395–1406, 2004.
11. Csaki, F.: Nonlinear Optimal and Adaptive Systems. Akademiai Kiado, Budapest, 1972 (in Russian).
12. Davis, G. and Spanos, P.: Developing force-limited random vibration test specifications for nonlinear systems using the frequency-shift method. In: Proc. 40th AIAA/SDM Conference, pp. 1688–1698, St.Louis, MO, 4/12-15/99, 1999.
13. Dimentberg, M.: Vibration of system with nonlinear cubic characteristic under narrow band random excitation. Izw.AN SSSR, MTT **2**, 1971 (in Russian).
14. Elishakoff, I.: Method of stochastic linearization: Revisited and improved. In: S. P. D. and C. Brebbia (eds.), Computational Stochastic Mechanics, pp. 101–111. Computational Mechanics Publication and Elsevier Applied Science, London, 1991.
15. Elishakoff, I.: Some results in stochastic linearization of nonlinear systems. In: K. W. H. and N. Namachchivaya (eds.), Nonlinear Dynamics and Stochastic Mechanics, pp. 259–281. CRC Press, Boca Raton, 1995.
16. Elishakoff, I.: Multiple combinations of the stochastic linearization criteria by the moment approach. J. Sound Vib. **237**, 550–559, 2000.
17. Elishakoff, I. and Bert, C.: Complementary energy criterion in nonlinear stochastic dynamics. In: R. Melchers and M. Steward (eds.), Application of Stochastic and Probability, pp. 821–825. A. A. Balkema, Rotterdam, 1999.
18. Elishakoff, I. and Colombi, P.: Successful combination of the stochastic linearization and Monte Carlo methods. J. Sound Vib. **160**, 554–558, 1993.

19. Elishakoff, I. and Falsone, G.: Some recent developments in stochastic linearization method. In: A.-D. Cheng and C. Yang (eds.), Computational Stochastic Mechanics, pp. 175–194. Computational Mechanics Publication, Southampton, New York, 1993.

20. Elishakoff, I., Fang, J., and Caimi, R.: Random vibration of a nonlinearly deformed beam by a new stochastic linearization method. Int. J. Solids Struct. **32**, 1571–1584, 1995.

21. Elishakoff, I. and Zhang, R.: Comparison of the new energy-based version of the stochastic linearization method. In: N. Bellomo and F. Casciati (eds.), Nonlinear Stochastic Mechanics, pp. 201–212. Springer, Berlin, 1992.

22. Elishakoff, I. and Zhang, X.: An appraisal of different stochastic linearization criteria. J. Sound Vib. **153**, 370–375, 1992.

23. Evlanov, L. and Kazakov, I. E.: Statistical investigation of nonlinear auto-oscillatory systems with non-steady state modes. Avtomatika i Telemekhanika **31**, 45–53, 1970.

24. Evlanov, L. and Kazakov, I. E.: Statistical investigation of nonlinear oscillatory systems in the steady states modes. Avtomatika i Telemekhanika **30**, 1902–1910, 1969.

25. Falsone, G.: An extension of the Kazakov relationship for non-Gaussian random variables and its use in the non-linear stochastic dynamics. Probab. Eng. Mech. **20**, 45–56, 2005.

26. Falsone, G. and Elishakoff, I.: Modified stochastic linearization method for coloured noise excitation of Duffing oscillator. Int. J. Non-Linear Mech. **29**, 65–69, 1994.

27. Fang, J., Elishakoff, I., and Caimi, R.: Nonlinear response of a beam under stationary random excitation by improved stochastic linearization method. Appl. Math. Modeling **19**, 106–111, 1995.

28. Ferrante, F., Arwade, S., and Graham-Brady, L.: A translation model for non-Gaussian random processes. Probab. Eng. Mech. **20**, 215–228, 2005.

29. Friedrich, H.: On a method of equivalent statistical linearization. J. Tech. Phys. **16**, 205–212, 1975.

30. Grigoriu, M.: Equivalent linearization for Poisson white noise input. Probab. Eng. Mech. **10**, 45–51, 1995.

31. Grigoriu, M.: Linear and nonlinear systems with non-Gaussian white noise input. Probab. Eng. Mech. **10**, 171–180, 1995.

32. Grigoriu, M.: Parametric models of nonstationary Gaussian processes. Probab. Eng. Mech. **10**, 95–102, 1995.

33. Grigoriu, M.: Applied non-Gaussian Processes: Examples, Theory, Simulation, Linear Random Vibration, and MATLAB Solutions. Prentice Hall, Englewood Cliffs, 1995.

34. Grigoriu, M.: Equivalent linearization for systems driven by Levy white noise. Probab. Eng. Mech. **15**, 185–190, 2000.

35. Gurley, K., Kareem, A., and Tognarelli, M.: Simulation of a class of non-normal random process. Int. J. Non-Linear Mech. **31**, 601–617, 1996.

36. Hou, Z., Noori, M., Wang, Y., and Duval, L.: Dynamic behavior of Duffing oscillators under a disordered periodic excitation. In: B. Spencer and E. Johnson (eds.), Stochastic Structural Dynamics, pp. 93–98. Balkema, Rotterdam, 1999.

37. Indira, R., et al.: Diffusion in a bistable potential: A comparartive study of different methods of solution. J. Stat. Phys. **33**, 181–194, 1983.

38. Isserlis, L.: On a formula for the product-moment coefficient in any number of variables. Biometrica **12**, 134–139, 1918.
39. Iwan, W. and Mason, A.: Equivalent linearization for systems subjected to non-stationary random excitation. Int. J. Non-Linear Mech. **15**, 71–82, 1980.
40. Iyengar, R.: Response of nonlinear systems to narrow band excitation. Struct. Safety **6**, 177–185, 1989.
41. Iyengar, R. N. and Jaiswal, O. R.: A new model for non-Gaussian random excitations. Probab. Eng. Mech. **8**, 281–287, 1993.
42. Jahedi, A. and Ahmadi, G.: Application of Wiener-Hermite expansion to nonstationary random vibration of a Duffing oscillator. Trans. ASME J. Appl. Mech. **50**, 436–442, 1987.
43. Jamshidi, M.: An overview on the solutions of the algebraic matrix Riccati equation and related problems. Large Scale Syst. **1**, 167–192, 1980.
44. Kazakov, I. E.: An approximate method for the statistical investigation for non-linear systems. Trudy VVIA im Prof. N. E. Zhukovskogo **394**, 1–52, 1954 (in Russian).
45. Kazakov, I. E.: Approximate probabilistic analysis of the accuracy of the operation of essentially nonlinear systems. Avtomatika i Telemekhanika **17**, 423–450, 1956.
46. Kazakov, I. E.: Statistical analysis of systems with multi-dimensional nonlinearities. Avtomatika i Telemekhanika **26**, 458–464, 1965.
47. Kazakov, I. E.: An extension of the method of statistical linearization. Avtomatika i Telemekhanika **59**, 220–224, 1998.
48. Kazakov, I.: Statistical Theory of Control Systems in State Space. Nauka, Moskwa, 1975 (in Russian).
49. Kazakov, I. and Dostupov, B.: Statistical Dynamic of Nonlinear Automatic Systems. Fizmatgiz, Moskwa, 1962 (in Russian).
50. Kimura, K. and Sakata, M.: Non-stationary responses of a non-symmetric nonlinear system subjected to a wide class of random excitation. J. Sound Vib. **76**, 261–272, 1981.
51. Lee, J.: Improving the equivalent linearization method for stochastic Duffing oscillators. J. Sound Vib. **186**, 846–855, 1995.
52. Legras, J.: Méthodes et Techniques de l'analyse numérique. Dudon, Paris, 1971.
53. Lyon, R., Heckl, M., and Hazlegrove, C.: Narrow band excitation of the hardspring oscillator. J. Acoust. Soc. Am. **33**, 1404–1411, 1961.
54. Masri, S., Smyth, A., and Traina, M.: Probabilistic representation and transmission of nonstationary processes in multi-degree-of-freedom systems. Trans. ASME J. Appl. Mech. **65**, 398–409, 1998.
55. Mel'ts, I., Pukhova, T., and Uskov, G.: Multidimensional statistical linearization of functions containing factors of the power, exponential and trigonometric types and also $\delta$ functions. Avtomatika i Telemekhanika **28**, 1871–1880, 1967.
56. Morosanov, J.: Practical methods for the calculation of the linearization factors for arbitrary nonlinearities. Avtomatika i Telemekhanika **29**, 81–86, 1968.
57. Naess, A., Galeazzi, F., and Dogliani, M.: Extreme response predictions of nonlinear compliant offshore structures by stochastic linearization. Appl. Ocean Res. **14**, 71–81, 1992.
58. Naumov, B.: The Theory of Nonlinear Automatic Control Systems. Frequency Methods. Nauka, Moscow, 1972 (in Russian).

59. Orabi, I. and Ahmadi, G.: A functional series expansion method for response analysis of nonlinear random systems subjected to random excitations. Int. J. Nonlinear Mech. **22**, 451–465, 1987.

60. Orabi, I. and Ahmadi, G.: An iterative method for nonstationary response analysis of nonlinear random systems. J. Sound Vib. **119**, 145–157, 1987.

61. Orabi, I. and Ahmadi, G.: Nonstationary response analysis of a Duffing oscillator by the Wiener-Hermite expansion method. Trans. ASME J. Appl. Mech. **54**, 434–440, 1987.

62. Pervozwanskii, A.: Random Processes in Nonlinear Automatic Control Systems. Fizmatgiz, Moskwa, 1962 (in Russian).

63. Pradlwarter, H.: Nonlinear stochastic response distributions by local statistical linearization. Int. J. Non-Linear Mech. **36**, 1135–1151, 2001.

64. Priestly, M.: Spectral Analysis and Time Series, 5th edn. Academic Press, New York, 1981.

65. Pugachev, V. and Sinitsyn, I.: Stochastic Differential Systems. Wiley, Chichester, 1987.

66. Pugachev, V.: Theory of Random Functions. Gostkhizdat, Moskwa, 1957 (in Russian).

67. Ricciardi, G.: A non-Gaussian stochastic linearization method. Probab. Eng. Mech. **22**, 1–11, 2007.

68. Roberts, J. and Spanos, P.: Random Vibration and Statistical Linearization. John Wiley and Sons, Chichester, 1990.

69. Romanov, V. and Khuberyan, B. K.: Approximate moment dynamics method using statistical linearization. Awtomatika i Telemechanika **37**, 29–35, 1978.

70. Sakata, M. and Kimura, K.: Calculation of the non-stationary mean square response of a non-linear system subjected to non-white excitation. J. Sound Vib. **73**, 333–343, 1980.

71. Samorodnitsky, G. and Taqqu, M.: Stable non-Gaussian Random Processes, Stochastic Models with Infinite Variance. Chapman and Hall, London, 1994.

72. Sawargai, Y., Sugai, N., and Sunahara, Y.: Statistical Studies of Nonlinear Control Systems. Nippon Printing and Publishing Co., Osaka, 1962.

73. Sinitsyn, I.: Methods of statistical linearization (survey). Awtomatika i Telemechanika **35**, 765–776, 1974.

74. Smyth, A. and Masri, S.: Nonstationary response of nonlinear systems using equivalent linearization with a compact analytical form of the excitation process. Probab. Eng. Mech. **17**, 97–108, 2002.

75. Smyth, A. and Masri, S.: The robustness of an efficient probabilistic data-based tool for simulating the nonstationary response of nonlinear systems. Int. J. Non-Linear Mech. **39**, 1453–1461, 2004.

76. Sobiechowski, C.: Statistical linearization of dynamical systems under parametric delta-correlated excitation. Z. Angew. Math. Mech. **79**, 315–316, 1999 S2.

77. Sobiechowski, C. and Socha, L.: Statistical linearization of the Duffing oscillator under non-Gaussian external excitation. In: P. D. Spanos (ed.), Proc. Int. Conf. on Computational Stochastic Mechanics, pp. 125–133. Balkema, Rotterdam, 1998.

78. Sobiechowski, C. and Socha, L.: Statistical linearization of the Duffing oscillator under non-Gaussian external excitation. J. Sound Vib. **231**, 19–35, 2000.

79. Socha, L.: Equivalent linearization for dynamical systems excited by colored noise. Z. Angew. Math. Mech. **70**, 738–740, 1990.

80. Socha, L.: Moment Equations in Stochastic Dynamic Systems. PWN, Warszawa, 1993 (in Polish).
81. Socha, L.: Statistical and equivalent linearization methods with probability density criteria. J. Theor. Appl. Mech. **37**, 369–382, 1999.
82. Socha, L.: Probability density statistical and equivalent linearization methods. Int. J. Syst. Sci. **33**, 107–127, 2002.
83. Spanos, P.: Numerical simulations of Van der Pol oscillator. Comput. Math. Appl. **6**, 135–145, 1980.
84. Steinwolf, A., Ferguson, N., and White, R.: Variations in steepness of probability density function of beam random vibration. Eur. J. Mech. A/Solids **19**, 319–341, 2000.
85. Steinwolf, A. and Ibrahim, R.: Numerical and experimental studies of linear systems subjected to non-Gaussian random excitations. Probab. Eng. Mech. **14**, 289–299, 1999.
86. Traina, M., R.K., M., and Masri, S.: Orthogonal decomposition and transmission of nonstationary random process. Probab. Eng. Mech. **1**, 136–149, 1986.
87. Tylikowski, A. and Marowski, W.: Vibration of a non-linear single degree of freedom system due to Poissonian impulse excitation. Int. J. Non-Linear Mech. **21**, 229–238, 1986.
88. Valsakumar, M., Murthy, K.P., and Ananthakrishna, G.: On the linearization of nonlinear langevin-type stochastic differential equations. J. Stat. Phys. **30**, 617–631, 1983.
89. Wang, G. and Dai, M.: Equivalent linearization method based on energy-to cth -power difference criterion in nonlinear stochastic vibration analysis of multi-degree-of-freedom systems. Appl. Math. Mech. **22**, 947–955, 2001 (English edn).
90. Wicher, J.: On the coefficients of statistical linearization in E. D. Zaydenberg's method. Nonlinear Vib. Probl. **12**, 147–154, 1971.
91. Zaydenberg, E. D.: A third mehod for the statistical linearization of a class of nonlinear differential equations. Avtomatika i Telemekhanika **25**, 195–200, 1964.
92. Zhang, R.: Work/energy-based stochastic equivalent linearization with optimized power. J. Sound Vib. **230**, 468–475, 2000.
93. Zhang, R., Elishakoff, I., and Shinozuka, M.: Analysis of nonlinear sliding structures by modified stochastic linearization methods. Nonlinear Dyn. **5**, 299–312, 1994.
94. Zhang, X., Elishakoff, I., and Zhang, R.: A new stochastic linearization method based on minimum mean-square deviation of potential energies. In: Y. Lin and I. Elishakoff (eds.), Stochastic Structural Dynamics – New Theoretical Developments, pp. 327–338. Springer Verlag, Berlin, 1991.
95. Zhang, X. and Zhang, R.: Energy-based stochastic equivalent linearization with optimized power. In: B. Spencer and E. Johnson (eds.), Stochastic Structural Dynamics, pp. 113–117. Balkema, Rotterdam, 1999.
96. Zhang, X., Zhang, R., and Xu, Y.: Analysis on control of flow-induced vibration by tuned liquid damper with crossed tube-like containers. J. Wind Eng. Ind. Aerodyn. **50**, 351–360, 1993.

# 6

# Equivalent Linearization of Stochastic Dynamic Systems Under External Excitation

## 6.1 Introduction

As was mentioned in Chap. 1 a different philosophy of the replacement of a nonlinear dynamic system under random excitation by an equivalent linear dynamic system under the same excitation has been first considered by Caughey who called his approach *equivalent linearization* (EL). This approach was generalized to multi–degree–of–freedom systems by Foster [49] and next by Atalik and Utku [6]. Socha and Pawleta [97] have observed that in many papers and chapters of books, the derivation of linearization coefficients by the differentiation of the mean-square criterion was incorrect. It was also independently stated by Elishakoff and Colojani [45]. This opinion has initiated a sharp discussion in the literature regarding the existence of the error and the "fathers" of this error, see, for instance, [39, 35, 79, 93, 94]. A short report of this discussion is given in Sect. 6.7.

We begin our considerations in this chapter with presentation of the differences between statistical and equivalent linearization given first in [99].

## 6.2 Moment Criteria

We consider again the nonlinear stochastic dynamic system with external excitation in the form of Gaussian white noises described by the Ito vector stochastic differential equation

$$d\mathbf{x} = \mathbf{\Phi}(\mathbf{x}, t)dt + \sum_{k=1}^{M} \mathbf{G}_{k0}(t)d\xi_k, \quad \mathbf{x}(t_0) = \mathbf{x_0}, \tag{6.1}$$

where $\mathbf{x} = [x_1, \ldots, x_n]^T$ is the vector state, $\mathbf{\Phi} = [\Phi_1, \ldots, \Phi_n]^T$ is a nonlinear vector function, $\mathbf{G}_{k0} = [\sigma_{k0}^1, \ldots, \sigma_{k0}^n]^T$, $k = 1, \ldots, M$, are deterministic vectors of intensities of noise, $\xi_k$, $k = 1, \ldots, M$, are independent standard Wiener

L. Socha: *Equivalent Linearization of Stochastic Dynamic Systems Under External Excitation*, Lect. Notes Phys. **730**, 147–210 (2008)
DOI 10.1007/978-3-540-72997-6_6

processes; the initial condition $\mathbf{x}_0$ is a vector random variable independent of $\xi_k(t)$, $k = 1, \ldots, M$. We assume that the solution of (6.1) exists.

The linearized system has the form

$$dx = [\mathbf{A}(t)\mathbf{x} + \mathbf{C}(t)]dt + \sum_{k=1}^{M} \mathbf{G}_{k0}(t)d\xi_k, \quad \mathbf{x}(t_0) = \mathbf{x_0} , \qquad (6.2)$$

where $\mathbf{A}(t) = [a_{ij}(t)]$, $i, j = 1, \ldots, n$ and $\mathbf{C}(t) = [C_1(t), \ldots, C_n(t)]^T$ are time-dependent matrix and vector of linearization coefficients, respectively.

### 6.2.1 SPEC Linearization Alternative

As in the case of statistical linearization we consider two basic moment criteria.

**Criterion 1-EL.** Equality of mean values and second-order moments of the response of nonlinear and linearized systems [64] (Kozin Criterion)

$$E[\mathbf{x}_N(t)] = E[\mathbf{x}_L(t)] , \qquad (6.3)$$

$$E[\mathbf{x}_N(t)\mathbf{x}_N(t)^T] = E[\mathbf{x}_L(t)\mathbf{x}_L(t)^T] , \qquad (6.4)$$

where $\mathbf{x}_N(t)$ and $\mathbf{x}_L(t)$ are solutions of (6.1) and (6.2), respectively.

Then we obtain a system of $n^2 + n$ equations with $n^2 + n$ unknowns $a_{ij}$ and $C_i$, $i, j = 1, \ldots, n$, where the linearization coefficients depend on moments of solutions of nonlinear systems (6.1). As in statistical linearization, moments of solutions of a nonlinear system are approximated by the corresponding moments of linearized system (6.2).

**Criterion 2-EL.** The mean-square error of approximation.

Consider the mean-square error of approximation of the function $y = \Phi(\mathbf{x}, t)$ by its linearized form (6.2), i.e.,

$$\delta_{EL} = E\left[(\Phi(\mathbf{x}, t) - \mathbf{A}(t)\mathbf{x} - \mathbf{C}(t))^2\right] , \qquad (6.5)$$

where the expectation appearing in equality (6.5) can be generated by the solution of a nonlinear or a linearized system. We denote them by $E_N[.]$ and $E_L[.]$, respectively.

The necessary conditions for minimum of Criterion $2 - EL$ have the form

$$\frac{\partial \delta_{EL}}{\partial a_{ij}} = 0, \frac{\partial \delta_{EL}}{\partial C_i} = 0, \quad i, j = 1, \ldots, n . \qquad (6.6)$$

The different definitions of averaging operations in Criterion $2 - EL$ (6.5) also give different necessary conditions for minimum of criterion. Shortly, one can say that if the averaging operation is with respect to the measure defining the solution of a nonlinear system or with respect to a measure independent of a linearized system, then we obtain the same linearization coefficients as in the case of statistical linearization, otherwise the linearization coefficients are different. We illustrate this idea on two examples of dynamic systems.

*Example 6.1.* Consider a scalar nonlinear Ito stochastic differential equation

$$dx = -f(x)dt + qd\xi(t), \quad x(t_0) = x_0 , \qquad (6.7)$$

where $f(x)$ is a nonlinear function, $f(0) = 0$, $q > 0$ is a constant, $\xi(t)$ is the standard Wiener process.

The equivalent linearized equation has the form

$$dx = -kxdt + qd\xi(t), \quad x(t_0) = x_0 , \qquad (6.8)$$

where $k$ is a linearization coefficient.

The stationary solutions of (6.7) and (6.8) are characterized by the corresponding stationary probability density functions

$$g_N(x) = \frac{1}{c_N} \exp\left\{ -\frac{1}{q^2} \int_0^x f(s)ds \right\} , \qquad (6.9)$$

$$g_L(x) = \frac{1}{c_L \sigma_L} \exp\left\{ -\frac{x^2}{2\sigma_L^2} \right\} , \qquad (6.10)$$

$$\sigma_L^2 = \frac{q^2}{2k} , \qquad (6.11)$$

where $c_N$ and $c_L$ are normalized constants, $\sigma_L^2$ is the variance of the stationary solution of linearized equation (6.8). We note that the mean values of the stationary solutions for both equations are equal to zero, $E[x] = 0$.

First we determine the linearization coefficients by statistical linearization. An equivalent linear element for a nonlinear element $y = f(\eta)$ has the form $y = k\eta$, where $\eta$ is a stationary zero-mean input process with $E_\eta[\eta] = 0$. Then from Criterion $1 - SL$ the following equality is obtained

$$E_\eta[f^2(\eta)] = k^2 E_\eta[\eta^2] , \qquad (6.12)$$

where $E_\eta[.]$ is the averaging operation with respect to the measure defined by the process $\eta$ (independent of the linearization coefficient). Hence

$$k_1 = \sqrt{\frac{E_\eta[f^2(\eta)]}{E_\eta[\eta^2]}} . \qquad (6.13)$$

In the case of the mean-square criterion (Criterion $2 - SL$) we have

$$I = E_\eta[(f(\eta) - k\eta)^2] . \qquad (6.14)$$

Then from the necessary conditions for the minimum of Criterion $2 - SL$ it follows that the linearization coefficient $k_2$ has the form

$$k_2 = \frac{E_\eta[f(\eta)\eta]}{E_\eta[\eta^2]} . \qquad (6.15)$$

We show that if the averaging operation $E_\eta[.]$ with respect to the input process $\eta$ in Criterion 2 − $SL$ is replaced by the averaging operation with respect to the output process from linearized element $y = k\eta$, and $\eta$ is assumed to be a Gaussian process, then the linearization coefficient is also defined by algebraic relation (6.15).

We denote the probability density functions corresponding to stationary solutions of the stationary input process $\eta(t)$ and the stationary output process of linearized element $y = k\eta$ by

$$g_\eta(x) = \frac{1}{c_\eta \sigma_\eta} \exp\left\{-\frac{x^2}{2\sigma_\eta^2}\right\}, \tag{6.16}$$

$$g_y(y) = g_{k\eta}(y) = \frac{1}{c_y \sigma_y} \exp\left\{-\frac{y^2}{2\sigma_y^2}\right\}, \tag{6.17}$$

where $c_\eta$ and $c_y$ are normalized constants, $\sigma_\eta^2$ and $\sigma_y^2$ are the variances of the input process $\eta(t)$ and the output process $y(t)$, respectively.

$$\sigma_\eta^2 = \frac{q^2}{2}, \qquad \sigma_y^2 = k^2 \sigma_\eta^2 = k^2 \frac{q^2}{2}. \tag{6.18}$$

The corresponding mean-square criteria have the form

$$I_\eta = \int_{-\infty}^{+\infty} (f(x) - kx)^2 \frac{1}{c_\eta \sigma_\eta} \exp\left\{-\frac{x^2}{2\sigma_\eta^2}\right\} dx, \tag{6.19}$$

$$I_y = \int_{-\infty}^{+\infty} \left[f\left(\frac{y}{k}\right) - y\right]^2 \frac{1}{c_\eta k \sigma_\eta} \exp\left\{-\frac{y^2}{2k^2\sigma_\eta^2}\right\} dy. \tag{6.20}$$

Substituting formally the variable $y = kx$ into equality (6.20) we obtain

$$I_y = \int_{-\infty}^{+\infty} [f(x) - kx]^2 \frac{1}{c_\eta k \sigma_x} \exp\left\{-\frac{k^2 x^2}{2k^2\sigma_x^2}\right\} k dx \tag{6.21}$$

$$= \int_{-\infty}^{+\infty} [f(x) - kx]^2 \frac{1}{c_\eta \sigma_x} \exp\left\{-\frac{x^2}{2\sigma_x^2}\right\} dx.$$

By differentiating $I_\eta$ and $I_y$ with respect to the coefficient $k$ and equating the corresponding derivatives to zero, we find

$$\frac{\partial I_\eta}{\partial k} = \frac{\partial I_y}{\partial k} = 2 \int_{-\infty}^{+\infty} [f(x) - kx](-x) \frac{1}{c_\eta \sigma_\eta} \exp\left\{-\frac{x^2}{2\sigma_\eta^2}\right\} dx = 0. \tag{6.22}$$

From (6.22) it follows that in both cases the linearization coefficient $k$ has the form (6.15). Furthermore one can show that this property is also satisfied for any convex combination of stationary Gaussian input and output processes.

Now, we present the application of equivalent linearization to the determination of the same response characteristics. In the case of Criterion $1 - EL$, we obtain the following relation

$$\int_{-\infty}^{+\infty} x^2 g_N(x) dx = \int_{-\infty}^{+\infty} x^2 g_L(x) dx = \frac{q^2}{2k} , \qquad (6.23)$$

where $g_N$ and $g_L$ are probability density functions of stationary response of nonlinear (6.9) and linearized equation (6.10), respectively. We note that for both equations the mean values of stationary solutions are equal to zero, $E[x] = 0$. Hence we find

$$k_{1EL} = \frac{q^2}{2q^2 E[x_N^2]} . \qquad (6.24)$$

In the case of the mean-square criterion (Criterion $2 - EL$) we have

$$I_{[.]} = E_{[.]}[(f(\eta) - k\eta)^2] , \qquad (6.25)$$

where the averaging operation $E_{[.]}$ can be defined by probability density function $g_N(x)$ as well as by $g_L(x)$. Hence we find

$$I_N = \int_{-\infty}^{+\infty} [f(x) - kx]^2 \frac{1}{c_N} \exp\left\{ -\frac{1}{q^2} \int_0^x f(s) ds \right\} dx , \qquad (6.26)$$

$$I_L = \int_{-\infty}^{+\infty} \left[ f\left(\frac{y}{k}\right) - y \right]^2 \frac{1}{c_L \sigma_L} \exp\left\{ -\frac{y^2}{2\sigma_L^2} \right\} dy , \qquad (6.27)$$

where $c_N$ and $c_L$ are the corresponding normalized constants, $\sigma_L$ is determined by (6.11). After transformation we obtain

$$I_L(k) = \int_{-\infty}^{+\infty} [f(x) - kx]^2 \frac{1}{c_L} \frac{\sqrt{2k}}{q} \exp\left\{ -\frac{k^2 x^2}{2\frac{q^2}{2k}} \right\} k dx \qquad (6.28)$$

$$= \int_{-\infty}^{+\infty} [f(x) - kx]^2 \frac{1}{c_L} \frac{1}{\sqrt{\frac{q^2}{2k^3}}} \exp\left\{ -\frac{x^2}{2\frac{q^2}{2k^3}} \right\} dx .$$

The necessary conditions for minimum of both criteria have the form

$$\frac{\partial I_N}{\partial k} = \frac{\partial I_y}{\partial k} = 2 \int_{-\infty}^{+\infty} [f(x) - kx](-x) \frac{1}{c_N} \exp\left\{ -\frac{1}{q^2} \int_0^x f(s) ds \right\} dx = 0 ,$$
$$(6.29)$$

$$\frac{\partial I_{L(k)}}{\partial k} = 2 \int_{-\infty}^{+\infty} [f(x) - kx](-x)$$

$$\times \frac{1}{c_L} \frac{\sqrt{2k^3}}{q} \exp\left\{ -\frac{x^2}{2\frac{q^2}{2k^3}} \right\} dx + 3 \int_{-\infty}^{+\infty} [f(x) - kx]^2$$

$$\times \left( \frac{1}{2k} - \frac{x^2 k^2}{q^2} \right) \frac{1}{c_L} \frac{\sqrt{2k^3}}{q} \exp\left\{ -\frac{x^2}{2\frac{q^2}{2k^3}} \right\} dx = 0 . \qquad (6.30)$$

In a particular case, if we assume that the function $f(x)$ has the form

$$f(x) = \lambda_0^2 x + \varepsilon x^3 , \qquad (6.31)$$

where $\lambda_0^2$ and $\varepsilon$ are positive constants, then the linearization coefficient obtained by statistical linearization and by equivalent linearization with averaging operation generated by the stationary solution of the nonlinear equation is the same and is calculated from relation (6.29), i.e.,

$$k = k_2 = \lambda_0^2 + \varepsilon \frac{E_N[x^4]}{E_N[x^2]} . \qquad (6.32)$$

In the case of equivalent linearization with averaging operation generated by the solution of the linearized equation, the linearization coefficient $k = k_{2L}$ is calculated from equality (6.30). After integration one can obtain the following algebraic equation for the linearization coefficient $k$

$$4k^8 - 16\lambda_0^2 k^7 + 12\lambda_0^4 k^6 - 60\varepsilon q^2 k^4 + 72\varepsilon\lambda_0^2 q^2 k^3 + 135\varepsilon^2 q^4 = 0 . \qquad (6.33)$$

The solution of (6.33) can be found only numerically.

Another mean-square criterion for equivalent linearization can be considered in the following form:

$$I_{L_1}(k) = \int_{-\infty}^{+\infty} [f(x) - kx]^2 \frac{1}{c_{L_1}} \frac{1}{\sqrt{\frac{q^2}{2k}}} \exp\left\{ -\frac{x^2}{2\frac{q^2}{2k}} \right\} dx , \qquad (6.34)$$

where $c_{L_1}$ is a normalized constant and $\sigma_x^2 = \sigma_L^2 = \frac{q^2}{2k}$ is the variance of the stationary solution of (6.8).

The differentiation of criterion (6.34) with respect to the linearization coefficient $k$ yields the necessary condition for minimum of $I_{L_1}(k)$

$$\frac{\partial I_{L_1}(k)}{\partial k} = 2 \int_{-\infty}^{+\infty} [f(x) - kx] (-x)$$

$$\times \frac{1}{c_{L_1}} \frac{\sqrt{2k}}{q} \exp\left\{ -\frac{x^2}{2\frac{q^2}{2k}} \right\} dx + \int_{-\infty}^{+\infty} [f(x) - kx]^2 \left( \frac{1}{\sqrt{2k}} - \frac{x^2}{q^2} \right)$$

$$\times \frac{1}{c_{L_1}} \frac{\sqrt{2k}}{q} \exp\left\{ -\frac{x^2}{2\frac{q^2}{2k}} \right\} dx = 0 . \qquad (6.35)$$

Assuming that the nonlinear function $f(x)$ is defined by (6.31), one can show by similar calculations to the previous case that the linearization coefficient $k$ satisfies the following algebraic equation:

$$(2\sqrt{2}+4)k^4 - (8-4\sqrt{2})\lambda_0^2 k^3 + [(4-6\sqrt{2})\lambda_0^4 + (18\sqrt{2}-12)\varepsilon q^2]k^2$$

$$- (30\sqrt{2}-12)\varepsilon\lambda_0^2 q^2 k + (105\frac{\sqrt{2}}{2}-15)\varepsilon^2 q^4 = 0. \qquad (6.36)$$

The difference between statistical and equivalent linearization is even more visible in the case of two-dimensional nonlinear dynamic system that will be shown in the next example.

*Example 6.2.* Consider a nonlinear oscillator described by the Ito vector stochastic differential equation

$$d\mathbf{x} = \mathbf{F}(\mathbf{x})dt + \mathbf{G}d\xi(t), \quad \mathbf{x}(t_0) = \mathbf{x_0}, \qquad (6.37)$$

where

$$\mathbf{x} = \begin{bmatrix} x_1 \\ x_2 \end{bmatrix}, \quad \mathbf{F}(\mathbf{x}) = \begin{bmatrix} x_2 \\ -2h\lambda_0 x_2 - \lambda_0^2 x_1 - \phi(x_1) \end{bmatrix},$$

$$\mathbf{G} = \begin{bmatrix} 0 \\ q \end{bmatrix}, \quad \mathbf{x_0} = \begin{bmatrix} x_{10} \\ x_{20} \end{bmatrix}, \qquad (6.38)$$

$\phi(x)$ is a nonlinear function, $\phi(0) = 0$, $\lambda_0^2$, $h$ and $q$ are positive constants, $\xi(t)$ is the standard Wiener process, the initial conditions $x_{10}$ and $x_{20}$ are random variables independent of $\xi(t)$.

The linearized vector stochastic differential equation has the form

$$d\mathbf{x} = \mathbf{B}\mathbf{x}dt + \mathbf{G}d\xi(t), \qquad (6.39)$$

where

$$\mathbf{B} = \begin{bmatrix} 0 & 1 \\ -k & -2h\lambda_0 \end{bmatrix}. \qquad (6.40)$$

First, we determine the linearization coefficient by statistical linearization. We limit our considerations to the mean-square criterion (Criterion $2 - SL$). For nonlinear element $y = \phi(\eta)$ the equivalent linear element has the form $y = k\eta$, where $\eta$ is a stationary zero mean input process $E[\eta] = 0$. Then Criterion $2 - SL$ has the form

$$I_1 = E_{n1}[(\phi(\eta) - k\eta)^2], \qquad (6.41)$$

where the averaging operation $E_{n1}$ can be defined by the probability density function $g_{n1}(x)$ of a one-dimensional non-Gaussian random variable, see Fig. 6.1, i.e.

$$E_{n1}[.] = \int_{-\infty}^{+\infty} [.] g_{n1}(x_1)dx_1. \qquad (6.42)$$

Hence, finding the necessary conditions for minimum of Criterion $2 - SL$ we calculate the linearization coefficient

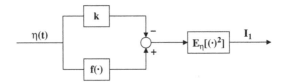

**Fig. 6.1.** Schematic of statistical linearization

$$k_{2SL} = \lambda_0^2 + \frac{E_{n1}[\phi(x_1)x_1]}{E_{n1}[x_1^2]} .$$ (6.43)

In the case of equivalent linearization, the mean-square criterion and the linearization coefficient have the same form as in the case of statistical linearization, i.e., defined by relations (6.41) and (6.43), respectively, where the averaging operation is defined by the probability density function $g_{N1}(x)$ of the two-dimensional random variable $\mathbf{x} = [x_1, x_2]^T$ being the stationary solution of nonlinear dynamic system (6.37). The averaging operation has the form

$$E_{N2}[.] = \int_{-\infty}^{+\infty} \int_{-\infty}^{+\infty} [.] g_{N2}(x_1, x_2) dx_1 dx_2 ,$$ (6.44)

where

$$g_{N2}(x_1, x_2) = \frac{1}{c_{N2}} \exp\left\{ -\frac{2h\lambda_0}{q^2} \left( \lambda_0^2 x_1^2 + \int_0^{x_1} \phi(s)ds + x_2^2 \right) \right\} ,$$ (6.45)

$c_{N2}$ is a normalized constant.

Hence, finding the necessary conditions for minimum of Criterion $2 - EL$ we calculate the linearization coefficient

$$k_{2EL} = \lambda_0^2 + \frac{E_{N2}[\phi(x_1)x_1]}{E_{N2}[x_1^2]} .$$ (6.46)

To obtain approximate moments of stationary solutions one should in both methods derive moment equations and apply in iterative procedures, where the averaging operations $E_{n1}[.]$ and $E_{N2}[.]$ with respect to non-Gaussian probability density functions defined by $g_{n1}$ and $g_{N2}$, respectively, are replaced by the averaging operation with respect to the corresponding Gaussian probability density functions, $E_{l1}[.]$ and $E_{L2}[.]$.

In the case of statistical linearization $E_{n1}[.]$ is replaced by $E_{l1}[.]$ defined as

$$E_{l1}[.] = \int_{-\infty}^{+\infty} [.] g_{l1}(x_1) dx_1 ,$$ (6.47)

where the probability density function of Gaussian random variable has the form

$$g_{l1}(x_1) = \frac{1}{\sqrt{2\pi}\sigma_{x_1}} \exp\left\{ -\frac{x_1^2}{2\sigma_{x_1}^2} \right\} .$$ (6.48)

In the case of equivalent linearization, the averaging operation $E_{N2}$ is replaced by operation $E_{L2}$ defined by the stationary solution of the linearized system as

$$E_{L2}[.] = \int_{-\infty}^{+\infty} \int_{-\infty}^{+\infty} [.] g_{L1}(x_1, x_2, k) dx_1 dx_2 , \qquad (6.49)$$

where the probability density function of the two-dimensional Gaussian random variable has the form

$$g_{L2}(x_1, x_2, k) = \frac{1}{c_{L2}} \frac{4h\lambda_0 \sqrt{k}}{q^2} \exp \left\{ -\frac{2h\lambda_0}{q^2}(kx_1^2 + x_2^2) \right\} . \qquad (6.50)$$

$c_{L2}$ is a normalized constant.

See Figs. 6.2 and 6.3

Criterion $I_2$ appearing in Figs. 6.2 and 6.3 has the form (6.41) with corresponding averaging operations $E_N$ and $E_L$ and the nonlinear function $\Phi(.) = \phi(.)$. The moment equations for stationary solutions of linearized equation (6.39) are

$$2E_{L2}[x_1 x_2] = 0 , \qquad (6.51)$$

$$E_{L2}[x_2^2] - kE_{L2}[x_1^2] - 2h\lambda_0 E_{L2}[x_1 x_2] = 0 , \qquad (6.52)$$

$$-2kE_{L2}[x_1 x_2] - 4h\lambda_0 E_{L2}[x_2^2] + q^2 = 0 . \qquad (6.53)$$

In the case of statistical linearization the iterative procedure has the form
**Procedure $EXAMPLE_{L1}$**

(1) Substitute $k = \lambda_0^2$ in moment equations (6.51–6.53).
(2) Replace the moments in equality (6.43) by the corresponding solutions of (6.51–6.53) and determine new linearization coefficient $k$ using Gaussian closure.
(3) Substitute the coefficient $k$ into moment equations (6.51–6.53) and solve them.
(4) Repeat steps (2) and (3) until convergence.

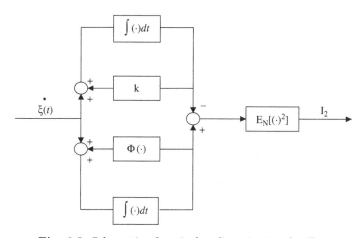

**Fig. 6.2.** Schematic of equivalent linearization for $E_{N2}$

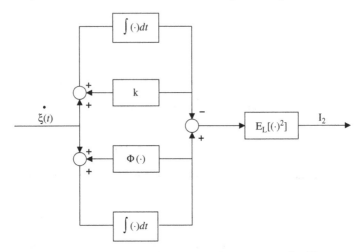

**Fig. 6.3.** Schematic of equivalent linearization for $E_{L2}$

In the case of equivalent linearization the iterative procedure is the same except step (2), that should be replaced by the following one.

(2') Replace moments in equality (6.46) by the corresponding moments being the solutions of moment equations (6.51–6.53) and determine the new linearization coefficient $k$ using Gaussian closure.

At the end we consider the case when the averaging operation in mean-square criterion (6.41) is replaced by $E_{L2}[.]$ defined by (6.49) and (6.50), i.e., the mean-square criterion has the form

$$
I_{L2} = E_{L2}[(\phi\left(\frac{y}{k}\right) - y)^2]
$$
$$
= \int_{-\infty}^{+\infty} \int_{-\infty}^{+\infty} \left[(\phi\left(\frac{y}{k}\right) - y)^2\right]
$$
$$
\times \frac{1}{c_{L2}} \frac{4h\lambda_0\sqrt{k}}{q^2} \exp\left\{-\frac{2h\lambda_0}{q^2}(ky^2 + x_2^2)\right\} dy\, dx_2, \qquad (6.54)
$$

where $c_{L2}$ is a normalized constant. Substituting formally $y = kx_1$ we obtain

$$
I_{L2} = \int_{-\infty}^{+\infty} \int_{-\infty}^{+\infty} \left[(\phi(x_1) - kx_1)^2\right]
$$
$$
\times \frac{1}{c_{L2}} \frac{4h\lambda_0\sqrt{k^3}}{q^2} \exp\left\{-\frac{2h\lambda_0}{q^2}(k^3x_1^2 + x_2^2)\right\} dx_1\, dx_2 \ . \qquad (6.55)
$$

The necessary condition for minimum of functional $I_{L2}$ has the form

$$
\frac{\partial I_{L2}}{\partial k} = 2 \int_{-\infty}^{+\infty} \int_{-\infty}^{+\infty} \left[(\phi(x_1) - kx_1)\right](-x_1)
$$

$$\times \frac{1}{c_{L2}} \frac{4h\lambda_0\sqrt{k^3}}{q^2} \exp\left\{-\frac{2h\lambda_0}{q^2}(k^3x_1^2 + x_2^2)\right\} dx_1\, dx_2$$

$$+ \int_{-\infty}^{+\infty}\int_{-\infty}^{+\infty} [(\phi(x_1) - kx_1)^2]$$

$$\times \frac{\partial}{\partial k}\left(\frac{1}{c_{L2}}\frac{4h\lambda_0\sqrt{k^3}}{q^2}\exp\left\{-\frac{2h\lambda_0}{q^2}(k^3x_1^2 + x_2^2)\right\}\right) dx_1\, dx_2 = 0.$$

$$(6.56)$$

Since the integrating operation with respect to $x_2$ does not depend on linearization coefficient $k$, relation (6.56) reduces to the following form:

$$\frac{\partial I_{L2}}{\partial k} = 2\bar{I}_L \int_{-\infty}^{+\infty}[(\phi(x_1) - kx_1)]\,(-x_1)k\bar{g}_L(x_1, k)dx_1$$

$$+\bar{I}_L \int_{-\infty}^{+\infty}[(\phi(x_1) - kx_1)^2]\left(\frac{3}{2} - \frac{6h\lambda_0 k^3 x_1^2}{q^2}\right)$$

$$\times \bar{g}_L(x_1, k)dx_1 = 0, \qquad (6.57)$$

where

$$\bar{I}_L = \int_{-\infty}^{+\infty}\frac{1}{c_{L12}}\frac{4h\lambda_0}{q^2}\exp\left\{-\frac{2h\lambda_0}{q^2}x_2^2\right\}dx_2,$$

$$\bar{g}_L(x_1, k) = \frac{\sqrt{k}}{c_{L11}}\exp\left\{-\frac{2h\lambda_0}{q^2}k^3x_1^2\right\}. \qquad (6.58)$$

where $\bar{c}_{L_{11}}$ and $\bar{c}_{L_{12}}$ are normalized constants.

Further simplifications are possible, for instance, for the function $\phi(x_1)$ in the form of the odd-order polynomial of $x_1$ and using properties of Gaussian random variables.

To obtain linearization coefficient $k$ and response statistics, we may use Procedure $EXAMPLE_{L1}$ where step (2) should be replaced by the following one.

(2") Replace moments calculated from condition (6.57) by the corresponding moments being the solutions of moment equations (6.51–6.53) and determine the new linearization coefficient $k$ using Gaussian closure.

We also note that like scalar equation, one can consider another mean-square criterion defined by

$$\bar{I}_{L2} = E_{L2}[(\phi(x) - kx)^2]$$

$$= \int_{-\infty}^{+\infty}\int_{-\infty}^{+\infty}[(\phi(x_1) - kx_1)^2]$$

$$\frac{1}{c_{L2}}\frac{4h\lambda_0\sqrt{k}}{q^2}\exp\left\{-\frac{2h\lambda_0}{q^2}(kx_1^2 + x_2^2)\right\}dx_1\, dx_2, \qquad (6.59)$$

where $c_{L2}$ is a normalized constant. Applying similar procedure as for criterion $I_{L2}$ including a modification of Procedure $EXAMPLE_{L1}$ one can derive an algebraic equation for the linearization coefficient. The numerical calculations show that linearization coefficient and response moments obtained by application of statistical linearization and equivalent linearization with criterion $I_{L2}$ or $\bar{I}_{L2}$ yield different results.

The presented two examples show that the different definitions of averaging operations in the mean-square criterion (6.5) imply different necessary conditions for minimum of criterion (6.5), i.e., conditions (6.6).

In the case when the averaging operation is with respect to the measure defined by the stationary solution of nonlinear system (6.1) or with respect to a measure independent of the stationary solution of linearized system (6.2), then this linearization method is called *standard equivalent linearization.*

In the case when the averaging operation is with respect to the measure defined by the stationary solution of linearized system (6.2), for instance, criteria $I_{L2}$ or $\bar{I}_{L2}$, then this linearization method was called by Crandall [36] *SPEC Alternative* after the first letters in the names of the four authors: Socha, Pawleta, Elishakoff, and Colajanni who in their papers [92, 45, 44, 31, 31, 93] have removed the inconsistency in the derivation of linearization coefficients by many authors in the literature.

### 6.2.2 Higher-Order Moment Criteria

In many linearization approaches, the response moments are not directly taken into account in the considered criteria of linearization. Usually, the response moments are calculated for linearized system and substituted into the linearization coefficients in an iterative procedure. An opposite idea was proposed by Socha [93] who calculated linearization coefficients taking into account the proposed criterion approximate higher-order moments of an original nonlinear system. The considered original and linearized systems are described by the corresponding Ito vector differential equations

$$d\mathbf{x} = \mathbf{\Phi}(\mathbf{x},t)dt + \sum_{k=1}^{M} \mathbf{G}_{k0}(t)d\xi_k, \quad \mathbf{x}(t_0) = \mathbf{x_0} , \qquad (6.60)$$

$$d\mathbf{y} = [\mathbf{A}(t)\mathbf{y} + \mathbf{C}(t)]dt + \sum_{k=1}^{M} \mathbf{G}_{k0}(t)d\xi_k, \quad \mathbf{y}(t_0) = \mathbf{x_0} , \qquad (6.61)$$

where $\mathbf{x} \in R^n$ and $\mathbf{y} \in R^n$ are the vector states of the original nonlinear system and linearized system, respectively; $\mathbf{\Phi}(\mathbf{x},t)$ is a vector nonlinear function, $\mathbf{G}_{k0}(t)$ are vectors of intensities of noises, $\mathbf{A}(t) = [a_{ij}(t)], i, j = 1, \ldots, n$, and $\mathbf{C}(t) = [C_1(t), \ldots, C_n(t)]^T$ are the time-dependent matrix and vector of linearization coefficients, respectively. $\xi_k, k = 1, \ldots, M$, are standard independent Wiener processes; the initial condition $\mathbf{x}_0$ is a vector random variable independent of $\xi_k(t), k = 1, \ldots, M$.

The corresponding moment equations can be obtained in a closed form. In the case of nonlinear system (6.60) by application of the cumulant closure technique

$$\frac{dE[\mathbf{x}^{[p]}]}{dt} = \mathbf{F}_p \left( E[\mathbf{x}(t)], E[\mathbf{x}^{[2]}], \ldots, E[\mathbf{x}^{[p]}] \right) , \qquad (6.62)$$

where $1 \leq p \leq N$ is the multiindex, $\mathbf{x}^{[p]} = x_1^{p_1} x_2^{p_2} \ldots x_n^{p_n}$, $\sum_{i=1}^n p_i = p$ and in the case of linearized system (6.61) by direct application of the Ito formula and the averaging operation

$$\frac{dE[\mathbf{y}^{[p]}]}{dt} = \mathbf{A}_{[p]} E[\mathbf{y}^{[p]}(t)] + \mathbf{C}_{[p]} E[\mathbf{y}^{[p-1]}(t)] + \mathbf{H}_{[p]} E[\mathbf{y}^{[p-2]}(t)]\mathbf{1},$$

$$p = 1, 2, \ldots, N , \qquad (6.63)$$

where $\mathbf{A}_{[p]}, \mathbf{C}_{[p]}$ and $\mathbf{H}_{[p]}$, $p = 1, 2, \ldots, N$, are matrices determined by elements of $\mathbf{A}, \mathbf{C}$, and $\mathbf{G}_{k0}$,

$$\mathbf{1} = \begin{cases} \mathbf{I} \ for \ p \geq 2 \\ 0 \ for \ p = 1 \end{cases} . \qquad (6.64)$$

The objective of moment equivalent linearization is to find the elements of $\mathbf{A}$ and $\mathbf{C}$ that minimize the following criterion

$$I_N = \sum_{p=1}^N \alpha_p \left| E[\mathbf{x}^{[p]}] - E[\mathbf{y}^{[p]}] \right|^2 , \qquad (6.65)$$

where $\alpha_p > 0$ are weighting coefficients. The proposed moment linearization was illustrated by Socha [93] in the Duffing oscillator.

## 6.2.3 Higher-Order Equivalent Linear Systems

In all considered linearization methods in application to a nonlinear dynamic system the equivalent linearized system was a linear oscillator with proper linearization coefficients. It is clear that the corresponding power spectral density functions will have different properties, namely in the power spectral density function of a linear oscillator, response will never appear at higher harmonics. The application of a linearization method to a nonlinear oscillator leads to a single-degree-of-freedom (SDOF) linear system that can oscillate at only one frequency. To improve the power spectral density function of a linearized system, one can increase the degree of freedom of the equivalent system. This idea was developed by Iyengar [56]. We discuss this approach on a class of oscillators with a nonlinear spring, namely for the vector Ito equation

$$d\mathbf{x} = [\mathbf{A}\mathbf{x} + \mathbf{B}(\mathbf{x})]dt + \mathbf{G}d\xi(t) , \qquad (6.66)$$

where

$$\mathbf{x} = \begin{bmatrix} x_1 \\ x_2 \end{bmatrix}, \quad \mathbf{A} = \begin{bmatrix} 0 & x_2 \\ -2\zeta\lambda_0 x_2 & -\lambda_0^2 x_1 \end{bmatrix}, \quad \mathbf{B(x)} = \begin{bmatrix} 0 \\ -\phi(x_1) \end{bmatrix}, \quad \mathbf{G} = \begin{bmatrix} 0 \\ q \end{bmatrix},$$
(6.67)

$\phi(x)$ is a nonlinear function, $\phi(0) = 0$, $\lambda_0^2$, $\zeta$, and $q$ are positive constants, $\xi$ is the standard Wiener process.

Introducing a new variable $z_1$ defined by

$$z_1 = \phi(x_1)$$
(6.68)

the Ito formula gives

$$dz_1 = \frac{\partial\phi(x_1)}{\partial x_1} x_2 dt .$$
(6.69)

Now if we introduce the next variable $z_2$ defined by

$$z_2 = \frac{\partial\phi(x_1)}{\partial x_1} x_2 ,$$
(6.70)

then the Ito formula gives

$$dz_2 = \left[ \frac{\partial^2\phi(x_1)}{\partial x_1^2} x_2^2 - \frac{\partial\phi(x_1)}{\partial x_1} z_1 - 2\zeta\lambda_0 z_2 - \lambda_0^2 \frac{\partial\phi(x_1)}{\partial x_1} x_1 \right] dt + \frac{\partial\phi(x_1)}{\partial x_1} q d\xi(t) .$$
(6.71)

In order to linearize equation (6.71), Iyengar proposed to apply Gaussian closure approximation. In the particular case for the Duffing oscillator, i.e., for $\phi(x_1) = \varepsilon x_1^3$, the extended linearized vector stochastic differential equation has the form

$$dx_1 = x_2 dt ,$$
$$dx_2 = [-2\zeta\lambda_0 x_2 - \lambda_0^2 x_1 - \varepsilon z_1] dt + q d\xi(t) ,$$
$$dz_1 = z_2 dt ,$$
$$dz_2 = [-2\zeta\lambda_0 z_2 - 3\lambda_0^2 k_z z_1 - k_x x_1] dt + 3\sigma_{x_1}^2 d\xi(t),$$
(6.72)

where $k_z$ and $k_x$ are linearization coefficients defined by

$$k_z = 1 + 10\varepsilon\sigma_{x_1}^2, \quad k_x = 6\sigma_{x_2}^2 + 45\varepsilon\lambda_0^2\sigma_{x_1}^4 ,$$
(6.73)

$$\sigma_{x_1}^2 = E[x_1^2], \quad \sigma_{x_2}^2 = E[x_2^2] .$$
(6.74)

Another application of the extension of dimension of equivalent linear system to the determination of response characteristics for externally excited nonlinear system was considered by Benfratello [10]. He used pseudoforce theory proposed by Molnar et al. [69] to replace a scalar nonlinear system

$$\dot{x} = \alpha x + \varepsilon g(x, t) + qw(t) ,$$
(6.75)

by a system of linear equations

$$\begin{aligned}
\dot{x}_0 &= & \alpha x_0 + q w(t) \,, \\
\dot{x}_1 &= & \alpha x_1 + \varepsilon g(x_0, t) + q w(t) \,,
\end{aligned}$$

$$\vdots \tag{6.76}$$

$$\dot{x}_n = \alpha x_n + \varepsilon g(x_{n-1}, t) + q w(t) \,,$$

where $\alpha$, $\varepsilon$, and $q$ are real constant parameters, $g(x, t)$ is a nonlinear function, and $w(t)$ is a Gaussian white noise.

System (6.76) is a cascade system, namely the right-hand side of the $n$th equation depends only on the $n$th and $(n-1)$th state variables; hence starting from the first equation (in a recursive way), the whole system can be solved. According to Benfratello [10], the convergence of this procedure is ensured for small values of the parameter $\varepsilon$. Benfratello has also shown that by a suitable choice of new variables, system (6.76) for the nonlinear function $g(x, t) = x^3$ can be transformed to a linear system with parametric excitations

$$\dot{\mathbf{Z}} = \mathbf{A}_0 \mathbf{Z} + (\mathbf{R}\mathbf{Z} + \mathbf{Q}) w(t) \,, \tag{6.77}$$

where $\mathbf{Z}$ is a new vector state, $\mathbf{A}_0$ and $\mathbf{R}$ are constant matrices and $\mathbf{Q}$ is a constant vector.

### 6.2.4 Energy Criteria

As in the case of statistical linearization, one can introduce energy criteria for equivalent linearization. This idea was first discussed by Elishakoff and his coworkers. We define energy criteria and illustrate their applications for the scalar dynamic system considered in Example 6.1.

**Criterion 3-*EL*.** Criterion of equality of first two order moments of potential energies of displacements of nonlinear and linearized dynamic systems

$$E_N[U^2(x)] = E_L[U^2(x)] \,, \tag{6.78}$$

where $U(x)$ is the potential energy of the displacement, $E_N$ and $E_L$ are the averaging operations defined by probability density functions of stationary solutions of nonlinear and linearized systems $g_N(x)$ and $g_L(x)$, respectively.

**Criterion 4-*EL*.** Criterion of the mean–square error for potential energies of responses of nonlinear and linearized dynamic systems

$$I_{[.]} = E_{[.]}\left[\left(U(\eta) - \frac{1}{2}k\eta^2\right)^2\right] \,, \tag{6.79}$$

where $U(\eta)$ is the potential energy of the nonlinear element $f(\eta)$, the averaging operation $E_{[.]}$ can be defined by probability density $g_N(x)$ as well as by $g_L(x)$. Hence we find

$$I_N = \int_{-\infty}^{+\infty} \left[ U(x) - \frac{1}{2}kx^2 \right]^2 \frac{1}{c_N} \exp\left\{ -\frac{1}{q^2} \int_0^x f(s)ds \right\} dx , \qquad (6.80)$$

$$I_{L1} = \int_{-\infty}^{+\infty} \left[ U\left(\frac{y}{k}\right) - \frac{1}{2}k\left(\frac{y}{k}\right)^2 \right]^2 \frac{1}{c_{L1}\sigma_L} \exp\left\{ -\frac{y^2}{2\sigma_L^2} \right\} dy , \qquad (6.81)$$

where $c_N$ and $c_{L_1}$ are normalized constants $\sigma_L$ is determined by (6.11). After transformation we obtain

$$I_{L1}(k) = \int_{-\infty}^{+\infty} \left[ U(x) - \frac{1}{2}kx^2 \right]^2 \frac{1}{c_{L1}} \frac{\sqrt{2k}}{q} \exp\left\{ -\frac{k^2x^2}{2\frac{q^2}{2k}} \right\} k dx$$

$$= \int_{-\infty}^{+\infty} \left[ U(x) - \frac{1}{2}kx^2 \right]^2 \frac{1}{c_{L1}} \frac{1}{\sqrt{\frac{q^2}{2k^3}}} \exp\left\{ -\frac{x^2}{2\frac{q^2}{2k^3}} \right\} dx . \qquad (6.82)$$

The necessary conditions for minimum of both criteria have the form

$$\frac{\partial I_N}{\partial k} = 2 \int_{-\infty}^{+\infty} \left[ U(x) - \frac{1}{2}kx^2 \right] \left( -\frac{x^2}{2} \right)$$

$$\times \frac{1}{c_N} \exp\left\{ -\frac{1}{q^2} \int_0^x f(s)ds \right\} dx = 0 , \qquad (6.83)$$

$$\frac{\partial I_{L1(k)}}{\partial k} = 2 \int_{-\infty}^{+\infty} \left[ U(x) - \frac{1}{2}kx^2 \right] \left( -\frac{x^2}{2} \right)$$

$$\times \frac{1}{c_{L1}} \frac{\sqrt{2k^3}}{q} \exp\left\{ -\frac{x^2}{2\frac{q^2}{2k^3}} \right\} dx$$

$$+ 3 \int_{-\infty}^{+\infty} \left[ U(x) - \frac{1}{2}kx^2 \right]^2 \left( \frac{1}{2k} - \frac{x^2k^2}{q^2} \right)$$

$$\times \frac{1}{c_{L1}} \frac{\sqrt{2k^3}}{q} \exp\left\{ -\frac{x^2}{2\frac{q^2}{2k^3}} \right\} dx = 0 . \qquad (6.84)$$

We note that from condition (6.83) one can derive the linearization coefficient $k = k_{eq}$ in the form (5.65), while condition (6.84) can be treated as an application of *SPEC Alternative* to energy criteria.

In a particular case if we assume that function $f(x)$ has the form (6.31), then

$$U(x) = \frac{1}{2}\lambda_0^2 x^2 + \frac{1}{4}\varepsilon x^4 , \qquad (6.85)$$

where $\lambda_0^2$ and $\varepsilon$ are positive constants, then the linearization coefficient obtained by statistical linearization and equivalent linearization of mean-square criterion for potential energies with averaging operation generated by the stationary solution of the nonlinear equation is the same and is calculated from relation (5.65), i.e.,

$$k = \lambda_0^2 + \varepsilon \frac{E_N[x^6]}{2E_N[x^4]} . \tag{6.86}$$

In the case of equivalent linearization with averaging operation generated by the solution of the linearized equation, the linearization coefficient $k = k_{4L}$ is calculated from equality (6.84). After integration using properties of Gaussian processes, one can obtain the following algebraic equation for the linearization coefficient $k$:

$$96k^8 - 192\lambda_0^2 k^7 + 72\lambda_0^4 k^6 - 227\varepsilon q^2 k^4 + 540\varepsilon\lambda_0^2 q^2 k^3 + 315\varepsilon^2 q^4 = 0 . \tag{6.87}$$

The solution of (6.87) can be found only numerically.

Another mean-square criterion for *SPEC Alternative* of equivalent linearization can be considered in the following form:

$$I_{L2}(k) = \int_{-\infty}^{+\infty} \left[ U(x) - \frac{1}{2}kx^2 \right]^2 \frac{1}{c_{L2}} \frac{1}{\sqrt{\frac{q^2}{2k}}} \exp\left\{ -\frac{x^2}{2\frac{q^2}{2k}} \right\} dx , \tag{6.88}$$

where $c_{L2}$ is a normalized constant and $\sigma_x^2 = \sigma_L^2 = \frac{q^2}{2k}$ is the variance of the stationary solution of (6.8).

The differentiation of criterion (6.34) with respect to the linearization coefficient $k$ yields the necessary condition for minimum of $I_{L2}(k)$

$$\frac{\partial I_{L2}(k)}{\partial k} = 2 \int_{-\infty}^{+\infty} \left[ U(x) - \frac{1}{2}kx^2 \right] \left( -\frac{x^2}{2} \right)$$

$$\times \frac{1}{c_{L2}} \frac{\sqrt{2k}}{q} \exp\left\{ -\frac{x^2}{2\frac{q^2}{2k}} \right\} dx$$

$$+ \int_{-\infty}^{+\infty} \left[ U(x) - \frac{1}{2}kx^2 \right]^2 \left( \frac{1}{\sqrt{2k}} - \frac{x^2}{q^2} \right)$$

$$\times \frac{1}{c_{L2}} \frac{\sqrt{2k}}{q} \exp\left\{ -\frac{x^2}{2\frac{q^2}{2k}} \right\} dx = 0 . \tag{6.89}$$

Assuming that the nonlinear function $f(x)$ is defined by (6.31) and using properties of Gaussian processes one can show by similar calculations to the previous case that the linearization coefficient $k$ satisfies the following algebraic equation:

$$(96 - 48\sqrt{2})k^4 - (96 - 48\sqrt{2})\lambda_0^2 k^3 + [(180\sqrt{2} - 240)\varepsilon q^2 - (240\sqrt{2} - 96)\lambda_0^4]k^2$$

$$-(420\sqrt{2} - 240)\varepsilon\lambda_0^2 q^2 k - (945\sqrt{2} - 210)\varepsilon^2 q^4 = 0. \tag{6.90}$$

The solution of (6.90) can also be found by numerical calculation.

At the end of this section we note that in the case of equivalent linearization, the determination of linearization coefficients and response characteristics requires an iterative procedure similar to Procedure $SL - MC$. We formulate this procedure for nonlinear system (6.1), linearized system (6.2), and mean-square criterion (6.5).

**Procedure** $EL - MC$

(1) Substitute initial values for linearization coefficients $\Phi_0$, $k_i$, $i = 1, \ldots, n$, and calculate response moments defined by stationary solutions of moment equations (5.31) and (5.32) for linearized system.

(2) Calculate new linearization coefficients $\Phi_0$, $k_i$, $i = 1, \ldots, n$, using, for instance, conditions (6.6).

(3) Substitute the coefficient $\Phi_0$, $k_i$, $i = 1, \ldots, n$, into moment equations (5.31 and 5.32) and solve them.

(4) Repeat steps (2) and (3) until convergence.

As in examples, the difference between statistical and equivalent linearization methods is in step 2 of the iterative procedures $SL - MC$ and $EL - MC$. Conditions (6.6) take different form for different criteria.

### 6.2.5 Moment Criteria with Non-Gaussian Distributions

Similar to statistical linearization one can consider the mean-square criterion with the averaging operation defined by a non-Gaussian probability density function. Such approach was discussed by a few authors, for instance Izumi, Zaiming, and Kimura [61] and Crandall [37, 38]. We will present shortly both methods. First, we discuss application of semi-Gaussian distribution in Criterion $2 - SL$ proposed by Izumi et al. [61] on a scalar nonlinear dynamic system (5.99), i.e.,

$$dx(t) = -f(x(t))dt + qd\xi , \tag{6.91}$$

where the nonlinear function $f(x)$ has the form

$$f(x) = cx + \varepsilon x^3 , \tag{6.92}$$

where $c$ and $\varepsilon$ are positive constants.

The averaging operation in (5.95) is defined by

$$E[.] = \frac{1}{\sqrt{2\pi}\sigma_w} \int_{-\infty}^{+\infty} [.] \exp\left\{-\frac{(x - m_x)^2}{2\sigma_w^2}\right\} dx , \tag{6.93}$$

where $\sigma_w$ is a modified standard deviation defined in a different way for different nonlinear functions, i.e.,

$$\sigma_w = \frac{2}{\sqrt{5}}\sigma_x \quad \text{for strong nonlinearity}(c > 0, \varepsilon > 0) , \tag{6.94}$$

$$\sigma_w = \frac{4}{\sqrt{15}}\sigma_x \quad \text{for soft nonlinearity}(c > 0, \varepsilon < 0) . \tag{6.95}$$

Like the calculation of linearization coefficient for Gaussian statistical linearization one can find that

$$k_1 = c + 3\sigma_w^2 .$$ (6.96)

It means that for nonlinear function defined by parameters $c, \varepsilon$ the linearization coefficients are

$$k_1 = c + \frac{4}{\sqrt{5}}\sigma_x^2 \quad \text{for strong nonlinearity}(c > 0, \varepsilon > 0) ,$$ (6.97)

$$k_1 = c + \frac{16}{\sqrt{15}}\sigma_x^2 \quad \text{for soft nonlinearity}(c > 0, \varepsilon < 0) .$$ (6.98)

Crandall [37, 38] used two types of distributions, namely rectangular distribution and the family of probability density functions

$$g(x) = \frac{\exp\left\{-\left(\frac{x}{a}\right)^m\right\}}{\int_{-\infty}^{+\infty} \exp\left\{-\left(\frac{x}{a}\right)^m\right\} dx} ,$$ (6.99)

where $a > 0$ and $m > 0$ are constant parameters.

In the particular case of the Duffing oscillator described by

$$dx_1(t) = x_2(t)dt ,$$
$$dx_2(t) = [-2\zeta\lambda_0 x_2(t) - f(x_1(t))]dt + \sqrt{2\pi S_0}d\xi(t),$$ (6.100)

where

$$f(x_1) = \lambda_0^2 x_1 + \varepsilon x_1^3 ,$$ (6.101)

$\zeta, \lambda_0, \varepsilon$, and $S_0$ are constant positive parameters.

Crandall [37, 38] proposed to consider the approximate probability density function in the following form:

$$g(z) = \frac{\exp\left\{-\left[b\frac{z^2}{a^2} + \frac{z^4}{a^4}\right]\right\}}{\int_{-\infty}^{+\infty} \exp\left\{-\left[b\frac{z^2}{a^2} + \frac{z^4}{a^4}\right]\right\} dz} ,$$ (6.102)

where $a$ and $b$ are parameters depending on response characteristics. Then the linearization coefficient $k$ can be evaluated from relation (5.95)

$$k = \lambda_0^2 + \varepsilon a^2 \frac{B(4)}{B(2)} ,$$ (6.103)

where

$$B(n) = \int_{-\infty}^{+\infty} v^n \exp\left\{-(bv^2 + v^4)\right\} dv .$$ (6.104)

The parameter $a$ can be determined from the equality of approximated second-order moments defined by (6.102) for nonlinear and linearized dynamic system. Then we obtain

$$a = \frac{\lambda_0^2}{\varepsilon} \left[ \sqrt{\sigma \frac{B(0)}{B(4)} + \left( \frac{B(2)}{2B(4)} \right)^2} - \frac{B(2)}{2B(4)} \right] , \tag{6.105}$$

where

$$\sigma = \frac{\pi S_0 \varepsilon}{2 \zeta \lambda_0^5} . \tag{6.106}$$

The parameter $b$ can be chosen in the form

$$b = w \frac{1}{\sqrt{\sigma}} , \tag{6.107}$$

where $w$ is an arbitrary chosen weighting coefficient. In the particular case when $w = 1$ we obtain true linearization.

A modification of equivalent linearization using a non-Gaussian probability density function has been proposed by Chen [28]. He divided the problem of approximation of the response of a nonlinear system under Gaussian excitations into two parts: Gaussian and non-Gaussian. Chen considered a general second-order nonlinear system described by the equation

$$\boldsymbol{\Phi}(\ddot{\mathbf{x}}(t), \dot{\mathbf{x}}(t), \mathbf{x}(t)) = \mathbf{F}(t) , \tag{6.108}$$

where $\boldsymbol{\Phi}$ is a nonlinear function of the displacement $\mathbf{x}(t)$, velocity $\dot{\mathbf{x}}(t)$, and acceleration $\ddot{\mathbf{x}}(t)$; $\mathbf{F}(t)$ is a Gaussian external excitation. Application of standard equivalent linearization leads to the following linearized equation:

$$\mathbf{M}\ddot{\mathbf{z}}(t) + \mathbf{C}\dot{\mathbf{z}}(t) + \mathbf{K}\mathbf{z}(t) = \mathbf{F}(t) , \tag{6.109}$$

where $\mathbf{M}, \mathbf{C},$ and $\mathbf{K}$ are the mass, damping, and stiffness matrices of linearization.

To improve the accuracy of linearization, Chen [28] proposed to divide the response of the nonlinear system into two parts

$$\mathbf{x}(t) = \mathbf{z}(t) + \Delta \mathbf{x}(t) , \tag{6.110}$$

where $\mathbf{z}(t)$ is the solution of (6.109). The nonlinear function $\boldsymbol{\Phi}(\ddot{\mathbf{x}}(t), \dot{\mathbf{x}}(t), \mathbf{x}(t))$ is expanded to the first-order approximation as a Taylor series

$$\boldsymbol{\Phi}(\ddot{\mathbf{x}}(t), \dot{\mathbf{x}}(t), \mathbf{x}(t)) = \boldsymbol{\Phi}(\ddot{\mathbf{z}}(t), \dot{\mathbf{z}}(t), \mathbf{z}(t)) + \overline{\mathbf{M}}\Delta\ddot{\mathbf{x}}(t) + \overline{\mathbf{C}}\Delta\dot{\mathbf{x}}(t) + \overline{\mathbf{K}}\Delta\mathbf{x}(t) , \tag{6.111}$$

where $\overline{\mathbf{M}} = \partial \boldsymbol{\Phi}/\partial \ddot{\mathbf{x}}, \overline{\mathbf{C}} = \partial \boldsymbol{\Phi}/\partial \dot{\mathbf{x}},$ and $\overline{\mathbf{K}} = \partial \boldsymbol{\Phi}/\partial \mathbf{x}.$

Hence, to reduce the approximation error $\epsilon_1$ defined by

$$\epsilon_1 = \mathbf{M}\ddot{\mathbf{z}}(t) + \mathbf{C}\dot{\mathbf{z}}(t) + \mathbf{K}\mathbf{z}(t) - \boldsymbol{\Phi}(\ddot{\mathbf{x}}(t), \dot{\mathbf{x}}(t), \mathbf{x}(t)) , \tag{6.112}$$

the process $\Delta \mathbf{x}(t)$ must satisfy

$$\overline{\mathbf{M}}\Delta\ddot{\mathbf{x}}(t) + \overline{\mathbf{C}}\Delta\dot{\mathbf{x}}(t) + \overline{\mathbf{K}}\Delta\mathbf{x}(t) = \epsilon_2 , \tag{6.113}$$

where

$$\epsilon_2 = \mathbf{M}\ddot{\mathbf{z}}(t) + \mathbf{C}\dot{\mathbf{z}}(t) + \mathbf{K}\mathbf{z}(t) - \mathbf{\Phi}(\ddot{\mathbf{z}}(t), \dot{\mathbf{z}}(t), \mathbf{z}(t)) \ . \tag{6.114}$$

The coefficient matrices $\mathbf{M}, \mathbf{C}$, and $\mathbf{K}$ are determined by minimizing the mean-square error $E[\epsilon_2^T \epsilon_2]$. Equation (6.114) represents the secondary linear system introduced in the cascade linearization (C–L) [28] and describes a random linear system with the coefficient matrices of non-Gaussian distributions. Hence it follows that the solution $\Delta\mathbf{x}(t)$ is also a non-Gaussian process. Chen [28] has applied this approach to one-dimensional systems, static and dynamic ones with a cubic nonlinearity.

## 6.3 Criteria in Probability Density Space

An application of criteria in probability density functions space is also possible for equivalent linearization. It was first proposed by Socha in [92] and further results are summarized in [95]. In contrast to statistical linearization where two criteria (probabilistic metrics) were considered in equivalent linearization, there are two groups of methods. In the first one, similar to methods in statistical linearization called *direct methods*, probabilistic metrics are used, while in the second one the Fokker–Planck–Kolmogorov equations. To discuss both approaches we consider again nonlinear system (6.1) and linearized one (6.2).

### 6.3.1 Direct Methods

The objective of the probability density equivalent linearization is to find the elements $a_{ij}$ and $C_i$ that minimize the following criterion

$$I = \int_{R^n} w(\mathbf{x})\Psi(g_N(\mathbf{x}) - g_L(\mathbf{x}))d\mathbf{x} \ , \tag{6.115}$$

where $\Psi$ is a convex function, $w(\mathbf{x})$ is a weighting function, $g_N(\mathbf{x})$ and $g_L(\mathbf{x})$ are joint probability density functions of stationary solutions of nonlinear (6.1) and linearized (6.2) systems, respectively. As in the case of statistical linearization, in what follows two basic criteria are considered

(a) Probabilistic square metric

$$I_{1e} = \int_{R^n} (g_N(\mathbf{x}) - g_L(\mathbf{x}))^2 d\mathbf{x} \ . \tag{6.116}$$

(b) Pseudomoment metric

$$I_{2e} = \int_{R^n} |\mathbf{x}|^{2l} |g_N(\mathbf{x}) - g_L(\mathbf{x})| d\mathbf{x}, \quad l = 1, 2, \dots \ . \tag{6.117}$$

It means that the discussed equivalent linearization method is also made in the space of probability density functions. To apply the proposed criteria

(6.116) or (6.117) we have to find the probability density functions for stationary solutions of the nonlinear system $g_N(\mathbf{x})$ and the linearized one $g_L(\mathbf{x})$. Unfortunately, except for some special cases, it is impossible to find the $g_N$ function in an analytical form. However, it can be done by approximation methods, for instance, the Gram–Charlier expansion or by simulations. For an $n$-dimensional system, the one-dimensional probability density function has the following truncated form [82]:

$$g_N(\mathbf{x}) = \frac{1}{c_N} g_G(\mathbf{x}) \left[ 1 + \sum_{k=3}^{N} \sum_{\sigma(\nu)=k} \frac{c_\nu H_\nu(\mathbf{x} - \mathbf{m_x})}{\nu_1! \ldots \nu_n!} \right] , \qquad (6.118)$$

where $g_G(\mathbf{x})$ is the probability density function of a vector Gaussian random variable $\mathbf{x}$

$$g_G(\mathbf{x}) = [(2\pi)^n |\mathbf{K}_G|]^{1/2} \exp\left\{ -\frac{1}{2}(\mathbf{x} - \mathbf{m_x})^T \mathbf{K}_G^{-1}(\mathbf{x} - \mathbf{m_x}) \right\} , \qquad (6.119)$$

where $\mathbf{m_x} = E[\mathbf{x}]$ and $\mathbf{K}_G = E[(\mathbf{x} - \mathbf{m_x})(\mathbf{x} - \mathbf{m_x})^T]$ are the mean value and the covariance matrix of the vector variable $\mathbf{x}$, respectively, $\nu$ is the multiindex, $\nu = [\nu_1, \ldots, \nu_n]^T$, $\sigma(\nu) = \sum_{i=1}^{n} \nu_i$, $|\mathbf{K}_G|$ is the determinant of the matrix $\mathbf{K}_G$, $N$ is the number of elements in the truncation series, and $c_\nu = E[G_\nu(\mathbf{x} - \mathbf{m_x})]$, $\nu = 3, 4, \ldots, N$, are quasi-moments; $c_N$ is a normalized constant, $H_\nu(\mathbf{x})$ and $G_\nu(\mathbf{x})$ are Hermite polynomials defined by (2.17) and (2.18), i.e.,

$$H_\nu(\mathbf{x} = (-1)^{\sigma(\nu)} \exp\left\{ \frac{1}{2}\mathbf{x}^T \mathbf{K}^{-1}\mathbf{x} \right\} \frac{\partial^{\sigma(\nu)}}{\partial x_1^{\nu_1} \ldots \partial x_n^{\nu_n}} \exp\left\{ -\frac{1}{2}\mathbf{x}^T \mathbf{K}^{-1}\mathbf{x} \right\} , \qquad (6.120)$$

$$G_\nu(\mathbf{x}) = (-1)^{\sigma(\nu)} \exp\left\{ \frac{1}{2}\mathbf{x}^T \mathbf{K}^{-1}\mathbf{x} \right\}$$

$$\times \left[ \frac{\partial^{\sigma(\nu)}}{\partial y_1^{\nu_1} \ldots \partial y_n^{\nu_n}} \exp\left\{ -\frac{1}{2}\mathbf{y}^T \mathbf{K}^{-1}\mathbf{y} \right\} \right]_{\mathbf{y}=\mathbf{K}^{-1}\mathbf{x}} , \qquad (6.121)$$

where $\mathbf{K}$ is a real positive-definite matrix. To obtain quasi-moments $c_\nu$, first we derive moment equations for the system (6.1), which can be closed, for instance, by cumulant or quasi-moment closure technique and next we use the algebraic relationships between quasi-moments and moments.

The probability density function of the stationary solution of linearized system (6.2) is known and has the analytical form that can be expressed as follows:

$$g_L(\mathbf{x}) = [(2\pi)^n |\mathbf{K}_L|]^{1/2} \exp\left\{ -\frac{1}{2}(\mathbf{x} - \mathbf{m_x})^T \mathbf{K}_L^{-1}(\mathbf{x} - \mathbf{m_x}) \right\} , \qquad (6.122)$$

where $\mathbf{m_x} = E[\mathbf{x}]$, and $\mathbf{K}_L = E[(\mathbf{x} - \mathbf{m_x})(\mathbf{x} - \mathbf{m_x})^T]$ are the mean and the covariance matrix of the stationary solution of linearized system (6.2), respectively. The vector $\mathbf{m_x}$ and the matrix $\mathbf{K}_L$ satisfy equations

$$\frac{d\mathbf{m_x}(t)}{dt} = \mathbf{A}(t)\mathbf{m_x}(t) + \mathbf{C}(t), \quad \mathbf{m_x}(t_0) = E[\mathbf{x_0}], \qquad (6.123)$$

$$\frac{d\mathbf{K}_L(t)}{dt} = \mathbf{K}_L(t)\mathbf{A}^T(t) + \mathbf{A}(t)\mathbf{K}_L(t) + \sum_{k=1}^{M} \mathbf{G}_k(t)\mathbf{G}_k^T(t), \qquad (6.124)$$

$$\mathbf{K}_L(t_0) = E[(\mathbf{x_0} - E[\mathbf{x_0}])(\mathbf{x_0} - E[\mathbf{x_0}])^T].$$

Since the probability density function $g_L(\mathbf{x})$ of linearized system (6.2) is a function of linearization coefficients $a_{ij}$ and $C_i$, then the necessary conditions for minimization of functional $I_{1e}$ have the form

$$\frac{\partial I_{1e}(t)}{\partial a_{ij}} = 0, \quad \frac{\partial I_{1e}(t)}{\partial C_i} = 0, \quad i,j = 1,\ldots,n. \qquad (6.125)$$

Using necessary conditions (6.125) and moment equations for nonlinear and linearized systems, one can determine linearization coefficients by an iterative procedure. It is illustrated for the criterion $I_{1e}$, rewritten as follows:

$$I_{1e} = \int_{-\infty}^{+\infty} \cdots \int_{-\infty}^{+\infty} [g_N(x_1,\ldots,x_n) - g_L(x_1,\ldots,x_n)]^2 dx_1 \ldots dx_n. \qquad (6.126)$$

The proposed iterative procedure has the following form.

**Procedure** $EL - PD$

*Step 1:* Substitute initial values of linearization coefficients $a_{ij}, C_i,\ i,j = 1,\ldots,n$, and calculate moments of the stationary response of linearized systems (6.123) and (6.124).

*Step 2:* Calculate approximate moments of the stationary response of nonlinear system (6.1).

*Step 3:* Calculate probability density functions $g_L(x_1,\ldots,x_n)$ and $g_N(x_1,\ldots,x_n)$ of the stationary response of nonlinear (6.1) and linearized (6.2) systems, respectively, using the moments obtained in Step 1 and Step 2, respectively, and relations between moments of the corresponding stationary responses and Hermite polynomials defined by (6.119–6.121).

*Step 4:* Choose any criterion, for instance, $I_{1e}$ and find the linearization coefficients $a_{ij_{\min}}, C_{i_{\min}},\ i,j = 1,\ldots,n$, that minimize $I_{1e}$ (jointly for whole nonlinear dynamic system) and substitute $a_{ij} = a_{ij_{\min}}, C_i = C_{i_{\min}}$.

*Step 5:* Calculate the stationary moments of linearized systems (6.123) and (6.124).

*Step 6:* Redefine the probability density function for the linearized system by substituting the moments obtained in Step 5 into the general formula defined by (6.118–6.121).

*Step 7:* Repeat Steps 3–6 until $a_{ij}$ and $C_i,\ i,j = 1,\ldots,n$, converge.

We note that in the case of criterion $I_{2e}$ the direct numerical minimization has to be used, because the function $|g_N(x) - g_L(x)|$ is not differentiable.

### 6.3.2 The Method of Fokker–Planck–Kolmogorov Equations in Linearization

When the probability density function of the nonlinear system is unknown and for some reason, the direct optimization technique cannot be applied, the present author proposed [94] instead of state equations (6.1 and 6.2) to consider the corresponding reduced Fokker–Planck–Kolmogorov equations (for stationary probability density functions)

$$\frac{\partial g_N}{\partial t} = -\sum_{i=1}^{n} \frac{\partial}{\partial x_i}[\Phi_i(\mathbf{x},t)g_N] + \frac{1}{2}\sum_{i=1}^{n}\sum_{j=1}^{n}\frac{\partial^2}{\partial x_i \partial x_j}[b_{Nij}g_N] = 0 , \quad (6.127)$$

and

$$\frac{\partial g_L}{\partial t} = -\sum_{i=1}^{n} \frac{\partial}{\partial x_i}[(\mathbf{A}_i^T\mathbf{x}+C_i)g_L] + \frac{1}{2}\sum_{i=1}^{n}\sum_{j=1}^{n}\frac{\partial^2}{\partial x_i \partial x_j}[b_{Lij}g_L] = 0 , \quad (6.128)$$

where $\mathbf{A}_i^T$ is $i$th row of matrix $\mathbf{A}$; $\mathbf{B}_N = [b_{Nij}]$ and $\mathbf{B}_L = [b_{Lij}]$ are the diffusion matrices

$$b_{Nij} = b_{Lij} = \sum_{k=1}^{M} G_{ki}G_{kj} . \quad (6.129)$$

If we denote

$$p_1 = g_N, p_{2i} = \frac{\partial g_N}{\partial x_i}, \quad q_1 = g_L, q_{2i} = \frac{\partial g_L}{\partial x_i} , \quad (6.130)$$

then (6.127) and (6.128) can be transformed to the following two-dimensional vector systems:

$$\frac{\partial p_1}{\partial x_i} = p_{2i},$$

$$\sum_{i=1}^{n}\left[\frac{\partial \Phi_i}{\partial x_i}p_1 + \Phi_i p_{2i}\right] - \frac{1}{2}\sum_{i=1}^{n}\sum_{j=1}^{n}\left[\frac{\partial^2 b_{ij}}{\partial x_i \partial x_j}p_1 + \frac{\partial b_{ij}}{\partial x_j}p_{2i} + \frac{\partial b_{ij}}{\partial x_i}p_{2j}\right.$$

$$\left. + b_{ij}\frac{\partial p_{2j}}{\partial x_i}\right] = 0, \quad (6.131)$$

$$i = 1,\ldots,n,$$

and

$$\frac{\partial q_1}{\partial x_i} = q_{2i},$$

$$\sum_{i=1}^{n}\left[a_{ii}q_1 + (A_i^T\mathbf{x}+C_i)q_{2i}\right] - \frac{1}{2}\sum_{i=1}^{n}\sum_{j=1}^{n}\left[\frac{\partial^2 b_{ij}}{\partial x_i \partial x_j}q_1 + \frac{\partial b_{ij}}{\partial x_j}q_{2i} + \frac{\partial b_{ij}}{\partial x_i}q_{2j}\right.$$

$$\left. + b_{ij}\frac{\partial q_{2j}}{\partial x_i}\right] = 0,$$

$$i = 1,\ldots,n, \quad (6.132)$$

where $b_{ij} = b_{Nij}$ for system (6.131) and $b_{ij} = b_{Lij}$ for system (6.132). Comparing the system equations (6.131) with (6.132) we find that $g_N$ and $\partial g_N/\partial x_i$ are approximated by $g_L$ and $\partial g_L/\partial x_i$, respectively. Then the following joint criterion is proposed [94]

$$I_3 = \int_{R^n} \epsilon_1{}^2(\mathbf{x})d\mathbf{x} \ , \tag{6.133}$$

where

$$\epsilon_1(\mathbf{x}) = \sum_{i=1}^{n} \frac{\partial}{\partial x_i} \left[ (\Phi_i - A_i^T \mathbf{x} - C_i)g_L \right] \ . \tag{6.134}$$

Then the necessary conditions for minimum of criterion $I_3$ have the form (6.125), and the linearization coefficients are determined by an iterative procedure similar to Procedure $EL - PD$. An application of the direct method and the Fokker–Planck–Kolmogorov equation method is illustrated in the following example.

*Example 6.3.* Consider again the scalar nonlinear Ito stochastic differential equation (6.7) for $f(x) = \lambda_0^2 x + \phi_1(x)$, i.e.,

$$dx = -[\lambda_0^2 x + \phi_1(x)]dt + qd\xi(t) \ , \tag{6.135}$$

where $\phi_1(x)$ is a nonlinear function, $\phi_1(0) = 0$, $\lambda_0^2$ and q are positive constants, and $\xi$ is a standard Wiener process.

The linearized system corresponding to (6.135) is described by

$$dx = -kxdt + qd\xi(t) \ , \tag{6.136}$$

where $k > 0$ is a linearization coefficient.

The stationary solutions of (6.134) and (6.135) are characterized by the corresponding stationary probability density functions

$$g_N(x) = \frac{1}{c_N} \exp\left\{ -\frac{2}{q^2} \int_0^x [\lambda_0^2 s + \phi_1(s)]ds \right\} \ , \tag{6.137}$$

$$g_L(x) = \frac{1}{c_L \sigma_L} \exp\left\{ -\frac{x^2}{2\sigma_L^2} \right\} \ , \tag{6.138}$$

$$\sigma_L^2 = \frac{q^2}{2k} \ , \tag{6.139}$$

where $c_N$ and $c_L$ are normalized constants, $\sigma_L^2$ is the variance of stationary solution of linearized system (6.135). Here, for both systems the mean values of stationary solutions are equal to zero, $E[x] = 0$.

First, we determine the linearization coefficient by the direct method for criterion $I_{1e}$. The necessary conditions for minimum of $I_{1e}$ are

$$\frac{\partial}{\partial k} \int_{-\infty}^{+\infty} (g_N(x) - g_L(x))^2 dx = -2 \int_{-\infty}^{+\infty} (g_N(x) - g_L(x)) \frac{\partial g_L(x)}{\partial k} dx = 0 .$$

$$(6.140)$$

Hence, we find

$$\int_{-\infty}^{+\infty} (g_N(x) - g_L(x)) \left( \frac{1}{2k} - \frac{x^2}{q^2} \right) g_L(x) dx = 0 , \qquad (6.141)$$

where $g_N(x)$ and $g_L(x)$ are defined by (6.137–6.139). The linearization coefficient $k$ can be found numerically.

Next we consider the application of the Fokker–Planck–Kolmogorov equations approach to the system (6.134–6.135). Then the linearization criterion has the form (6.133), i.e.,

$$I_3 = \int_{-\infty}^{+\infty} \left\{ \frac{\partial}{\partial x} [(\lambda_0^2 x + \phi_1(x) - kx) g_L(x)] \right\}^2 dx , \qquad (6.142)$$

where $g_L(x)$ is defined by (6.138–6.139).

The necessary condition for minimum of criterion $I_3$ can be found by equating to zero the derivative of $I_3$ with respect to the linearization coefficient $k$

$$\frac{\partial}{\partial k} \int_{-\infty}^{+\infty} \left\{ \frac{\partial}{\partial x} [(\lambda_0^2 x + \phi_1(x) - kx) g_L(x)] \right\}^2 dx = 0 . \qquad (6.143)$$

Hence, we calculate

$$\int_{-\infty}^{+\infty} \left\{ \left( \frac{\partial f}{\partial x} - k \right) g_L + (f(x) - kx) \frac{\partial g_L}{\partial x} \right\} \qquad (6.144)$$

$$\times \left\{ \left( \frac{\partial f}{\partial x} - k \right) \frac{\partial g_L}{\partial k} + (f(x) - kx) \frac{\partial^2 g_L}{\partial x \partial k} - \frac{\partial}{\partial x} (x g_L) \right\} dx = 0,$$

where

$$f(x) = \lambda_0^2 x + \phi_1(x), \quad g_L = g_L(x, k), \quad \frac{\partial g_L}{\partial x} = \frac{2kx}{q^2} g_L,$$

$$\frac{\partial}{\partial x} (x g_L) = \left( 1 - \frac{2kx^2}{q^2} \right) g_L, \quad \frac{\partial g_L}{\partial k} = \left( \frac{1}{2k} - \frac{x^2}{q^2} \right) g_L, \qquad (6.145)$$

$$\frac{\partial}{\partial k} \left( \frac{\partial g_L}{\partial x} \right) = \left( -3 + \frac{2kx^2}{q^2} \right) \frac{x}{q^2} g_L .[6pt]$$

Further calculations are possible only for some class of nonlinear functions, for instance, if it is assumed that the function $f(x)$ is an odd polynomial of third order, i.e.,

$$f(x) = \lambda_0^2 x + \varepsilon x^3 . \qquad (6.146)$$

where $\varepsilon > 0$.

Then condition (6.144) takes the form

$$
\int_{-\infty}^{+\infty} \left\{ [3\varepsilon x^2 + (\lambda_0^2 - k)] + [\varepsilon x^3 + (\lambda_0^2 - k)x] \left( -\frac{2kx}{q^2} \right) \right\}
$$
$$
\times \left\{ [3\varepsilon x^2 + (\lambda_0^2 - k)] \left( \frac{1}{2k} - \frac{x^2}{q^2} \right) \right.
$$
$$
+ [\varepsilon x^3 + (\lambda_0^2 - k)x] \left( -3 + \frac{2kx^2}{q^2} \right) \frac{x}{q^2}
$$
$$
\left. - \left( 1 - \frac{2kx^2}{q^2} \right) \right\} \frac{2k}{c_L q^2} \exp\left\{ -\frac{2kx^2}{q^2} \right\} dx = 0. \tag{6.147}
$$

Next, using the properties of Gaussian process we calculate from (6.147) the integrals of polynomials with weighting function $\exp\{-2kx^2/q^2\}$. In paper [95] it was shown that the linearization coefficient $k$ satisfies the following algebraic equation:

$$
48k^3(\lambda_0^2 - k)^2 - 192k^4(\lambda_0^2 - k) - 891k\varepsilon^2 q^4 - 48\varepsilon k^3 q^2 - 24(\lambda_0^2 - k)k^2 q^2 = 0 . \tag{6.148}
$$

To illustrate the obtained results, we show a comparison of the relative errors of the response variances $E[x^2]$ and six–order moments $E[x^6]$ versus the ratio of parameters $r = \varepsilon/\lambda_0^2$. In these comparisons we consider the moments obtained by standard statistical linearization (SSL), potential energy statistical linearization (PESL), true equivalent linearization (TEL), second-order pseudomoment statistical linearization (PMSL-2), six-order pseudomoment statistical linearization (PMSL-6), square metric probability density statistical linearization (SMSL), Fokker–Planck–Kolmogorov equation linearization (F-PEL), second-order pseudomoment equivalent linearization (PMEL-2), six–order pseudomoment equivalent linearization (PMEL–6), and square metric probability density equivalent linearization (SMEL). The numerical results for parameters $\lambda_0^2 = 0.5$, $\varepsilon = i \times 0.5$, $i = 1, \ldots, 10$, and $q^2 = 0.2$ are presented in Figs. 6.4 and 6.5.

Figures 6.4 and 6.5 show that for second-order moments, the relative error obtained by TEL is equal to zero and the errors obtained by PESL, PMSL-2, and PMEL-2 are almost zero, while the errors obtained by the other methods are significantly greater. The opposite situation is observed for six-order moments of the response. Here, the preferable accuracy of the F-PEL, PMSL-2, and PMSL-6 is clearly seen.

We note that although statistical as will as equivalent linearization with criteria in probability density function space was applied to dynamic systems under Gaussian white noise excitation, both approaches can be extended to dynamic systems under Gaussian colored noise. The difference is in the calculation of moments of stationary solutions.

At the end of this section, we note that in both the direct method and the Fokker–Planck–Kolmogorov equations method, the proposed criteria usually are not convex with respect to the linearization coefficients. This fact

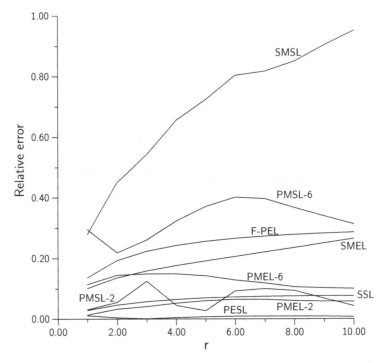

**Fig. 6.4.** Comparison of the relative errors of the response variances $E[x^2]$ versus the ratio of parameters $r = \varepsilon/\lambda_0^2$ for $\lambda_0^2 = 0.5$, $\varepsilon = i \times 0.5$, $i = 1, ..., 10$ and $q^2 = 0.2$, reproduced from [94]

requires application of complex optimization methods to the determination of linearization coefficients. Therefore, both the approaches are not recommended for nonlinear multi–degree-of-freedom systems.

## 6.4 Criteria in Spectral Density Space

As was mentioned in Chap. 1, the idea of linearization of nonlinear elements in the frequency domain for deterministic systems and signals is known in the literature as *harmonic balance* or *describing function*. It was used mainly by control engineers for the analysis of nonlinear automatic control systems. The mathematical background for this approach was given by Russian researchers Krylov (1934) and Bogoliuboff (1937) partially presented in [66] and developed by many authors (see for details and references, for instance, in [62, 104, 72]). The first attempts of application of describing function method to stochastic dynamic systems were made in 1965 by Pupkov [83] and later in the 1980s and 1990s in [5, 4, 3]. The derivation of formulas for linearization coefficients in the frequency domain is much more complicated than in the time domain. The

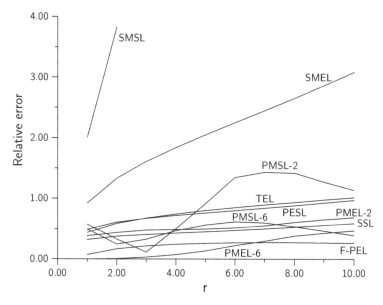

**Fig. 6.5.** Comparison of the relative errors of the response six–order moments $E[x^6]$ versus the ratio of parameters $r = \varepsilon/\lambda_0^2$ for $\lambda_0^2 = 0.5$, $\varepsilon = i \times 0.5$, $i = 1, ..., 10$ and $q^2 = 0.2$, reproduced from [94]

basic difficulty was the problem ("good approximation") of the power spectral density function of the stationary response of a nonlinear dynamic system. At the same time it has been found that in contrast to linearization criteria in time domain where a nonlinear element is replaced by a linear one in the frequency domain, such replacement always leads to incorrect approximation. For instance, the response power spectral density function for the Duffing oscillator denoted by $S_r(\lambda)$ consists of at least two modes, while the response power spectral density function for the corresponding linear oscillator has the only one mode (see Fig. 6.6).

Therefore, it was necessary to approximate even a simple nonlinear dynamic system with stochastic external excitation by a multi-degree-of-freedom system with deterministic coefficients and with the same external excitation or by a simple linear dynamic system with random parameters.

The linearization procedure of stochastic dynamic systems in frequency domain consists of four basic stages:

(i) The determination of the approximate power spectral density function $S_N(\lambda)$ of the stationary solution of the considered nonlinear dynamic system for a given set of frequencies $\lambda \in \Lambda$.

(ii) The determination of an approximate operational transfer function $H(s)$ that defines a linear system such that its response power spectral density function $S_L(\lambda)$ approximates the power spectral density function $S_N(\lambda)$

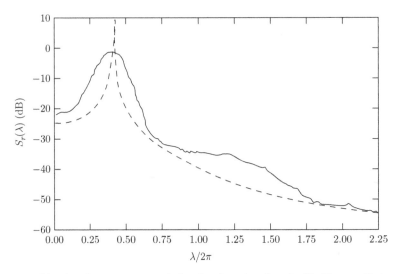

**Fig. 6.6.** Simulated power spectral density function for the Duffing oscillator $(-)$ and its linearized form 1-DOF model $(- - -)$, from [18] with permission

for $\lambda \in \Lambda$, for instance, by relation $S_L(\lambda_k) = S_N(\lambda_k)$ for a given set of $\lambda_k \in \Lambda$, $k = 1, \ldots, M$, where $M$ is a given number. This step is called *spectral decomposition.*

(iii) Identification by ARMA models, i.e., the determination of an approximate ARMA model for the function $H(e^{-i\lambda}) = \frac{Q}{P}(e^{-i\lambda})$, where $Q$ and $P$ are some polynomials.

(iv) The determination of an approximate ARMA model for a linear mechanical system, i.e., an approximate model for the function $\frac{Q}{P}(e^{-i\lambda})$ by ARMA model for a linear mechanical system.

In what follows we discuss shortly all proposed steps, starting from approximate methods of the determination of the power spectral density function of the response of a simple nonlinear dynamic system.

### 6.4.1 The Determination of the Power Spectral Density Function of the Stationary Response of a Nonlinear Dynamic System $S_N(\lambda)$

In the first works concerning the determination of the power spectral density functions of the stationary response of simple nonlinear dynamic systems under external excitation in the form of Gaussian white noise, the perturbation methods were used, for instance, by Crandall [33, 34], Morton and Corrisin [71], Manning [68] and an iterative operator method by Adomian [1]. Since the obtained results were not enough, satisfying new approaches were proposed, for instance, [87, 65, 23]. Roy and Spanos [87] have proposed to expand in series the approximated power spectral density function, where the coefficients of this expansion are spectral moments that are determined by a recursive method. Krenk and Roberts [65] used the stochastic averaging for the total energy $E$ to the determination of the conditional correlation function $R_x(\tau|E)$ and next the conditional power spectral density function $S_x(\lambda|E)$. It has the form of a series whose components represent oscillators with damping and different resonance frequencies.

Based on the knowledge of an exact probability density function for the total energy, $p(E)$ that is known as an exact solution of the FPK equation corresponding to the averaged stochastic differential equation for the total energy, one can calculate the sought power spectral density function of stationary response of nonlinear system in the form

$$S_x(\lambda) = \int_0^{+\infty} S_x(\lambda|E)p(E)dE . \tag{6.149}$$

One of the new proposed methods of the determination of an approximate power spectral density of the stationary solution of a nonlinear oscillator is an application of cumulant closure technique to the differential equations for correlation moments by Cai and Lin [23] and next the application of Fourier transform to the obtained differential equations for correlation moments in the closed form. Following [23] we illustrate this approach in an example of the Duffing oscillator using cumulant closure technique to the including sixth order.

*Example 6.4.* [23] Determine the power spectral density function of the stationary solution of the Duffing oscillator

$$d\mathbf{x}(t) = [\mathbf{A}\mathbf{x} + \mathbf{B}(\mathbf{x})]dt + \mathbf{G}d\xi , \tag{6.150}$$

where

$$\mathbf{x} = \begin{bmatrix} x_1 \\ x_2 \end{bmatrix}, \mathbf{A} = \begin{bmatrix} 0 & 1 \\ -\lambda_0^2 & -2h \end{bmatrix}, \mathbf{B}(\mathbf{x}) = \begin{bmatrix} 0 \\ \varepsilon x_1^3 \end{bmatrix}, \mathbf{G} = \begin{bmatrix} 0 \\ \delta \end{bmatrix}, \tag{6.151}$$

$\lambda_0, h, \varepsilon$, and $\delta$ are positive constant parameters; $\xi(t)$ is the standard Wiener process. In Example 4.2 were defined the correlation moments

$$\rho_{lj}(\tau) = E[x_1^l(t)x_2^j(t)x_1(t-\tau)], p_{lj}(\tau) = E[x_1^l(t)x_2^j(t)x_2(t-\tau)], l, j = 1,\ldots,n , \tag{6.152}$$

$$R_{11}(\tau) = E[x_1(t)x_1(t-\tau)] = \rho_{10}(\tau) ,$$
$$R_{12}(\tau) = E[x_1(t)x_2(t-\tau)] = p_{10}(\tau) ,$$
$$R_{22}(\tau) = E[x_2(t)x_2(t-\tau)] = p_{01}(\tau) . \tag{6.153}$$

Using relations (4.21) and applying cumulant closure technique to moment correlation equations we obtain:

**Approximation by Cumulant Closure Second Order (Gaussian Closure Technique)**

$$\rho_{30} = E[x_1^3(t)x_1(t-\tau)] = 3E[x_1^2]E[x_1(t)x_1(t-\tau)] = 3m_{20}\rho_{10} , \quad (6.154)$$

where $m_{lj} = E[x_1^l x_2^j]$. For stationary solutions of system (6.150–6.151) $m_{lj}$ are constant. Then the correlation moment equations of first order (4.21) take the form

$$\frac{d\rho_{10}}{d\tau} = \rho_{01},$$

$$\frac{d\rho_{01}}{d\tau} = -(\lambda_0^2 + 3\varepsilon m_{20})\rho_{10} - 2h\rho_{01}, \quad (6.155)$$

with initial conditions

$$\rho_{10}(0) = m_{20}, \quad \rho_{01}(0) = m_{11} = 0 . \quad (6.156)$$

By solving the system of equations (6.155) with initial conditions (6.156) we determine the autocorrelation function $R_{x_1 x_1}(\tau)$ in the form

$$R_{x_1 x_1}(\tau) = R_{11}\tau = \rho_{10}(\tau) = m_{20}e^{-h\tau}\left(\cos\lambda_l\tau + \frac{h}{\lambda_l}\sin\lambda_l\tau\right) , \quad (6.157)$$

where

$$\lambda_l^2 = \lambda_0^2 + 3\varepsilon m_{20} - h^2 . \quad (6.158)$$

To find the power spectral density function corresponding to the correlation function (6.157), one should directly apply the Fourier transform to autocorrelation function or to system (6.155). For convenience we use an integral transformation $\mathcal{F}^*$ introduced by Cai and Lin [23], which is connected with the Fourier transform $\mathcal{F}$ as follows:

$$\bar{\rho}_{lj}(\lambda) = \mathcal{F}^*[\rho_{lj}(\tau)] = \frac{1}{\pi}\int_0^{+\infty}\rho_{lj}(\tau)e^{-i\lambda\tau}\,d\tau , \quad (6.159)$$

The power spectral density $\Phi_{x_1 x_1}(\lambda)$ can be obtained as

$$\Phi_{x_1 x_1}(\lambda) = \Phi_{11}(\lambda) = \frac{1}{2\pi}\int_{-\infty}^{+\infty}R_{11}(\tau)e^{-i\lambda\tau}\,d\tau = Re[\bar{R}_{11}(\lambda)] = Re[\bar{\rho}_{10}(\lambda)] . \quad (6.160)$$

We note that the integral transform (6.159) for the derivative of correlation moment has the following form:

$$\mathcal{F}^*\left[\frac{d\rho_{lj}}{d\tau}(\tau)\right] = i\lambda\mathcal{F}^*[\rho_{lj}(\tau)] - \frac{1}{\pi}\rho_{lj}(0) = i\lambda\bar{\rho}_{lj}(\lambda) - \frac{1}{\pi}m_{l+1j} . \quad (6.161)$$

Applying Fourier transform to system (6.155) we obtain

$$i\lambda\bar{\rho}_{10}(\lambda) - 2m_{20} = \bar{\rho}_{01}$$

$$i\lambda\bar{\rho}_{01}(\lambda) = -(\lambda_0^2 + 3\varepsilon m_{20})\bar{\rho}_{10} - 2h\bar{\rho}_{01} . \qquad (6.162)$$

By solving (6.162) we obtain

$$\bar{\rho}_{10}(\lambda) = \frac{(i\lambda + 2h)2m_{20}}{\pi[-\lambda^2 + \lambda_0^2 + 3\varepsilon m_{20} + 2ih\lambda]} . \qquad (6.163)$$

Then the power spectral density function of the process $x_1(t)$ is equal to

$$\Phi_{x_1 x_1}(\lambda) = \Phi_{11}(\lambda) = Re[\bar{\rho}_{10}(\lambda)] = \frac{2hm_{20}(\lambda_0^2 + 3\varepsilon m_{20})}{\pi[(\lambda_0^2 + 3\varepsilon m_{20} - \lambda_0^2)^2 + 4h^2\lambda^2]} . \quad (6.164)$$

## Approximation by Cumulant Closure Technique of Fourth Order

By equating the cumulants of fourth order to zero we obtain the following approximations:

$$\rho_{50} = 10m_{20}\rho_{30} + 5c_1\rho_{10},$$
$$\rho_{41} = 6m_{20}\rho_{21} + 5c_1\rho_{01},$$
$$\rho_{32} = m_{02}\rho_{30} + 3m_{20}\rho_{12} - 3m_{22}\rho_{10} , \qquad (6.165)$$

where

$$c_1 = m_{40} - 6m_{20}^2 . \qquad (6.166)$$

Substituting equalities (6.165) to correlation moment equations of third order (4.21) we find

$$\frac{d\rho_{30}}{d\tau} = 3\rho_{21},$$

$$\frac{d\rho_{21}}{d\tau} = 2\rho_{12} - (\lambda_0^2 + 10\varepsilon m_{20})\rho_{30} - 2h\rho_{21} - 5\varepsilon c_1\rho_{10} ,$$

$$\frac{d\rho_{12}}{d\tau} = \rho_{03} - (2\lambda_0^2 + 12\varepsilon m_{20})\rho_{21} - 4h\rho_{12} - 2\varepsilon c_1\rho_{01} + \delta^2\rho_{10} , \qquad (6.167)$$

$$\frac{d\rho_{03}}{d\tau} = 3\varepsilon m_{02}\rho_{30} - (3\lambda_0^2 + 9\varepsilon m_{20})\rho_{12} - 6h\rho_{03} + 9\varepsilon m_{22}\rho_{10} + 3\delta^2\rho_{01} ,$$

with initial conditions

$$\rho_{10}(0) = m_{20}, \quad \rho_{01}(0) = 0, \quad \rho_{30}(0) = m_{40} ,$$
$$\rho_{21}(0) = 0, \quad \rho_{12}(0) = m_{22}, \quad \rho_{03}(0) = 0. \qquad (6.168)$$

The equations for transform in frequency domain are created as follows. We consider equations of the first-order correlation moments (4.21) without applying cumulant closure and the third-order equations (4.21) with this closure, i.e., system of equations (6.167) with the corresponding initial conditions. After applying the integral transform to both systems of equations we obtain

$$i\lambda\bar\rho_{10}(\lambda) - \tfrac{1}{\pi}m_{20} = \bar\rho_{01},$$

$$i\lambda\bar\rho_{01}(\lambda) = -\lambda_0^2\bar\rho_{10} - \varepsilon\bar\rho_{30} - 2h\bar\rho_{01}, \tag{6.169}$$

$$i\lambda\bar\rho_{30}(\lambda) - \tfrac{1}{\pi}m_{40} = 3\bar\rho_{21}, \tag{6.170}$$

$$i\lambda\bar\rho_{21}(\lambda) = -(\lambda_0^2 + 10\varepsilon m_{20})\bar\rho_{30} - 2h\bar\rho_{21} + 2\bar\rho_{12} - 5\varepsilon c_1\bar\rho_{10},$$

$$i\lambda\bar\rho_{12}(\lambda) - \tfrac{1}{\pi}m_{22} = -(2\lambda_0^2 + 12\varepsilon m_{20})\bar\rho_{21} - 4h\bar\rho_{12} + \bar\rho_{03} - 2\varepsilon c_1\bar\rho_{12} + \delta^2\bar\rho_{10},$$

$$i\lambda\bar\rho_{03}(\lambda) = -3\varepsilon m_{02}\bar\rho_{30} - (3\lambda_0^2 + 9\varepsilon m_{20})\bar\rho_{12} - 6h\bar\rho_{03} + 9\varepsilon m_{22}\bar\rho_{10} + 3\delta^2\bar\rho_{01}.$$

From linear system of equations (6.169) and (6.171) we calculate $\bar\rho_{10}$, and next we determine $\Phi_{x_1x_1}(\lambda) = \Phi_{11}(\lambda) = Re[\bar\rho_{10}(\lambda)]$.

## Approximation by Six-Order Cumulant Closure Technique

By equating six-order cumulants to zero we obtain the following approximations:

$$\rho_{70} = 21m_{20}\rho_{50} - 35c_1\rho_{30} + 7c_3\rho_{10},$$

$$\rho_{61} = 15m_{20}\rho_{41} - 15c_1\rho_{21} + c_3\rho_{01},$$

$$\rho_{52} = m_{02}\rho_{50} + 10m_{20}\rho_{32} - 10m_{22}\rho_{30} + 5c_1\rho_{12} - 5m_{02}c_1\rho_{10},$$

$$\rho_{43} = 3m_{02}\rho_{41} + 6m_{20}\rho_{23} - 18m_{22}\rho_{21} + c_1\rho_{03} - 3m_{02}c_1\rho_{01},$$

$$\rho_{34} = 6m_{02}\rho_{32} + 3m_{20}\rho_{14} + c_2\rho_{30} - 18m_{22}\rho_{12} - 3m_{02}c_2\rho_{10}, \tag{6.171}$$

where $c_1$ is determined by (6.166), and coefficients $c_2$ and $c_3$ are as follows:

$$c_2 = m_{40} - 6m_{02}^2, \quad c_3 = m_{60} - 30m_{02}(m_{40} - 3m_{20}^2). \tag{6.172}$$

Further procedure is similar to the case when fourth-order cumulant closure was applied, i.e., we create the system of equations for the first- and third-order correlation moments (4.21) without application of cumulant closure technique and the system of fifth-order correlation moments closed by cumulant closure technique (6.171) with the corresponding initial conditions. The obtained system of differential equations is a linear one and can be transformed by the integral transform $\mathcal{F}^*$ defined in (6.159). Then we obtain the following system of transformed equations:

$$i\lambda\bar\rho_{10}(\lambda) - \frac{1}{\pi}m_{20} = \bar\rho_{01},$$

$$i\lambda\bar\rho_{01}(\lambda) = -\lambda_0^2\bar\rho_{10} - \varepsilon\bar\rho_{30} - 2h\bar\rho_{01}, \tag{6.173}$$

$$i\lambda\bar\rho_{30}(\lambda) - \frac{1}{\pi}m_{40} = 3\bar\rho_{21},$$

$$i\lambda\bar{\rho}_{21}(\lambda) = -\lambda_0^2\bar{\rho}_{30} - 2h\bar{\rho}_{21} + 2\bar{\rho}_{12} - \varepsilon\bar{\rho}_{50} ,$$

$$i\lambda\bar{\rho}_{12}(\lambda) - \frac{1}{\pi}m_{22} = -2\lambda_0^2\bar{\rho}_{21} - 4h\bar{\rho}_{12} + \bar{\rho}_{03} - 2\varepsilon\bar{\rho}_{41} + \delta^2\bar{\rho}_{10} ,$$

$$i\lambda\bar{\rho}_{03}(\lambda) = -3\varepsilon\bar{\rho}_{32} - 3\lambda_0^2\bar{\rho}_{12} - 6h\bar{\rho}_{03} + 3\delta^2\bar{\rho}_{01}, \tag{6.174}$$

$$i\lambda\bar{\rho}_{50}(\lambda) - \frac{1}{\pi}m_{60} = 5\bar{\rho}_{41},$$

$$i\lambda\bar{\rho}_{41}(\lambda) = -(\lambda_0^2 + 21\varepsilon m_{20})\bar{\rho}_{50} - 2h\bar{\rho}_{41} + 4\bar{\rho}_{32}$$
$$-35\varepsilon c_1\bar{\rho}_{30} - 7\varepsilon c_3\bar{\rho}_{10} ,$$

$$i\lambda\bar{\rho}_{32}(\lambda) - \frac{1}{\pi}m_{42} = -(2\lambda_0^2 + 15\varepsilon m_{20})\bar{\rho}_{41} - 4h\bar{\rho}_{12} + 3\bar{\rho}_{23}$$
$$-30\varepsilon c_1\bar{\rho}_{21} - 2\varepsilon c_3\rho_{01} + \delta^2\bar{\rho}_{30} ,$$

$$i\lambda\bar{\rho}_{23}(\lambda) = -3\varepsilon m_{02}\bar{\rho}_{50} - 3(\lambda_0^2 + 10\varepsilon m_{20})\bar{\rho}_{32} - 6h\bar{\rho}_{23} + 2\bar{\rho}_{14}$$
$$+30\varepsilon m_{22}\bar{\rho}_{30} + 2\delta^2\bar{\rho}_{21} - 15\varepsilon c_1\bar{\rho}_{12} + 15\varepsilon m_{02}c_1\bar{\rho}_{10} ,$$

$$i\lambda\bar{\rho}_{14}(\lambda) - \frac{1}{\pi}m_{24} = -12\varepsilon m_{02}\bar{\rho}_{41} - 4(\lambda_0^2 + 6\varepsilon m_{20})\bar{\rho}_{23} - 8h\bar{\rho}_{14} + \bar{\rho}_{05}$$
$$+72\varepsilon m_{22}\bar{\rho}_{21} + 6\delta^2\bar{\rho}_{12} - 4\varepsilon c_1\bar{\rho}_{03} + 12\varepsilon m_{02}c_1\bar{\rho}_{01},$$

$$i\lambda\bar{\rho}_{05}(\lambda) = -30\varepsilon m_{02}\bar{\rho}_{32} - 5(\lambda_0^2 + 3\varepsilon m_{20})\bar{\rho}_{14} - 10h\bar{\rho}_{05} + 90\varepsilon m_{22}\bar{\rho}_{12}$$
$$+10\delta^2\bar{\rho}_{03} - 5\varepsilon c_2\bar{\rho}_{30} + 15\varepsilon m_{20}c_2\bar{\rho}_{10}. \tag{6.175}$$

As in the previous case we calculate $\bar{\rho}_{10}$ from the system of linear equations (6.174–6.175), and next we determine $\Phi_{x_1x_1}(\lambda) = \Phi_{11}(\lambda) = Re[\bar{\rho}_{10}(\lambda)]$.

### 6.4.2 Spectral Decomposition

One of the simplest methods of the approximation of a given power spectral density function $S_N(\lambda)$ by the power spectral density function of a stationary solution of a linear continuous time dynamic system $S_L(\lambda, \mathbf{r})$ depending on a vector of constant parameters $\mathbf{r}$ is the analysis of the minimum of a square criterion defined, for instance, by

$$I(\mathbf{r}) = \int_{-\infty}^{+\infty} (S_N(\lambda) - S_L(\lambda, \mathbf{r}))^2 d\lambda , \tag{6.176}$$

or application of this criterion to the approximation of $S_N(\lambda)$ by the power spectral density function of a stationary solution of a linear discrete-time dynamic system defined by an ARMA$(p, q)$ model

$$x(t) + \sum_{l=1}^{p} a_i x(t - l) = \sum_{j=0}^{q} b_j v(t - j) , \tag{6.177}$$

where $v(t)$ is a discrete-time white noise. The power spectral density function of the stationary solution of such ARMA process is a periodic function over the interval $[-\pi, \pi]$ defined by

$$S_A(\lambda) = \left|\frac{MA(e^{-i\lambda})}{AR(e^{-i\lambda})}\right|^2 , \tag{6.178}$$

where

$$AR(z) = 1 + \sum_{l=1}^{p} a_l z^l, \quad MA(z) = 1 + \sum_{j=0}^{q} b_j z^j \tag{6.179}$$

are two polynomials characterizing ARMA model (AR, auto regressive; MA, moving average), $a_l$ and $b_j$ are parameters that should be found from conditions of minimization of functional (6.176), where instead of $S_L(\lambda, \mathbf{r})$ we substitute $S_A(\lambda, a_l, b_j)$ defined by (6.178), i.e.,

$$I(a_l, b_j) = \int_{-\infty}^{+\infty} (S_N(\lambda) - S_A(\lambda, a_l, b_j))^2 d\lambda . \tag{6.180}$$

Unfortunately, both the criteria (6.176) and (6.180) are nonconvex with respect to both types of parameters: elements of vector $\mathbf{r}$ for continuous time model and $a_l, b_j$ for discrete time model. This fact causes that the optimization process is very complicated.

The mentioned difficulties were omitted by Bernard [12] who proposed a modification of a linearization method in frequency domain. His idea was developed by Bernard and Taazount [13] and Ismaili [54]. We discuss their approaches.

If we restrict the obtained power spectral density function of a scalar stationary response process of a nonlinear system to the frequency interval $[-\lambda_c, +\lambda_c]$, where $\lambda_c$ is the cutoff pulsation, then by a suitable change of variables $\bar{\lambda} = \frac{\lambda\pi}{\lambda_c}$ the range $[-\lambda_c, +\lambda_c]$ is transformed to $[-\pi, +\pi]$. Further, we consider the obtained power spectral density function of the stationary solution of a nonlinear system $S_N(\bar{\lambda})$ for $\bar{\lambda} \in [-\pi, +\pi]$.

If we assume that the function $\log S(\bar{\lambda})$ is Lebesgue integrable and is restricted to the unit circle of a complex-valued function that is analytic inside the unit circle, the spectral density function can be represented as follows [80, 54]:

$$S(\bar{\lambda}) = e^{c_0} H(e^{-i\bar{\lambda}}) H^*(e^{-i\bar{\lambda}}), \quad \forall \bar{\lambda} \in [-\pi, +\pi] , \tag{6.181}$$

where

$$H(z) = \exp\left\{\sum_{k=1}^{+\infty} c_k z^k\right\}, \quad c_k \in \mathbf{R}, \quad \sum_{k\geq 0} |c_k|^2 < \infty , \tag{6.182}$$

$$c_k = \frac{1}{2\pi} \int_{-\pi}^{+\pi} \log(S(\bar{\lambda})) e^{ik\bar{\lambda}} d\bar{\lambda}, \quad k \geq 0 . \tag{6.183}$$

The coefficients $c_k$ can be determined by fast Fourier transform (FFT) based on the knowledge of values of the function $S_N(\bar\lambda)$ in $2N + 1$ points in the range $[-\pi, \pi]$. Denoting

$$\bar\lambda_j = \frac{\pi j}{N} \quad for \quad j = 0, \ldots, N , \tag{6.184}$$

we obtain approximate coefficients $c_k$ for $k = 0, \ldots, \frac{N}{2}$,

$$c_k = \frac{1}{N} \left( \sum_{j=0}^{N-1} \log\left(S(\bar\lambda_j)\right) \right) e^{ik\bar\lambda_j} . \tag{6.185}$$

The power spectral density function in the form (6.181) can be treated as a power spectral density function of an output process of a linear filter excited by a Gaussian white noise.

### 6.4.3 Identification by ARMA Models

For the obtained function $H(e^{-i\lambda})$ in (6.182) in Sect. 6.4.2 we find an approximate ARMA model by minimizing the following square criterion:

$$I(a_l, b_j) = \int_{-\pi}^{+\pi} |H(e^{-i\bar\lambda})P(e^{-i\bar\lambda}) - Q(e^{-i\bar\lambda})|^2 d\lambda , \tag{6.186}$$

where $Q$ and $P$ are polynomials

$$P(z) = 1 + \sum_{l=1}^{p} a_l z^l, \quad Q(z) = 1 + \sum_{j=0}^{q} b_j z^j, q < p . \tag{6.187}$$

After discretization of criterion (6.186) we obtain

$$I(a_l, b_j) = \sum_{k=1}^{N-1} |P(e^{-i\bar\lambda_k})H(e^{-i\bar\lambda_k}) - Q(e^{-i\bar\lambda_k})|^2 . \tag{6.188}$$

Using the necessary conditions for minimum of criterion (6.188)

$$\frac{\partial I}{\partial a_l} = 0, \ l = 1, \ldots, p; \quad \frac{\partial I}{\partial b_j} = 0, \ j = 1, \ldots, q , \tag{6.189}$$

we obtain the system of linear algebraic equations for coefficients $a_l$ and $b_j$ in the form

$$\mathbf{FY} = \mathbf{G} , \tag{6.190}$$

where $\mathbf{Y}^T = [\mathbf{a}^T \mathbf{b}^T]$, $\mathbf{a}^T = [a_1, \ldots, a_p]$, $\mathbf{b}^T = [b_0, \ldots, b_q]$; elements of matrix $\mathbf{F} = [F_{lj}]$ and vector $\mathbf{G} = [G_1, \ldots, G_{p+q+1}]$ are the following [13]:

$$
F_{lj} = \begin{cases} Re < HH^* z^{j-1}, 1 >, & l = 1,\ldots,p & j = 1,\ldots,p \\ -Re < H^* z^{j-l-p-1}, 1 >, & l = 1,\ldots,p & j = p+1,\ldots,p+q+1 \\ -Re < H z^{j-l+p+1}, 1 >, & l = p+1,\ldots,p+q+1 & j = 1,\ldots,p \\ N\delta_{lj}, & l = p+1,\ldots,p+q+1 & j = p+1,\ldots,p+q+1 \end{cases},
$$

$$
G_l = \begin{cases} -Re < HH^* z^{-l}, 1 >, & l = 1,\ldots,p \\ Re < H z^{-l+p+1}, 1 >, & l = p+1,\ldots,p+q+1 \end{cases}, \tag{6.191}
$$

where $\delta_{lj}$ is the Kronecker delta, $< f,g >= \sum_{k=0}^{N-1} f(k)g^*(k)$ and $z = e^{-2i\pi k/N}$. Hence, from (6.190) we find the linearization coefficients $a_l$, $l = 1,\ldots,p$, and $b_j$, $j = 0,\ldots,q$.

### 6.4.4 Mechanical ARMA Model Approximation

The approximation of the power spectral density function of a scalar stationary response process of a nonlinear system presented in Sect. 6.4.3 gives good results. However, the corresponding linear system does not represent any model of a real mechanical system. Therefore Ismaili [54] proposed the following modification of the discussed procedure.

Consider again the function $H(e^{-i\bar\lambda})$ in the form

$$
H(e^{-i\bar\lambda}) = \frac{Q}{P}(e^{-i\bar\lambda}) \quad \forall \bar\lambda \in [-\pi, \pi] . \tag{6.192}
$$

If some poles $z_i$ of function $H$ are real, then they represent aperiodic terms and cannot be used in the modeling of mechanical oscillators. Therefore, Ismaili [54] has made a decomposition of polynomial $P$ as a product of two polynomials

$$
P = P_1 P_2 , \tag{6.193}
$$

where the roots of $P_1$ are nonreal and the roots of $P_2$ are real. In further consideration we take into account only the polynomial $P_1$ and we are interested in the approximation of the function $H(e^{-i\bar\lambda})$ by the ratio $\frac{Q_1}{P_1}(e^{-i\bar\lambda})$, i.e., we have to find the coefficients of the polynomial $Q_1$ that minimize the following criterion:

$$
I = \int_{-\pi}^{+\pi} |H(e^{-i\bar\lambda})P_1(e^{-i\bar\lambda}) - Q_1(e^{-i\bar\lambda})|^2 d\bar\lambda , \tag{6.194}
$$

where the polynomial $P_1$ and the degree of the polynomial $Q_1$ are fixed, i.e., if the degree of the polynomial $P_1$ is equal to $2n_1$, $n_1 = 1,\ldots,N_1$, $N_1$ is a given number, then the degree of the polynomial $Q_1$ is equal to $2n_1 - 1$; the operational transfer function $H$ has the form (6.192).

To obtain a spectrally equivalent linear mechanical model, the denominator $P_1$ of the rational function $\frac{Q_1}{P_1}$ will be decomposed into irreducible quadratic factors. Then the function $\frac{Q_1}{P_1}$ can be decomposed as follows:

$$\frac{Q_1}{P_1}(e^{-i\bar{\lambda}}) = \sum_{j=1}^{n_1} \alpha_j \frac{MA_j}{AR_j}(e^{-i\bar{\lambda}}) \quad \forall \lambda \in [-\pi, \pi] , \qquad (6.195)$$

where $[\alpha_1, \ldots, \alpha_{n_1}]^T \in R^{n_1}$, $\frac{MA_j}{AR_j}(e^{-i\bar{\lambda}})$ is an output of the filter of an ARMA $(2,1)$ process; the polynomials $MA_j$ and $AR_j$ define linear oscillators described by the following equations

$$\ddot{x}_j(t) + 2\eta_j \lambda_{0j} \dot{x}_j(t) + \lambda_{0j}^2 x_j(t) = q_j \dot{\xi}(t), \quad j = 1, \ldots, n_1 . \qquad (6.196)$$

i.e.,

$$MA_j(z) = A_j z + B_j ,$$

$$A_j^2 + B_j^2 = \frac{q_j^2}{4\pi\eta_j\gamma_j\lambda_{0j}^3}[\gamma_j(1 - e^{-4\eta_j\lambda_{0j}h}) - 2\eta_j e^{-2\eta_j\lambda_{0j}h} \sin 2\lambda_{dj}h] ,$$

$$A_j B_j = \frac{q_j^2}{4\pi\eta_j\gamma_j\lambda_{0j}^3}[\gamma_j(e^{-3\eta_j\lambda_{0j}h} - e^{-\eta_j\lambda_{0j}h})\cos \lambda_{dj}h$$

$$+\eta_j(e^{-3\eta_j\lambda_{0j}h} + e^{-\eta_j\lambda_{0j}h})\sin \lambda_{dj}h], \qquad (6.197)$$

$$AR_j(z) = 1 + a_{1j}z + a_{2j}z^2,$$
$$a_{1j} = -2e^{-\eta_j\lambda_{0j}h}\cos(\lambda_{dj}h),$$
$$a_{2j} = e^{-2\eta_j\lambda_{0j}h}, \qquad (6.198)$$

where $0 < \lambda_{0j}, 0 < \eta_j < 1$ and $q_j$ $(j = 1, \ldots, n_1)$ are constant parameters; $\dot{\xi}(t)$ is a standard scalar white noise, $h$ is the discrete time step,

$$\gamma_j = \sqrt{1 - \eta_j^2}, \quad \lambda_{dj} = \gamma_j \lambda_{0j} . \qquad (6.199)$$

Ismaili [54] has shown that the coefficients $\eta_j$ and $\lambda_{0j}$ defining $j$th oscillator can be calculated from the following relation:

$$\eta_j = \sqrt{\frac{K^2}{K^2 + 1}}, \quad \lambda_{0j} = \frac{\log(|z_j|)}{\eta_j h} , \qquad (6.200)$$

where

$$K = \frac{\log(|z_j|)}{arg(z_j)} . \qquad (6.201)$$

Then, criterion (6.194) can be rewritten as follows:

$$I_1(\alpha_1, \ldots, \alpha_{n_1}) = \int_{-\pi}^{+\pi} |H(e^{-i\bar{\lambda}}) - \sum_{j=1}^{n_1} \alpha_j \frac{MA_j}{AR_j}(e^{-i\bar{\lambda}})|^2 d\bar{\lambda}$$
$$= \int_{-\pi}^{+\pi} |H(e^{-i\bar{\lambda}}) - \sum_{j=1}^{n_1} \alpha_j H_j(e^{-i\bar{\lambda}})|^2 d\bar{\lambda}, \quad (6.202)$$

where

$$H_j(e^{-i\bar{\lambda}}) = \frac{MA_j}{AR_j}(e^{-i\bar{\lambda}}) \quad \forall \bar{\lambda} \in [-\pi, \pi], \quad (6.203)$$

with the linear constraints

$$\alpha_j \geq 0, j = 1, \ldots, n_1. \quad (6.204)$$

From (6.202) it follows that the operational transfer function $H(e^{-i\bar{\lambda}})$ is approximated by a linear combination of the operational transfer functions of linear oscillators. This constitutes a mechanical $ARMA(2n_1, 2n_1 - 1)$ model for the nonlinear oscillator, where $n_1$ is the order of linear combination.

For the convenience of solving the constrained criterion (6.202–6.204) Ismaili [54] proposed the following two-step procedure:

(1) First, for the coefficients of $MA_j$ that are chosen fulfilling $A_j \geq 0$, $B_j \geq 0$ the criterion (6.202–6.203) is minimized without constraints.

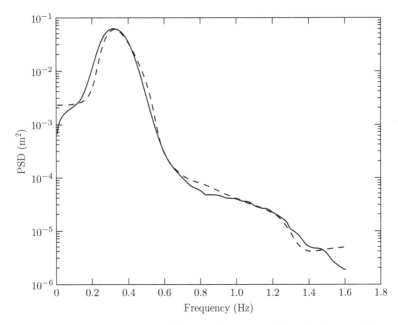

**Fig. 6.7.** Simulated power spectral density function of the Duffing oscillator $(-)$ and its mechanical ARMA model $(- - -)$, from [54] with permission

Then one can obtain a linear system of equations for unknowns $\alpha_j \geq 0$; $j = 1, \ldots, n_1$, i.e.,

$$\mathbf{A}_L \alpha = \mathbf{B}_L , \qquad (6.205)$$

where

$$\alpha = [\alpha_1, \ldots, \alpha_{n_1}]^T , \quad \mathbf{A}_L = [\mathbf{A}_{L_{ij}}], \; \mathbf{B}_L = [\mathbf{B}_{L_1}, \ldots \mathbf{B}_{L_{n1}}]^T \qquad (6.206)$$

$$\mathbf{A}_{L_{lj}} = \; < H_l, H_j >, \quad for \quad l, j = 1, \ldots, n_1 , \qquad (6.207)$$

$$\mathbf{B}_{L_j} = Re < H, H_l >, \quad for \quad l = 1, \ldots, n_1 . \qquad (6.208)$$

If $\alpha_j < 0$, then the signs of $A_j$ and $B_j$ are changed.

(2) The second step of the procedure is to increase the degrees $p$ and $q$ of the polynomials $P$ and $Q$ in the first ARMA identification step by step until satisfactory approximation of the spectrum is obtained. Hence it follows that it must increase the number $n_1$ of linear oscillators approximating the given nonlinear oscillator.

Ismaili [54] has shown that in the case of the Duffing oscillator, the equivalent linear system consists of three oscillators. The good approximation of the corresponding power spectral density functions is illustrated in Fig. 6.7.

However, the convexity of a criterion with respect to $p$ and $q$ was not discussed. We also note that further generalization of Bernard and Taazount [13] method was done by Daucher et al. [40], who used a vector ARMA formulation. In their approximation model of a nonlinear oscillator under stochastic excitation, the identified parameters of the ARMA structure were $p = 39$ and $q = 38$.

## 6.5 Multi-criterial Linearization Methods

The application of different linearization criteria to nonlinear dynamic systems gives an opportunity for finding approximate linear dynamic systems having similar response characteristics or probability density functions or power spectral density functions. Since the number of different linearization criteria is large ($> 100$), it is necessary to consider a methodology that takes into account a group of criteria. See, for instance, a survey paper [100] or [44]. It is clear that the choice of a group of criteria of linearization depends on the considered problem, but in one group, for instance, in the space of moments of stochastic processes, the problem is open and the choice requires an additional analysis. To study multi-criterial problems, special approaches called multi-criteria optimization methods were developed in the literature. In the field of mechanics it was reviewed in [102] and recently studied for linear stochastic systems, for instance, in [19, 88].

In this section, we show an application of two basic multi-criterial optimization methods, i.e., scalarization method and Pareto optimal solution in application to the determination of the response characteristics. First, we quote basic definitions and facts from this approach [89, 90].

### 6.5.1 Poli-Criteria Optimization Methods

**Definition 6.1.** *A subset $\Theta$ of a linear space $B$ is called a convex cone if and only if*

$$\forall \alpha_1 \geq 0 \ \forall \alpha_2 \geq 0 \ \forall \mathbf{x}^1 \mathbf{x}^2 \in \Theta (\alpha_1 \mathbf{x}^1 + \alpha_2 \mathbf{x}^2) \in \Theta . \tag{6.209}$$

To every convex cone $\Theta$, there corresponds the ordering relation $R$ in $B$ defined by

$$\mathbf{x}^1 \leq_\Theta \mathbf{x}^2 <=> \mathbf{x}^2 - \mathbf{x}^1 \in \Theta . \tag{6.210}$$

The relations of partial order induced by convex cones are generalization of the natural order in $R^n$ defined as follows:

$$\mathbf{x} \leq \mathbf{y} <=> \forall i = 1, \ldots, n, x_i \leq y_i , \tag{6.211}$$

where $\mathbf{x} = [x_1, \ldots, x_n]^T$ and $\mathbf{y} = [y_1, \ldots, y_n]^T$.

This is equivalent to the relation $\mathbf{y} - \mathbf{x} \in R_+^n$. The positive orthant $R_+^n$ satisfies all properties of convex cones.

The general problem of multi-criteria optimization is

$$(F : U_d \rightarrow B) \rightarrow \min(\Theta) , \tag{6.212}$$

where the set of the admissible strategies (controls) $U_d$ is a subset of a linear space $U$, the goal space $B$ is partially ordered Banach space with a closed convex cone $\Theta$. Moreover, it is assumed that the admissible set $F(U_d)$ is nonempty and closed.

**Pareto Optimal Approach**

**Definition 6.2.** *A strategy (control) $u_{\text{opt}}$ is said to be nondominated or Pareto optimal, or $\Theta$ optimal if and only if*

$$(F(u_{\text{opt}}) - \Theta) \cap F(U_d) = \{F(u_{\text{opt}})\} . \tag{6.213}$$

The condition (6.213) means that no element of admissible set is better than $u_{\text{opt}}$ in the sense of the partial order relation.

The relation (6.213) plays the fundamental role in the classical problems of multi-criteria optimization, which can be reduced to the simultaneous minimization of scalar functions

$$(F_1, F_2, \ldots, F_m) \rightarrow \min . \tag{6.214}$$

**Scalarization Methods**

The most frequently used scalarization method for the problem

$$(F : U_d \rightarrow R^n) \rightarrow \min(R_+^n) \tag{6.215}$$

is the positive convex combination of the criteria

$$F_w(u) = \sum_{i=1}^{N} w_i F_i(u) , \tag{6.216}$$

where $u \in U_d$, $w_i > 0$ for $1 \le i \le N$, and $\sum_{i=1}^{N} w_i = 1$. The parameters $w_i > 0$, $1 \le i \le N$, are weighting coefficients.

**The Scalarization by Distance**

$$F_d(u) = d(q, F(u)) , \tag{6.217}$$

where $d$ is a metric in the goal space, $q$ is a fixed unattainable element of the goal space that dominates at least one point from $F(U_d)$, for instance,

$$F_p(u) = \|q - F(u)\|_p^p , \tag{6.218}$$

where $\|.\|_p^p$ is a $p$th power of the norm in $L_p$ space. For instance, as the scalarizing family for the finite-dimensional multi-criteria optimization problem with respect to the natural partial order in $R^n$, one can consider the following family of functionals:

$$N_p(u, w) = \sum_{i=1}^{N} w_i (F_i(u) - q_i)^p, \quad w \in R_+^n \setminus \{0\}, \quad 1 \le p \le \infty . \tag{6.219}$$

In the particular case when $N = 2$, $U_d = R^1 = \{k : -\infty < k < +\infty\}$, and $F_j = I_j$, $j = 1, 2$, an illustration of a nondominated point $q$ and relation (6.217) is given in Fig. 6.8.

### 6.5.2 Applications for Single-Degree-of-Freedom Systems

Consider the single-degree-of-freedom system described by

$$dx_1 = x_2 dt,$$
$$dx_2 = [-f(x_1) - 2hx_2]dt + \sigma d\xi(t), \tag{6.220}$$

where $h > 0$ and $\sigma > 0$ are constant parameters, $f$ is a nonlinear function such that $f(0) = 0$, and $\xi(t)$ is the standard Wiener process. Then the mean value of the stationary solution is equal to zero, i.e., $E[x_1] = 0$.

An equivalent linearized system has the form

$$dx_1 = x_2 dt,$$
$$dx_2 = [-kx_1 - 2hx_2]dt + \sigma d\xi(t), \tag{6.221}$$

where $k$ is a linearization coefficient.

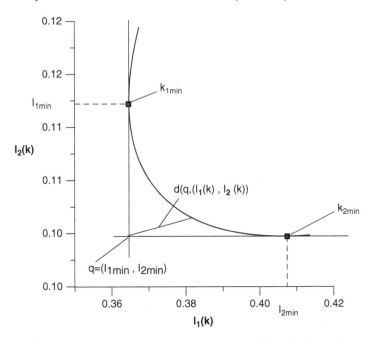

**Fig. 6.8.** A geometric illustration of condition (6.217), from [95] with permission

The most frequently used scalarization method is the positive convex combination of the considered criteria, i.e.,

$$I_{\text{opt}}(k) = \sum_{i=1}^{N} \alpha_i I_i(k) , \qquad (6.222)$$

where $I_{\text{opt}}$ and $I_i, i = 1, \ldots, N$, are multiobjective criterion of linearization and partial criteria of linearization, respectively, $\alpha_i > 0$, $i = 1, \ldots, N$, are weighting coefficients such that $\sum_{i=1}^{N} \alpha_i = 1$.

The idea of the Pareto optimal solution is to find a nondominated point $\mathbf{q}$ whose coordinates are defined by minimal values of the considered criteria, i.e.,

$$\mathbf{q}(k) = \mathbf{q}(\mathbf{I}_{i_{\min}}(k)), \quad i = 1, \ldots, N , \qquad (6.223)$$

where

$$I_{i_{\min}}(k) = \min_{k} I_i(k), \quad i = 1, \ldots, N . \qquad (6.224)$$

The scalarization distance $d_w$ is defined, for instance, by

$$d_w = \left( \sum_{i=1}^{N} \alpha_i (I_i(k) - I_{i_{\min}}(k)) \right)^{1/2} , \qquad (6.225)$$

where $\alpha_i > 0, i = 1, \ldots, N$, are weighting coefficients such that $\sum_{i=1}^{N} \alpha_i = 1$.

To illustrate an application of Pareto optimal approach and scalarization method in the determination of response characteristics, we use two criteria of statistical linearization ($N = 2$). In further consideration we analyze two cases of the moment criteria and the criteria in probability density space.

The corresponding criteria and linearization coefficients have the following forms.

### 6.5.3 Moment Criteria

· **Mean-Square Criterion**

$$I_{MS} = \frac{E[f^2(x_1)]E[x_1^2] - (E[f(x_1)x_1])^2}{E[x_1^2]} , \tag{6.226}$$

$$k_{MS} = \frac{E[f(x_1)x_1]}{E[x_1^2]} . \tag{6.227}$$

**Energy Criterion**

$$I_E = \frac{E\left[\left(\int_0^{x_1} f(s)ds\right)^2\right]\frac{1}{4}E[x_1^4]}{\frac{1}{4}E[x_1^4]} - \frac{\left(E\left[\frac{x_1^2}{2}\int_0^{x_1} f(s)ds\right]\right)^2}{\frac{1}{4}E[x_1^4]}, \tag{6.228}$$

$$k_E = \frac{E\left[\frac{x_1^2}{2}\int_0^{x_1} f(s)ds\right]}{\frac{1}{4}E[x_1^4]} . \tag{6.229}$$

### 6.5.4 Probability Density Criteria

**Pseudomoment Metric**

$$I_{PMM} = \int_{-\infty}^{+\infty}\int_{-\infty}^{+\infty} x_1^{2p}x_2^{2q}|g_N(x_1, x_2) - g_L(x_1, x_2)|dx_1 dx_2, \tag{6.230}$$

where $p + q = r$, $p, q = 0, 1, \dots$.

**Square Probability Metric**

$$I_{PSM} = \int_{-\infty}^{+\infty}\int_{-\infty}^{+\infty} [g_N(x_1, x_2) - g_L(x_1, x_2)]^2 dx_1 dx_2 , \tag{6.231}$$

where $g_N(\mathbf{x})$ and $g_L(\mathbf{x})$ are probability density functions of solutions of nonlinear (6.220) and linearized systems (6.221), respectively.

$g_N(x_1, x_2)$ is defined by the Gram–Charlier expansion [82]

$$g_N(x_1, x_2) = g_G(x_1, x_2) \left[ 1 + \sum_{k=3}^{N} \sum_{\sigma(\nu)=k} \frac{c_{\nu_1 \nu_2} H_{\nu_1 \nu_2}(x_1, x_2)}{\nu_1! \nu_2!} \right], \quad (6.232)$$

where

$$g_G(x_1, x_2) = \frac{1}{2\pi \sqrt{k_{11} k_{22} - k_{12}^2}} \exp \left\{ -\frac{k_{11}(x_2)^2 - 2k_{12} x_1 x_2 + k_{22}(x_1)^2}{2(k_{11} k_{22} - k_{12}^2)} \right\},$$

$$(6.233)$$

$c_{\nu_1 \nu_2} = E[G_{\nu_1 \nu_2}(x_1, x_2)]$ are quasi-moments, $\nu_1, \nu_2 = 0, 1, \ldots, N$, $\nu_1 + \nu_2 = 3, 4, \ldots, N$, $H_{\nu_1 \nu_2}(x_1, x_2)$, and $G_{\nu_1 \nu_2}(x_1, x_2)$ are Hermite polynomials; $k_{ij} = E[x_i x_j]$, $i, j = 1, 2$.

For stationary probability density function, the corresponding moments are as follows:

$$k_{12} = 0, k_{22} = \frac{1}{v_{22}}, v_{12} = 0, k_{11} = \frac{1}{v_{11}}. \quad (6.234)$$

The moment $k_{11}$ has to be found from moment equations.

We illustrate the derived formulas for the application of Pareto optimal approach and scalarization method to an example.

*Example 6.5.* Consider the Duffing oscillator described by

$$dx_1 = x_2 dt,$$
$$dx_2 = [-\lambda_0^2 x_1 - \varepsilon x_1^3 - 2h x_2] dt + \sigma d\xi(t), \quad (6.235)$$

where $\lambda_0^2, \varepsilon, h$, and $\sigma$ are positive constant parameters, $\xi(t)$ is the standard Wiener process. The parameters selected for calculations are: $\lambda_0^2 = 0.5, \varepsilon = 0.1, h = 0.05$, and $\sigma^2 = 0.2$.

The corresponding linearized system has the form (6.221).

**Moment Criteria in Statistical Linearization**

**Mean-Square Criterion**

$$I_{MS} = (\lambda_0^2 - k)^2 E[x_1^2] + 15\varepsilon^2 (E[x_1^2])^3 + 6\varepsilon(\lambda_0^2 - k)(E[x_1^2])^2 . \quad (6.236)$$

**Energy Criterion**

$$I_E = \frac{3}{4}(\lambda_0^2 - k)^2 (E[x_1^2])^2 + 105 \left(\frac{\varepsilon}{4}\right)^2 (E[x_1^2])^4$$

$$+ \frac{15}{4}\varepsilon(\lambda_0^2 - k)(E[x_1^2])^3, \quad (6.237)$$

where

$$E[x_1^2] = \frac{\sigma^2}{4hk} . \quad (6.238)$$

Then the set of dominating points is presented in Fig. 6.9

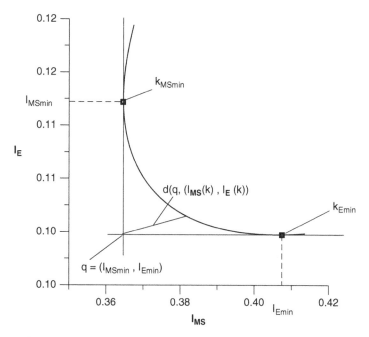

**Fig. 6.9.** A graphical illustration of the set of dominating points determined by (6.236–6.238), from [95] with permission

## Convex Combination

$$I_{\text{opt}} = \alpha I_{MS}(k) + (1 - \alpha)I_E, 0 \le \alpha \le 1 . \tag{6.239}$$

The characteristics $k_{\min} = k_{\min}(\alpha)$ and $I_{\text{opt}} = I_{\text{opt}}(\alpha)$ determined by relation (6.239) are shown in Figs. 6.10 and 6.11, respectively.

The characteristic $d_E = d_E(k)$ obtained by the *scalarization by distance* has the form

$$d_E(k) = \sqrt{(I_{MS_{\min}} - I_{MS}(k))^2 + (I_{E_{\min}} - I_E(k))^2} . \tag{6.240}$$

A graphical illustration of this characteristic is given in Fig. 6.12.

## Probability Density Linearization Criteria in Equivalent Linearization

## Pseudomoment Metric

$$I_{PMM} = \int_{-\infty}^{+\infty} \int_{-\infty}^{+\infty} x_1^{2p} x_2^{2q} |g_N(x_1, x_2) - g_L(x_1, x_2)| dx_1 dx_2, \tag{6.241}$$

where $p + q = r$, $p, q = 0, 1, \ldots$.

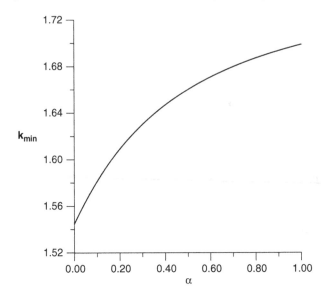

**Fig. 6.10.** The characteristics $k_{min} = k_{min}(\alpha)$ determined by relation (6.239), from [95] with permission

**Square Probability Metric**

$$I_{PSM} = \int_{-\infty}^{+\infty} \int_{-\infty}^{+\infty} (g_N(x_1, x_2) - g_L(x_1, x_2))^2 dx_1 dx_2 , \tag{6.242}$$

where $g_N(x_1, x_2)$ and $g_L(x_1, x_2)$ are probability density functions defined by

$$g_N(x_1, x_2) = \frac{1}{c_N} exp\left\{-\frac{2h}{\sigma^2}(\lambda_0^2 x_1^2 + \frac{\varepsilon}{2}x_1^4 + x_2^2)\right\}, \tag{6.243}$$

$$g_L(x_1, x_2, k) = \frac{1}{c_L}\frac{4h\sqrt{k}}{\sigma^2}exp\left\{-\frac{2h}{\sigma^2}(kx_1^2 + x_2^2)\right\}, \tag{6.244}$$

where $c_N$ and $c_L$ are normalized constants. Then the set of dominating points for parameters $\lambda_0^2 = 0.5$, $h = 0.05$, $\varepsilon = 0.1$, $\sigma^2 = 0.2$, $p = 1$, $q = 0$ is presented in Fig. 6.13.

## 6.6 Special Linearization Methods

In this section we present some important new methods that are interesting from theoretical point of view.

### 6.6.1 Exact Equivalent Linearization

The idea of exact linearization for deterministic systems described, for instance, by Isidori [53] was developed by Socha [91] for stochastic systems. By suitable choice of nonlinear transformations it has been shown for a few classes of nonlinear systems the corresponding exact linearized higher-order systems. For instance, for the scalar nonlinear system

$$dx = \left[ A_1 x + A_2 x^{\frac{n-1}{n}} + a_{12}^2 \frac{n-1}{2n} x^{\frac{n-2}{n}} \right] dt + \left[ a_{11} x + a_{12} x^{\frac{n-1}{n}} \right] d\xi(t) , \tag{6.245}$$

where $x \in R^1$ is the scalar state variable, $A_1, A_2, a_{11}$, and $a_{12}$ are constants; $n = 1, 2, \ldots$, $\xi(t)$ is the standard Wiener process.

The equivalent linear system is a multi-dimensional system that has the form

$$d\mathbf{x} = (\mathbf{A}\mathbf{x} + \mathbf{B})dt + (\mathbf{C}\mathbf{x} + \mathbf{D})d\xi(t) , \tag{6.246}$$

where $\mathbf{x} = [x_1, \ldots, x_n]^T$, $x_k = x^{\frac{n+1-k}{n}}$, $k = 1, \ldots, n$ and

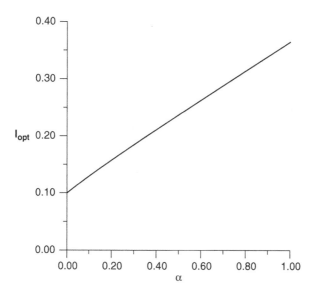

**Fig. 6.11.** The characteristics $I_{\mathrm{opt}} = I_{\mathrm{opt}}(\alpha)$ determined by relation (6.239), from [95] with permission

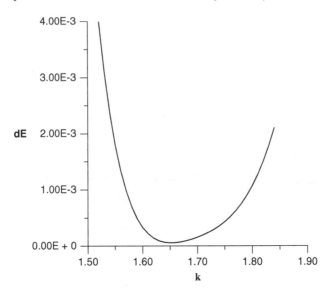

**Fig. 6.12.** The characteristics $d_E = d_E(k)$ determined by relation (6.240), from [95] with permission

$$\mathbf{A} = \begin{bmatrix} A_{11} & A_{12} & A_{13} & 0 & \cdots & 0 \\ 0 & A_{22} & A_{23} & A_{24} & \cdots & 0 \\ \cdot & & \cdot & \cdot & \cdot & \cdot \\ \cdot & & & \cdot & \cdot & \cdot \\ \cdot & & & & \cdot & \cdot \\ 0 & 0 & 0 & 0 & 0 & A_{nn} \end{bmatrix}, \quad \mathbf{C} = \begin{bmatrix} c_{11} & c_{12} & 0 & 0 & \cdots & 0 \\ 0 & c_{22} & c_{23} & 0 & \cdots & 0 \\ \cdot & & \cdot & \cdot & \cdot & \cdot \\ \cdot & & & \cdot & \cdot & \cdot \\ \cdot & & & & \cdot & \cdot \\ 0 & 0 & 0 & 0 & 0 & c_{nn} \end{bmatrix} \quad (6.247)$$

$$\mathbf{B} = [0, \ldots, 0, B_{n-1}, B_n]^T, \quad \mathbf{D} = [0, \ldots, 0, D_n]^T, \quad B_{n-1} = A_{n-1,n+1},$$
$$B_n = A_{n,n+1},$$
$$D_n = c_{n,n+1}, A_{kk} = \frac{n+1-k}{n}\left(A_1 - \frac{k-1}{2n}a_{11}^2\right),$$
$$A_{k,k+1} = \frac{n+1-k}{n}\left(A_2 - \frac{k-1}{n}a_{11}a_{12}\right), \quad \text{for } k = 1, \ldots, n,$$
$$A_{k,k+2} = \frac{(n+1-k)(n-k)}{2n^2}a_{12}^2, \text{ for } k = 1, \ldots, n-1,$$
$$c_{kk} = \frac{n+1-k}{n}a_{11}, c_{k,k+1} = \frac{n+1-k}{n}a_{12}, \text{ for } k = 1, \ldots, n.$$

Unfortunately, the proposed exact linearization is valid only for positive variables $x_i > 0$ and, therefore is not useful for random vibration analysis.

### 6.6.2 Conditional Equivalent Linearization

A new idea of conditional linearization for the Duffing oscillator under narrowband excitation has been proposed by Iyengar and Roy [58]. The system is described by

$$\ddot{x} + 2\zeta\lambda_0\dot{x} + \lambda_0^2 x + \varepsilon\lambda_0^2 x^3 = \eta(t), \tag{6.248}$$

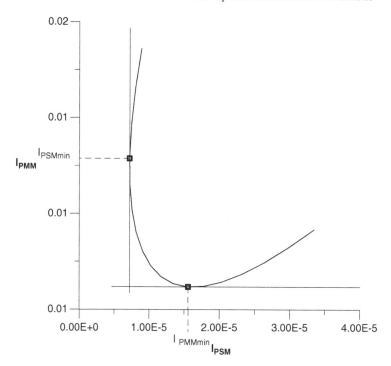

**Fig. 6.13.** A graphical illustration of the set of dominating points determined by (6.241–6.242), from [95] with permission

where $\zeta$, $\lambda_0$, and $\varepsilon$ are positive constant parameters, $\eta(t)$ is a narrowband noise (a filtered noise). For small $\varepsilon$ and damping filter parameter $\zeta$, the response is expected to be at the frequency $\lambda$ in the following approximation form:

$$x(t) = x_m \sin(\lambda t - \theta) , \tag{6.249}$$

where $x_m$ and $\theta$ are random variables. Based on the results of Spanos and Iwan [101] one can show that the linear system corresponding to the solution (6.249) has the form

$$\ddot{x}(t) + 2\zeta\lambda_0\dot{x}(t) + \lambda_0^2(1 + 0.75\varepsilon x_m^2)x(t) = \eta(t) . \tag{6.250}$$

Here $x_m$ is the unknown random peak amplitude. Although system (6.248) is still nonlinear, the corresponding conditional system (6.250) for $(x|x_m)$ is already linear and Gaussian. Iyengar and Roy [58] proposed an iterative procedure for the determination of the probability density function $g(x_m)$. Using this function, one can calculate the response characteristics of (6.250) that approximate the corresponding ones for original nonlinear system (6.248). This is an example of an equivalent linear oscillator with random parameters.

The idea of conditional linearization was also used by Iyengar and Roy [60, 59] to develop a numeric–analytical phase-space linearization (PSL)

method for the analysis and the simulation of nonlinear oscillators under deterministic excitation. The concept of PSL is essentially a conditional and local
linearization principle that works by reducing a given nonlinear dynamic system into a set of linear ones. The derivation of these linearized equations is
based on a locally defined least-squares error minimization on the nonlinear
part of the vector field. An extension of the PSL procedure for a nonlinear oscillator under deterministic and stochastic excitations was proposed by
Roy [85].

The considered nonlinear oscillator is given by

$$\ddot{x} + 2\pi\varepsilon_1\dot{x} + 4\pi^2\varepsilon_2 g(x)x = 4\pi^2\varepsilon_3 \cos(2\pi t) + \sigma(t)w(t) , \qquad (6.251)$$

where $\varepsilon_k, k = 1, 2, 3$, are positive constant parameters, $g(x)$ is a nonlinear
function, $\sigma(t) \in L_2[0, +\infty)$, and $w(t)$ is a zero-mean Gaussian white noise.

The idea of the explicit linearization procedure is to use a short time-
averaged Ito–Taylor expansion for replacing the nonlinear terms in (6.251)
with linear ones. Then, the conditionally linear oscillator is given by

$$\ddot{y} + 2\pi\varepsilon_1\dot{y} + 4\pi^2\varepsilon_2\beta(x(t_i), \dot{x}(t_i), t_i, h_i)y = 4\pi^2\varepsilon_3 \cos(2\pi t) + \sigma(t)w(t) , \quad (6.252)$$

where $t_0 < t_1 < t_2 < \ldots < t_i < t_{i+1}$, $h_i = t_{i+1} - t_i$, and the linearization
coefficient $\beta$ has the form

$$\beta = g(x(t_i)) + x(t_i)\dot{x}(t_i)h_i + \frac{2}{h_i}\pi G_0 x(t_i) \int_{t_i}^{t_i+h_i} (w(t) - w(t_i))dt$$

$$+ \frac{1}{3}[\dot{x}^2(t_i) - 2\pi\varepsilon_1 x(t_i)\dot{x}(t_i) - 4\pi^2\varepsilon_2 x^4(t_i)]$$

$$+ 4\pi^2\varepsilon_3 x(t_i) \cos(2\pi t_i)]h_i^2 + O(h_i^{2.5}). \qquad (6.253)$$

This approach was called "pathwise stochastic linearization" by Roy [85].


### 6.6.3 Discrete Time Equivalent Linearization

An application of statistical linearization to nonlinear discrete time systems
was developed in Soviet Union parallel with continuous time systems in the
context of control engineering, (see, for instance, [75]). The systematic derivation of stochastic difference equations and statistical linearization for discrete
time systems can be found in Kazakov and Maltchikov's book [63]. We shortly
discuss the main idea of this approach following Uchino et al.[105] who considered a discrete time nonlinear elements

$$y_k = g(z_k), \quad k = 0, 1, \ldots , \qquad (6.254)$$

where $g(z)$ is a nonlinear function, $z_k$ and $y_k$ are the input and the output of
the nonlinear $k$th element, respectively. The linearized variable is assumed to
be in the form

$$y_k = g(E[z_k]) + K_{n_k}(z_k - E[z_k]) , \qquad (6.255)$$

where $K_{n_k}$ is a linearization coefficient. Uchino et al.[105] have proposed the linearization criterion as follows:

$$E[(z_k - E[z_k])^n \epsilon_k] = 0 , \qquad (6.256)$$

where

$$\epsilon_k = g(z_k) - K_{n_k}(z_k - E[z_k]) - g(E[z_k]) . \qquad (6.257)$$

The determination of linearization coefficients is possible in a similar way as for the continuous time case by using an iterative procedure.

### 6.6.4 Equivalent Linear System with Random Coefficients

The idea of equivalent linear system with random coefficient introduced by Bouc [18] in the context of power spectral density criteria was modified by Ismaili and Bernard [55] who proposed a new local linearization approach. These authors considered a nonlinear two-wells Duffing oscillator described by

$$\ddot{x}(t) + 2\zeta\dot{x}(t) + k(-x + \varepsilon x^3) = \sigma w(t) , \qquad (6.258)$$

where $\zeta, k, \varepsilon$, and $\sigma$ are positive constant parameters and $w(t)$ is a standard white noise. This oscillator has two stable equilibria at $x_{c1} = -x_{c2} = 1/\sqrt{\varepsilon}$. The existence and uniqueness of the stationary solution of (6.258) follows from Lyapunov method. The idea of Ismaili and Bernard [55] was to find a locally linear oscillator with random coefficients that coincides over the domain attracted to each equilibrium position with a linearization in the neighborhood of this equilibrium position. That is, over the $D(x^i)$ domain attracted to the asymptotically equilibrium position $x_c^i$, they substitute into the nonlinear solution $x(t)$ of (6.258) the solution $x^i(t)$ of the following linear system

$$\ddot{x}^i(t) + c_i\dot{x}^i(t) + k_i(x^i - m_i) = \sigma w(t) , \qquad (6.259)$$

where $c_i, k_i > 0$, and $m_i \in D(x^i)$ are unknown constants to be determined by minimizing the cross entropy between the invariant measure of the nonlinear system (6.258) and the equivalent linear system (6.259). The strict mathematical derivation of this entropy approach is given by Bernard and Wu [11]. In this paper, which to our knowledge is the first in the literature where the entropy criteria are used to the determination of linearization coefficients, the authors discussed four linearization methods including "Gaussian linearization" and "true linearization".

### 6.6.5 Equivalent Linear System with Non-Gaussian Excitation

Falsone and Pirrotta [47] proposed a new version of equivalent linearization where a nonlinear system excited by a Gaussian white noise is replaced by a linear system excited by a non-normal delta-correlated process characterized by intensity coefficients up to a fixed order $k$. We note that a stochastic process $w(t)$ is delta-correlated if its correlation functions have the following expressions;

$$K_1(w(t_1)) = q_1(t_1),$$

$$K_r(w(t_1), \ldots, w(t_r)) = q_r(t_1)\delta(t_2 - t_1) \ldots \delta(t_r - t_1), \qquad r \geq 2,$$

(6.260)

where $q_r(t_1)$, $r \geq 2$, are the intensity functions of the process $w(t)$.

Falsone and Pirrotta [47] considered a one-dimensional system described by

$$\dot{x} = f(x) + w(t) ,$$

(6.261)

where $f(x)$ is a nonlinear function and $w(t)$ is a Gaussian white noise process. In the case of standard equivalent linearization, the linearized system has the form

$$\dot{x} = Ax + a + w(t) ,$$

(6.262)

while in the Falsone and Pirrotta's approach, the linearized system has the form

$$\dot{x} = Ax + a + \bar{w}(t) ,$$

(6.263)

where $\bar{w}(t)$ is a non-Gaussian the functions delta-correlated process whose intensities $q_r(t_1)$, $r \geq 2$, are unknown. The coefficients $A$, $a$, and the functions $\bar{q}_r(t)$ have to be evaluated in order to minimize the difference between systems (6.260) and (6.262). The differential formulas in the sense of Di Paola and Falsone [41] corresponding to these systems are

$$\Delta\phi(x,t) = \left[\frac{\partial\phi(x,t)}{\partial t} + \frac{\partial\phi(x,t)}{\partial x}f(x) + \frac{\partial^2\phi(x,t)}{\partial x^2}q_2(t)\right] dt ,$$

(6.264)

$$\Delta\phi(x,t) = \left[\frac{\partial\phi(x,t)}{\partial t} + \frac{\partial\phi(x,t)}{\partial x}(Ax + a) + \sum_{i=2}^{\infty}\frac{1}{i!}\frac{\partial^i\phi(x,t)}{\partial x^i}\bar{q}_i(t)\right] dt .$$

(6.265)

Falsone and Pirrotta [47] have shown that by substituting the polynomial into (6.264) and (6.265) and by minimizing the mean difference between these results with respect to $\bar{q}_r(t)$ after algebraic manipulations one obtains

$$\bar{q}_r(t) = r\left\{E[f(x)x^{r-1}] - AE[x^r] - aE[x^{r-1}]\right\}$$

$$-\sum_{j=3}^{r-1}\left\{\frac{1}{j!}\left[\prod_{i=0}^{j-1}(r-1)\right]E[x^{r-j}]\bar{q}_j(t)\right\} .$$

(6.266)

The iterative relation (6.266) was used for the determination of a new closure technique. Using the differential formula (6.265) one can derive the moment equations for the linearized system (6.262). They depend on intensity coefficients $\bar{q}_r(t)$ determined by (6.266). Therefore, to close the moments equation, an iterative procedure and a standard closure technique have to be used.

# Bibliography Notes

### Ref.6.2

As was mentioned in the introduction to this chapter, the observation done by Socha and Pawleta [97] and Elishakoff and Colojani [45] has initiated a sharp discussion in the literature. These first two papers were written in the spirit that equivalent linearization presented in the literature was incorrectly derived and the correct derivation is given in [97, 45].

Since the statement that equivalent linearization was incorrectly derived has not been accepted by "Mechanical and Structural Engineering Society", Socha and Pawleta [98] tried to find a compromise between an already existing correctly derived Kazakov's statistical linearization and an "incorrectly" or "not precisely" derived equivalent linearization. This compromise was the idea of input and output stochastic linearization in application to statistical, equivalent, and true linearization. The authors have introduced new definitions of input and output equivalent linearization, and they studied the influence of the averaging operation in the mean-square criteria of linearization. Showing the difference between these approaches for equivalent linearization, the authors wanted to show indirectly the error in previous derivations. A defense of standard equivalent linearization and critical discussion of the thesis presented in [97, 45, 32, 31] was given by Crandall [36] and Proppe et al. [81]. Crandall [36] quoted his paper [35] from 1980, where the formulas for linearization coefficients in the standard algebraic forms were derived correctly. We denote this Crandall's derivation by CD approach. In his opinion [37] the derivations were correct in all pioneers' papers also. The conclusion following from Crandall's paper [36] one can reduce to the following sentence.

"Since equivalent linearization has been correctly derived in the literature, the approaches proposed by Socha and Pawleta [97] and Elishakoff and Colojani [45] are useless."

Also Proppe et al. [81] concluded their critical comments in the following words.

"Socha and Pawleta [97] and Elishakoff and Colojani [45] proposed a linearization method that gives worse results than those obtained from the usual mean-square criterion. Unfortunately, by calling the proposed method as corrected linearization [97] and the usual method as erroneous [45] the authors gave the impression that the usual method is false."

The answer of Socha has appeared in [96]. He has written:

"Both conclusions seem to be sharp for the following reason. One can find the incorrect or not explained derivations of linearization coefficients in hundreds of papers in journals and proceedings and chapters of books. Usually the authors rewrite the derivation of linearization coefficients from books or some basic papers where it was done imprecisely (without information about expectation in the mean-square criterion). These authors do not quote the Crandall's derivation given in [35]. Also after publishing Crandall's paper [36] still appear papers with incorrect derivations of linearization coefficients.

Therefore, in our opinion, the papers by Socha and Pawleta [97] and by Elishakoff and Colojani [45] attracted some researchers' attention (unfortunately
not all) on this problem and showed incorrect derivation of linearization coefficients. This is the main advantage of these papers. The words used by Socha
and Pawleta [97] "corrected linearization" should be understood as "the correction of the often used derivation of linearization coefficients", it means that
the derivation of the "usual method" presented in many papers was false."

Recently, a new view on this discussion was presented by Crandall in [39],
who stated that equivalent linearization was correctly proposed by the "pioneers" of linearization methods for stochastic dynamic systems. Crandall has
observed that a kind of so called inconsistency in derivation formulas for linearization coefficient was introduced by Lin in his text [67] on p. 284. This
inconsistency involved in overlooking Lin's warning went unnoticed by anyone
in the field for 27 years. Crandall has stated, that Socha, Pawleta, Elishakoff
and Colojani supposed that they have discovered an error that has been introduced particularly by Caughey. In fact this misunderstanding was done by
Lin in [67]. In general, one can confirm this observation by a careful analysis
of papers with included list of references that appeared after 1967. However,
there is a paper written by Busby and Weingarten [22] where the mentioned
inconsistency in the derivation of formulas for linearization coefficient appeared and the authors did not cite Lin's book. It may suggest (however,
not necessary) that not only Lin but also other authors have problems with
correct understanding of the true idea of equivalent linearization. The reason
is probably in fact that Caughey has written almost at the same time three
papers [27, 26, 25] where the versions of derivation of linearization coefficients
are a little different.

At the end of this very academic discussion, it should be stressed that
in all cases of statistical linearization, CD approach, and incorrectly derived
equivalent linearization, the linearization coefficient $k_{eq}$ for one-dimensional
nonlinear function and mean-square criterion has the same form. However,
in statistical linearization and CD approach, the expectation $E$ in the mean-
square criterion is replaced by $E_N$, while in incorrect equivalent linearization
it is replaced by $E_L$. This difference is eliminated in the determination of
approximation of response moments by an iterative procedure.

Similar to statistical linearization, equivalent linearization has also been
used jointly with harmonic linearization for the analysis of response of nonlinear dynamic systems under joint harmonic and stationary Gaussian excitation.
Budgor [20] and Bulsara et al. [21] used this approach to investigate the Duffing oscillator under external excitation in the form $\xi(t) + A \sin \Lambda t$. Budgor [20]
considered the case in which the spectrum of the narrowband Gaussian noise
$\xi(t)$ lies outside the bandwidth associated with the linear part of the system
and presented a modified equivalent–harmonic linearization. Bulsara et al. [21]
considered a sequential procedure, applying first equivalent linearization and
then harmonic linearization. Although the solution of the linearized oscillator
equation is simply a superposition of the displacements caused by the ran-

dom and deterministic excitations, these excitations are nonlinearly coupled through the linearization coefficients. Some other applications and modifications of equivalent and harmonic linearization were considered by Windrich et al. [107], Hagedorn and Walaschek [51], Windrich [106].

Casciati et al. [24] proposed a different idea of the determination of linearization coefficients on two examples: the Duffing oscillator and the hysteretic oscillator. In the case of the Duffing oscillator, the authors derived analytically expressions for the expected uncrossing rate at the level $u$, the exact one for the linearized system and the approximated one for the original nonlinear system. By comparing these quantities they have found the approximated expressions for linearization coefficients.

An application of a linearization criterion with the averaging operation defined by a non-Gaussian distribution was proposed by Pradlwarter et al. [79] and developed by Pradlwarter [76, 77, 78]. Non-Gaussian properties of response vectors $y$ are taken into account by utilizing nonlinear transformations $y(x)$ between nonlinear response vector $y$ and its linearized (Gaussian) state vector $x$. The consideration of non-Gaussian properties significantly improves the estimates for the first two moments and makes equivalent linearization suitable for strongly nonlinear systems where the joint distribution deviates considerably from a jointly normal one. The non-Gaussian joint distribution of $y$ is represented by its marginal distribution and by modified correlation coefficients applicable to Gaussian distribution. The approximate shape of all marginal distributions is found by Monte Carlo simulation of non-Gaussian initial response vectors $y$ and numerical integration over a short time interval.

### Ref.6.4

The problem of the determination of the response power spectral density function for nonlinear dynamic systems under zero-mean Gaussian or sine wave excitation was considered by many authors. Bendat and Piersol [9] proposed an approach based on bispectral analysis for finite memory square law systems. This approach was developed by the same and other authors [29, 8, 84]. Another methodology was proposed by Billings and Tsang [15, 16, 14, 17]. Their method consists of estimating the parameters in a NARMAX (nonlinear autoregressive moving average) model description of the system and then computing the generalized frequency response functions directly from the estimated model. An interesting example of a nonlinear oscillator under white noise excitation with the exact known response power spectral density function was shown by Dimentberg et al. [42]. Also Iourtchenko [52] obtained an exact, closed-form analytical expression for a response power spectral density function for a class of linear systems subjected to external and parametric non-Gaussian stationary (delta-correlated) processes. An inverse problem was presented by Iyengar and Basak [57]. They studied the Duffing oscillator under random excitation with the prescribed power spectral density function and unspecified non-Gaussian probability density function and obtained sufficient conditions on these excitation under which the response can be a Gaussian process.

**Ref.6.5**
Pareto optimality approach was used in the study of linear stochastic dynamic system, for instance in [7, 43, 103].

**Ref.6.6**
An application of algebraic methods to linearization of stochastic systems was proposed by Wong [108] and next by Roy and Spanos [86]. They have developed the so-called "Carleman linearization" introduced for deterministic systems and applied to stochastic ones. This approach is in fact a kind of bilinearization method. Pan [74] has obtained the coordinate-free necessary and sufficient conditions for the solvability of the feedback complete linearization problem for SISO stochastic nonlinear systems. It was shown by suitable transformation rule to the equivalent problem of the solvability of the feedback complete linearization for the deterministic uncertain nonlinear system that is associated with the considered nonlinear stochastic system.

Statistical linearization was applied to discrete time systems mainly in the field of control engineering, for instance, [30, 73, 70].

The idea of the approximation of nonlinear oscillator with non-Gaussian excitations by a linear oscillator with equivalent Gaussian excitation was proposed by Floris [48].

Another interesting criterion of linearization was introduced by Anh and Shiehlen [2]

$$P\{-\rho < \epsilon(x) < \rho\} \longrightarrow \max_{c,k}, \tag{6.267}$$

where $\epsilon = f(x, \dot{x}) - c\dot{x} - kx$ is an error of linearization for function $f(x, \dot{x})$.

At the end we note a new, non-parametric linearization method for nonlinear stochastic dynamic systems proposed by Fujimura and Der Kiureghian [50]. In the considered models of nonlinear and linearized systems, the stochastic excitation is used in a discrete time representation. The main idea of this approach is to find the equivalent linear system that is defined by matching the "design points" of the linear and nonlinear responses in the space of the standard Gaussian random variables obtained from the discretization of the excitation. Hence, it follows that the first-order approximation of tail probability of the nonlinear system is equal to the tail probability of the linear system for the specified threshold.

# References

1. Adomian, G. and Sibul, L.: Symmetrized solutions for nonlinear stochastic differential equations. Int. J. Math. Sci. Appl. **4**, 529–542, 1981.
2. Anh, N. and Shiehlen, W.: New criterion for Gaussian equivalent linearization. Eur. J. Mech. A/Solids **16**, 1025–1039, 1997.

3. Apetaur, M.: Linearization function of a complex nonlinear two-force-element. Veh. Syst. Dyn. **20**, 309–320, 1991.

4. Apetaur, M.: Modified second order linearization procedure – problems encountered and their solution. Veh. Syst. Dyn. **17**, 255–265, 1988.

5. Apetaur, M. and Opicka, F.: Linearization of nonlinear stochastically excited dynamic systems. J. Sound Vib. **86**, 563–585, 1983.

6. Atalik, T. and Utku, S.: Stochastic linearization of multi-degree-of-freedom nonlinear systems. Earthquake Eng. Struct. Dyn. **4**, 411–420, 1976.

7. Basar, T.: Decentralized multi-criteria optimization of linear stochastic systems. IEEE Trans. Autom. Control **23**, 233–243, 1978.

8. Bendat, J. and Piersol, A.: Decomposition of wave forces into linear and nonlinear components. J. Sound Vib. **106**, 391–408, 1986.

9. Bendat, J. and Piersol, A.: Spectral analysis for nonlinear systems involving square law operations. J. Sound Vib. **81**, 199–213, 1982.

10. Benfratello, S.: Pseudo-force method for a stochastic analysis of nonlinear systems. Probab. Eng. Mech. **11**, 113–123, 1996.

11. Bernard, P. and Wu, L.: Stochastic linearization: The theory. J. Appl. Probab. **35**, 718–730, 1998.

12. Bernard, P.: About stochastic linearization. In: N. Bellomo and F. Casciati (eds.), Nonlinear Stochastic Mechanics, pp. 61–70. Springer, Berlin, 1991.

13. Bernard, P. and Taazount, M.: Random dynamics of structures with gaps: Simulation and spectral linearization. Nonlinear Dyn. **5**, 313–335, 1994.

14. Billings, S. and Tsang, K.: Spectral analysis for nonlinear systems Part III: Case study examples. Mech. Sys. Signal Proc. **4**, 3–21, 1990.

15. Billings, S. and Tsang, K.: Spectral analysis for nonlinear systems Part I: Parametric non-linear spectral analysis. Mech. Sys. Signal Proc. **3**, 319–339, 1989.

16. Billings, S. and Tsang, K.: Spectral analysis for nonlinear systems Part II: Interpretation of non-linear frequency response functions. Mech. Sys. Signal Proc. **3**, 341–359, 1989.

17. Billings, S. and Tsang, K.: Spectral analysis of block structured nonlinear systems. Mech. Sys. Signal Proc. **4**, 117–130, 1990.

18. Bouc, R.: The power spectral density of response for a strongly non-linear random oscillator. J. Sound Vib. **175**, 317–331, 1994.

19. Brown, A., Ankireddi, S., and Yang, H.: Actuator and sensor placement for multiobjective control of structures. J. Struct. Eng. **125**, 757–765, 1999.

20. Budgor, A.: Studies in nonlinear stochastic processes III. Approximate solutions of nonlinear stochastic differential equations excited by Gaussian noise and harmonic disturbances. J. Stat. Phys. **17**, 21–44, 1977.

21. Bulsara, A., Lindberg, K., and Schuler, K.: Spectral analysis of a nonlinear oscillator driven by random and periodic forces, I. linearized theory. J. Stat. Phys. **27**, 787–808, 1982.

22. Busby, H. and Weingarten, V.: Response of nonlinear beam to random excitation. ASCE J. Eng. Mech. **99**, 55–67, 1973.

23. Cai, G. Q. and Lin, Y.: Response spectral densities of strongly nonlinear systems under random excitation. Probab. Eng. Mech. **12**, 41–47, 1997.

24. Casciati, F. Faravelli, L., and Hasofer, A.: A new philosophy for stochastic equivalent linearization. Probab. Eng. Mech. **8**, 179–185, 1993.

25. Caughey, T. and Dienes, J.: Analysis of a nonlinear first order system with a white noise input. J. Appl. Phys. **23**, 2476–2479, 1961.

26. Caughey, T.: Random excitation of a system with bilinear hysteresis. Trans. ASME J. Appl. Mech. **27**, 649–652, 1960.
27. Caughey, T.: Response of Van der Pol's oscillator to random excitations. Trans. ASME J. Appl. Mech. **26**, 345–348, 1959.
28. Chen, G.: Cascade linearization of nonlinear system subjected to Gaussian excitations. In: B. Spencer and E. Johnson (eds.), Stochastic Structural Dynamics, pp. 69–76. Balkema, Rotterdam, 1999.
29. Choi, D., Miksad, R., and Powers, E.: Application of digital cross-bispectral analysis techniques to model the nonlinear response of a moored vessel system in random seas. J. Sound Vib. **99**, 309–326, 1985.
30. Cohen, G.: Analysis of a class of nonlinear sampled-data systems with a Gaussian input signal. IEEE Trans. Aerospace Electron. Syst. **7**, 71–91, 1971.
31. Colojani, P. and Elishakoff, I.: A new look at the stochastic linearization method for hyperbolic tangent oscillator. Chaos Solitons Fractals **9**, 1611–1623, 1998.
32. Colojani, P. and Elishakoff, I.: A subtle error in conventional stochastic linearization methods. Chaos Solitons Fractals **9**, 479–491, 1998.
33. Crandall, S.: Perturbation techniques for random vibration of nonlinear systems. J. Acoust. Soc. Am. **35**, 1700–1705, 1963.
34. Crandall, S.: Correlations and spectra of nonlinear system response. Probl. Nonlinear Vib. **14**, 39–53, 1973.
35. Crandall, S.: On statistical linearization for nonlinear oscillators. In: J. Rammath, R. V. Hedrick and H. Paynter (eds.), Nonlinear System Analysis and Synthesis, pp. 199–209. American Society of Mechanical Engineers, New York, 1980.
36. Crandall, S.: Is stochastic equivalent linearization a subtly flawed procedure. Probab. Eng. Mech. **16**, 169–176, 2001.
37. Crandall, S.: On using non-Gaussian distributions to perform statistical linearization. In: W. Zhu, G. Cai, and R. Zhang (eds.), Advances in Stochastic Structural Dynamics, pp. 49–62. CRC Press, Boca Raton, 2003.
38. Crandall, S.: On using non-Gaussian distributions to perform statistical linearization. Int. J. Non-Linear Mech. **39**, 1395–1406, 2004.
39. Crandall, S.: A half-century of stochastic equivalent linearization. Struct. Control Health Monit. **13**, 27–40, 2006.
40. Daucher, D., Fogli, M., and Clair, D.: Modelling of complex dynamical behaviours using a state representation technique based on a vector ARMA approach. Probab. Eng. Mech. **21**, 73–80, 2006.
41. Di Paola, M. and Falsone, G.: Stochastic dynamics of nonlinear systems driven by non-normal delta-correlated processes. Trans. ASME J. Appl. Mech. **60**, 141–148, 1993.
42. Dimentberg, M., Hou, Z., and Noori, M.: Spectral density of a non-linear single-degree-of-freedom system's response to white noise random excitation: A unique case of an exact solution. Int. J. Non-Linear Mech. **30**, 673–676, 1995.
43. Eberhard, P., Schiehlen, W., and Bestle, D.: Some advantages of stochastic methods in multi-criteria optimization of multibody systems. Arch. Appl. Mech. **69**, 543–554, 1999.
44. Elishakoff, I.: Multiple combinations of the stochastic linearization criteria by the moment approach. J. Sound Vib. **237**, 550–559, 2000.
45. Elishakoff, I. and Colojani, P.: Stochastic linearization critically re-examined. Chaos Solitons Fractals **8**, 1957–1972, 1997.

46. Elishakoff, I. and Colojani, P.: Booton's problem re-examined. J. Sound Vib. **210**, 683–691, 1998.

47. Falsone, G. and Pirrotta, A.: A new stochastic linearization approach. In: S. P. D. (ed.), Computational Stochastic Mechanics, pp. 105–112. A. A. Balkema, Rotterdam/Brookfield, 1995.

48. Floris, C.: Equivalent Gaussian process in stochastic dynamics with application to along-wind response of structures. Int. J. Non-Linear Mech. **31**, 779–794, 1996.

49. Foster, E.: Semi-linear random vibrations in discrete systems. Trans. ASME J. Appl. Mech. **35**, 560–564, 1968.

50. Fujimura, K. and Der Kiureghian, A.: Tail-equivalent linearization method for nonlinear random vibration. Probab. Eng. Mech. **22**, 63–76, 2007.

51. Hagedorn, P. and Wallaschek, J.: On equivalent harmonic and stochastic linearization of nonlinear shock-absorbers in nonlinear stochastic dynamic engineering systems. In: F. Ziegler and G. Schuëller (eds.), Nonlinear Stochastic Dynamic Engineering Systems, IUTAM Symposium, Innsbruck/ Igls Austria, pp. 23–32. Springer-Verlag, Berlin, 1988.

52. Iourtchenko, D.: Response spectral density of linear systems with external and parametric non-Gaussian, delta-correlated excitation. Probab. Eng. Mech. **18**, 31–36, 2003.

53. Isidori, A.: Nonlinear Control Systems. An Introduction, 2nd edn. Springer, Berlin, 1989.

54. Ismaili, M.: Design of a system of linear oscillators spectrally equivalent to a non-linear one. Int. J. Non-Linear Mech. **31**, 573–580, 1996.

55. Ismaili, M. and Bernard, P.: Asymptotic analysis and linearization of the randomly perturbed two-wells Duffing oscillator. Probab. Eng. Mech. **12**, 171–178, 1997.

56. Iyengar, R.: Higher order linearization in nonlinear random vibration. Int. J. Non-Linear Mech. **23**, 385–391, 1988.

57. Iyengar, R. and Basak, B.: Investigation of a non-linear system under partially prescribed random excitation. Int. J. Non-Linear Mech. **40**, 1102–1111, 2005.

58. Iyengar, R. and Roy, D.: Conditional linearization in nonlinear random vibration. ASCE J. Eng. Mech. **119**, 197–200, 1996.

59. Iyengar, R. and Roy, D.: Extensions of the phase space linearization method. J. Sound Vib. **211**, 877–906, 1998.

60. Iyengar, R. and Roy, D.: New approaches for the study of nonlinear oscillators. J. Sound Vib. **211**, 843–875, 1998.

61. Izumi, M., Zaiming, L., and Kimura, M.: A stochastic linearization method and its application to response analysis of nonlinear system based on weighted least–square minimization. AIJ J. Struct. Constr. Eng. **395**, 72–81, 1989.

62. Kazakov, I. and Dostupov, B.: Statistical Dynamic of Nonlinear Automatic Systems. Fizmatgiz, Moskwa, 1962 (in Russian).

63. Kazakov, I. and Malczikow, S.: Analysis of Stochastic Systems in State Space. Nauka, Moskwa, 1975 (in Russian).

64. Kozin, F.: The method of statistical linearization for non-linear stochastic vibration. In: F. Ziegler and G. Schueller (eds.), Nonlinear Stochastic Dynamic Engineering Systems, pp. 45–56. Springer, Berlin, 1989.

65. Krenk, S. and Roberts, J.: Local similarity in nonlinear random vibration. Trans. ASME J. Appl. Mech. **66**, 225–235, 1999.

66. Krilov, N. and Bogoliuboff, N.: Introduction to Non-linear Mechanics. Princeton University Press, Princeton, 1943.

67. Lin, Y.: Probabilistic Theory of Structural Dynamics. McGraw Hill, New York, 1967.

68. Manning, J.: Response spectra for nonlinear oscillator. ASCE J. Eng. Ind. **97**, 1223–1226, 1975.

69. Molnar, A., Vaschi, K., and Gay, C.: Application of normal mode theory of pseudo force methods to solve problems with nonlinearities. J. Press. Vessel Tech. **98**, 151–156, 1976.

70. Moore, J., Zhou, X., and Lim, A.: LQG controls with control dependent noise. Syst. Control. Lett. **36**, 199–206, 1999.

71. Morton, J. and Corrisin, S.: Consolidated expansions for estimating the response of a randomly driven nonlinear oscillator. J. Stat. Phys. **2**, 153–194, 1970.

72. Naumov, B.: The Theory of Nonlinear Automatic Control Systems. Frequency Methods. Nauka, Moscow, 1972 (in Russian).

73. Pakshin, P.: Approximate synthesis of discrete-time stochastic saturating systems. Probab. Control. Inform. Theory **8**, 327–339, 1979.

74. Pan, Z.: Differential geometric condition for feedback complete linearization of stochastic nonlinear systems. Automatica **37**, 145–149, 2001.

75. Perov, V.: Statistical Synthesis of Sampled-Data Systems. Sovet-skoe Radio, Moscau, 1959 (in Russian).

76. Pradlwarter, H.: Non-Gaussian linearization: An efficient tool to analyse nonlinear MDOF-systems. In: Proc. 10th Int. Conf. on Structural Mechanics in Reactor Technology, pp. 19–30, California, 1989.

77. Pradlwarter, H.: Consideration of non-Gaussian hysteretic response properties of MDOF-systems by use equivalent linearization. In: A. Ang, M. Shinozuka, and G. Schuëller (eds.), Structural Safety and Reliability, pp. 1333–1340. ASCE, New York, 1990.

78. Pradlwarter, H.: On the existence of "True" equivalent linear systems for the evaluation of the nonlinear stochastic response due to nonstationary Gaussian excitation. Int. J. Non-Linear Mech. **32**, 1990.

79. Pradlwarter, H., Schuëller, G., and Chen, X.: Consideration of non-Gaussian response properties by use of stochastic equivalent linearization. In: M. Petyt, H. Wolfe, and C. Mei (eds.), Proc. Third Int. Conf. on Recent Advances in Structural Dynamics, pp. 737–752. Wright-Patterson Air Force, Southhampton, 1988.

80. Priestly, M.: Spectral Analysis and Time Series, 5th edn. Academic Press, New York, 1981.

81. Proppe, C., Pradlwarter, H., and Schueller, G.: Equivalent linearization and Monte Carlo simulation in stochastic dynamics. Probab. Eng. Mech. **18**, 1–15, 2003.

82. Pugachev, V. and Sinitsyn, I.: Stochastic Differential Systems. Wiley, Chichester, 1987.

83. Pupkov, K.: Statistical Calculations of Nonlinear Automatic Control Systems. Mashinostroyene, Moscau, 1965 (in Russian).

84. Rice, H. and Fitzpatrick, J.: A generalised technique for spectral analysis for nonlinear systems. Mech. Sys. Signal Proc. **2**, 195–207, 1988.

85. Roy, D.: Exploration of the phase–space linearization method for deterministic and stochastic nonlinear dynamical systems. Nonlinear Dyn. **23**, 225–258, 2000.

86. Roy, R. and Spanos, P.: Wiener–Hermite functional representation of nonlinear stochastic systems. Stochastics **6**, 187–202, 1989.

87. Roy, R. and Spanos, P.: Power spectral density of nonlinear system response: The recursive method. Trans. ASME J. Appl. Mech. **60**, 358–365, 1993.

88. Sarkar, A.: Linear stochastic dynamical system under uncertain load: inverse reliability analysis. ASCE J. Eng. Mech. **129**, 665–671, 2003.

89. Sawaragi, Y., Nakayama, H., and Tanino, T.: Theory Multiobjective Optimization. Academic Press, New York, 1985.

90. Skulimowski, A.: Decision Support Systems Based on Reference Sets Theory Multiobjective Optimization. AGH-Press, Krakow, 1996.

91. Socha, L.: Some remarks on exact linearization of a class of stochastic dynamical systems. IEEE Trans. Autom. Control **39**, 1980–1984, 1994.

92. Socha, L.: Application of probability metrics to the linearization and sensitivity analysis of stochastic dynamic systems. In: Proc. Int. Conf. on Nonlinear Stochastic Dynamics, pp. 193–202. Hanoi, Vietnam, 1995.

93. Socha, L.: Moment equivalent linearization method. Z. Angew. Math. Mech. **75**, 577–578, 1995 SII.

94. Socha, L;, Probability density statistical and equivalent linearization techniques. Int. J. Syst. Sci. **33**, 107–127, 2002.

95. Socha, L.: Application of poli-criterial linearization for control problem of stochastic dynamic systems, J. Theor. Appl. Mech., **43**, 675–693, 2005.

96. Socha, L.: Linearization in analysis of nonlinear stochastic systems – recent results. Part I. Theory. ASME Appl. Mech. Rev. **58**, 178–205, 2005.

97. Socha, L. and Pawleta, M.: Corrected equivalent linearization. Mach. Dyn. Probl. **7**, 149–161, 1994.

98. Socha, L. and Pawleta, M.: Some remarks on equivalent linearization of stochastic systems. In: B. Spencer and E. Johnson (eds.), Stochastic Structural Dynamics, pp. 105–112. Balkema, Rotterdam, 1999.

99. Socha, L. and Pawleta, M.: Are statistical linearization and standard equivalent linearization the same methods in the analysis of stochastic dynamic systems. J. Sound Vib. **248**, 387–394, 2001.

100. Socha, L. and Soong, T.: Linearization in analysis of nonlinear stochastic systems. Appl. Mech. Rev. **44**, 399–422, 1991.

101. Spanos, P. and Iwan, W.: Harmonic analysis of dynamic systems with nonsymmetric nonlinearities. J. Dyn. Syst. **101**, 31–36, 1979.

102. Stadler, W.: multi-criteria optimization in mechanics (a survey). Appl. Mech. Rev. **37**, 277–286, 1984.

103. Starzec, P. and Anderson, J.: Application of two-level factorial design to sensitivity analysis of keyblock statistics from fracture geometry. Int. J. Rock Mech. Min. **39**, 243–255, 2002.

104. Thaler, G. and Pastel, M.: Analysis and Design of Nonlinear Feedback Control Systems. MacGraw-Hill, New York, 1962.

105. Uchino, E., Ohta, M., and Takata, H.: A new state estimation method for a quantized stochastic sound system based on a generalized statistical linearization. J. Sound Vib. **160**, 193–203, 1993.

106. Windrich, H.: Application of statistical linearization and averaging methods for limit cycle systems. Prikladnaya Mekhanika **24**, 122–126, 1988 (in Russian).

107. Windrich, H., Müller, P., and Popp, K.: Approximate analysis of limit cycles in the presence of stochastic excitations. In: F. Ziegler and G. Schuëller (eds.), Nonlinear Stochastic Dynamic Engineering Systems, IUTAM Symposium, Innsbruck/Igls Austria, pp. 33–44. Springer Verlag, Berlin, 1988.

108. Wong, W.: Carleman linearization and moment equations of nonlinear stochastic equations. Stochastics **9**, 77–101, 1983.

# 7

# Nonlinearization Methods

## 7.1 Introduction

The idea of finding an equivalent model for a nonlinear system has not been limited to a linear model. It was first extended to equivalent nonlinear models by Lutes [13], who used an energy-dependent system with a nonlinear restoring force to replace a hysteretic system subjected to a white noise excitation. In general, this idea is the replacement of a nonlinear oscillator by another nonlinear oscillator with known exactly response probability density function. This approach is often called *equivalent nonlinearization* (EN) *method.* Maltchikov [14, 15] replaced a multidimensional system with parametric and external excitations by one-dimensional model whose output has the same distribution as that of the original system. Caughey [3, 4] developed an equivalent nonlinear differential equation model for quasilinear stochastic systems subjected to an external white noise excitation specified by

$$\ddot{x} + f(x,\dot{x})sgn(\dot{x}) + x = \dot{\xi}(t) , \qquad (7.1)$$

where

$$f(x,\dot{x}) = f_{00} + f_{01}|\dot{x}| + f_{10}|x| + f_{11}|x\dot{x}| + \ldots + f_{ij}|x^i\dot{x}^j| + \ldots, \quad 0 \le i,j \le N \qquad (7.2)$$

$f_{ij}$ are constant parameters and $\dot{\xi}(t)$ is a white noise, $E[\dot{\xi}(t_1)\dot{\xi}(t_2)] = 2\sigma^2\delta(t_1 - t_2)$, and $\sigma$ is a positive constant.

This system is then replaced by the following nonlinear model:

$$\ddot{x} + c(H)\dot{x} + x + e(x,\dot{x}) = \dot{\xi}(t) , \qquad (7.3)$$

where

$$H = \frac{1}{2}(x^2 + \dot{x}^2) , \qquad (7.4)$$

$$c(H) = a_{00}c_{00}(H) + a_{01}c_{01}(H) + a_{10}c_{10}(H) \qquad (7.5)$$

L. Socha: *Nonlinearization Methods*, Lect. Notes Phys. **730**, 211–249 (2008)
DOI 10.1007/978-3-540-72997-6_7

and $e(x, \dot{x})$ is the error term. The "nonlinearization coefficients" $a_{ij}$ and $c_{ij}(H)$ can be determined indirectly from the conditions of minimization of mean-square error $E[e^2(x, \dot{x})]$. To calculate these coefficients, one may use the polar coordinates $\rho$ and $\theta$, i.e.,

$$x = \rho \cos \theta , \quad \dot{x} = \rho \sin \theta , \tag{7.6}$$

then

$$E[e^2(\rho \cos \theta, \rho \sin \theta)] = E[e^2(\rho, \theta)] = \frac{1}{2\pi} \int_0^{2\pi} \int_0^{+\infty} e^2(\rho, \theta) g_e(\rho) d\rho d\theta , \tag{7.7}$$

where $g_e(\rho)$ has the form

$$g_e(\rho) = b_0 exp\left\{ -\frac{1}{\sigma} \int_0^H c(z) dz \right\} \int_0^\rho \frac{dx}{\dot{x}(\rho, x)} , \tag{7.8}$$

$b_0$ is a normalizing constant.

Minimizing of $E[e^2(\rho, \theta)]$ with respect to $a_{ij}$ yields $N^2$ equations of the form

$$E[c_{ij}(H) \rho \sin \theta e(\rho, \theta)] = 0 . \tag{7.9}$$

Caughey [3] proposed to select the coefficients $c_{ij}(H)$ in the form

$$c_{ij}(H) = (2H)^{(i+j-1)/2} = \rho^{i+j-1} , \tag{7.10}$$

then (7.9) becomes:

$$E\left[ \rho^{(i+j)}[f_{ij}| \cos^i \theta \sin^{(j+1)}(\theta)| - a_{ij} \sin^2 \theta] \right] = 0 \tag{7.11}$$

and to select

$$a_{ij} = f_{ij} \frac{\int_0^{\pi/2} \cos^i \theta \sin^{(j+1)}(\theta) d\theta}{\int_0^{\pi/2} \sin^2 \theta d\theta} = \frac{2 f_{ij} \Gamma(\frac{i+2}{2}) \Gamma(\frac{i+1}{2})}{\pi \Gamma(\frac{i+j+3}{2})} . \tag{7.12}$$

In this case, the $N^2$ equations (7.9) are satisfied and the steady-state envelope probability density function is given by

$$g_e(\rho) = b_1 exp\left\{ -\frac{1}{\sigma} \int_0^{\frac{1}{2}\rho^2} c(z) dz \right\} , \tag{7.13}$$

where $b_1$ is a normalizing constant. The steady-state probability density function, in polar coordinates, is given by

$$g_e(\rho, \theta) = \frac{1}{2\pi} g_e(\rho) . \tag{7.14}$$

*Example 7.1.* [3] Consider a nonlinear oscillator with Coulomb damping excited by standard Gaussian white noise

$$\ddot{x} + f\,sgn(\dot{x}) + x = \dot{\xi}(t) \ . \tag{7.15}$$

This is a special case of (7.1) with $f_{00} = f$, $f_{ij} = 0$, $0 < i, j \leq N$.

We replace (7.15) by the equivalent nonlinear differential equation

$$\ddot{x} + ac(H)\dot{x} + x = \dot{\xi}(t) \ . \tag{7.16}$$

Using the previous derivations one can select

$$c(H) = \frac{1}{\sqrt{2H}} \tag{7.17}$$

and

$$a = \frac{2f\Gamma(1)\Gamma(\frac{1}{2})}{\pi\Gamma(\frac{3}{2})} = \frac{4f}{\pi} \ , \tag{7.18}$$

$$g_e(\rho) = \left(\frac{4f}{\sigma\pi}\right)^2 \rho \exp\left\{-\frac{4f\rho}{\pi\sigma}\right\} \ . \tag{7.19}$$

One can calculate using the probability density function (7.19)

$$E[x^2] = E[\dot{x}^2]\frac{1}{2}E[\rho^2] = 3\left(\frac{\sigma\pi}{4f}\right)^2 \ . \tag{7.20}$$

We note that the probability density function of stationary solution of (7.3) is known and has the following form:

$$g(x, \dot{x}) = b_1 \exp\left\{-\frac{4f}{\pi\sigma}\sqrt{x^2 + \dot{x}^2}\right\} \ , \tag{7.21}$$

where $b_1$ is a normalized constant.

In general case *equivalent nonlinearization* can be treated as a generalization of equivalent linearization. It means that if we consider the nonlinear stochastic dynamic system with external excitation in the form of Gaussian white noises described by the Ito vector stochastic differential equation

$$d\mathbf{x} = \mathbf{\Phi}(\mathbf{x}, t)dt + \sum_{k=1}^{M} \mathbf{G}_{k0}(t)d\xi_k \ , \tag{7.22}$$

where $\mathbf{\Phi}(\mathbf{x}, t)$ is a nonlinear vector function, $\mathbf{G}_{k0}(t)$ are vectors of intensity of noises, $\xi_k$, $k = 1, \ldots, M$, are independent standard Wiener processes, then the equivalent nonlinear system has the form

$$d\mathbf{x} = \mathbf{\Phi}_{EN}(\mathbf{x}, \mathbf{p}_{en}, t)dt + \sum_{k=1}^{M} \mathbf{G}_{k0}(t)d\xi_k \ , \tag{7.23}$$

where $\boldsymbol{\Phi}_{EN}(\mathbf{x}, \mathbf{p}_{en}, t)$ is a nonlinear vector function corresponding to the equivalent nonlinear system, $\mathbf{p}_{en}$ is a vector of nonlinearization coefficients, that is determined by application of a criterion of nonlinearization.

Since the exact solutions of nonlinear dynamic systems are mainly known for one- and two-dimensional systems the class of considered nonlinear dynamic systems was reduced to one-degree-of-freedom systems. Similar to equivalent linearization methods the determination of equivalent nonlinear systems has been developed by many authors using a few basic approaches, namely, statistical linearization of nonlinear damping, solvable Fokker–Planck–Kolmogorov equation, and generalized stochastic averaging. Mainly the authors considered the replacement of a nonlinear oscillator by another nonlinear oscillator with known response probability density function. We show the four basic approaches mentioned and some possible generalizations.

## 7.2 Moment Criteria

### 7.2.1 Partial Linearization

Statistical linearization is as well the convenient mathematical tool in approximation of a nonlinear oscillator by a linear one as by a simpler nonlinear one. This fact was first observed by Elishakoff and Cai [6], who considered the following original nonlinear system:

$$dx_1 = x_2 dt$$
$$dx_2 = [-f(x_1, x_2) - r(x_1)]dt + \sqrt{2q}d\xi(t) , \qquad (7.24)$$

where $f(x_1, x_2)$ and $r(x_1)$ are nonlinear functions, $\xi(t)$ is the standard Wiener process, and $q$ is a positive constant. They proposed to apply statistical linearization only to nonlinear damping and called this approach by "partial linearization," i.e., they replaced the nonlinear damping force $f(x_1, x_2)$ by an equivalent linear one and obtained the equivalent nonlinear system in the form

$$dx_1 = x_2 dt ,$$
$$dx_2 = [-2h_e x_2 - r(x_1)]dt + \sqrt{2q}d\xi(t) . \qquad (7.25)$$

To obtain the linearization coefficient $2h_e$, the authors proposed the criterion of equality of the average energy dissipation for both systems, namely,

$$E[x_2 f(x_1, x_2)] = 2h_e E[x_2^2] . \qquad (7.26)$$

This approach can be generalized by looking for equivalent nonlinear system in the form

$$dx_1 = x_2 dt$$
$$dx_2 = [-2h_e x_2 - k_e x_1 - r(x_1)]dt + \sqrt{2q}d\xi(t) , \qquad (7.27)$$

where $h_e$ and $k_e$ are linearization coefficients that can be determined, for instance, from the mean-square criterion,

$$I_N = E_N[(f(x_1, x_2) - 2h_e x_2 - k_e x_1)^2] , \qquad (7.28)$$

i.e., from the corresponding necessary conditions

$$\frac{\partial I_N}{\partial h_e} = 0 , \quad \frac{\partial I_N}{\partial k_e} = 0 . \qquad (7.29)$$

If we assume that the averaging operation $E_N[.]$ is defined by the probability density function that does not depend on linearization coefficients (it can be, for instance, the probability density of stationary solution of the original nonlinear system (7.24)), then we find from conditions (7.29)

$$E_N[x_2 f(x_1, x_2)] - 2h_e E_N[x_2^2] - k_e E_N[x_1 x_2] = 0 ,$$

$$E_N[x_1 f(x_1, x_2)] - 2h_e E_N[x_1 x_2] - k_e E_N[x_1^2] = 0 . \qquad (7.30)$$

The moment equations for the original nonlinear system (7.24) have the form

$$\frac{dE_N[x_1^2]}{dt} = E_N[x_2^2] ,$$

$$\frac{dE_N[x_1 x_2]}{dt} = E_N[x_2^2] - E_N[x_1 f(x_1, x_2)] - E_N[x_1 r(x_1)] ,$$

$$\frac{dE_N[x_2^2]}{dt} = -2E_N[x_2 f(x_1, x_2)] - 2E_N[x_2 r(x_1)] + 2q^2 . \qquad (7.31)$$

For the stationary solution we find that

$$E_N[x_1 x_2] = 0 , \quad E_N[x_2 r(x_1)] = 0 , \quad E_N[x_2 f(x_1, x_2)] = q^2 , \qquad (7.32)$$

hence,

$$2h_e = \frac{E_N[x_2 f(x_1, x_2)]}{E_N[x_2^2]} = \frac{q^2}{E_N[x_2^2]} ,$$

$$k_e = \frac{E_N[x_1 f(x_1, x_2)]}{E_N[x_1^2]} = \frac{E_N[x_2^2] - E_N[x_1 r(x_1)]}{E_N[x_1^2]} . \qquad (7.33)$$

Here we note that $h_e$ and $k_e$ have been found as the linearization coefficients for the nonlinear function (static element) $f(x_1, x_2)$ and at the same time they can be treated as the nonlinearization coefficients for the dynamic system (7.25).

To calculate the linearization coefficients, we have to approximate the moments defined by the averaging operation $E_N[.]$ by the corresponding moments obtained from stationary moment equations for equivalent nonlinear system

(7.27). Then the averaging operation $E_N[.]$ in equalities (7.33) is replaced by $E_n[.]$ defined by the probability density function of the stationary solution of (7.27), which is known in the analytical form

$$g_n(x_1, x_2) = \frac{1}{c_n} \exp\left\{ -\frac{2h_e}{q^2} \left[ \frac{x_2^2}{2} + \frac{k_e x_1^2}{2} + \int_0^{x_1} r(u)du \right] \right\} , \qquad (7.34)$$

where $c_n = c_n(h_e, k_e)$ is a normalized constant being a function of linearization coefficients, i.e.,

$$c_n(h_e, k_e) = \int_{-\infty}^{+\infty} \int_{-\infty}^{+\infty} \exp\left\{ -\frac{2h_e}{q^2} \left[ \frac{x_2^2}{2} + \frac{k_e x_1^2}{2} + \int_0^{x_1} r(u)du \right] \right\} dx_1 dx_2 . \qquad (7.35)$$

Taking into account (7.33–7.35) and applying an iterative procedure similar to Procedure $EXAMPLE_{L1}$, one can calculate as well nonlinearization coefficients $h_e$ and $k_e$ as approximate response characteristics for a given original nonlinear system (7.24).

## Procedure $EXAMPLE_{NL1}$

(1) Substitute the initial values for the nonlinearization coefficients $h_e$ and $k_e$ and calculate response moments defined by the joint probability density function (7.34) and (7.35).
(2) Calculate the new nonlinearization coefficients $h_e$ and $k_e$ by formulas (7.33) replacing the moments defined by the averaging operation $E_N[.]$ by the corresponding moments defined by the joint probability density function (7.34) and (7.35).
(3) Substitute the coefficients $h_e$ and $k_e$ into the joint probability density function (7.34) and (7.35) and calculate the new response moments.
(4) Repeat steps (2) and (3) until convergence.

Here it is important to stress what kind of substitution in Procedure $EXAMPLE_{NL1}$ is applied. If we use the substitution

$$2h_e = \frac{q^2}{E_N[x_2^2]} \cong \frac{q^2}{E_n[x_2^2]} ,$$

$$k_e = \frac{E_N[x_2^2] - E_N[x_1 r(x_1)]}{E_N[x_1^2]} \cong \frac{E_n[x_2^2] - E_n[x_1 r(x_1)]}{E_n[x_1^2]} , \qquad (7.36)$$

then the nonlinearization coefficients do not depend on function $f(x_1, x_2)$ and therefore the proposed approximation is unacceptable. If we apply the substitution

$$2h_e = \frac{E_N[x_2 f(x_1, x_2)]}{E_N[x_2^2]} \cong \frac{E_n[x_2 f(x_1, x_2)]}{E_n[x_2^2]} ,$$

$$k_e = \frac{E_N[x_1 f(x_1, x_2)]}{E_N[x_1^2]} \cong \frac{E_n[x_1 f(x_1, x_2)]}{E_n[x_1^2]} , \tag{7.37}$$

then of course both nonlinearization coefficients depend on function $f(x_1, x_2)$ and the proposed approximation (7.37) is acceptable. We show it with an example.

*Example 7.2.* Consider Van der Pol–Duffing oscillator under external Gaussian white noise excitation described by

$$dx_1 = x_2 dt ,$$
$$dx_2 = -[(\alpha + \beta x_1^2)x_2 + \gamma x_1 + \delta x_1^3]dt + \sqrt{2q}d\xi(t) \tag{7.38}$$

where $\alpha, \beta, \gamma, \delta$ and $q$ are positive constants, and the equivalent nonlinear system has the form

$$dx_1 = x_2 dt ,$$
$$dx_2 = -[2h_e x_2 + k_e x_1 + \gamma x_1 + \delta x_1^3]dt + \sqrt{2q}d\xi(t) . \tag{7.39}$$

From (7.38) and (7.39) it follows that

$$f(x_1, x_2) = (\alpha + \beta x_1^2)x_2, \quad r(x_1) = \gamma x_1 + \delta x_1^3 \tag{7.40}$$

and the probability density function of the stationary solution of (7.39), which is known in the analytical form

$$g_n(x_1, x_2) = \frac{1}{c_n} \exp\left\{ -\frac{2h_e}{q^2} \left[ \frac{x_2^2}{2} + \frac{(k_e + \gamma)x_1^2}{2} + \frac{\delta x_1^4}{4} \right] \right\} , \tag{7.41}$$

where $c_n = c_n(h_e, k_e)$ is a normalized constant being a function of linearization coefficients, i.e.,

$$c_n(h_e, k_e) = \int_{-\infty}^{+\infty} \int_{-\infty}^{+\infty} \exp\left\{ -\frac{2h_e}{q^2} \left[ \frac{x_2^2}{2} + \frac{(k_e + \gamma)x_1^2}{2} + \frac{\delta x_1^4}{4} \right] \right\} dx_1 dx_2 . \tag{7.42}$$

Applying substitutions (7.37) we obtain

$$2h_e = \frac{E_N[x_2 f(x_1, x_2)]}{E_N[x_2^2]} = \frac{E_N[(\alpha + \beta x_1^2)x_2^2]}{E_N[x_2^2]} \cong \frac{E_n[(\alpha + \beta x_1^2)x_2^2]}{E_n[x_2^2]}$$
$$= \alpha + \beta \frac{E_n[x_1^2 x_2^2]}{E_n[x_2^2]} ,$$
$$k_e = \frac{E_N[(\alpha + \beta x_1^2)x_1 x_2]}{E_N[x_1^2]} \cong \frac{E_n[x_1 x_2(\alpha + \beta x_1^2)]}{E_n[x_1^2]}$$
$$= \frac{\alpha E_n[x_1 x_2] + \beta E_n[x_1^3 x_2]}{E_n[x_1^2]} = 0 . \tag{7.43}$$

The calculation of stationary response moments and nonlinearization co-efficient $h_e$ is possible by application of Procedure $EXAMPLE_{NL1}$, where in step (2) the nonlinearization coefficient $h_e$ is defined by (7.43).

### 7.2.2 SPEC Nonlinearization Alternative

We may also extend the SPEC Alternative to original nonlinear system (7.24) and equivalent nonlinear system (7.27) and the criterion is defined by (7.28) where the averaging operation is replaced by $E_n[.]$. Then the determination of linearization coefficients is more complicated, because of algebraic calculations. The necessary conditions of minimum of criterion

$$I_n = E_n[(f(x_1, x_2) - 2h_e x_2 - k_e x_1)^2]$$

$$= \frac{1}{c_n} \int_{-\infty}^{+\infty} \int_{-\infty}^{+\infty} (f(x_1, x_2) - 2h_e x_2 - k_e x_1)^2 \qquad (7.44)$$

$$\times \exp\left\{-\frac{2h_e}{q^2}\left[\frac{x_2^2}{2} + \frac{k_e x_1^2}{2} + \int_0^{x_1} r(u)du\right]\right\} dx_1 dx_2,$$

corresponding to criterion (7.28) are the following:

$$\frac{\partial I_n}{\partial h_e} = 0, \quad \frac{\partial I_n}{\partial k_e} = 0. \qquad (7.45)$$

After differentiation, (7.45) yield

$$E_n\left[2(f(x_1, x_2) - 2h_e x_2 - k_e x_1)(-2x_2)\right]$$

$$+ E_n\left[(f(x_1, x_2) - 2h_e x_2 - k_e x_1)^2\left(-\frac{2}{q^2}\right)\left(\frac{x_2^2}{2} + \frac{k_e x_1^2}{2} + \int_0^{x_1} r(u)du\right)\right]$$

$$- E_n\left[\left(\frac{2}{q^2}\right)\left(\frac{x_2^2}{2} + \frac{k_e x_1^2}{2} + \int_0^{x_1} r(u)du\right)\right]$$

$$E_n\left[(f(x_1, x_2) - 2h_e x_2 - k_e x_1)^2\right] = 0, \qquad (7.46)$$

$$E_n\left[2(f(x_1, x_2) - 2h_e x_2 - k_e x_1)(-x_1)\right]$$

$$+ E_n\left[(f(x_1, x_2) - 2h_e x_2 - k_e x_1)\left(-\frac{h_e x_1^2}{q^2}\right)\right]$$

$$+ E_n\left[\frac{h_e x_1^2}{q^2}\right] E_n\left[(f(x_1, x_2) - 2h_e x_2 - k_e x_1)^2\right] = 0. \qquad (7.47)$$

Then the corresponding iterative procedure has the form

## Procedure $EXAMPLE_{NL2}$

(1) Substitute the initial values for the nonlinearization coefficients $h_e$ and $k_e$ and calculate the response moments defined by the joint probability density function (7.34) and (7.35).

(2) Calculate the new nonlinearization coefficients $h_e$ and $k_e$ using coupled algebraic nonlinear equations defined by formulas (7.46) and (7.47) and the required moments obtained from stationary moments defined by the joint probability density function (7.34) and (7.35).

(3) Substitute the coefficients $h_e$ and $k_e$ into the joint probability density function (7.34) and (7.35) and calculate the new response moments.

(4) Repeat steps (2) and (3) until convergence.

### 7.2.3 Entropy Criteria

An interesting idea of the application of entropy criteria for stochastic nonlinearization have been proposed by Ricciardi and Elishakoff [21]. The authors considered the original nonlinear oscillator

$$\ddot{x}(t) + c\dot{x}(t) + f(x) = \sigma w(t) \tag{7.48}$$

and for the first criterion the equivalent nonlinear oscillator

$$\ddot{x}(t) + c\dot{x}(t) + kx(t) + \alpha \, sgn(x(t)) = \sigma w(t) , \tag{7.49}$$

where $f(x)$ is a nonlinear function, $w(t)$ is a zero-mean white noise process with correlation function $E[w(t)w(t + \tau)] = q\delta(\tau)$, $\sigma = \sqrt{2c}$, $c$ and $q$ are constant parameters, $k$ and $\alpha$ are nonlinearization coefficients.

The first criterion was the mean-square criterion

$$\int_{-\infty}^{+\infty} (f(x) - kx - \alpha \, sgn(x))^2 g_{ent1}(x)dx , \tag{7.50}$$

where $g_{ent1}(x)$ is the maximum entropy probability density function defined by

$$g_{ent1} = \frac{1}{C(k,\alpha)} \exp\left\{-\frac{1}{q}\left(\alpha|x| + \frac{1}{2}kx^2\right)\right\} , \tag{7.51}$$

where $C(k,\alpha)$ is a normalized constant.

In the second case, the equivalent nonlinear oscillator and the corresponding criterion have the form

$$\ddot{x} + c\dot{x} + k_*x + \alpha \, sgn(x) = \sigma w(t) , \tag{7.52}$$

$$\int_{-\infty}^{+\infty} (f(x) - k_*x - \alpha \, sgn(x))^2 g_{ent2}(x)dx , \tag{7.53}$$

where $k_*$ is the equivalent linearization coefficient obtained from the application of the standard equivalent linearization or by Kazakov's criterion of equality second-order moments of nonlinear and linearized elements. $\alpha$ is a nonlinearization coefficient and $g_{ent2}(x)$ is the corresponding minimum cross-entropy probability density function defined by

$$g_{ent2} = \frac{1}{C_2(\alpha)} \exp\left\{-\frac{1}{q}\left(\alpha|x| + \frac{1}{2}k_* x^2\right)\right\} , \tag{7.54}$$

where $C_2(\alpha)$ is a normalized constant.

### 7.2.4 Polynomial Energy Approximation

Another generalization of Elishakoff and Cai result was done by To [28], who considered the same original nonlinear system (7.24) and equivalent nonlinear system in the form

$$dx_1 = x_2 dt ,$$

$$dx_2 = [-\tilde{h}(H)x_2 - r(x_1)]dt + \sqrt{2}q d\xi(t) , \tag{7.55}$$

where $\tilde{h}(H)$ is the equivalent nonlinear damping coefficient of the system and $H$ is the energy function defined by

$$H = H(x_1, x_2) = \frac{1}{2}x_2^2 + \int_0^{x_1} r(s)ds . \tag{7.56}$$

To determine the equivalent nonlinear damping coefficient, To [28] introduced new variables

$$x_1 = \sqrt{2H}\cos\theta , \quad x_2 = \sqrt{2H}\sin\theta \tag{7.57}$$

and proposed the following criterion:

$$I = \int_0^{2\pi} [f(\sqrt{2H}\cos\theta , \sqrt{2H}\sin\theta) - \tilde{h}(H)\sqrt{2H}\sin\theta]^2 d\theta . \tag{7.58}$$

The necessary condition of minimum of criterion (7.57) under assumption $\tilde{h}(H) = h$ has the form

$$\frac{\partial I}{\partial h}\Big|_{h=\tilde{h}(H)} = \int_0^{2\pi} 2[f(\sqrt{2H}\cos\theta , \sqrt{2H}\sin\theta) - \tilde{h}(H)\sqrt{2H}\sin\theta]\sqrt{2H}\sin\theta d\theta = 0 . \tag{7.59}$$

Equality (7.59) yields

$$\tilde{h}(H) = \frac{1}{\pi\sqrt{2H}}\int_0^{2\pi} f(\sqrt{2H}\cos\theta, \sqrt{2H}\sin\theta)\sin\theta d\theta \tag{7.60}$$

and the second derivative

$$\frac{\partial^2 I}{\partial h^2}\Big|_{h=\tilde{h}(H)} = 4\pi H > 0 \; . \tag{7.61}$$

The next generalization of Elishakoff and Cai approach was proposed by Polidori and Beck [18], who considered also the original nonlinear system (7.24) and the equivalent nonlinear system in the form

$$dx_1 = x_2 dt \; ,$$

$$dx_2 = [-\tilde{f}(H)x_2 - r(x_1)]dt + \sqrt{2q}d\xi(t) \; , \tag{7.62}$$

where $\tilde{f}(H)$ was assumed in a polynomial form

$$\tilde{f}(H) = \sum_{i=0}^{N} \alpha_i H^i \; , \tag{7.63}$$

where $H = H(x_1, x_2)$ is defined by (7.56), and $\alpha_i$, $i = 1, \ldots, N$, are constant coefficients that can be determined by minimization of a criterion, $N$ is chosen arbitrarily.

We note that the probability density function of the stationary solution of (7.24) is known exactly and has the form

$$g_{EN}(x_1, x_2, \alpha_0, \ldots, \alpha_N) = \frac{1}{c_{EN}} \exp\left\{ -\frac{1}{q^2} \sum_{i=0}^{N} \frac{\alpha_i}{i+1} H^{i+1}(x_1, x_2) \right\} \; , \tag{7.64}$$

where $c_{EN}$ is a normalized constant defined by

$$c_{EN} = \int_{-\infty}^{+\infty} \int_{-\infty}^{+\infty} \exp\left\{ -\frac{1}{q^2} \sum_{i=0}^{N} \frac{\alpha_i}{i+1} H^{i+1}(x_1, x_2) \right\} dx_1 dx_2 \; . \tag{7.65}$$

For instance, coefficients $\alpha_i$, $i = 1, \ldots, N$, can be determined by the minimization of the following square criterion:

$$\min_{\alpha_i} \| f(x_{1_{eqln}}, x_{2_{eqln}}) - \left( \sum_{i=0}^{N} \alpha_i H^i \right) x_{2_{eqln}} \|_\rho \; , \tag{7.66}$$

where $\|.\|_\rho$ is the $L_2(R^2)$ norm with some weighting function $\rho(x_1, x_2)$, $\rho(x_1, x_2) > 0$ almost everywhere.

Polidori and Beck [18] proposed to use the weighting function $\rho(x_1, x_2) = g_{lin}(x_1, x_2)$ or $\rho(x_1, x_2) = g_{lin}^2(x_1, x_2)$ in (7.66), where $g_{lin}$ is computed for linearized system, for instance by statistical linearization.

Further generalization of a nonlinearization method for a general nonlinear oscillator

$$dx_1 = x_2 dt \; ,$$

$$dx_2 = -f(x_1, x_2)dt + \sqrt{2q}d\xi(t) \tag{7.67}$$

was proposed by Wang et al. [31], who considered the equivalent nonlinear system in the form

$$dx_1 = x_2 dt ,$$

$$dx_2 = -[\tilde{f}_2(H)x_2 + \tilde{r}_5(x_1)]dt + qd\xi(t) , \qquad (7.68)$$

where $f(x_1, x_2)$ is a nonlinear function, $q > 0$ is a constant parameter

$$\tilde{f}_2(H) = \sum_{i=0}^{2} \alpha_i H^i , \qquad (7.69)$$

$$\tilde{r}_5(x_1) = \sum_{j=0}^{5} \beta_j x_1^j , \qquad (7.70)$$

where $\alpha_i$, $i = 0, 1, 2$, and $\beta_j$, $j = 0, 1, \dots, 5$, are constant coefficients. One can write the energy function $H$ of (7.67) and (7.68) in two different forms

$$H_1 = \frac{1}{2}x_2 + f_0 x_1 + \frac{1}{2}f_x x_1^2 + \frac{1}{6}f_{xx}x_1^3 + \frac{1}{24}f_{xxx}x_1^4 + \frac{1}{120}f_{xxxx}x_1^5 + \frac{1}{720}f_{xxxxx}x_1^6 + c , \qquad (7.71)$$

and

$$H_2 = \frac{1}{2}x_2 + \int_0^{x_1} \tilde{r}_5(s)ds + c = \frac{1}{2}x_2 + \sum_{j=0}^{5} \frac{\beta_j}{j+1}x_1^{j+1} + c , \qquad (7.72)$$

where $c$ is an arbitrary constant,

$$f_0 = f(0,0) , \quad f_x = \frac{\partial f(x_1, x_2)}{\partial x_1}\Big|_{(0,0)} , \quad f_{xx} = \frac{\partial^2 f(x_1, x_2)}{\partial x_1^2}\Big|_{(0,0)}, \dots \quad (7.73)$$

To find coefficients $\alpha_i$, $i = 0, 1, 2$, and $\beta_j$, $j = 0, 1, \dots, 5$, Wang et al. [31] have used two groups of conditions. The first group of conditions follows from the equality of energy functions, i.e.,

$$H_1 = H_2 = H . \qquad (7.74)$$

Hence we find

$$\beta_0 = f_0, \beta_1 = f_x, \beta_2 = \frac{1}{2}f_{xx}, \beta_3 = \frac{1}{6}f_{xxx}, \beta_4 = \frac{1}{24}f_{xxxx}, \beta_5 = \frac{1}{120}f_{xxxxx} . \qquad (7.75)$$

To obtain the second group of conditions, we first define the error of approximation

$$\Delta f = \sum_{i=0}^{2} \alpha_i H_2^i + \sum_{j=0}^{5} \beta_i x^j - f(x_1, x_2) , \qquad (7.76)$$

where the nonlinear function $f(x_1, x_2)$ in equality (7.76) is replaced by its Taylor series until fifth order. Then we obtain

$$\Delta f = \sum_{i=0}^{8} \sum_{j=0}^{5} w_{ij} x_1^i x_2^j , \qquad (7.77)$$

where $w_{ij}$ are coefficients depending on $\alpha_i$, $i = 0, 1, 2$, $\beta_i$, $i = 0, 1, \ldots, 5$, and partial derivatives

$$\frac{\partial^{p_1 + p_2} f(x_1, x_2)}{\partial x_1^{p_1} \partial x_2^{p_2}} \Big|_{(0,0)}, \quad for \quad p_1 + p_2 \le 5.$$

The corresponding square criterion has the form

$$I_f = \sum_{i=0}^{8} \sum_{j=0}^{5} w_{ij}^2 . \qquad (7.78)$$

The second group of conditions that Wang et al. [31] have used to find coefficients $\alpha_i$, $i = 0, 1, 2$, are

$$\frac{\partial I_f}{\partial \alpha_i} = 0, \quad i = 0, 1, 2 . \qquad (7.79)$$

Using the coefficients $\alpha_i$, $i = 0, 1, 2$, and $\beta_j$, $j = 0, 1, \ldots, 5$, obtained from conditions (7.79) and (7.74), respectively, using the results of [17] one gets the exact stationary joint probability density function of $x_1$ and $x_2$ in the form

$$g(x_1, x_2) = \frac{1}{c_n} \exp \left\{ -\frac{1}{q^2} \sum_{i=0}^{2} \frac{\alpha_i}{i+1} H^{i+1}(x_1, x_2) \right\} , \qquad (7.80)$$

where $H = H_2$ is defined by (7.72) and $c_n$ is a normalized constant.

Using the results obtained by Elishakoff and Cai [6], Polidori and Beck [18], and Wang et al. [31], we can formulate the natural generalization of their approaches for system (7.67). The equivalent nonlinear system has the form (7.68) defined by modified functions $\tilde{f}_2(H)$ and $\tilde{r}_5(H)$, i.e.,

$$\tilde{f}_2(H) = \sum_{i=0}^{N_\alpha} \alpha_i H^i , \qquad (7.81)$$

$$\tilde{r}_5(x_1) = \sum_{j=0}^{N_\beta} \beta_j x_1^j , \qquad (7.82)$$

where the coefficients $\alpha_i$, $i = 0, 1, 2, \ldots, N_\alpha$, and $\beta_j$, $j = 0, 1, \ldots, N_\beta$ for arbitrarily chosen numbers $N_\alpha$ and $N_\beta$ can be found from minimization of the following criterion:

$$\min_{\alpha_i, \beta_j} \| f(x_{1_{eqln}}, x_{2_{eqln}}) - \left( \sum_{i=0}^{N_\alpha} \alpha_i H^i \right) x_{2_{eqln}} - \sum_{j=0}^{N_\beta} \beta_j x_1^j \|_\rho , \qquad (7.83)$$

where $\|.\|_\rho$ is the $L_2(R^2)$ norm with a weighting function $\rho(x_1, x_2)$, for instance, according to Polidori and Beck [18] $\rho(x_1, x_2) = g_{lin}(x_1, x_2)$ or $\rho(x_1, x_2) = g_{lin}^2(x_1, x_2)$ in (7.66), where $g_{lin}$ is computed for linearized system, for instance, by application of statistical linearization.

This criterion can be simplified for the case when the equivalent nonlinear system has the form (7.68–7.70) and the coefficients $\beta_j$, $j = 0, 1, \ldots, 5$, are found from condition (7.74) and the coefficients $\alpha_i$, $i = 0, 1, 2$, are found from minimization of the simplified version of criterion (7.83), i.e.,

$$\min_{\alpha_i} \| f(x_{1_{eqln}}, x_{2_{eqln}}) - \left( \sum_{i=0}^{2} \alpha_i H^i \right) x_{2_{eqln}} \|_\rho , \qquad (7.84)$$

where similar to previous cases $\rho(x_1, x_2) = p_{lin}(x_1, x_2)$ or $\rho(x_1, x_2) = p_{lin}^2(x_1, x_2)$, $p_{lin}$ is computed for linearized system, for instance, by application of statistical linearization.

In this case, one can also use $\rho(x_1, x_2) = p_{parlin}(x_1, x_2)$, where $p_{parlin}(x_1, x_2)$ is the stationary joint probability density function of $x_1$ and $x_2$ defined by partial linearization of system (7.67), i.e., when system (7.67) is replaced by the equivalent nonlinear system

$$dx_1 = x_2 dt ,$$

$$dx_2 = -[2h_e x_2 + \tilde{r}_5(x_1)]dt + \sqrt{2q}d\xi(t) , \qquad (7.85)$$

where $h_e$ is a coefficient of linearized damping obtained by a modified Elishakoff and Cai approach [6] and a simplified Wang et al. approach [31].

The stationary joint probability density function for system (7.85) is known and has the form

$$g_{parlin}(x_1, x_2) = \frac{1}{c_{parlin}} \exp \left\{ -\frac{2h_e}{q^2} \left[ \frac{x_2^2}{2} + \int_0^{x_1} \tilde{r}_5(u)du \right] \right\} , \qquad (7.86)$$

where $c_{parlin}$ is a normalized constant. We note that the coefficients $\beta_j$, $j = 0, 1, \ldots, 5$, are found from condition (7.75). It means that the stationary joint probability density function $g_{parlin}(x_1, x_2)$ can be presented as follows:

$$g_{parlin}(x_1, x_2) = \frac{1}{c_{parlin}} \exp \left\{ -\frac{2h_e}{q^2} \left[ \frac{x_2^2}{2} + \sum_{j=0}^{5} \frac{f_{x_1}^{(j)}(x_1, x_2)|_{(0,0)}}{j!} (x_1)^{j+1} \right] \right\} . \qquad (7.87)$$

where $f_{x_1}^{(j)}(x_1, x_2) = \frac{\partial^j f(x_1, x_2)}{\partial x_1^j}$. We illustrate this approach by an example.

*Example 7.3.* Consider the Van der Pol–Duffing oscillator under external Gaussian white noise excitation described by

$$dx_1 = x_2 dt ,$$

$$dx_2 = -[(\alpha + \beta x_1^2)x_2 + \gamma x_1 + \delta x_1^3]dt + \sqrt{2q}d\xi(t) \qquad (7.88)$$

and the equivalent nonlinear system

$$dx_1 = x_2 dt \, ,$$

$$dx_2 = -[2h_e x_2 + k_e x_1 + \tilde{r}_5(x_1)]dt + \sqrt{2q}d\xi(t) \, . \qquad (7.89)$$

where $\alpha, \beta, \gamma, \delta$, and $q$ are positive constants, $\xi(t)$ is the standard Wiener process, $h_e$ and $k_e$ are nonlinearization coefficients, and $\tilde{r}_5(x_1)$ was defined by (7.82).

From (7.70) and (7.75), it follows that

$$f(x_1, x_2) = (\alpha + \beta x_1^2)x_2 + \gamma x_1 + \delta x_1^3 \, , \quad \tilde{r}_5(x_1) = \sum_{j=0}^{5} \beta_j x_1^j \, , \qquad (7.90)$$

and

$$\begin{aligned}
f(0,0) &= 0 &\Rightarrow& \quad \beta_0 = 0 \, , \\
f_{x_1}^{(1)}(x_1, x_2) &= 2\beta x_1 x_2 + \gamma + 3\delta x_1^2 &\Rightarrow& \quad \beta_1 = \gamma, \\
f_{x_1}^{(2)}(x_1, x_2) &= 2\beta x_2 + 6\delta x_1 &\Rightarrow& \quad \beta_2 = 0 \, , \\
f_{x_1}^{(3)}(x_1, x_2) &= 6\delta &\Rightarrow& \quad \beta_3 = \delta, \\
f_{x_1}^{(4)}(x_1, x_2) &= f_{x_1}^{(5)}(x_1, x_2) = 0 &\Rightarrow& \quad \beta_4 = \beta_5 = 0 \, .
\end{aligned} \qquad (7.91)$$

Hence

$$\tilde{r}_5(x_1) = \gamma x_1 + \delta x_1^3 \, , \qquad (7.92)$$

and the corresponding criterion of nonlinearization has the form

$$\begin{aligned}
I &= E_N \left[ \left( (\alpha + \beta x_1^2)x_2 + \gamma x_1 + \delta x_1^3 - \gamma x_1 - \delta x_1^3 - 2h_e x_2 - k_e x_1 \right)^2 \right] \\
&= E_N \left[ \left( \alpha x_2 + \beta x_1^2 x_2 - 2h_e x_2 - k_e x_1 \right)^2 \right] \rightarrow \min_{h_e, k} \, . \qquad (7.93)
\end{aligned}$$

The necessary conditions of minimum of criterion (7.93) yield

$$\alpha E_N[x_2^2] + \beta E_N[x_1^2 x_2^2] - 2h_e E_N[x_2^2] - k_e E_N[x_1 x_2] = 0 \, ,$$

$$\alpha E_N[x_1 x_2] + \beta E_N[x_1^3 x_2] - 2h_e E_N[x_1 x_2] - k_e E_N[x_1^2] = 0 \, . \qquad (7.94)$$

Since $E_N[x_1 x_2] = 0$ and $E_N[x_1^3 x_2] = 0$, we find the formula for nonlinearization coefficients

$$2h_e = \alpha + \beta \frac{E_N[x_1^2 x_2^2]}{E_N[x_2^2]} \cong \alpha + \beta \frac{E_n[x_1^2 x_2^2]}{E_n[x_2^2]}, \quad k_e = 0 \, , \qquad (7.95)$$

where the averaging operation $E_n[.]$ is defined by

$$g_{parlin}(x_1, x_2) = \frac{1}{c_{parlin}} \exp \left\{ -\frac{2h_e}{q^2} \left[ \frac{x_2^2}{2} + \frac{\gamma x_1^2}{2} + \frac{\delta x_1^4}{4} \right] \right\} \, . \qquad (7.96)$$

To find nonlinearization coefficients and response statistics, one can apply a modified Procedure $EXAMPLE_{NL1}$.

## 7.3 Probability Density Criteria

### 7.3.1 Direct Methods

The natural generalization of equivalent linearization with criteria in the probability density functions space for equivalent nonlinear systems was proposed by Socha [24]. We will shortly discuss this approach below.

The objective of the probability density equivalent nonlinearization is to find the elements $\alpha_i$, which minimize the following criterion:

$$I = \int_{R^n} w(\mathbf{x})\Psi(g_N(\mathbf{x}) - g_{EN}(\mathbf{x}, \boldsymbol{\alpha}))d\mathbf{x} , \qquad (7.97)$$

where $\Psi$ is a convex function, $w(\mathbf{x})$ is a weighting function, and $g_N(\mathbf{x})$ and $g_{EN}(\mathbf{x}, \boldsymbol{\alpha})$ are the probability density functions of stationary solutions of the original and equivalent nonlinear systems, respectively. $\boldsymbol{\alpha} = [\alpha_0, \alpha_1, \ldots, \alpha_N]^T$ is a vector of nonlinearization coefficients, $\alpha_i \in R$, $i = 0, 1, \ldots, N$.

As in the case of statistical and equivalent linearization, in what follows, two basic criteria are considered

(a) Probabilistic square metric

$$I_{1en} = \int_{R^n} (g_N(\mathbf{x}) - g_{EN}(\mathbf{x}, \boldsymbol{\alpha}))^2 d\mathbf{x} . \qquad (7.98)$$

(b) Pseudomoment metric

$$I_{2en} = \int_{R^n} |\mathbf{x}|^{2l} |g_N(\mathbf{x}) - g_{EN}(\mathbf{x}, \boldsymbol{\alpha})| d\mathbf{x} . \qquad (7.99)$$

To apply the proposed criterion, we have to find the probability density functions for stationary solutions of the original system $g_N(\mathbf{x})$ and the equivalent nonlinear one $g_{EN}(\mathbf{x}, \boldsymbol{\alpha})$. Usually, the equivalent nonlinear system is looked for the class of nonlinear systems with the known probability density function in the exact form. Unfortunately, except for some special cases, the probability density function $g_N(\mathbf{x})$ can be found only by approximation methods, for instance, the Gramm–Charlier expansion or by simulations. For an $n$-dimensional system, the one-dimensional density has the truncated form [19] given by (6.118–6.121). In what follows, we consider the case when $\mathbf{x} = [x_1, x_2]^T$.

The probability density function of the stationary solution of equivalent nonlinear system (7.62) is known and has the analytical form (7.64), i.e.,

$$g_{EN}(\mathbf{x}, \boldsymbol{\alpha}) = g_{EN}(x_1, x_2, \alpha_0, \ldots, \alpha_N)$$

$$= c_{EN}^{-1} \exp\left\{ -\frac{1}{q^2} \sum_{i=0}^{N} \frac{\alpha_i}{i+1} H^{i+1}(x_1, x_2) \right\} , \qquad (7.100)$$

where for convenience of further calculations the normalized constant $c_{EN}$ is defined by

$$c_{EN} = c_{EN}(\boldsymbol{\alpha}) = \int_{-\infty}^{+\infty} \int_{-\infty}^{+\infty} \exp\left\{-\frac{1}{q^2}\sum_{i=0}^{N}\frac{\alpha_i}{i+1}H^{i+1}(x_1, x_2)\right\}dx_1 dx_2 .$$

$$(7.101)$$

In the case when the function $\Psi$ in criterion (7.97) is differentiable, the necessary conditions of minimization one can find, for instance, from conditions

$$\frac{\partial I}{\partial \alpha_i} = 2\int_{R^2} w(\mathbf{x})\frac{\partial \Psi(g_N, g_{EN})}{\partial g_{EN}}\frac{\partial g_{EN}(\mathbf{x}, \boldsymbol{\alpha})}{\partial \alpha_i}d\mathbf{x} = 0, \ i = 0,\ldots,N , \quad (7.102)$$

where

$$\frac{\partial g_{EN}(\mathbf{x}, \boldsymbol{\alpha})}{\partial \alpha_i} = \frac{1}{c_{EN}^2}\exp\left\{-\frac{1}{q^2}\frac{H^{i+1}(\mathbf{x})}{i+1}\right\}$$

$$-\frac{1}{c_{EN}^2}\exp\left\{-\frac{1}{q^2}\sum_{i=0}^{N}\frac{\alpha_i}{i+1}H^{i+1}(\mathbf{x})\right\}\frac{\partial c_{EN}}{\partial \alpha_i}, \quad (7.103)$$

$$\frac{\partial c_{EN}}{\partial \alpha_i} = \int_{-\infty}^{+\infty}\int_{-\infty}^{+\infty}\exp\left\{-\frac{1}{q^2}\frac{H^{i+1}(x_1, x_2)}{i+1}\right\}dx_1 dx_2, \ i = 0,\ldots,N .$$

$$(7.104)$$

In the particular case for criterion $I_{1en}$, one can show that the necessary conditions of minimization have the form

$$\int_{R^2}(g_N(\mathbf{x}) - g_{EN}(\mathbf{x}))\left(E_{EN}[H^{i+1}(\mathbf{x})] - H^{i+1}(\mathbf{x})\right)g_{EN}(\mathbf{x})d\mathbf{x} = 0 ,$$

$$for \quad i = 0,\ldots,N , \quad (7.105)$$

where

$$E_{EN}[.] = \int_{R^2}[.]g_{EN}(\mathbf{x})d\mathbf{x} . \quad (7.106)$$

Using necessary conditions (7.105) and moment equations for the original nonlinear system and directly calculated for the equivalent nonlinear system, one can determine nonlinearization coefficients $\alpha_i$ for $i = 0,\ldots,N$ by an iterative procedure that is similar to Procedure $EL-PD$ presented in Sect. 6.4.2.

## 7.3.2 The Fokker–Planck–Kolmogorov Equation in Equivalent Nonlinearization

The second approach used in equivalent linearization with criteria in the probability density functions space was the application of the reduced Fokker–Planck–Kolmogorov equations (for stationary probability density functions). This approach can be extended to the creation of an equivalent nonlinear systems. For instance, if we consider the original and equivalent nonlinear systems in the form (7.24) and (7.62), respectively, then the corresponding reduced Fokker–Planck–Kolmogorov equations have the form

$$\frac{\partial g_N}{\partial t} = \frac{\partial g_N}{\partial x_1} x_2 + \frac{\partial}{\partial x_2}[(-f(x_1, x_2) - g(x_1))g_N] + q^2 \sum_{i=1}^{2} \sum_{j=1}^{2} \frac{\partial^2 g_N}{\partial x_i \partial x_j} = 0$$

(7.107)

and

$$\frac{\partial g_{EN}}{\partial t} = \frac{\partial g_{EN}}{\partial x_1} x_2 + \frac{\partial}{\partial x_2}[(-\tilde{f}(H)x_2 - g(x_1))g_{EN}] + q^2 \sum_{i=1}^{2} \sum_{j=1}^{2} \frac{\partial^2 g_{EN}}{\partial x_i \partial x_j} = 0 .$$

(7.108)

If we denote

$$p_1 = g_N , \quad p_2 = \frac{\partial g_N}{\partial x_1} , \quad q_1 = g_{EN} , \quad q_2 = \frac{\partial g_{EN}}{\partial x_1} ,$$

(7.109)

then (7.107) and (7.108) can be transformed to the following two-dimensional vector systems:

$$\frac{\partial p_1}{\partial x_1} = p_2 ,$$

$$p_2 x_2 + \frac{\partial}{\partial x_2}[(-f(x_1, x_2) - g(x_1))p_1] + q^2 \sum_{i=1}^{2} \sum_{j=1}^{2} \frac{\partial^2 p_1}{\partial x_i \partial x_j} = 0$$

(7.110)

and

$$\frac{\partial q_1}{\partial x_1} = q_2 ,$$

$$q_2 x_2 + \frac{\partial}{\partial x_2}[(-\tilde{f}(H)x_2 - g(x_1))q_1] + q^2 \sum_{i=1}^{2} \sum_{j=1}^{2} \frac{\partial^2 q_1}{\partial x_i \partial x_j} = 0 .$$

(7.111)

Comparing the system equations (7.110) with (7.111), we find that $g_N$ and $\partial g_N/\partial x_i$ are approximated by $g_{EN}$ and $\partial g_{EN}/\partial x_i$, respectively.

Then the following joint criterion can be proposed:

$$I_{3ne}(\alpha) = \int_{-\infty}^{+\infty} \int_{-\infty}^{+\infty} \epsilon_{EN}^2(x_1, x_2, \alpha) dx_1 dx_2 ,$$

(7.112)

where

$$\epsilon_{EN}(x_1, x_2, \alpha) = \frac{\partial}{\partial x_2}\left[(f(x_1, x_2) - \tilde{f}(H)x_2)g_{EN}(x_1, x_2, \alpha)\right] .$$

(7.113)

Then the necessary conditions of minimum for criterion $I_{3ne}$ have the form

$$\frac{\partial I_{3ne}}{\partial \alpha_i} = 2 \int_{-\infty}^{+\infty} \int_{-\infty}^{+\infty} \epsilon_{EN}(x_1, x_2, \boldsymbol{\alpha}) \left[ \frac{\partial}{\partial x_2} \left[ (-H^i(x_1, x_2)x_2)g_{EN}(x_1, x_2, \boldsymbol{\alpha}) \right] \right.$$

$$\left. + \left( f(x_1, x_2) - \tilde{f}(H)x_2 \right) \frac{\partial g_{EN}(x_1, x_2, \boldsymbol{\alpha})}{\partial \alpha_i} \right] dx_1 dx_2 = 0 ,$$

$$i = 0, \dots, N ,$$

$$(7.114)$$

where $\frac{\partial g_{EN}(\mathbf{x}, \boldsymbol{\alpha})}{\partial \alpha_i}, i = 0, \dots, N$ is defined by (7.103) and (7.104).

The nonlinearization coefficients $\alpha_i$, $i = 0, \dots, N$, are determined from conditions (7.114) by an iterative procedure similar to Procedure $EL-PD$.

A different criterion followed from the two Fokker–Planck–Kolmogorov equations (7.110) and (7.111) was proposed by Polidori and Beck [18]

$$I_{4ne} = \int_{-\infty}^{+\infty} \int_{-\infty}^{+\infty} \left[ f(x_1, x_2)g_{EN}(x_1, x_2, \boldsymbol{\alpha}) + \frac{1}{q^2} \frac{\partial g_{EN}(x_1, x_2, \boldsymbol{\alpha})}{\partial x_2} \right]^2 dx_1 dx_2 .$$

$$(7.115)$$

It can be treated as criterion $I_{3ne}$ with the error of approximation defined by

$$\epsilon_{EN}(x_1, x_2, \boldsymbol{\alpha}) = \frac{\partial}{\partial x_2} \left[ (f(x_1, x_2) - \tilde{f}(H)x_2)g_{EN}(x_1, x_2, \boldsymbol{\alpha}) \right] . \qquad (7.116)$$

## 7.4 Application of the Generalized Stationary Potential Approach

The most important problem for application of moment or probability density functions criteria in a nonlinearization procedure is to find a class of nonlinear oscillators for which the exact solution, even for stationary case is known in an analytical form. The wide class of such systems was defined by *the generalized stationary potential method* proposed by Cai and Lin [2] and described in Sect. 2.11.

To recall these results, we consider a nonlinear system described by the Stratonovich equation [2]

$$dx_1 = x_2 dt ,$$

$$dx_2 = -f(x_1, x_2)dt + \sum_{k=1}^{M} \sigma_k(x_1, x_2)d\xi_k , \qquad (7.117)$$

where $f(.)$ and $\sigma_k(.)$ are nonlinear functions and $\xi_k(t)$, $k = 1, \dots, M$, are zero-mean Wiener processes with the variance of derivatives given by

$$E[d\xi_k d\xi_l] = 2\pi K_{kl} dt , \quad k, l = 1, \dots, M . \qquad (7.118)$$

where $[K_{kl}]$ is a matrix of intensity of noises.

Let the function $f(x_1, x_2)$ satisfies condition (2.268), i.e.,

$$
f(x_1, x_2) = \pi \sum_{k=1}^{M} \sum_{l=1}^{M} x_2 K_{kl} \sigma_k(x_1, x_2) \sigma_l(x_1, x_2) \left[ \lambda_y \frac{d\phi_0(\lambda)}{d\lambda} - \frac{\lambda_{yy}}{\lambda_y} \right]
$$
$$
- \pi \sum_{k=1}^{M} \sum_{l=1}^{M} K_{kl} \sigma_k(x_1, x_2) \frac{\partial \sigma_l(x_1, x_2)}{\partial x_2} + \frac{\lambda_x}{\lambda_y} + \frac{1}{\lambda_y} \frac{dD_1(x_1)}{dx_1} e^{\phi_0(\lambda)} ,
$$
$$
(7.119)
$$

where $x = x_1, y = \frac{1}{2}(x_2)^2$, $\lambda(x, y)$ is a proposed function twice differentiable with respect to $x$ and $y$, $\lambda_x = \partial \lambda / \partial x$, $\lambda_y = \partial \lambda / \partial y$, $\lambda_{yy} = \partial^2 \lambda / \partial y^2$ and $\phi_0(\lambda)$, $D_1(x_1)$ are arbitrary functions of one variable.

Then the stationary solution of (7.117) is defined by (2.251) and (2.265) and has the form

$$
p_s(x_1, x_2) = \frac{1}{c_s \lambda_y} e^{-\phi_0} , \tag{7.120}
$$

where $c_s$ is a normalized constant.

The considered class of nonlinear oscillators for which one can find an exact stationary solution can be extended to a class of nonlinear oscillators having the same corresponding Fokker–Planck–Kolmogorov equation. For instance, Lin and Cai [12] have shown that one of such classes of the Stratonovich stochastic differential equations is

$$
dx_1 = x_2 dt, \quad dx_2 = [-2hx_2 - f(x_1)]dt + \sqrt{2\pi K_{11}} d\xi_1 ,
$$
$$
dx_1 = x_2 dt, \quad dx_2 = [-2hx_1^2 x_2 - f(x_1)]dt + \sqrt{2\pi K_{11}} x_1 d\xi_1 ,
$$
$$
dx_1 = x_2 dt, \quad dx_2 = [\pi K_{11} x_2 - 2hx_2^3 - f(x_1)]dt + \sqrt{2\pi K_{11}} x_2 d\xi_1 ,
$$
$$
dx_1 = x_2 dt, \quad dx_2 = \left[ -2hx_2 \left( 1 + \frac{K_{22}}{K_{11}} x_1^2 \right) - f(x_1) \right] dt
$$
$$
+ \sqrt{2\pi K_{11}} d\xi_1 + \sqrt{2\pi K_{22}} x_1 d\xi_2 ,
$$
$$
dx_1 = x_2 dt, \quad dx_2 = \Bigg[ -(2h - \pi K_{33}) x_2
$$
$$
-2hx_2 \left( \frac{K_{22}}{K_{11}} x_1^2 + \frac{K_{33}}{K_{11}} x_2^2 \right) - f(x_1) \Bigg] dt
$$
$$
+ \sqrt{2\pi k_{11}} d\xi_1 + \sqrt{2\pi K_{22}} x_1 d\xi_2 + \sqrt{2\pi K_{33}} x_2 d\xi_3 ,
$$
$$
(7.121)
$$

where $h$ and $K_{ii}, i = 1, 2, 3$, are constant positive parameters, $f(x_1)$ is a nonlinear function, $f(0) = 0$, and $\xi_i, i = 1, 2, 3$, are independent standard Wiener processes with the power spectral densities equal to $K_{ii}$, respectively.

The corresponding probability density function of stationary solutions of (7.121) is the same for all equations and has the form

$$
p_s(x_1, x_2) = \frac{1}{c_s} \exp \left\{ -\frac{h}{\pi K_{11}} \left[ (x_2)^2 + 2 \int_0^{x_1} f(u) du \right] \right\} . \tag{7.122}
$$

where $c_s$ is a normalized constant. A wider class of nonlinear oscillators described by (7.117) that the exact stationary solution has the form similar to (7.120) was proposed by To and Li [29], i.e.,

$$p_s(x_1, x_2) = \frac{1}{c_s M(x_1, x_2)} e^{-\phi_0(r)} , \tag{7.123}$$

where $M = M(x_1, x_2)$, $\phi_0(r)$ are arbitrary functions, $c_s$ is a normalized constant, and $r$ is a constant defined by

$$r = \int_{-\infty}^{x_1} \int_{-\infty}^{x_2} \left\{ \left[ -C(s_1) - \int_0^{x_2} s_2 \frac{\partial M(s_1, s_2)}{\partial s_1} ds_2 \right] ds_1 + M(s_1, s_2) s_2 ds_2 \right\} , \tag{7.124}$$

where $C(x_1)$ is an arbitrary function.

Then the function $f(x_1, x_2)$ satisfies similar condition to (7.119), i.e.,

$$f(x_1, x_2) = \pi \sum_{k=1}^{M} \sum_{l=1}^{M} K_{kl} \sigma_k(x_1, x_2) \sigma_l(x_1, x_2) \left( -\frac{1}{M} \frac{\partial M(s_1, s_2)}{\partial x_2} + \frac{d\phi_0(r)}{dr} \frac{\partial r}{\partial x_2} \right)$$

$$- \pi \sum_{k=1}^{M} \sum_{l=1}^{M} K_{kl} \sigma_k(x_1, x_2) \frac{\partial \sigma_l(x_1, x_2)}{\partial x_2}$$

$$- \frac{1}{M(x_1, x_2)} \left( C(x_1) - \int_0^{x_2} s_2 \frac{\partial M(s_1, s_2)}{\partial s_1} ds_2 \right) + \frac{dD_1(x_1)}{dx_1} \frac{e^{\phi_0}}{M} , \tag{7.125}$$

where $D_1(x_1)$ is an arbitrary function of one variable. If we assume that $M = \lambda_y(x, y)$, where $\lambda(x, y)$ is an arbitrary function with continuous second-order derivatives with respect to $x$ and $y$, where $x = x_1$, $y = \frac{1}{2} x_2^2$, and $C(x_1) = 0$, then one can show that $r = \lambda(x, y)$ and the function $f(x_1, x_2)$ satisfies the condition (7.125). It means that the result obtained by To and Li [29] is a generalization of the derivation given by Lin and Cai [12].

The obtained class of stochastic differential equations was used by To and Li [29] to the determination of an equivalent nonlinear oscillator with the corresponding solvable reduced Fokker–Planck–Kolmogorov equation.

To and Li [29] considered original nonlinear oscillator in the form (7.117) and (7.118) with two additional assumptions on functions $f(x_1, x_2)$ and $\sigma_k(x_1, x_2)$, $k = 1, \ldots, M$, namely, there exist functions $u_0(x_1, x_2)$ and $u_1(x_1, x_2)$ such that

$$f(x_1, x_2) = f_0(x_1) + u_0(x_1, x_2) , \tag{7.126}$$

$$\pi \sum_{k=1}^{M} \sum_{l=1}^{M} K_{kl} \sigma_k(x_1, x_2) \frac{\partial \sigma_l(x_1, x_2)}{\partial x_2} = f_1(x_1) + u_1(x_1, x_2) . \tag{7.127}$$

An equivalent nonlinear system was proposed in the form

$$dx_1 = x_2 dt ,$$

$$dx_2 = -h(x_1, x_2)dt + \sum_{k=1}^{M} \sigma_k(x_1, x_2)d\xi_k , \qquad (7.128)$$

where

$$h(x_1, x_2) = \pi \sum_{k=1}^{M} \sum_{l=1}^{M} K_{kl} x_2 \sigma_k(x_1, x_2)\sigma_l(x_1, x_2)\phi_0'(\lambda)$$

$$-\pi \sum_{k=1}^{M} \sum_{l=1}^{M} K_{kl}\sigma_k(x_1, x_2)\frac{\partial\sigma_l(x_1, x_2)}{\partial x_2} + g(x_1), \qquad (7.129)$$

$$g(x_1) = f_0(x_1) + f_1(x_1) , \qquad (7.130)$$

$$\lambda = \lambda(x_1, x_2) = \frac{1}{2}x_2^2 + \int_0^{x_1} g(s_1)ds_1 , \qquad (7.131)$$

$\phi_0'$ is an unknown function of $\lambda$ that can be further determined.

Since in this case $y = \frac{1}{2}x_2^2$, then $\lambda_y = 1$ and the probability density function of the stationary solution of (7.128) depends only on $\lambda$ and has the form

$$p_s(x_1, x_2) = c_s \exp\{-\phi_0(\lambda)\} . \qquad (7.132)$$

The criterion of approximation of the original nonlinear oscillator (7.117) by an equivalent nonlinear one (7.126) was proposed as a mean-square error in the form

$$I(\phi_0') = E[\epsilon^2(x_1, x_2)] = \int_{-\infty}^{+\infty} \int_{-\infty}^{+\infty} \epsilon^2(x_1, x_2)p_s(x_1, x_2)dx_1 dx_2 , \qquad (7.133)$$

where

$$\epsilon(x_1, x_2) = f(x_1, x_2) - h(x_1, x_2)$$

$$= u_0(x_1, x_2) + u_1(x_1, x_2) - \pi \sum_{k=1}^{M} \sum_{l=1}^{M} K_{kl} x_2 \sigma_k(x_1, x_2)\sigma_l(x_1, x_2)\phi_0'(\lambda).$$

$$(7.134)$$

Let

$$x_2^+ = \left(2\lambda - 2\int_0^{x_1} g(s_1)ds_1\right)^{1/2} , \qquad x_2^- = -\left(2\lambda - 2\int_0^{x_1} g(s_1)ds_1\right)^{1/2} .$$

$$(7.135)$$

Then criterion (7.133) can be presented in the form

$$I(\phi_0') = \int_0^{+\infty} p_s(\lambda)d\lambda \int_{\lambda_1}^{\lambda_2} \left( \frac{\epsilon^2}{x_2 \big|_{x_2^+}} - \frac{\epsilon^2}{x_2 \big|_{x_2^-}} \right) dx_1 , \qquad (7.136)$$

where $\lambda_i$, $i = 1, 2$, are found from the following conditions:

$$\lambda = \int_0^{\lambda_i} g(x_1)dx_1, \quad i = 1, 2 . \qquad (7.137)$$

Denoting by

$$I_\lambda(\lambda, \phi_0') = p_s(\lambda) \int_{\lambda_1}^{\lambda_2} \left( \frac{\epsilon^2}{x_2 \big|_{x_2^+}} - \frac{\epsilon^2}{x_2 \big|_{x_2^-}} \right) dx_1 \qquad (7.138)$$

and using the Euler equation

$$\frac{\partial I_\lambda}{\partial \phi_0'} = 0 \qquad (7.139)$$

one can find that

$$p_s(\lambda) \int_{\lambda_1}^{\lambda_2} \left( \frac{2\epsilon \, \epsilon^2}{x_2 \, x_2 \big|_{x_2^+}} - \frac{2\epsilon \, \epsilon^2}{x_2 \, x_2 \big|_{x_2^-}} \right) dx_1 = 0 . \qquad (7.140)$$

Taking into account that

$$\frac{\partial \epsilon}{\partial \phi_0'} = -\pi \sum_{k=1}^{M} \sum_{l=1}^{M} K_{kl} x_2 \sigma_k(x_1, x_2) \sigma_l(x_1, x_2) , \qquad (7.141)$$

we find

$$\int_{\lambda_1}^{\lambda_2} \left[ \sum_{k=1}^{M} \sum_{l=1}^{M} K_{kl} \sigma_k(x_1, x_2) \sigma_l(x_1, x_2) \epsilon(x_1, x_2) \big|_{x_2^+} \right.$$
$$\left. - \sum_{k=1}^{M} \sum_{l=1}^{M} K_{kl} \sigma_k(x_1, x_2) \sigma_l(x_1, x_2) \epsilon(x_1, x_2) \big|_{x_2^-} \right] dx_1 = 0 . \quad (7.142)$$

One can calculate substituting (7.134) into (7.142)

$$\phi_0'(\lambda) = \frac{\int_{\lambda_1}^{\lambda_2} \left[ \sum_{k=1}^{M} \sum_{l=1}^{M} K_{kl} \sigma_k \sigma_l (u_0 + u_1) \big|_{x_2^+} - \sum_{k=1}^{M} \sum_{l=1}^{M} K_{kl} \sigma_k \sigma_l (u_0 + u_1) \big|_{x_2^-} \right] dx_1}{\pi \int_{\lambda_1}^{\lambda_2} \left[ \sum_{k=1}^{M} \sum_{l=1}^{M} x_2 (K_{kl} \sigma_k \sigma_l)^2 \big|_{x_2^+} - \sum_{k=1}^{M} \sum_{l=1}^{M} x_2 (K_{kl} \sigma_k \sigma_l)^2 \big|_{x_2^-} \right] dx_1} .$$
$$(7.143)$$

In the case of additive excitations, i.e., when $\sigma_k(x_1, x_2) = \sigma_k(t)$, then equality (7.143) reduces to

$$\phi_0'(\lambda) = \frac{\int_{\lambda_1}^{\lambda_2} \left[ (u_0 + u_1)|_{x_2^+} - (u_0 + u_1)|_{x_2^-} \right] dx_1}{\pi \sum_{k=1}^M \sum_{l=1}^M K_{kl} \sigma_k \sigma_l \int_{\lambda_1}^{\lambda_2} \left[ x_{2|_{x_2^+}} - x_{2|_{x_2^-}} \right] dx_1} . \tag{7.144}$$

*Example 7.4.* [29] Consider the nonlinear oscillator described by the Stratonovich stochastic differential equation

$$dx_1 = x_2 dt ,$$

$$dx_2 = (-\alpha_1 x_2 - \alpha_2 x_1^2 x_2 + x_1 - x_1^3) dt + x_1 \sqrt{2\pi q} d\xi(t) , \tag{7.145}$$

where $\alpha_i$, $i = 1, 2$ and q are real positive constants and $\xi(t)$ is the standard Wiener process. One can establish using the obtained results the following relations:

$$f_0(x_1) = -x_1 + x_1^3 , \quad u_0(x_1, x_2) = \alpha_1 x_2 + \alpha_2 x_1^2 x_2 ,$$
$$f_1(x_1) = 0 , \quad\quad\quad u_1(x_1, x_2) = 0 . \tag{7.146}$$

Hence

$$g(x_1) = f_0(x_1) = -x_1 + x_1^3 ,$$
$$\lambda = \frac{1}{2}x_2^2 - \frac{1}{2}x_1^2 + \frac{1}{4}x_1^4 , \tag{7.147}$$

and

$$f(x_1, x_2) = \pi q x_2 x_1^2 \phi_0'(\lambda) - x_1 + x_1^3 . \tag{7.148}$$

The relation (7.137) has the form

$$-\frac{1}{2}\lambda_i^2 + \frac{1}{4}\lambda_i^4 - \lambda = 0 . \tag{7.149}$$

Hence, the positive root of (7.149) is

$$\lambda_i^2 = 1 + \sqrt{1 + 4\lambda}, \quad i = 1, 2 , \tag{7.150}$$

and

$$x_2^+ = \sqrt{2\lambda + x_1^2 - \frac{1}{2}x_1^4}, \quad x_2^- = -\sqrt{2\lambda + x_1^2 - \frac{1}{2}x_1^4} \tag{7.151}$$

Substituting all the above equations into (7.143), we obtain

$$\phi_0'(\lambda) = \frac{q \int_{\lambda_1}^{\lambda_2} \left[ x_1^2 u_0|_{x_2^+} - x_1^2 u_0|_{x_2^-} \right] dx_1}{\pi \int_{\lambda_1}^{\lambda_2} \left[ x_2 q^2 x_1^4|_{x_2^+} - x_2 q^2 x_1^4|_{x_2^-} \right] dx_1}$$

$$= \frac{\alpha_2}{\pi q} + \frac{\alpha_1 \int_0^{\sqrt{1+\sqrt{1+4\lambda}}} x_1^2 \sqrt{2\lambda + x_1^2 - \frac{1}{2}x_1^4} dx_1}{\pi q \int_0^{\sqrt{1+\sqrt{1+4\lambda}}} x_1^4 \sqrt{2\lambda + x_1^2 - \frac{1}{2}x_1^4} dx_1} . \tag{7.152}$$

The equivalent nonlinear system and the corresponding probability density function of its stationary solution have the form

$$dx_1 = x_2 dt \ ,$$
$$dx_2 = (-\pi q x_1^2 \phi_0'(\lambda) + x_1 - x_1^3)dt + x_1\sqrt{2\pi q}d\xi(t) \ , \qquad (7.153)$$

$$p_s(x_1, x_2) = c_s \exp\left\{ -\int_0^\lambda \phi_0'(\beta)d\beta \right\} \ . \qquad (7.154)$$

where $c_s$ is a normalized constant.

# 7.5 Application of Stochastic Averaging Approach

Since the stochastic averaging approach is one of the strongest mathematical tools in the determination of exact stationary solutions of stochastically excited (externally and/or parametrically) and dissipated integrable and partially integrable Hamiltonian systems, Zhu and his coworkers have used this approach to the determination of an equivalent nonlinear system for non-resonant case (see, for instance, [35, 34]. This approach was developed and the general method for $n$-DOF stochastically excited and dissipated partially integrable Hamiltonian systems was proposed by Zhu et al. [33] for the non-resonant and resonant cases and by Zhu and Deng [32] for the resonant case only, respectively. We will show only the nonresonant case presented in [33] below.

First we remember the meaning of partially integrable Hamiltonian systems in nonresonant case. We consider a deterministic $n$-DOF Hamiltonian system governed by the following $n$ pairs of Hamilton's equations:

$$\dot{q}_i = \frac{\partial H}{\partial p_i}, \quad i = 1, \ldots, n \ , \qquad (7.155)$$

$$\dot{p}_i = -\frac{\partial H}{\partial q_i}, \quad i = 1, \ldots, n \ , \qquad (7.156)$$

where $q_i$ and $p_i$ are the generalized displacements and momenta, respectively; $H = H(\mathbf{q}, \mathbf{p})$ is a Hamiltonian function with continuous first-order derivatives, $\mathbf{q} = [q_1, \ldots, q_n]^T$, $\mathbf{p} = [p_1, \ldots, p_n]^T$.

A Hamiltonian system of $n$-DOF (7.155) is said to be integrable if there exists a set of canonical transformations

$$I_i = I_i(\mathbf{q}, \mathbf{p}), \quad i = 1, \ldots, n \ , \qquad (7.157)$$

$$\theta_i = \theta_i(\mathbf{q}, \mathbf{p}), \quad i = 1, \ldots, n \ , \qquad (7.158)$$

such that the new Hamilton's equations are of the following simplest canonical form:

$$\dot{I}_i = -\frac{\partial}{\partial \theta_i} H(\mathbf{I}) = 0, \quad i = 1, \ldots, n, \tag{7.159}$$

$$\dot{\theta}_i = \frac{\partial}{\partial I_i} H(\mathbf{I}) = \lambda_i(\mathbf{I}), \quad i = 1, \ldots, n, \tag{7.160}$$

where $I_i$ and $\lambda_i$ are action variables and frequencies, respectively, connected with corresponding degrees of freedom; $\theta_i$ are angle variables adjoint with $I_i$; $H(\mathbf{I})$ is the new Hamiltonian independent of $\theta_i$.

Then the solutions of (7.159) and (7.160) are

$$I_i = const, \quad i = 1, \ldots, n, \tag{7.161}$$

$$\theta_i = \lambda_i(\mathbf{I})t + \kappa_i, \quad i = 1, \ldots, n, \tag{7.162}$$

where $\kappa_i$ are constants of integration.

A Hamiltonian system is called nonintegrable if there exists no integral of motion other than the Hamiltonian itself. A Hamiltonian system is called partially integrable Hamiltonian system if there exist $r$ integrals of motion $H_1, H_2, \ldots H_r, (1 < r < n)$ which are in involution, it means it can be converted into one consisting of an integrable and a nonintegrable Hamiltonian subsystems by using canonical transformation, i.e., the Hamiltonian function can be presented as follows:

$$H(\mathbf{q}, \mathbf{p}) = \sum_{\alpha=1}^{r-1} H_\alpha(I_\alpha) + H_r(q_r, \ldots, q_n, p_r, \ldots, p_n). \tag{7.163}$$

An integrable Hamiltonian system of $n$-DOF is called resonant if frequencies $\lambda_\alpha$ are rationally related, i.e.,

$$k_\alpha^\mu \lambda_\alpha = 0, \quad \alpha = 1, \ldots, n, \quad \mu = 1, \ldots, \upsilon, \tag{7.164}$$

where $k_i^\mu$ are integers. The system is called completely resonant if $\upsilon = n - 1$, partially resonant if $1 \le \upsilon \le n - 1$, and it is nonresonant if $\upsilon = 0$.

Similar to integrable Hamiltonian systems, partially integrable Hamiltonian systems can also be resonant or partially resonant or nonresonant depending on whether the integrable part of Hamiltonian is resonant or partially resonant or nonresonant, respectively.

Now we consider an $n$-DOF stochastically excited and dissipated Hamiltonian system. The equations of motion treated as Stratonovich equations after transformation to the Ito form with adding correction terms are

$$dQ_i = \frac{\partial H}{\partial P_i} dt,$$

$$dP_i = -\left(\frac{\partial H}{\partial P_i} + M_{ij}\frac{\partial H}{\partial P_i}\right) dt + \sum_{k=1}^{m} \sigma_{ik} d\xi_k, \quad i, j = 1, 2, \ldots, n, \tag{7.165}$$

where $Q_i$ and $P_i$ are generalized displacements and momenta, respectively, $H = H(\mathbf{Q}, \mathbf{P})$ is the twice differentiable Hamiltonian, $M_{ij} = M_{ij}(\mathbf{Q}, \mathbf{P})$ and $\sigma_{ik} = \sigma_{ik}(\mathbf{Q}, \mathbf{P})$ are the differentiable and twice differentiable, respectively, and $\xi_k(t)$ are the Wiener processes with intensities calculated from zero-mean white noises Gaussian processes $w_k(t)$ modeling the equation of motion of the original nonlinear system with $E[w_k(t+\tau)w_l(t)] = 2\pi K_{kl}\delta(\tau), k, l = 1, \ldots, m$.

If the exact stationary solution of system (7.165) (stationary probability density of the solution) is not obtainable, then Zhu et al. [33] proposed to replace system (7.165) by the following one:

$$dQ_i = \frac{\partial H}{\partial P_i}dt \ ,$$

$$dP_i = -\left(\frac{\partial H}{\partial P_i} + m_{ij}\frac{\partial H}{\partial P_i}\right)dt + \sum_{k=1}^{m}\sigma_{ik}d\xi_k, i, j = 1, 2, \ldots, n \ . \quad (7.166)$$

whose exact stationary solution is obtainable and whose behavior is closest to that of system (7.165) in some statistical sense.

As it was mentioned earlier we will discuss only partially integrable Hamiltonian nonresonant systems. Then the exact stationary solution of system (7.166) is known of the form [33]

$$g_{enon}(\mathbf{q}, \mathbf{p}) = \frac{1}{C}\exp\left\{-\phi(H_1, \ldots, H_r)\right\}|_{H_i = H_i(\mathbf{q},\mathbf{p})} \ , \quad (7.167)$$

where $H_i$ are the components of the Hamiltonian function, i.e., $H(\mathbf{q}, \mathbf{p}) = \sum_{i=1}^{r} H_i(\mathbf{q}, \mathbf{p})$, $C$ is a normalization constant, and $\phi$ is a function of $H_i$ satisfying the following set of $n$ first-order linear partial differential equations:

$$\sum_{j=1}^{n} m_{ij}\frac{\partial H}{\partial p_j} + \frac{1}{2}\frac{\partial b_{ij}}{\partial p_j} - \sum_{j=1}^{n}\sum_{s=1}^{r}\frac{1}{2}b_{ij}\frac{\partial H_s}{\partial p_j}\frac{\partial \phi}{\partial H_s} = 0,$$

$$i = 1, 2, \ldots, n \ . \quad (7.168)$$

If a set of consistent $\frac{\partial \phi}{\partial H_s}$ can be determined from (7.168) and they satisfy the conditions

$$\frac{\partial^2 \phi}{\partial H_{s_1}\partial H_{s_2}} = \frac{\partial^2 \phi}{\partial H_{s_2}\partial H_{s_1}} \quad s_1, s_2 = 1, 2, \ldots, r \ , \quad (7.169)$$

where $b_{ij} = 2(\sigma\sigma^T)_{ij}$ are diffusion coefficients, $\sigma = [\sigma_{lk}], i, j, l, k = 1, \ldots, r, \ldots$ then according to [33], the solution of (7.168) is

$$\phi(H_1, \ldots, H_r) = \phi_0 + \sum_{s=1}^{r}\int_{0}^{H_1, \ldots, H_r}\frac{\partial \phi}{\partial H_s}dH_s \ , \quad (7.170)$$

where $\phi_0 = \phi(0, \ldots, 0)$. Substituting the obtained function $\phi(H_1, \ldots, H_r)$ into equality (7.167), we find the solution of system (7.165).

We note that the only difference between systems (7.165) and (7.166) is in the damping coefficients $M_{ij}$ and $m_{ij}$. The choice of the coefficients $m_{ij}$ can be replaced by the choice of functions $h_s(\mathbf{q}, \mathbf{p})$ defined by

$$h_s(\mathbf{q}, \mathbf{p}) = \frac{\partial \phi}{\partial H_s}\Big|_{H_s=H_s(\mathbf{q},\mathbf{p})}, \quad s = 1, \ldots, r \tag{7.171}$$

and satisfying (7.168) and (7.169).

To determine functions $h_s(\mathbf{q}, \mathbf{p})$ Zhu, et al. [33] proposed three criteria. The first criterion is the mean-square error of damping forces

$$I_{stav1} = E\left[\sum_{i=1}^{n} \epsilon_i^2\right], \tag{7.172}$$

where

$$\epsilon_i = \sum_{j=1}^{n}(M_{ij} - m_{ij})\frac{\partial H}{\partial P_j} = \sum_{j=1}^{n} M_{ij}\frac{\partial H}{\partial P_j} + \frac{1}{2}\frac{\partial b_{ij}}{\partial P_j} - \frac{1}{2}\sum_{j=1}^{n}\sum_{s=1}^{r} b_{ij}\frac{\partial H_s}{\partial P_j}h_s,$$

$$i, j = 1, 2, \ldots, n, \quad s = 1, 2, \ldots, r \tag{7.173}$$

and the averaging operation

$$E[.] = \int_{R^{2n}} [.]g(\mathbf{q}, \mathbf{p})d\mathbf{q}d\mathbf{p}, \tag{7.174}$$

the probability density function $g(\mathbf{q}, \mathbf{p})$ is assumed to be independent of $h_s$.

Then the necessary conditions of minimum of $I_{stav1}$ have the form

$$\int_{R^{2n}}\sum_{i=1}^{n}\epsilon_i\frac{\partial \epsilon_i}{\partial h_s}g(\mathbf{q}, \mathbf{p})d\mathbf{q}d\mathbf{p} = 0, \quad s = 1, 2, \ldots, r. \tag{7.175}$$

Since the exact probability density function $g(\mathbf{q}, \mathbf{p})$ for original nonlinear system (7.165) is unknown, we replace it by $g_{en}(\mathbf{q}, \mathbf{p})$ for equivalent nonlinear system (7.166) that is already defined by (7.167). However, some parameters are unknown at this stage. Next, using the transformation from variables $\mathbf{q}, \mathbf{p}$ to $H_1, \ldots, H_r, \theta_1, \ldots, \theta_{r-1}, q_r, \ldots, q_n,$ and $p_{r+1}, \ldots, p_n$, equality (7.175) can be rewritten as follows:

$$\int_0^{+\infty} g_{enonH}(H_1, \ldots, H_r)dH_1 \ldots dH_r \int_0^{2\pi}\int_\Lambda\left(\sum_{i=1}^{n}\epsilon_i\frac{\partial \epsilon_i}{\partial h_s}\sum_{j=1}^{n} b_{ij}\frac{\partial H_s}{\partial p_j}/\left|\prod_{\gamma=1}^{r}\frac{\partial H_\gamma}{\partial p_\gamma}\right|\right)$$

$$\times dq_r \ldots dq_n dp_{r+1} \ldots dp_n d\theta_1 \ldots d\theta_{r-1} = 0, \quad s = 1, 2, \ldots, r,$$

$$\tag{7.176}$$

where the domain of integration $\Lambda$ is defined by

$$\Lambda = \{(q_r \ldots q_n; p_{r+1} \ldots p_n) | H_r(q_r \ldots q_n; 0, p_{r+1} \ldots p_n) \leq H_r\}, \qquad (7.177)$$

$|\prod_{\gamma=1}^{r} \frac{\partial H_\gamma}{\partial p_\gamma}|$ is the absolute value of the Jacobian determinant for the transformation $\mathbf{q}, \mathbf{p}$ to $H_1, \ldots, H_r, \theta_1, \ldots, \theta_{r-1}, q_r, \ldots, q_n$, and $p_{r+1}, \ldots, p_n$.

Since the function $g_{enonH}(H_1, \ldots, H_r)$ is unknown, we simplify the condition (7.176) assuming that for every combination of $H_1, \ldots, H_r$ holds

$$\int_0^{2\pi} \int_\Lambda \left( \sum_{i=1}^n \epsilon_i \frac{\partial \epsilon_i}{\partial h_s} / | \prod_{\gamma=1}^r \frac{\partial H_\gamma}{\partial p_\gamma} | \right) \times dq_r \ldots dq_n dp_{r+1} \ldots \qquad (7.178)$$

$$dp_n d\theta_1 \ldots d\theta_{r-1} = 0, \quad s = 1, 2, \ldots, r.$$

Inserting of $\epsilon_i$ defined by equality (7.173) into (7.178) gives

$$\int_0^{2\pi} \int_\Lambda \sum_{i=1}^n \left[ \left( \sum_{j=1}^n M_{ij} \frac{\partial H}{\partial p_j} + \frac{1}{2} \frac{\partial b_{ij}}{\partial p_j} - \frac{1}{2} \sum_{j=1}^n \sum_{s=1}^r b_{ij} \frac{\partial H_s}{\partial p_j} h_s \right) \sum_{j=1}^n b_{ij} \frac{\partial H_s}{\partial p_j} | \prod_{\gamma=1}^r \frac{\partial H_\gamma}{\partial p_\gamma} | \right]$$

$$\times dq_r \ldots dq_n dp_{r+1} \ldots dp_n d\theta_1 \ldots d\theta_{r-1} = 0,$$

$$(7.179)$$

The second criterion for obtaining $h_s$ is minimizing the mean-square difference in the dissipating energies of the original and equivalent systems, i.e.,

$$I_{stav1} = E \left[ \left( \sum_{i=1}^n \epsilon_i \frac{\partial H}{\partial p_j} \right)^2 \right], \qquad (7.180)$$

where the averaging operation $E[.]$ is also assumed to be independent of the function $h_s = \frac{\partial \phi}{\partial H_s}$.

Then the similar derivation leads to the following necessary conditions of minimization of criterion (7.180):

$$\int_0^{2\pi} \int_\Lambda \sum_{i=1}^n \left[ \frac{\partial H}{\partial p_i} \left( \sum_{j=1}^n M_{ij} \frac{\partial H}{\partial p_j} + \frac{1}{2} \frac{\partial b_{ij}}{\partial p_j} - \frac{1}{2} \sum_{j=1}^n \sum_{s=1}^r b_{ij} \frac{\partial H_s}{\partial p_j} h_s \right) \right] \sum_{i=1}^n \sum_{j=1}^n \frac{\partial H}{\partial p_i} b_{ij} \frac{\partial H_s}{\partial p_j}$$

$$\times | \prod_{\gamma=1}^r \frac{\partial H_\gamma}{\partial p_\gamma} | dq_r \ldots dq_n dp_{r+1} \ldots dp_n d\theta_1 \ldots d\theta_{r-1} = 0,$$

$$(7.181)$$

The third criterion for obtaining $h_s$ is the equality of the averaged derivatives with respect to time of $r$ integrals of motion for the original and equivalent nonlinear systems, i.e.,

$$E\left[\frac{dH_s}{dt}\right]_{orig} = E\left[\frac{dH_s}{dt}\right]_{en} , \quad s = 1, 2, \ldots, r , \qquad (7.182)$$

where the averaging operation $E[.]$ is defined by the probability density function corresponding to stationary solution of equivalent nonlinear system (7.167), i.e.,

$$E[.] = \int_{R^{2n}} [.]g_{en}(\mathbf{q}, \mathbf{p})d\mathbf{q}d\mathbf{p} . \qquad (7.183)$$

We note that one can derive the Ito equation for Hamiltonian functions $H_{s|orig}$ and $H_{s|en}$ corresponding to systems (7.165) and (7.166) by applying Ito formula to both systems, respectively. Substituting them into (7.182), we find

$$\int_{R^{2n}} \sum_{i=1}^{n} \epsilon_i \frac{\partial H_s}{\partial p_i} g_{en}(\mathbf{q}, \mathbf{p})d\mathbf{q}d\mathbf{p} = 0, \quad s = 1, 2, \ldots, r . \qquad (7.184)$$

Then, following the derivation for the first criterion, one can find equations for determining $h_s$

$$\int_0^{2\pi} \int_\Lambda \left( \sum_{i=1}^{n} \epsilon_i \frac{\partial H_s}{\partial p_i} / | \prod_{\gamma=1}^{r} \frac{\partial H_\gamma}{\partial p_\gamma} | \right) dq_r \ldots dq_n \; dp_{r+1} \ldots dp_n \; d\theta_1 \ldots d\theta_{r-1} = 0 ,$$

$$s = 1, 2, \ldots, r. \qquad (7.185)$$

Inserting of $\epsilon_i$ defined by equality (7.173) into (7.185) gives

$$\int_0^{2\pi} \int_\Lambda \sum_{i=1}^{n} \left[ \frac{\partial H}{\partial p_i} \left( \sum_{j=1}^{n} M_{ij} \frac{\partial H}{\partial p_j} + \frac{1}{2} \frac{\partial b_{ij}}{\partial p_j} - \frac{1}{2} \sum_{j=1}^{n} \sum_{s=1}^{r} b_{ij} \frac{\partial H_s}{\partial p_j} h_s \right) | \prod_{\gamma=1}^{r} \frac{\partial H_\gamma}{\partial p_\gamma} | \right]$$

$$\times dq_r \ldots dq_n \; dp_{r+1} \ldots dp_n \; d\theta_1 \ldots d\theta_{r-1} = 0, \qquad (7.186)$$

The three presented criteria were compared with a 4-DOF system by Zhu et al. [33]. It was observed from numerical calculations that the three different criteria yield similar results for both the nonresonant and resonant cases, respectively.

## 7.6 Application of Volterra Functional Series Approach

When an original nonlinear system system can be approximated by an equivalent nonlinear dynamic system with nonlinear elements modeled by polynomials of finite order, then the Volterra series method can be used for the determination of the so-called "nonlinearization coefficients." This idea was first introduced by Donley and Spanos [5] and developed by these authors in [25] and [26] as well for SDOF as for MDOF systems. Using results given in [25], we will present this approach for a particular case of one-degree-of-freedom systems (original and equivalent) with the nonlinear function modeled by second-order polynomial. Then this approach is called *quadratization method*.

We consider the original nonlinear oscillator in the form

$$dx_1 = x_2 dt,$$
$$dx_2 = (-2hx_2 - \lambda_0^2 x_1 - f(x_1, x_2))dt + \eta(t)dt, \qquad (7.187)$$

where $h$ and $\lambda_0^2$ are positive constant coefficients, $\eta(t)$ is a stationary colored noise treated as an output of a linear filter with white noise input process, i.e.,

$$\eta(t) = \mu_\eta + \int_{-\infty}^{+\infty} h_\eta(\tau) w(t-\tau) d\tau , \qquad (7.188)$$

$h_\eta(\tau)$ is the impulse response (transfer) function of the linear filter, $\mu_\eta = E[\eta]$, $w(t)$ is a zero-mean Gaussian white noise process with the power spectral density $S_{ww}(\lambda) = 1$ or equivalently with autocorrelation function $E[w(t)w(t+\tau)] = 2\pi\delta(\tau)$; the nonlinear function $f(x_1, x_2)$ is nonsymmetric with respect to $(x_1, x_2)$, i.e.,

$$\forall x_1, x_2 \in R^1 \quad f(x_1, x_2) \neq -f(-x_1, -x_2) . \qquad (7.189)$$

The stationary solution of (7.187) is assumed in the form

$$x_1(t) = \mu_{x_1} + \tilde{x}_1(t) , \qquad (7.190)$$

where $\mu_{x_1} = E[x_1]$ can be determined by substitution of equality (7.190) into (7.187) and next by averaging. Then we obtain

$$\mu_{x_1} = \frac{\mu_\eta - E[f(\mu_{x_1} + \tilde{x}_1, \tilde{x}_2)]}{\lambda_0^2} . \qquad (7.191)$$

The equivalent nonlinear system is proposed as a quadratic system

$$
\begin{aligned}
d\tilde{x}_1 &= \tilde{x}_2 dt, \\
d\tilde{x}_2 &= [-2h\tilde{x}_2 - \lambda_0^2 \tilde{x}_1 - \alpha_1 \tilde{x}_1 + \alpha_2 \tilde{x}_2 - \alpha_3(\tilde{x}_1^2 - E[\tilde{x}_1^2]) \\
&\quad - \alpha_4(\tilde{x}_2^2 - E[\tilde{x}_2^2]) - \alpha_5 \tilde{x}_1 \tilde{x}_2 + \eta(t) - \mu_\eta] dt.
\end{aligned}
\qquad (7.192)
$$

The quadratization coefficients $\alpha_i$, $i = 1, \ldots, 5$, are determined by mean-square minimization of an error expression

$$
\begin{aligned}
\varepsilon &= \lambda_0^2 \mu_{x_1} + f(\mu_{x_1} + \tilde{x}_1, \tilde{x}_2) - \mu_\eta - \alpha_1 \tilde{x}_1 - \alpha_2 \tilde{x}_2 - \alpha_3(\tilde{x}_1^2 - E[\tilde{x}_2^2]) \\
&\quad - \alpha_4(\tilde{x}_2^2 - E[\tilde{x}_2^2]) - \alpha_5 \tilde{x}_1 \tilde{x}_2
\end{aligned}
\qquad (7.193)
$$

that yields a symmetric system of linear algebraic equations for the five unknowns $\alpha_i$.

$$\mathbf{A}_\alpha \alpha = \mathbf{b}_\alpha , \qquad (7.194)$$

where

$$\alpha = \begin{bmatrix} \alpha_1 \\ \alpha_2 \\ \alpha_3 \\ \alpha_4 \\ \alpha_5 \end{bmatrix}, \quad \mathbf{A}_\alpha = \begin{bmatrix} \mu_{\tilde{x}_1^2} & 0 & \mu_{\tilde{x}_1^3} & \mu_{\tilde{x}_1 \tilde{x}_2^2} & 0 \\ 0 & \mu_{\tilde{x}_2^2} & 0 & \mu_{\tilde{x}_2^3} & \mu_{\tilde{x}_1 \tilde{x}_2^2} \\ \mu_{\tilde{x}_1^3} & 0 & \mu_{\tilde{x}_1^4} - (\mu_{\tilde{x}_1^2})^2 & \mu_{\tilde{x}_1^2 \tilde{x}_2^2} - \mu_{\tilde{x}_1^2}\mu_{\tilde{x}_2^2} & 0 \\ \mu_{\tilde{x}_1 \tilde{x}_2} & \mu_{\tilde{x}_2^3} & \mu_{\tilde{x}_1^4} - (\mu_{\tilde{x}_1^2})^2 & \mu_{\tilde{x}_2^4} - (\mu_{\tilde{x}_2^2})^2 & \mu_{\tilde{x}_1 \tilde{x}_2^3} \\ 0 & \mu_{\tilde{x}_1 \tilde{x}_2^2} & 0 & \mu_{\tilde{x}_1 \tilde{x}_2^3} & \mu_{\tilde{x}_1^2 \tilde{x}_2^2} \end{bmatrix} ,$$

$$(7.195)$$

$$\mathbf{b}_\alpha = \begin{bmatrix} E[f(x_1,x_2)x_1] \\ E[f(x_1,x_2)x_2] \\ E[f(x_1,x_2)x_1^2] - E[f(x_1,x_2)]\mu_{\tilde{x}_1^2} \\ E[f(x_1,x_2)x_2^2] - E[f(x_1,x_2)]\mu_{\tilde{x}_2^2} \\ E[f(x_1,x_2)x_1 x_2] \end{bmatrix} , \quad \mu_{\tilde{x}_1^p \tilde{x}_2^q} = E[\tilde{x}_1^p \tilde{x}_2^q] . \quad (7.196)$$

To calculate the quadratic terms appearing in (7.196), one may apply the Voltera series method [23], where the equivalent nonlinear system is approximated by two linear systems. Using this approach, we rewrite (7.192) as follows:

$$d\tilde{x}_1 = \tilde{x}_2 dt,$$
$$d\tilde{x}_2 = [-2h_{eq}\tilde{x}_2 - \lambda_{0_{eq}}^2 \tilde{x}_1 - \alpha_1 \tilde{x}_1 + \alpha_2 \tilde{x}_2 - \alpha_3(\tilde{x}_1^2 - E[\tilde{x}_1^2])$$
$$- \alpha_4(\tilde{x}_2^2 - E[\tilde{x}_2^2]) - \alpha_5 \tilde{x}_1 \tilde{x}_2 + \gamma(\eta(t) - \mu_\eta)]dt, \qquad (7.197)$$

where $\gamma$ is a scalar coefficient and

$$2h_{eq} = 2h + \alpha_1, \quad \lambda_{0_{eq}}^2 = \lambda_0^2 + \alpha_2 . \qquad (7.198)$$

To use the Voltera series method, we assume that the solution of (7.197) can be presented as an infinite series in the following form:

$$\tilde{\mathbf{x}}(t) = \sum_{i=1}^\infty \gamma^i \tilde{\mathbf{x}}^{(i)}(t) , \qquad (7.199)$$

where $\tilde{\mathbf{x}} = [\tilde{x}_1, \tilde{x}_2]^T$, $\tilde{\mathbf{x}}^{(i)}(t)$ are unknown functions of $t$.

For further calculations, Spanos and Donley [25] have taken into account only the first two terms in infinite sum. Substituting the truncated series (7.199) into (7.197) and equating like powers of $\gamma$ leads to

$$d\tilde{x}_1^{(1)} = \tilde{x}_2^{(1)} dt,$$
$$d\tilde{x}_2^{(1)} = [-2h_{eq}\tilde{x}_2^{(1)} - \lambda_{0_{eq}}^2 \tilde{x}_1^{(1)} + \tilde{\eta}^{(1)}(t)]dt, \qquad (7.200)$$

$$d\tilde{x}_1^{(2)} = \tilde{x}_2^{(2)} dt,$$
$$d\tilde{x}_2^{(2)} = [-2h_{eq}\tilde{x}_2^{(2)} - \lambda_{0_{eq}}^2 \tilde{x}_1^{(2)} + \tilde{\eta}^{(0)} + \tilde{\eta}^{(2)}(t)]dt, \qquad (7.201)$$

where

$$\tilde{\eta}^{(0)} = \alpha_3 E[(\tilde{x}_1^{(1)})^2] + \alpha_4 E[(\tilde{x}_2^{(1)})^2], \quad \tilde{\eta}^{(1)}(t) = \eta(t) - \mu_\eta,$$
$$\tilde{\eta}^{(2)}(t) = -\alpha_3(\tilde{x}_1^{(1)})^2 - \alpha_4(\tilde{x}_2^{(1)})^2 - \alpha_5\tilde{x}_1^{(1)}\tilde{x}_2^{(1)}. \qquad (7.202)$$

In this case, the nonlinear system (7.197) is approximated by two linear systems (7.200) and (7.201), both of which have the same differential operator and different external excitations. One can obtain combining (7.200) and (7.201) one equivalent linear system in the following form:

$$d\tilde{x}_1 = \tilde{x}_2 dt,$$
$$d\tilde{x}_2 = [-2h_{eq}\tilde{x}_2 - \lambda_{0_{eq}}^2 \tilde{x}_1 + \tilde{\eta}(t)]dt, \qquad (7.203)$$

where

$$\tilde{\eta}(t) = \tilde{\eta}^{(0)} + \tilde{\eta}^{(1)}(t) + \tilde{\eta}^{(2)}(t) \qquad (7.204)$$

depends on the solution moments of (7.197).

From comparison of (7.188) and (7.202), it follows that $\tilde{\eta}^{(1)}(t)$ is the linear filtered white noise process and from (7.200) and (7.202) that $\tilde{\eta}^{(2)}(t)$ is a quadratic transform of white noise process. Hence, from the Volterra series theory, it follows that $\tilde{\eta}(t)$ can be expressed by equation

$$\tilde{\eta}(t) = \tilde{\eta}^{(0)} + \int_{-\infty}^{+\infty} h_{\tilde{\eta}}^{(1)}(\tau_1)w(t - \tau_1)d\tau_1$$
$$+ \int_{-\infty}^{+\infty}\int_{-\infty}^{+\infty} h_{\tilde{\eta}}^{(2)}(\tau_1, \tau_2)w(t - \tau_1)w(t - \tau_2)d\tau_1 d\tau_2, \quad (7.205)$$

where the functions $h_{\tilde{\eta}}^{(1)}(\tau_1)$ and $h_{\tilde{\eta}}^{(2)}(\tau_1, \tau_2)$ are the linear and quadratic impulse response (transfer) functions.

Taking into account that the differential operators defined by (7.200), (7.201), and (7.203) are linear, the stationary solution of (7.203) can be presented as follows:

$$\tilde{x}_1(t) = \int_{-\infty}^{+\infty} h(\tau)\tilde{\eta}(t - \tau)d\tau, \qquad (7.206)$$

where $h(\tau)$ is the impulse response function corresponding to dynamic system (7.203). Substituting (7.205) into (7.206), we obtain the following Volterra series representation for stationary solution of system ((7.203)):

$$\tilde{x}(t) = \tilde{x}_0 + \int_{-\infty}^{+\infty} h_{\tilde{x}_1}^{(1)}(\tau_1)w(t - \tau_1)d\tau_1$$
$$+ \int_{-\infty}^{+\infty}\int_{-\infty}^{+\infty} h_{\tilde{x}_1}^{(2)}(\tau_1, \tau_2)w(t - \tau_1)w(t - \tau_2)d\tau_1 d\tau_2, \quad (7.207)$$

where

$$\tilde{x}_0 = \tilde{\eta}^{(0)} \int_{-\infty}^{+\infty} h(\tau)d\tau , \qquad (7.208)$$

$$h_{\tilde{x}_1}^{(1)}(\tau_1) = \int_{-\infty}^{+\infty} h(\tau)h_{\tilde{\eta}}^{(1)}(\tau_1 - \tau)d\tau , \qquad (7.209)$$

$$h_{\tilde{x}_1}^{(2)}(\tau_1, \tau_2) = \int_{-\infty}^{+\infty} h(\tau)h_{\tilde{\eta}}^{(2)}(\tau_1 - \tau, \tau_2 - \tau)d\tau , \qquad (7.210)$$

$h_{\tilde{x}_1}^{(1)}(\tau_1)$ and $h_{\tilde{x}_1}^{(2)}(\tau_1, \tau_2)$ are the Volterra kernels of the first- and second-order zero-mean displacement response of system (7.203). We note that $h_{\tilde{\eta}}^{(1)}(\tau_1)$ is the linear impulse response function of the white noise filter, while $h_{\tilde{\eta}}^{(2)}(\tau_1, \tau_2)$ can be found if we substitute the linear part of (7.207) into the third equality of (7.202) and we take into account (7.205). Then we obtain

$$h_{\tilde{\eta}}^{(2)}(\tau_1, \tau_2) = -\alpha_3 h_{\tilde{x}_1}^{(1)}(\tau_1)h_{\tilde{x}_1}^{(1)}(\tau_2) - \alpha_4 h_{\tilde{x}_2}^{(1)}(\tau_1)h_{\tilde{x}_2}^{(1)}(\tau_2)$$
$$- \frac{1}{2}\alpha_5[h_{\tilde{x}_1}^{(1)}(\tau_1)h_{\tilde{x}_2}^{(1)}(\tau_2) + h_{\tilde{x}_2}^{(1)}(\tau_1)h_{\tilde{x}_1}^{(1)}(\tau_2)] , \qquad (7.211)$$

where $h_{\tilde{x}_2}^{(1)}(\tau)$ is the linear Volterra kernel for $\tilde{x}_2(t)$, i.e., $h_{\tilde{x}_2}^{(1)}(\tau) = \frac{d}{d\tau}h_{\tilde{x}_1}^{(1)}(\tau)$.

To calculate response statistics, we have to determine the moments appearing in vector $\mathbf{b}_\alpha$. It can be done only approximately because the exact probability distribution function of the response vector $\tilde{\mathbf{x}}(t)$ is unknown. Spanos and Donley proposed in [25] to approximate the joint probability density function by the truncating Gram–Charlier expansion up to the third order. For higher-order moments, they have used the Gaussian closure technique.

This approach was extended by Spanos and Donley [26] for MDOF systems and two modifications were proposed by Kareem and Zhao [8], Kareem et al. [9], and Tognarelli et al. [30]. In the first modification, the stationary solution expressed by the Voltera series decomposition of the entire response process has the form

$$\tilde{x}_1(t) = x_0 + \tilde{x}_1^{(1)}(t) + \frac{1}{2}\tilde{x}_1^{(2)}(t) = x_0 + \int_{-\infty}^{+\infty} h_{\tilde{x}_1}^{(1)}(\tau_1)\eta(t - \tau_1)d\tau_1$$

$$+ \frac{1}{2}\int_{-\infty}^{+\infty}\int_{-\infty}^{+\infty} h_{\tilde{x}_1}^{(2)}(\tau_1, \tau_2)\eta(t - \tau_1)\eta(t - \tau_2)d\tau_1 d\tau_2, \qquad (7.212)$$

where $x_0$ is the static response of the first-order system to be developed momentarily, and $h_{\tilde{x}_1}^{(1)}(\tau_1)$ and $h_{\tilde{x}_1}^{(2)}(\tau_1, \tau_2)$ are Volterra kernels of the first- and second-order displacement response. The nonlinearity $f(x_1, x_2)$ is first approximated by a Taylor series

$$f(x_1, x_2) \approx f(x_o + \tilde{x}_1^{(1)}, \tilde{x}_2^{(1)}) + \mu_{f x_1}\frac{\tilde{x}_1^{(2)}}{2} + \mu_{f x_2}\frac{\tilde{x}_2^{(2)}}{2} , \qquad (7.213)$$

where $\bar{x}_2^{(1)} = \dot{\bar{x}}_1^{(1)}$, $\bar{x}_2^{(2)} = \dot{\bar{x}}_1^{(2)}$,

$$\mu_{fx_1} = E\left[\frac{\partial f}{\partial x_1} f(x_o + \bar{x}_1^{(1)}, \bar{x}_2^{(1)})\right], \quad \mu_{fx_2} = E\left[\frac{\partial f}{\partial x_2} f(x_o + \bar{x}_1^{(1)}, \bar{x}_2^{(1)})\right]$$

(7.214)

and next proposed to replace by

$$f(x_o + \bar{x}_1^{(1)}, \bar{x}_2^{(1)}) \approx \frac{1}{2}[\beta_0 + \beta_1 \bar{x}_1^{(1)} + \beta_2 \bar{x}_2^{(1)} + \beta_3 (\bar{x}_1^{(1)})^2 + \beta_4 \bar{x}_1^{(1)} \bar{x}_2^{(1)} + \beta_5 (\bar{x}_2^{(1)})^2],$$

(7.215)

where $\beta_i$, $i = 0, \ldots, 5$, are quadratization coefficients.

Then the mean-square error $E[\epsilon^2]$ is defined for

$$\epsilon = f(x_o + \bar{x}_1, \dot{\bar{x}}_1) - \frac{1}{2}[\beta_0 + \beta_1 \bar{x}_1^{(1)} + \beta_2 \bar{x}_2^{(1)} + \beta_3 (\bar{x}_1^{(1)})^2 + \beta_4 \bar{x}_1^{(1)} \bar{x}_2^{(1)} + \beta_5 (\bar{x}_2^{(1)})^2].$$

(7.216)

The minimization of the mean-square error $E[\epsilon^2]$ leads to a symmetric system of linear algebraic equations for the six unknowns $\beta_i$, $i = 0, \ldots, 5$.

$$\mathbf{A}_\beta \beta = \mathbf{b}_\beta,$$

(7.217)

where

$$\mathbf{A}_\beta = \begin{bmatrix} 1 & 0 & 0 & E[(\bar{x}_1^{(1)})^2] & 0 & E[(\bar{x}_2^{(1)})^2] \\ 0 & E[(\bar{x}_1^{(1)})^2] & 0 & 0 & 0 & 0 \\ 0 & 0 & E[(\bar{x}_2^{(1)})^2] & 0 & 0 & 0 \\ E[(\bar{x}_1^{(1)})^2] & 0 & 0 & E[(\bar{x}_1^{(1)})^4] & 0 & E[(\bar{x}_1^{(1)})^2(\bar{x}_2^{(1)})^2] \\ 0 & 0 & 0 & 0 & E[(\bar{x}_1^{(1)})^2(\bar{x}_2^{(1)})^2] & 0 \\ E[(\bar{x}_2^{(1)})^2] & 0 & 0 & E[(\bar{x}_1^{(1)})^2(\bar{x}_2^{(1)})^2] & 0 & E[(\bar{x}_2^{(1)})^4] \end{bmatrix}$$

(7.218)

$$\beta = \begin{bmatrix} \beta_0 \\ \beta_1 \\ \beta_2 \\ \beta_3 \\ \beta_4 \\ \beta_5 \end{bmatrix}, \quad \mathbf{b}_\beta = 2\begin{bmatrix} E[f(x_o + \bar{x}_1, \bar{x}_2^{(1)})] \\ E[\bar{x}_1^{(1)} f(x_o + \bar{x}_1, \bar{x}_2^{(1)})] \\ E[\bar{x}_2^{(1)} f(x_o + \bar{x}_1, \bar{x}_2^{(1)})] \\ E[(\bar{x}_1^{(1)})^2 f(x_o + \bar{x}_1, \bar{x}_2^{(1)})] \\ E[\bar{x}_1^{(1)} \bar{x}_2^{(1)} f(x_o + \bar{x}_1, \bar{x}_2^{(1)})] \\ E[(\bar{x}_2^{(1)})^2 f(x_o + \bar{x}_1, \bar{x}_2^{(1)})] \end{bmatrix}.$$

(7.219)

Using equalities (7.212–7.215) one can rewrite and split (7.187) into the equivalent Volterra representation

$$d\bar{x}_1^{(1)} = \bar{x}_2^{(1)} dt,$$

$$d\bar{x}_2^{(1)} = \left[-\left(2h + \frac{\beta_2}{2}\right)\bar{x}_2^{(1)} - (\lambda_0^2 + \frac{\beta_1}{2})\bar{x}_1^{(1)} + \eta(t)\right] dt,$$

(7.220)

and

$$d\bar{x}_1^{(2)} = \bar{x}_2^{(2)} dt,$$
$$d\bar{x}_2^{(2)} = \left[ -(2h + \mu_{fx_2})\bar{x}_2^{(2)} - (\lambda_{0_{eq}}^2 + \mu_{fx_1})\bar{x}_1^{(2)} + \beta_3(\bar{x}_1^{(1)})^2 + \beta_4\bar{x}_1^{(1)}\bar{x}_2^{(1)} \right.$$
$$\left. + \beta_5(\bar{x}_2^{(1)})^2 \right] dt . \tag{7.221}$$

Similar to the equivalent quadratization approach, Tognarelli et al. [30] used (7.220) and (7.221) first for the determination of the impulse transfer functions and next exact integral formulas for response cumulants [1]. Application of the moment-based Hermite transformation to response cumulants yields the response probability density function. The Gram–Charlier series expansion appearing in the Spanos and Donley approach was replaced by the moment-based Hermite transformation model of the probability distribution of the response in order to eliminate a problem of negative probabilities in the tails. Also, Tognarelli et al. [30] have calculated the moments appearing in vector $\mathbf{b}_\beta$ up to the fourth order directly without Gaussian closure.

The third approach, proposed by Tognarelli et al. [30] is identical to the first, except for two changes:

(a) The Gram–Charlier series expansion appearing in first approach was replaced by the moment-based Hermite transformation model of the probability distribution of the response,
(b) the moments are computed up to fourth order rather than just to the third order.

## Bibliography Notes

Caughey's technique was generalized by Zhu and Yu [36] for nearly conservative systems subjected to external and parametric white noise excitations. This approach was based on the equivalence of the drift and diffusion coefficients of the averaged total energies of the given and equivalent systems. The same concept of nonlinearization has been applied by Manohar and Iyengar [16] in a slightly different way in two examples. In the first one, they replaced a nonlinear oscillator under Gaussian white noise excitation by equivalent nonlinear Van der Pol oscillator under the same excitation using mean-square criterion of approximation. In the second example, Manohar and Iyengar [16] extended this approach to a nonlinear system with combined periodic and Gaussian white noise excitation. In this case, the response is a nonstationary process and therefore a combination of stochastic averaging and equivalent nonlinearization has been proposed and developed.

The problem of the determination of an equivalent nonlinear system is connected with finding a class of nonlinear stochastic dynamic systems with known form of solution, for instance, stationary solution. Therefore, the results in the field of exact or very well approximated solution for a given class of nonlinear stochastic differential equations are valuable for the study of equivalent nonlinear system. One can find these results mainly in the literature concerning methods of the determination of exact or approximate solutions of Fokker–Planck–Kolmogorov equation. It was partially reviewed in Sect. 2.12.

The most popular—because of the simplicity of calculations in the study of mechanical and structural systems—are methods of finding equivalent nonlinear system with nonlinearity in polynomial form or in ratio power.

An application of the Donley and Spanos approach to the determination of stochastic response of offshore platforms by statistical quadratization and statistical cubicization was given by Quek et al. [20] and by Li et al. [11], respectively. Also Fatica and Floris [7] have shown an application of a nonlinearization method to the moment analysis of base-isolated buildings subjected to support motion. The authors approximated the function $sgn(x)$ by cubic or quintic polynomial and derived cubicization and pentization coefficients using mean-square criterion.

The idea of cubicization has been adopted by Spanos et al. [27] for the determination of approximate response spectrum of a nonlinear oscillator. First the authors approximate a nonlinear stiffness by a polynomial third order using square criterion (not mean-square) and next used the $N$-order Galerkin approximation. The obtained response spectral density function was compared with the corresponding one derived from the linearized equation by statistical linearization.

Another idea for the determination of equivalent nonlinear oscillator has been proposed by Krenk and Roberts [10] and developed by Rudinger and Krenk [22]. The authors have used the derivation of nonlinearization coefficient (damping) introduced by Caughey [4] by an assumption that this coefficient depends on a given energy level $\lambda$, which can be evaluated by considering free undamped vibration for $\lambda$.

# References

1. Bedrosian, E. and Rice, S.: The outpost properties of Volterra systems (nonlinear systems with memory) driven by harmonic and Gaussian inputs. Proc. IEEE **59**, 1688–1707, 1971.
2. Cai, G. and Lin, Y.: On exact stationary solutions of equivalent nonlinear stochastic systems. Int. J. Non-Linear Mech. **23**, 315–325, 1988.
3. Caughey, T.: On the response of nonlinear oscillators to stochastic excitation in random vibration. In: T. Huang and P. Spanos (eds.), Random Vibration, AMD 65, 9–14. American Society of Mechanical Engineers, New York, 1984.
4. Caughey, T.: On the response of nonlinear oscillators to stochastic excitation. Probab. Eng. Mech. **1**, 2–10, 1986.
5. Donley, M. and Spanos, P.: Dynamic analysis of non-linear structures by the method of statistical quadratization. In: Lecture Notes in Engineering, Vol. 37. 1–172, Springer, New York, 1990.

6. Elishakoff, I. and Cai, G.: Approximate solution for nonlinear random vibration problems by partial stochastic linearization. Probab. Eng. Mech. **8**, 233–237, 1993.
7. Fatica, G. and Floris, C.: Moment equation analysis of base-isolated buildings subjected to support motion. ASCE J. Eng. Mech. **129**, 94–106, 2003.
8. Kareem, A. and Zhao, J.: Stochastic response analysis tension leg platforms: A statistical quadratization and qubicization approach. In: (ed.), Proceedings of Offshore Mechanics and Arctic Engineers, Vol. I. American Society of Mechanical Engineers, New York, 1994.
9. Kareem, A., Zhao, J., and Tognarelli, M.: Surge response statistics of tension leg platforms under wind and wave loads: Statistical quadratization approach. Probab. Eng. Mech. **10**, 225–240, 1995.
10. Krenk, S. and Roberts, J.: Local similarity in nonlinear random vibration. Trans. ASME J. Appl. Mech. **66**, 225–235, 1999.
11. Li, X., Quek, S., and Koh, C.: Stochastic response of offshore platform by statistical cubization. ASCE J. Eng. Mech. **121**, 1056–1068, 1995.
12. Lin, Y. and Cai, G.: Equivalent stochastic systems. Trans. ASME J. Appl. Mech. **55**, 918–922, 1988.
13. Lutes, L.: Approximate technique for treating random vibration of hysteretic systems. J. Acoust. Soc. Am. **48**, 299–306, 1970.
14. Maltchikov, S.: An approximate method of statistical analysis of systems which contain nonlinearities of multiplicative type. Awtomatika i Telemechanika **34**, 1560–1565, 1973.
15. Maltchikov, S.: Determination of the distribution of the output variables of a multidimensional nonlinear system. Awtomatika i Telemechanika **34**, 1724–1729, 1973.
16. Manohar, C. and Iyengar, R.: Random vibration of a limit cycle system. Structural Engineering Laboratory. Department of Civil Engineering Indiana Institute of Science, Report No **1**, 1990.
17. Nigam, N.: Introduction to Random Vibrations. M.I.T. Press, Cambridge, Mass., 1983.
18. Polidori, D. and Beck, J.: Approximate solutions for nonlinear random vibration problems. Probab. Eng. Mech. **11**, 179–185, 1996.
19. Pugachev, V. and Sinitsyn, I.: Stochastic Differential Systems. Wiley, Chichester, 1987.
20. Quek, S., Li, X., and Koh, C.: Stochastic response of jack-up platform by the method for statistical quadratization. Appl. Ocean Res. **16**, 113–122, 1994.
21. Ricciardi, G. and Elishakoff, I.: A novel local stochastic linearization method via two extremum entropy principles. Int. J. Non-Linear Mech. **37**, 785–800, 2002.
22. Rudinger, F. and Krenk, S.: Spectral density of oscillator with power law damping excited by white noise. J. Sound Vib. **261**, 365–371, 2003.
23. Schetzen, M.: The Volterra and Wiener Theories of Nonlinear Systems. John Wiley and Sons, New York, 1980.
24. Socha, L.: Application of probabilistic metrics to the nonlinearization analysis. In: B. Spencer and E. Johnson (eds.), Stochastic Structural Dynamics, 99–104. A. A. Balkema, Rotterdam, 1999.
25. Spanos, P. and Donley, M.: Equivalent statistical quadratization for nonlinear system. ASCE J. Eng. Mech. **117**, 1289–1310, 1991.

26. Spanos, P. and Donley, M.: Non-linear multi-degree-of-freedom system random vibration by equivalent statistical quadratization. Int. J. Non-Linear Mech. **27**, 735–748, 1992.

27. Spanos, P., Di Paola, M., and Failla, G.: A Galerkin approach for power spectrum determination for nonlinear oscillators. Meccanica **37**, 51–65, 2002.

28. To, C.: A statistical nonlinearization technique in structural dynamics. J. Sound Vib. **161**, 543–548, 1993.

29. To, C. and Li, D.: Equivalent nonlinearization of nonlinear systems to random excitations. Probab. Eng. Mech. **6**, 184–192, 1991.

30. Tognarelli, M., Zhao, J., Rao, K., and Kareem, A.: Equivalent statistical quadratization and qubicization for nonlinear system. ASCE J. Eng. Mech. **123**, 512–523, 1997.

31. Wang, R., Kusumoto, S., and Zhang, Z.: A new equivalent non-linearization technique. Probab. Eng. Mech. **11**, 129–137, 1996.

32. Zhu, W. and Deng, M.: Equivalent non-linear system method for stochastically excited and dissipated integrable Hamiltonian systems-resonant case. J. Sound Vib. **274**, 1110–1122, 2004.

33. Zhu, W., Huang, Z., and Suzuki, Y.: Equivalent non-linear system method for stochastically excited and dissipated partially integrable Hamiltonian systems. Int. J. Non-Linear Mech. **36**, 773–786, 2001.

34. Zhu, W. and Lei, Y.: Equivalent nonlinear system method for stochastically excited and dissipated integrable Hamiltonian systems. Trans. ASME J. Appl. Mech. **64**, 209–216, 1997.

35. Zhu, W., T.T., S., and Lei, Y.: Equivalent nonlinear system method for stochastically excited integrable Hamiltonian systems. Trans. ASME J. Appl. Mech. **61**, 618–623, 1994.

36. Zhu, W. and Yu, J.: The equivalent non-linear system method. J. Sound Vib. **129**, 385–395, 1989.

# 8

## Linearization of Dynamic Systems with Stochastic Parametric Excitations

### 8.1 Introduction

For many real physical, chemical, biological, economical, and engineering systems, the corresponding mathematical models are dynamic, nonlinear with variable parameters. One can find many examples of such systems, for instance, in monographs by Evlanov and Konstantinov [7], Kazakov and Maltchikov [11], and Ibrahim [10]. In the case of dynamic deterministic systems, it is assumed that they are described by ordinary differential equations with parameters being deterministic time-dependent functions, while in the case of stochastic dynamic systems with stochastic parameters, they are described by stochastic differential equations.

### 8.2 Statistical Linearization

The generalization of statistical linearization for dynamic systems under stochastic parametric excitations was proposed by Kazakov and Maltchikov [11]. It consists of independent linearization procedures of deterministic and stochastic parts of a dynamic system that can be modeled by the Stratonovich stochastic differential equation. We discuss a modification of this approach using the equivalent Ito stochastic differential equation.

Consider a nonlinear stochastic dynamic system described by the vector Ito stochastic differential equation

$$dx(t) = \Phi(x, t)dt + \sigma(x, t)d\xi(t), \quad x(t_0) = x_0 , \tag{8.1}$$

where $\mathbf{x} = [x_1, \ldots, x_n]^T$ is the state vector, $\boldsymbol{\Phi} = [\Phi_1, \ldots, \Phi_n]^T$ is a nonlinear vector function, $\boldsymbol{\sigma} = [\sigma_{jk}]$, $j = 1, \ldots, n$, $k = 1, \ldots, M$ is a matrix with elements being nonlinear functions, $\boldsymbol{\sigma} : R^n \times R^+ \to R^n \times R^M$, $\boldsymbol{\xi} = [\xi_1, \ldots, \xi_M]^T$, $\xi_k$, $k = 1, \ldots, M$, are independent standard Wiener processes, and the initial condition $\mathbf{x}_0$ is a vector random variable independent of $\xi_k$, $k = 1, \ldots, M$.

L. Socha: *Linearization of Dynamic Systems with Stochastic Parametric Excitations*, Lect. Notes Phys. **730**, 251–280 (2008)
DOI 10.1007/978-3-540-72997-6_8      © Springer-Verlag Berlin Heidelberg 2008

We assume that the solution of (8.1) exists and the sufficient conditions of the exponential stability with probability 1 of equilibrium of this equation are satisfied.

In the case of parametric excitations, the objective of statistical linearization is the replacement as well the vector $\boldsymbol{\Phi}$ as the matrix $\boldsymbol{\sigma}$ by their linearized forms

$$\boldsymbol{\Phi}(\mathbf{x}, t) = \boldsymbol{\Phi}_0(\mathbf{m}_x, \boldsymbol{\Theta}_x, t) + \mathbf{K}_\phi(\mathbf{m}_x, \boldsymbol{\Theta}_x, t)\mathbf{x}^0 , \qquad (8.2)$$

$$\boldsymbol{\sigma}(\mathbf{x}, t) = \boldsymbol{\sigma}_0(\mathbf{m}_x, \boldsymbol{\Theta}_x, t) + \mathbf{K}_\sigma^T(\mathbf{m}_x, \boldsymbol{\Theta}_x, t)\mathbf{x}^0 \qquad (8.3)$$

or

$$\boldsymbol{\sigma}(\mathbf{x}, t) = \boldsymbol{\sigma}_0(\mathbf{m}_x, \boldsymbol{\Theta}_x, t) + \sum_{i=1}^{n} \mathbf{K}_{\sigma i}(\mathbf{m}_x, \boldsymbol{\Theta}_x, t)x_i^0 , \qquad (8.4)$$

where $\boldsymbol{\Phi}_0(\mathbf{m}_x, \boldsymbol{\Theta}_x, t)$, $\mathbf{K}_\phi(\mathbf{m}_x, \boldsymbol{\Theta}_x, t)$, $\mathbf{m}_x, \boldsymbol{\Theta}_x$, and $\mathbf{x}^0$ have the same meaning as in the case of statistical linearization for dynamic systems under external excitation,

$$\boldsymbol{\sigma}_0(\mathbf{m}_x, \boldsymbol{\Theta}_x, t) = [\sigma_{jk0}(\mathbf{m}_x, \boldsymbol{\Theta}_x, t)], \ j = 1, \ldots, n, \ k = 1, \ldots, M \qquad (8.5)$$

is an $n \times M$ matrix,

$$\mathbf{K}_\sigma = [\mathbf{K}_{\sigma 1}, \ldots, \mathbf{K}_{\sigma n}]^T \qquad (8.6)$$

is a block vector which elements are matrices defined by

$$\mathbf{K}_{\sigma i}(\mathbf{m}_x, \boldsymbol{\Theta}_x, t) = [h_{ijk}(\mathbf{m}_x, \boldsymbol{\Theta}_x, t)] = \begin{bmatrix} h_{i11} & \cdots & h_{i1M} \\ \vdots & & \vdots \\ h_{in1} & \cdots & h_{inM} \end{bmatrix},$$

$$i, j = 1, \ldots, n, \quad k = 1, \ldots, M. \qquad (8.7)$$

Linearization of the matrix $\boldsymbol{\sigma}(\mathbf{x}, t)$ is made similar to the linearization of the vector $\boldsymbol{\Phi}(\mathbf{x}, t)$, i.e., each element of the matrix $\boldsymbol{\sigma}$ is linearized separately by one of the criteria presented in Sect. 5.2. For instance, if we use the mean-square criterion defined by

$$\delta_{jk} = E\left[\left(\sigma_{jk}(x_1, \ldots, x_n, t) - \sigma_{jk0} - \sum_{i=1}^{n} h_{ijk}x_i^0\right)^2\right],$$

$$j = 1, \ldots, n, \quad k = 1, \ldots, M , \qquad (8.8)$$

then the linearization coefficients $\sigma_{jk0}$ and $h_{ijk}$, $i = 1, \ldots, n$, are determined from the necessary conditions of minimum of the considered criterion

$$\frac{\partial \delta_{jk}}{\partial \sigma_{jk0}} = 0, \quad \frac{\partial \delta_{jk}}{\partial h_{ijk}} = 0, \quad i, j = 1, \ldots, n, k = 1, \ldots, M . \qquad (8.9)$$

Similar to the derivation of linearization coefficients presented in Sect. 5.2, we obtain

$$\sigma_{jk0} = E[\sigma_{jk}] \quad j = 1, \ldots, n, \quad k = 1, \ldots, M .  \tag{8.10}$$

$$h_{ijk} = \frac{\partial \sigma_{jk0}}{\partial m_{x_i}}, \quad i, j = 1, \ldots, n, \quad k = 1, \ldots, M ,  \tag{8.11}$$

under the assumption that variables $x_i$, $i = 1, \ldots, n$, are Gaussian.

As in the case of application of statistical linearization to the vector function $\Phi(\mathbf{x}, t)$, also, the elements of linearized matrices $\sigma_0$ and $\mathbf{K}_\sigma$ are nonlinear functions of the mean value and covariance matrix of the state vector $\mathbf{x}$. One can find these nonlinear relations from (8.10) and (8.11). Since the determination of exact moments of the first and second order is impossible in a general case, they are replaced by the corresponding moments of the linearized system. It is obtained by substitution of relations (8.2) and (8.3) into (8.1), i.e.,

$$dx(t) = [\Phi_0(\mathbf{m}_x, \Theta_x, t) + \mathbf{K}_\phi(\mathbf{m}_x, \Theta_x, t)\mathbf{x}^0] dt$$

$$+ [\sigma_0(\mathbf{m}_x, \Theta_x, t) + \sum_{i=1}^{n} \mathbf{K}_{\sigma i}(\mathbf{m}_x, \Theta_x, t)\mathbf{x}_i^0] d\xi .  \tag{8.12}$$

Equation (8.12) can be rewritten in an equivalent form

$$dx(t) = [\mathbf{A}_0(t) + \mathbf{A}(t)\mathbf{x}(t)] dt + \sum_{k=1}^{M} [\mathbf{D}_k(t)\mathbf{x}(t) + \mathbf{G}_{k0}(t)] d\xi_k(t), \quad \mathbf{x}(t_0) = \mathbf{x}_0 ,  \tag{8.13}$$

where $\mathbf{A}_0 = \Phi_0 - \mathbf{K}_\phi \mathbf{m_x} = [A_{01}, \ldots, A_{0n}]^T$, $\mathbf{A} = \mathbf{K}_\phi = [a_{ij}]$, $i, j = 1, \ldots, n$,
$\mathbf{G}_{k0} = \sigma_{kj0} - \sum_{i=1}^{n} \mathbf{K}_{\sigma i} m_{xi} = [G_{1k0}, \ldots, G_{nk0}]^T$, $\mathbf{D}_k = [h_{ijk}]$, $i, j = 1, \ldots, n$, $k = 1, \ldots, M$, are vectors and matrices of linearization coefficients, respectively.

$$\sigma_{k0} = \begin{bmatrix} \sigma_{1k0} \\ \vdots \\ \sigma_{nk0} \end{bmatrix}, \quad \mathbf{D}_k = \begin{bmatrix} h_{11k} & \cdots & h_{1nk} \\ \vdots & & \vdots \\ h_{n1k} & \cdots & h_{nnk} \end{bmatrix},$$

$$i, j = 1, \ldots, n, \quad k = 1, \ldots, M.  \tag{8.14}$$

Using the Ito formula and averaging operation, one can derive the moment equations of the first and second order for the linearized equation (8.12)

$$\frac{d\mathbf{m_x}(t)}{dt} = \Phi_0(\mathbf{m_x}(t), \Theta_x(t), t), \quad \mathbf{m_x}(t_0) = E[\mathbf{x}_0] ,  \tag{8.15}$$

$$\frac{d\Theta_\mathbf{x}(t)}{dt} = \mathbf{K}_\phi(\mathbf{m_x}(t), \Theta_x(t), t)\Theta_\mathbf{x}(t) + \Theta_\mathbf{x}(t)\mathbf{K}_\phi^T(\mathbf{m_x}(t), \Theta_x(t), t)$$

$$+ \sigma_0(\mathbf{m_x}(t), \Theta_x(t), t)\sigma_0^T(\mathbf{m_x},(t)\Theta_x, (t)t) , \quad \Theta(t_0) = E[\mathbf{x}_0^0 \mathbf{x}_0^{0T}]  \tag{8.16}$$

or in an equivalent form for the linearized equation (8.13)

$$\frac{d\mathbf{m_x}(t)}{dt} = \mathbf{A}_0(t) + \mathbf{A}(t)\mathbf{m_x}(t), \quad \mathbf{m_x}(t_0) = E[\mathbf{x}_0], \tag{8.17}$$

$$\frac{d\mathbf{\Gamma}_{Lp}(t)}{dt} = \mathbf{m_x}(t)\mathbf{A}_0^T(t) + \mathbf{A}_0(t)\mathbf{m_x}^T(t) + \mathbf{\Gamma}_{Lp}(t)\mathbf{A}^T(t) + \mathbf{A}(t)\mathbf{\Gamma}_{Lp}(t)$$

$$+ \sum_{k=1}^{M}[\mathbf{G}_{k0}(t)\mathbf{G}_{k0}^T(t) + \mathbf{D}_k(t)\mathbf{m_x}(t)\mathbf{A}_0^T(t) + \mathbf{A}_0(t)\mathbf{m_x}^T(t)\mathbf{D}_k^T(t)$$

$$+ \mathbf{D}_k(t)\mathbf{\Gamma}_{Lp}(t)\mathbf{D}_k^T(t)], \tag{8.18}$$

$$\mathbf{\Gamma}_{Lp}(t_0) = E[\mathbf{x}_0\mathbf{x}_0^T],$$

where $\mathbf{m_x} = E[\mathbf{x}]$, $\mathbf{\Gamma}_{Lp} = E[\mathbf{xx}^T]$.

The moment equations (8.15) and (8.16) are nonlinear differential equations that can be solved by standard methods (usually numerical methods). To obtain stationary solutions, one should solve the system of algebraic equations obtained by equating to zero the right-hand sides of the moment equations (8.15) and (8.16) or (8.17) and (8.18) for $t \to \infty$ taking into consideration the nonlinear dependence of linearization coefficients with respect to response moments. Then, one can use the iterative procedure discussed in Sect. 5.2.

We note that this approach is valid for the white noise excitation. In the case of the colored noise excitation, a modification of this approach has to be done. The problem is connected with the solution of moment equations for linear systems with colored noise parametric excitations. It was discussed in Sect. 3.3.

In the case of statistical linearization with criteria in probability density functions space, the procedure is the same as in the case of moment criteria, i.e., we linearize separately each nonlinear element and next we determine moments of solutions using, for instance, a modified Procedure $SL$–$PD$, where the solution of moment equations for nonlinear system (5.31) and (5.32) is replaced by solutions of corresponding moments for linearized system (8.17) and (8.18).

The method of linearization described for dynamic systems with stochastic parametric excitations can be generalized to dynamic systems with stochastic external and parametric excitations. Then linearization coefficients have the same form (they are calculated separately) and moment equations for linearized system depend on external and parametric excitations. We illustrate this approach with an example.

*Example 8.1.* Consider the vector nonlinear stochastic Ito equation

$$d\mathbf{x}(t) = [\mathbf{Ax}(t) + \mathbf{B}(\mathbf{x}(t))]dt + [\mathbf{C}(\mathbf{x}(t)) + \mathbf{D}]d\mathbf{\xi}(t), \quad \mathbf{x}(t_0) = \mathbf{x}_0, \tag{8.19}$$

where

$$\mathbf{x} = \begin{bmatrix} x_1 \\ x_2 \end{bmatrix}, \mathbf{A} = \begin{bmatrix} 0 & 1 \\ -\lambda_0^2 & -2h \end{bmatrix}, \mathbf{B}(\mathbf{x}) = \begin{bmatrix} 0 \\ \varepsilon x_1^3 \end{bmatrix}, \mathbf{C}(\mathbf{x}) = \begin{bmatrix} 0 & 0 \\ \varepsilon x_1^3 & c|x_2|x_2 \end{bmatrix},$$

$$\mathbf{D} = \begin{bmatrix} 0 & 0 \\ q_1 & q_2 \end{bmatrix}, \boldsymbol{\xi} = \begin{bmatrix} \xi_1 \\ \xi_2 \end{bmatrix}, \mathbf{x} = \begin{bmatrix} x_{10} \\ x_{20} \end{bmatrix}, \quad (8.20)$$

where $\lambda_0, h, q_1, q_2, \varepsilon$, and $c$ are positive constant parameters satisfying sufficient conditions of exponential stability with probability 1; $\xi_1$ and $\xi_2$ are independent standard Wiener processes, $x_{i0}$, $i = 1, 2$, are random variables independent of $\xi_1$ and $\xi_2$.

The nonlinear functions $\phi_1(x_1) = \varepsilon x_1^3$ and $\phi_2(x_2) = c|x_2|x_2$ are approximated by the corresponding linearized forms, i.e.,

$$\phi_1(x_1) = \varepsilon x_1^3 \approx k_0 m_{x_1} + k_1(x_1 - m_{x_1}), \quad (8.21)$$

$$\phi_2(x_2) = c|x_2|x_2 \approx \sigma_0 m_{x_2} + h_1(x_2 - m_{x_2}), \quad (8.22)$$

where $m_{x_i} = E[x_i]$, $i = 1, 2$, $k_0, k_1, \sigma_0$, and $h_1$ are linearization coefficients. If we substitute the linearized functions to system (8.19) and (8.20), we then obtain a linear vector Ito stochastic differential equation

$$d\mathbf{x}(t) = [\mathbf{A_1}\mathbf{x(t)} + \mathbf{A_0}]dt + \sum_{k=1}^{2}[\mathbf{D_k}\mathbf{x(t)} + \mathbf{G_k}]d\xi_k(t), \quad \mathbf{x}(t_0) = \mathbf{x_0}, \quad (8.23)$$

where

$$\mathbf{A_0} = \begin{bmatrix} 0 \\ (k_0 - k_1)m_{x_1} \end{bmatrix}, \quad \mathbf{A_1} = \begin{bmatrix} 0 & 1 \\ -\lambda_0^2 + k_1 & -2h \end{bmatrix}, \quad \mathbf{D_1} = \begin{bmatrix} 0 & 0 \\ k_1 & 0 \end{bmatrix},$$

$$\mathbf{D_2} = \begin{bmatrix} 0 & 0 \\ 0 & h_1 \end{bmatrix}, \mathbf{G_1} = \begin{bmatrix} 0 \\ (k_0 - k_1)m_{x_1} + q_1 \end{bmatrix}, \mathbf{G_2} = \begin{bmatrix} 0 \\ (\sigma_0 - h_1)m_{x_2} + q_2 \end{bmatrix}.$$
$$(8.24)$$

Then the first-order moments satisfy the following vector differential equation:

$$\frac{d\mathbf{m_x}(t)}{dt} = \mathbf{A_1}\mathbf{m_x}(t) + \mathbf{A_0}, \quad \mathbf{m_x}(t_0) = E[\mathbf{x_0}]. \quad (8.25)$$

The stationary solution of (8.25) is equal to zero, i.e.,

$$m_{x_1}(\infty) = m_{x_2}(\infty) = 0. \quad (8.26)$$

Hence, it follows that linearization coefficients for stationary solution also are equal to zero

$$k_0 = \sigma_0 = 0, \quad (8.27)$$

while coefficients $k_1$ and $h_1$ in the case, for instance, of Gaussian excitations and the mean-square criterion have the form

$$k_1 = E\left[\frac{\partial(\varepsilon x_1^3)}{\partial x_1}\right] = 3\varepsilon E[x_1^2], \quad h_1 = E\left[\frac{\partial(c|x_2|x_2)}{\partial x_2}\right] = c\sqrt{\frac{8}{\pi}}E[x_2^2]. \quad (8.28)$$

Deriving the second-order moment equations for the linearized system (8.23) and taking into consideration equalities (8.27), we find relations for stationary solutions

$$2E[x_1 x_2] = 0, \quad (8.29)$$

$$E[x_2^2] + (-\lambda_0^2 + k_1)E[x_1^2] - 2hE[x_1 x_2] = 0, \quad (8.30)$$

$$2(-\lambda_0^2 + k_1)E[x_1 x_2] - 4hE[x_2^2] + q_1^2 + q_2^2 + k_1^2 E[x_1^2] + h_1^2 E[x_2^2] + 2k_1 h_1 E[x_1 x_2] = 0. \quad (8.31)$$

Hence, after substituting relations (8.28) into (8.29–8.31), we obtain

$$E[x_2^2] + (-\lambda_0^2 + 3\varepsilon E[x_1^2])E[x_1^2] = 0, \quad (8.32)$$

$$-4hE[x_2^2] + q_1^2 + q_2^2 + 9\varepsilon^2 E[x_1^2]^3 + \frac{8c^2}{\pi}E[x_2^2]^3 = 0. \quad (8.33)$$

Using notations

$$y = E[x_1^2], \quad a_1 = \frac{648\varepsilon^3 c^2}{\pi}, \quad a_2 = -\frac{216\varepsilon^2 c^2 \lambda_0^2}{\pi}, \quad a_3 = \frac{72\varepsilon c^2 \lambda_0^4}{\pi},$$

$$a_4 = -\left(\frac{8c^2 \lambda_0^6}{\pi} + 9\varepsilon^2\right), \quad a_5 = -12h\varepsilon, \quad a_6 = 4h\lambda_0^2, \quad a_7 = -q_1^2 - q_2^2$$

one can reduce the system of (8.32) and (8.33) to the following algebraic equation:

$$\sum_{i=1}^{7} a_i y^{7-i} = 0. \quad (8.34)$$

Hence, it follows that for the determination of linearization coefficients $k_1$ and $h_1$ for stationary solutions, it is necessary to solve numerically algebraic equation (8.34).

In the case of application of statistical linearization with a criterion in probability density functions space, for instance, Criterion 1 − $SLPD$, we assume that for each nonlinearity the input process is a stationary Gaussian process $x(t)$ with the probability density function given by

$$g_I(x) = \frac{1}{\sqrt{2\pi}\sigma_x} \exp\left\{-\frac{x^2}{2\sigma_x^2}\right\}, \quad (8.35)$$

where $\sigma_x^2 = E[X^2]$ is the variance of the process $x(t)$.

Then, the probability density function for stationary output processes $Y_i(t) = \phi_i(X(t))$, $i = 1, 2$, have the form

$$g_{Y_1}(y) = \frac{1}{\sqrt{2\pi}\sigma_x} \exp\left\{-\frac{v_1^2}{2\sigma_x^2}\right\} \frac{1}{3\varepsilon v_1^2}, \qquad (8.36)$$

$$g_{Y_2}(y) = \frac{1}{\sqrt{2\pi}\sigma_x} \exp\left\{-\frac{v_2^2}{2\sigma_x^2}\right\} \frac{1}{2c v_2}, \qquad (8.37)$$

where

$$v_1 = \left[\frac{y}{\varepsilon}\right]^{\frac{1}{3}}, \quad v_2 = \left[\frac{|y|}{c}\right]^{\frac{1}{2}}.$$

The probability density functions for linearized variables

$$Y_1 = k_1 x, \quad Y_2 = h_1 x \qquad (8.38)$$

have the form

$$g_{L_1}(y) = \frac{1}{\sqrt{2\pi}k_1\sigma_x(t)} \exp\left\{-\frac{y^2}{2k_1^2\sigma_x^2(t)}\right\}, \qquad (8.39)$$

$$g_{L_2}(y) = \frac{1}{\sqrt{2\pi}h_1\sigma_x(t)} \exp\left\{-\frac{y^2}{2h_1^2\sigma_x^2(t)}\right\}, \qquad (8.40)$$

respectively.

To obtain the linearization coefficients and approximate response stationary moments, we use a simplified Procedure $SL$–$PD$.

**Procedure $SSL$–$PD$**

*Step 1*: Substitute the initial values of linearization coefficients $k_1 = -\lambda_0^2$, $h_1 = 0$ and calculate the stationary moments of solutions of the linearized system (8.32) and (8.33), i.e.,

$$E[x_1^2] = \frac{q_1^2 + q_2^2}{(4h - h_1^2)(\lambda_0^2 - k_1) + k_1^2}, \qquad (8.41)$$

$$E[x_2^2] = \frac{(q_1^2 + q_2^2)(\lambda_0^2 - k_1)}{(4h - h_1^2)(\lambda_0^2 - k_1) + k_1^2}, \qquad (8.42)$$

under the exponential stability with probability one assumptions

$$\lambda_0^2 > k_1, \quad 4h > h_1. \qquad (8.43)$$

*Step 2*: For nonlinear and corresponding linearized elements (one dimensional) defined by (8.21) and (8.22), respectively, determine the probability density functions using $E[x_i] = 0$ and $E[x_i^2]$, $i = 1, 2$ calculated from the moments obtained in Step 1. They have the forms (8.36) and (8.37) for nonlinear elements and (8.39) and (8.40) for the corresponding linearized ones.

*Step 3*: Choose any criterion, for instance, $I_1$ and apply them to all non-linear elements separately, i.e.,

$$I_{1_j} = \int_{-\infty}^{+\infty} (g_{Y_j}(y_j) - g_{L_j}(y_j))^2 dy_j, \quad j = 1, 2 \tag{8.44}$$

where $g_{Y_j}(y_j)$ and $g_{L_j}(y_j)$, $j = 1, 2$, are probability density functions defined by (8.36), (8.37) and (8.39), (8.40) and next find the coefficients $k_{1_{min}}$ and $h_{1_{min}}$ that minimize criterion (8.44) separately for each nonlinear element and then substitute $k_1 = k_{1_{min}}$ and $h_1 = h_{1_{min}}$.

*Step 4*: If the stability conditions (8.43) are satisfied then calculate the stationary moments of linearized system (8.41) and (8.42), otherwise stop.

*Step 5*: For all nonlinear and linearized elements (one dimensional) defined by (8.21) and (8.22), determine the new stationary probability density functions using $E[x_i] = 0$, $E[x_i^2]$, $i = 1, 2$, calculated on the basis of stationary moments obtained in Step 4.

*Step 6*: Repeat Steps 3–5 until $k_1$, $h_1$, $E[x_i^2]$, $i = 1, 2$, converge with a given accuracy.

We note that if we treat (8.39) as the Stratonovich stochastic differential equation, then there are two possibilities of linearization. In the first one, that is similar to the one proposed by Kazakov and Maltchikov [11], first we linearize separately the deterministic and stochastic parts of the Stratonovich stochastic differential equation and next we transform the obtained linearized Stratonovich equation to the corresponding Ito equation and then we calculate moment equations. In the second approach, we first transform the Stratonovich equation to the corresponding Ito equation and next we linearize separately deterministic and stochastic part of Ito equation and we determine moment equations for linearized Ito equation. The results obtained by both the approaches are different.

## 8.3 Equivalent Linearization

### 8.3.1 Moment Criteria

Similar to statistical linearization, the method of equivalent linearization has been applied to models as well with parametric as external and parametric excitations. Two basic approaches of equivalent linearization have been proposed in the literature. The first one was given by Kottalam et al. [12] and independently by Young and Chang [21], while the second approach by Brückner and Lin [2]. We are going to show both the approaches.

Kottalam et al. [12] considered the nonlinear system with parametric excitations described by the vector Ito stochastic differential equation

$$d\mathbf{x}(t) = \mathbf{\Phi}(\mathbf{x})dt + \sum_{k=1}^{M} \boldsymbol{\sigma}_{\mathbf{k}}(x)d\xi_k(t), \quad \mathbf{x}(t_0) = \mathbf{x}_0, \tag{8.45}$$

where $\mathbf{x} = [x_1, \ldots, x_n]^T$ is a state vector, $\xi_k(t), k = 1, \ldots, M$, are standard Wiener processes, such that

$$E[d\xi_i(t)d\xi_j(t)] = \delta_{ij}dt , \qquad (8.46)$$

$[\delta_{ij}]$ is the matrix of intensity, the initial condition $\mathbf{x}_0$ is a vector random variable independent of $\xi_k(t), k = 1, \ldots, M$, $\mathbf{\Phi}(\mathbf{x})$ is a vector nonlinear function having components

$$\phi_i(\mathbf{x}) = f_i(\mathbf{x}) + \frac{1}{2}\sum_{j=1}^{n}\sum_{k=1}^{M}\sum_{l=1}^{M}\left[\frac{\partial}{\partial x_j}\sigma_{ik}(\mathbf{x})\right]\delta_{ij}\sigma_{jl}(\mathbf{x}), \quad i = 1, \ldots, n , \quad (8.47)$$

$\boldsymbol{\sigma}_k(x) = [\sigma_{1k}(\mathbf{x}), \ldots, \sigma_{nk}(\mathbf{x})]^T$, and $f_i(\mathbf{x})$ and $\sigma_{ik}(\mathbf{x})$ are nonlinear functions $i, j = 1, \ldots, n, k = 1, \ldots, M$.

Elements of the diffusion matrix $\mathbf{B}(\mathbf{x}) = [b_{ij}(\mathbf{x})]$ corresponding to (8.45) have the form

$$b_{ij}(\mathbf{x}) = \frac{1}{2}\sum_{k=1}^{M}\sum_{l=1}^{M}\sigma_{ik}(\mathbf{x})\delta_{kl}\sigma_{lj}(\mathbf{x}) . \qquad (8.48)$$

We replace (8.45) by a linear one with time-dependent coefficients and additive nonstationary excitations, i.e.,

$$d\mathbf{x}(t) = [\mathbf{A}(t)\mathbf{x}(t) + \mathbf{A}_0(t)]dt + \mathbf{G}(t)d\xi(t), \quad \mathbf{x}(t_0) = \mathbf{x}_0 , \qquad (8.49)$$

where $\mathbf{A} = [a_{ij}]$, $\mathbf{G} = [G_{ik}]$, $i, j = 1, \ldots, n$, $k = 1, \ldots, M$ and $\mathbf{A}_0 = [A_{01}, \ldots, A_{0n}]^T$ are the matrices and vector of time-dependent linearization coefficients, and $\boldsymbol{\xi}(t) = [\xi_1(t), \ldots, \xi_M(t)]^T$ is a standard vector Wiener process.

$$G_{ij}(t) = \sum_{k=1}^{M}\sum_{l=1}^{M}\gamma_{ik}(t)\delta_{kl}\gamma_{lj}(t) . \qquad (8.50)$$

The elements $a_{ij}, A_{0i}, \gamma_{ik}, i, j = 1, \ldots, n$, $k = 1, \ldots, M$ are determined from the necessary conditions of minimization of mean-square linearization criteria $I_1$ and $I_2$ defined by

$$I_1 = E\left[(\mathbf{\Phi}(\mathbf{x}) - \mathbf{A}(t)\mathbf{x} - \mathbf{A}_0(t))^T(\mathbf{\Phi}(\mathbf{x}) - \mathbf{A}(t)\mathbf{x} - \mathbf{A}_0(t))\right] , \qquad (8.51)$$

$$I_2 = E[\mathbf{G}(t)\mathbf{G}^T(t) - \mathbf{B}(\mathbf{x})] . \qquad (8.52)$$

Deriving equations for the first- and second-order moments for the nonlinear system (8.45) and the linearized one (8.49), and next equating the right-hand sides of these equations, we find

$$\mathbf{A}(t) = \left(E[\mathbf{\Phi}(\mathbf{x})\mathbf{x}^T] - E[\mathbf{\Phi}(\mathbf{x})]E[\mathbf{x}^T]\right)\left(E[\mathbf{x}\mathbf{x}^T] - E[\mathbf{x}]E[\mathbf{x}^T]\right)^{-1} , \qquad (8.53)$$

$$\mathbf{A}_0(t) = E[\mathbf{\Phi}(\mathbf{x})] - \mathbf{A}(t)E[\mathbf{x}] . \qquad (8.54)$$

To calculate the linearization coefficients and response moments one can use, as in the previous cases, an iterative procedure. The moment equations are linear with time-dependent coefficients and time-dependent excitations, i.e.,

$$\frac{d\mathbf{m}_x(t)}{dt} = \mathbf{A}(t)\mathbf{m}_x(t) + \mathbf{A}_0(t), \quad \mathbf{m}_\mathbf{x}(t_0) = E[\mathbf{x}_0] , \tag{8.55}$$

$$\frac{d\mathbf{\Gamma}_L(t)}{dt} = \mathbf{\Gamma}_L(t)\mathbf{A}^T(t) + \mathbf{A}(t)\mathbf{\Gamma}_L(t) + \mathbf{A}_0(t)\mathbf{m}_x^T(t) + \mathbf{m}_x(t)\mathbf{A}_0^T(t)$$

$$+ \boldsymbol{\mu}(t), \quad \mathbf{\Gamma}_L(t_0) = E[\mathbf{x}_0\mathbf{x}_0^T] , \tag{8.56}$$

where

$$\mathbf{m}_\mathbf{x}(t) = E[\mathbf{x}(t)], \quad \mathbf{\Gamma}_L(t) = E[\mathbf{x}(t)\mathbf{x}(t)^T], \quad \boldsymbol{\mu}(t) = E[\mathbf{B}(\mathbf{x}(t))] . \tag{8.57}$$

This approach was illustrated by Kottalam, Lindberg, and West for the Duffing oscillator subject to both additive and stiffness stochastic excitations.

*Example 8.2.* [12] Consider the Duffing oscillator with external and parametric excitation described by the vector Ito stochastic differential equation

$$dx(t) = [\mathbf{A}_1\mathbf{x} + \mathbf{F}(\mathbf{x})]dt + \sum_{k=1}^{2} \boldsymbol{\sigma}_k(\mathbf{x})d\xi_k(t), \quad \mathbf{x}(t_0) = \mathbf{x}_0 , \tag{8.58}$$

where

$$\mathbf{x} = \begin{bmatrix} x_1 \\ x_2 \end{bmatrix}, \quad \mathbf{A}_1 = \begin{bmatrix} 0 & 1 \\ -\lambda_0^2 & -\delta_{11} \end{bmatrix}, \quad \mathbf{F} = \begin{bmatrix} 0 \\ -\varepsilon x_1^3 - 2\delta_{12}x_1x_2 - \delta_{22}x_1^2x_2 \end{bmatrix},$$

$$\boldsymbol{\sigma}_1 = \begin{bmatrix} 0 \\ 1 \end{bmatrix}, \quad \boldsymbol{\sigma}_2 = \begin{bmatrix} 0 \\ x_1 \end{bmatrix}, \quad E[d\xi_i(t)d\xi_j(t)] = \delta_{ij}dt, \ i, \ j = 1, \ 2, \ \mathbf{x}_0 = \begin{bmatrix} x_{10} \\ x_{20} \end{bmatrix}. \tag{8.59}$$

The linearized system has the form (8.49), where

$$\mathbf{A} = [a_{ij}], \quad \mathbf{A}_0 = [C_1, C_2]^T, \quad \mathbf{G} = [G_1, G_2]^T,$$

$$a_{11} = 0, \quad a_{12} = 1, \quad C_1 = 0, \quad G_1 = 0,$$

$$a_{21} = -\frac{1}{\Delta} \{ E[Q(\mathbf{x})](E[x_1]E[x_2^2] - E[x_2]E[x_1x_2])$$

$$- E[x_1Q(\mathbf{x})](E[x_2^2] - (E[x_2])^2) + E[x_2Q(\mathbf{x})](E[x_1x_2] - E[x_1]E[x_2])\} ,$$

$$a_{22} = \frac{1}{\Delta}\{[E[Q(\mathbf{x})](E[x_1]E[x_1x_2] - E[x_2]E[x_1^2]) - E[x_1Q(\mathbf{x})]$$

$$\times(E[x_1x_2] - E[x_1]E[x_2]) + E[x_2Q(\mathbf{x})](E[x_1^2] - (E[x_1])^2\},$$

$$C_2 = \frac{1}{\Delta} \{ E[Q(\mathbf{x})](E[x_1^2]E[x_2^2] - (E[x_1x_2])^2)$$

$$-E[x_1Q(\mathbf{x})](E[x_1]E[x_2^2] - E[x_1x_2]E[x_2])$$

$$+E[x_2Q(\mathbf{x})](E[x_1]E[x_1x_2] - E[x_1^2]E[x_2])\},$$

$$G_2 = \sqrt{\frac{E[(\delta_{11} + 2\delta_{12}x_1 + \delta_{22}x_1^2)]}{\delta_{11}}},$$

$$\Delta = E[x_1^2]E[x_2^2] - (E[x_1x_2])^2 - (E[x_1])^2 E[x_2^2]$$

$$+2E[x_1]E[x_2]E[x_1x_2] - E[x_1^2](E[x_2])^2,$$

$$Q(\mathbf{x}) = -\lambda_0^2 x_1 - \varepsilon x_1^3 - (\delta_{11} + 2\delta_{12}x_1 + \delta_{22}x_1^2)x_2. \tag{8.60}$$

The solution of the first- and second-order moment equations for system (8.49) with the coefficients given by (8.60) is possible only by application, for instance, by cumulant closure technique of second order. Then, we obtain a system of nonlinear differential equations that can be solved directly or by the iterative procedure discussed in previous section. In the case of the application of the iterative procedure, one has to solve multiple times the system of linear differential equations.

An alternative approach was presented by Young and Chang [21] and Chang and Young [4], who considered a second-order stochastic nonlinear model in a special form of (8.45) when the nonlinear functions in deterministic and stochastic part of equation are the same, namely,

$$dx_1(t) = x_2(t)dt, \quad x_1(t_0) = x_{10},$$

$$dx_2(t) = -\sum_{k=1}^{M} a_k \Psi_k(x_1, x_2)dt$$

$$-\sum_{k=1}^{M} a_k \Psi_k(x_1, x_2)d\xi_k(t) + d\xi_0(t), x_2(t_0) = x_{20}, \tag{8.61}$$

where $a_k$ are constants, $\xi_k$ are independent standard Wiener processes, i.e., $E[d\xi_k(t)d\xi_l(t)] = \delta_{kl}dt$, $k, l = 0, 1, \ldots, M$, the initial conditions $x_{i0}$, $i = 1, 2$, are mutually independent random variables and independent of $\xi_k$, $k, l = 0, 1, \ldots, M$; $\Psi_k(x_1, x_2)$ are nonlinear functions.

The corresponding linearized system has the form

$$dx_1(t) = x_2(t)dt, \quad x_1(t_0) = x_{10},$$

$$dx_2(t) = -\sum_{k=1}^{M} a_k(b_k x_2 + c_k x_1)dt$$

$$-\sum_{k=1}^{M} a_k(b_k x_2 + c_k x_1)d\xi_k(t) + d\xi_0(t), \quad x_2(t_0) = x_{20}, \quad (8.62)$$

where the linearization coefficients $b_k, c_k, k = 1, \ldots, M$, are functions of moments of stationary solution of nonlinear system (8.61), i.e.,

$$b_k = b_k(m_{20}, m_{02}), \quad c_k = c_k(m_{20}, m_{02}), \quad k = 1, \ldots, M,$$

$m_{ij} = E[(x_1)^i(x_2)^j], \; i, j = 1, 2$, are stationary solutions of corresponding moment equations obtained from (8.61).

The linearization coefficients are selected to minimize the mean-square errors $I_k$ given by

$$I_k = E[(\Psi_k(x_1, x_2) - b_k x_2 - c_k x_1)^2], \quad k = 1, \ldots, M . \quad (8.63)$$

It is assumed that the averaging operation $E[.]$ appearing in criterion (8.63) does not depend on linearization coefficients $b_k, c_k, k = 1, \ldots, M$. Utilizing the necessary conditions of minimization of $I_k$, i.e.,

$$\frac{\partial I_k}{\partial b_k} = 0, \quad \frac{\partial I_k}{\partial c_k} = 0, \quad k = 1, \ldots, M \quad (8.64)$$

one can find that

$$b_k = \frac{m_{02}E[x_1\Psi_k(x_1, x_2)] - m_{11}E[x_2\Psi_k(x_1, x_2)]}{m_{20}m_{02} - m_{11}^2}, \quad k = 1, \ldots, M,$$

$$c_k = \frac{m_{20}E[x_2\Psi_k(x_1, x_2)] - m_{11}E[x_1\Psi_k(x_1, x_2)]}{m_{20}m_{02} - m_{11}^2}, \quad k = 1, \ldots, M, \quad (8.65)$$

where $m_{ij} = E[(x_1)^i(x_2)^j]$ and $E[x_i\Psi_k(x_1, x_2)], \; i, j = 1, 2$ appearing in equalities (8.65) are replaced by corresponding moments of stationary solutions of (8.62).

We note, that the linearized system (8.62) has the form of a linear system with stochastic coefficients which is equivalent in the sense of the same first- and second-order moments to the following linear system:

$$dx_1(t) = x_2(t)dt, \quad x_1(t_0) = x_{10},$$

$$dx_2(t) = -\sum_{k=1}^{M} a_k(b_k x_2 + c_k x_1)dt$$

$$-\left(\sum_{k=1}^{M} \delta_{kk}E[(b_k x_2 + c_k x_1)^2] + \delta_{00}\right)^{\frac{1}{2}} dw(t), \quad x_2(t_0) = x_{20},$$

$$(8.66)$$

where $w(t)$ is a standard (scalar) Wiener process, $x_{i0}$, $i = 1, 2$, are random variables independent of $w(t)$.

To find coefficients $b_k$ and $c_k$, it is necessary to apply one of the closure techniques in moment equations. In some particular cases when one can find an analytical solution of FPK equation corresponding to system (8.66) or to the following system:

$$dx_1(t) = x_2(t)dt, \quad x_1(t_0) = x_{10},$$

$$dx_2(t) = -\sum_{k=1}^{M} a_k(b_k x_2 + c_k x_1)dt$$

$$- \left( \sum_{k=1}^{M} E[\Psi_k^2(x_1, x_2)] + \delta_{00} \right)^{\frac{1}{2}} dw(t), \quad x_2(t_0) = x_{20}, \quad (8.67)$$

then the algorithm of the determination of solution characteristics proposed by Young and Chang [21] has the form

*Step 1*: Transform the nonlinear system (8.61) to the form (8.67).

*Step 2*: Determine the probability density function of stationary solution of system (8.67).

*Step 3*: Determine the stationary moments that appear in terms $E[\Psi_k^2(x_1, x_2)]$ indirectly.

*Step 4*: Find the moments of stationary solution appearing in Step 3 directly or by an iterative procedure.

*Step 5*: Determine the other moments of stationary solution.

We illustrate the above algorithm with an example.

*Example 8.3.* [21] Consider the vector nonlinear Ito stochastic differential equation

$$dx_1(t) = x_2(t)dt, \quad x_1(t_0) = x_{10},$$

$$dx_2(t) = [-\lambda_0^2 x_1^3 - 2hx_2]dt - x_1^2 d\xi_1(t)$$

$$- x_2 d\xi_2(t) + d\xi_0(t), \quad x_2(t_0) = x_{20}, \quad (8.68)$$

where $\lambda_0^2$ and $h$ are constant parameters and the independent Wiener processes satisfy the condition $E[d\xi_k(t)d\xi_l(t)] = \delta_{kl}dt$, $k, l = 0, 1, 2$.

Then the following procedure of the determination of linearization coefficients and response characteristics has the following form:

*Step 1*: The equivalent nonlinear system has the form

$$dx_1(t) = x_2(t)dt, \quad x_1(t_0) = x_{10},$$

$$dx_2(t) = [-\lambda_0^2 x_1^3 - (2h - \frac{1}{2}\delta_{22})x_2]dt + \sqrt{Q}dw(t), \quad x_2(t_0) = x_{20}, (8.69)$$

where

$$Q = \delta_{11}E[x_1^6] + \delta_{22}E[x_2^2] + \delta_{00} . \tag{8.70}$$

*Step 2*: The marginal probability density functions of stationary solution of nonlinear system (8.68) determined by FPK equations method have the form

$$g_1(x_1) = \frac{2(k_1)^{1/4}}{\Gamma(\frac{1}{4})} \exp\{-k_1 x_1^4\}, \quad k_1 = \frac{(2h - \frac{1}{2}\delta_{22})\lambda_0^2}{Q} , \tag{8.71}$$

$$g_2(x_2) = \sqrt{\frac{k_2}{\pi}} \exp\{-k_2 x_2^2\}, \quad k_2 = \frac{2k_1}{\lambda_0^2} . \tag{8.72}$$

*Step 3*: Using the probability density functions obtained in Step 2, we calculate the moments of the stationary solution of nonlinear system (8.68) $E[x_1^6]$ and $E[x_2^2]$

$$E[x_1^6] = \int_{-\infty}^{+\infty} x_1^6 g_1(x_1)dx_1 = 0.254 \left(\frac{Q}{(2h - \frac{1}{2}\delta_{22})\lambda_0^2}\right)^{\frac{3}{2}} , \tag{8.73}$$

$$E[x_2^2] = \int_{-\infty}^{+\infty} x_2^2 g_2(x_2)dx_2 = \frac{Q}{4h - \delta_{22}} . \tag{8.74}$$

*Step 4*: The obtained stationary moments are connected by nonlinear algebraic equations. Substituting (8.70) into (8.73) and (8.74) eliminating the variable $E[x_1^6]$, we obtain the third-order algebraic equation with respect to $E[x_2^2]$

$$4.113\frac{\delta_{11}^2}{\lambda_0^6}(E[x_2^2])^3 - 4(2h-\delta_{22})^2(E[x_2^2])^2 + 4\delta_{00}(2h-\delta_{22})E[x_2^2] - \delta_{00}^2 = 0 . \tag{8.75}$$

Equation (8.75) usually is solved numerically and next we calculate the second unknown $E[x_1^6]$. Then we have completely determined the variable $Q$ and the probability density functions $g_1(x_1)$ and $g_2(x_2)$.

*Step 5*: Using the probability density functions $g_1(x_1)$ and $g_2(x_2)$, we determine other characteristics, for instance,

$$E[x_1^2] = \int_{-\infty}^{+\infty} x_1^2 g_1(x_1)dx_1 = 0.676 \left(\frac{E[x_2^2]}{\lambda_0^2}\right)^{\frac{1}{2}} . \tag{8.76}$$

Chang [6] extended the described approach to a new practical non-Gaussian linearization method for a general stable $n$-dimensional nonlinear systems. He considered the system described by the vector Ito equation

$$d\mathbf{x}(t) = \mathbf{F}[\mathbf{x}(t)]dt + \boldsymbol{\sigma}[\mathbf{x}(t)]d\boldsymbol{\xi}(t) , \tag{8.77}$$

where $\mathbf{x} \in R^n$ is the vector of state process, $\mathbf{F}(\mathbf{x}) \in R^n$ is the vector nonlinear function of vector state, $\boldsymbol{\sigma}(\mathbf{x}) \in R^n \times R^m$ is the matrix nonlinear function of states, and $\boldsymbol{\xi} \in R^m$ is a vector Wiener process with intensity

$$E[d\boldsymbol{\xi}(t)d\boldsymbol{\xi}^T(t)] = \mathbf{Q}_\xi(t)dt \ . \tag{8.78}$$

The nonlinear functions $\mathbf{F}(\mathbf{x})$ and $\boldsymbol{\sigma}(\mathbf{x})$ are approximated by

$$\mathbf{F}[\mathbf{x}(t)] \cong \mathbf{A}_0(t) + \mathbf{K}(t)[\mathbf{x}(t) - \boldsymbol{\mu}(t)] \ , \tag{8.79}$$

$$\boldsymbol{\sigma}[\mathbf{x}(t)] \cong \mathbf{B}(t) + \sum_{i=1}^{n} \mathbf{L}_i(x_i(t) - \mu_i(t)) \ , \tag{8.80}$$

where $\mu = E[\mathbf{x}]$, $\mu_i = E[x_i]$, $i = 1, \ldots, n$. The unknown vector $\mathbf{A}_0$ and matrices $\mathbf{K}, \mathbf{B}$, and $\mathbf{L}_i, i = 1, \ldots, n$, are derived such that the mean-square errors defined by

$$I_F = E[\boldsymbol{\Delta}_1^T \boldsymbol{\Delta}_1], \quad I_G = E[\boldsymbol{\Delta}_{2i}^T \boldsymbol{\Delta}_{2i}], \ i = 1, \ldots, n \tag{8.81}$$

will be minimized, where

$$\boldsymbol{\Delta}_1 = \mathbf{F}[\mathbf{x}(t)] - \mathbf{A}_0(t) - \mathbf{K}(t)[\mathbf{x}(t) - \boldsymbol{\mu}(t)],$$

$$\boldsymbol{\Delta}_{2i} = \boldsymbol{\sigma}[\mathbf{x}(t)] - \mathbf{B}(t) - \sum_{l=1}^{n} \mathbf{L}_l(x_l(t) - \mu_l(t)) \ , \tag{8.82}$$

$\boldsymbol{\Delta}_{2i}$ is an $i$th partitioned column vector of $\boldsymbol{\Delta}_2$.

From the necessary conditions for the minimization of the mean-square error (8.81) and moment equations for linearized system, one can calculate the linearization coefficients $\mathbf{A_0}, \mathbf{K}, \mathbf{B}$, and $\mathbf{L}_i, i = 1, \ldots, n$. In these equations, the averaging operation $E[.]$ usually was defined by a Gaussian probability density function. Chang [6] proposed to apply a simply and practical multidimensional non-Gaussian density, which has been justified by physical arguments and can be constructed as the weighing sum of Gaussian densities with adjusted means and variances

$$g_k(\mathbf{x}) = \sum_{i=1}^{p} \alpha_i N_i(m_i, S_i) \ , \tag{8.83}$$

where $p$ is a given number and $\alpha_i$ satisfy

$$\sum_{i=1}^{p} \alpha_i = 1, \quad \alpha_i > 0 \tag{8.84}$$

and $N_i(m_i, S_i)$, $i = 1, \ldots, p$, are $n$-dimensional Gaussian probability density functions. The unknown parameters $\alpha_i, m_i$, and $S_i, i = 1, \ldots, p$, can be derived by solving a set of algebraic nonlinear moment equations for stationary solutions.

The second approach of linearization has been proposed by Brückner and Lin [2]. They considered a particular case of system (8.45), namely,

$$dx_1(t) = x_2(t)dt, \quad x_1(t_0) = x_{10},$$

$$dx_2(t) = -f(x_1, x_2)dt + \sum_{k=1}^{M} \Psi_k(x_1, x_2)d\xi_k(t), \quad x_2(t_0) = x_{20}, \quad (8.85)$$

where $x_i$, $i = 1, 2$, are elements of the state vector $\mathbf{x} = [x_1, x_2]^T$, $f$, $\Psi_k$ are scalar functions, $\xi_k$ are Wiener processes, $E[d\xi_{ik}(t)d\xi_{jl}(t)] = \delta_{ijkl}dt$, $k, l = 1, \ldots, M$, the initial conditions $x_{i0}$, $i = 1, 2$, are random variables independent of $\xi_k$, $k = 1, \ldots, M$.

Introducing a one-dimensional standard Wiener process $w(t)$, Brückner and Lin replaced system (8.85) by the following one:

$$dx_1(t) = x_2(t)dt, \quad x_1(t_0) = x_{10},$$

$$dx_2(t) = F(x_1, x_2)dt + \sigma(x_1, x_2)dw(t), \quad x_2(t_0) = x_{20}, \quad (8.86)$$

where

$$F(x_1, x_2) = -f(x_1, x_2) + \frac{1}{2}\sum_{k=1}^{M}\sum_{l=1}^{M} \frac{\partial \Psi_k(x_1, x_2)}{\partial x_2} \Psi_l(x_1, x_2)\delta_{kl}, \quad (8.87)$$

$$\sigma^2(x_1, x_2) = \sum_{k=1}^{M}\sum_{l=1}^{M} \Psi_k(x_1, x_2)\Psi_l(x_1, x_2)\delta_{kl}. \quad (8.88)$$

We add to (8.86) equations for second-order powers of coordinates $x_1$ and $x_2$, i.e.,

$$d(x_1^2)(t) = x_1 x_2 dt, \quad x_1^2(t_0) = x_{10}^2,$$

$$d(x_1 x_2)(t) = [x_2^2 + x_1 F(x_1, x_2)]dt + x_1\sigma(x_1, x_2)dw(t), \quad (x_1 x_2)(t_0) = x_{10}x_{20},$$

$$d(x_2^2)(t) = [2x_2 F(x_1, x_2) + \sigma^2(x_1, x_2)]dt$$

$$+2x_2\sigma(x_1, x_2)dw(t), x_2^2(t_0) = x_{20}^2. \quad (8.89)$$

The linearized system of equations has been assumed in the form

$$dx_1(t) = x_2(t)dt, \quad x_1(t_0) = x_{10},$$

$$dx_2(t) = (a_1 x_1 + a_2 x_2 + b)dt$$

$$+[c_{11}(x_1)^2 + c_{12}x_1 x_2 + c_{22}(x_2)^2 + d_1 x_1$$

$$+d_2 x_2 + e]^{1/2}dw(t), \quad x_2(t_0) = x_{20}. \quad (8.90)$$

where $a_i, b, c_{ij}, d_i$ and $e$ are the linearization coefficients $i, j = 1, 2$.

In a similar way, one can add to the system of (8.90), equations for second-order powers of coordinates $x_1$ and $x_2$, i.e.,

$$d(x_1)^2(t) = x_1 x_2 dt, \quad (x_1)^2(t_0) = x_{10}^2,$$

$$d(x_1 x_2)(t) = [(x_2)^2 + a_1(x_1)^2 + a_2 x_1 x_2 + b x_2]dt + x_1[c_{11}(x_1)^2 + c_{12}x_1 x_2$$

$$+c_{22}(x_2)^2 + d_1 x_1 + d_2 x_2 + e]^{1/2}dw(t), \quad (x_1 x_2)(t_0) = x_{10}x_{20},$$

$$d(x_2)^2(t) = [(2a_1 + c_{12})x_1 x_2 + (2a_2 + c_{22})(x_2)^2$$

$$+(2b + d_2)x_2 + c_{11}(x_1)^2 + d_1 x_1 + e]dt$$

$$+2x_2[c_{11}(x_1)^2 + c_{12}x_1 x_2 + c_{22}(x_2)^2$$

$$+d_1 x_1 + d_2 x_2 + e]^{1/2}dw(t), \quad (x_2)^2(t_0) = x_{20}^2, \tag{8.91}$$

where the linearization coefficients $a_i, b, c_{ij}, d_i, i, j = 1, 2$ and $e$ are determined from the necessary conditions for minimum of mean-square criteria $I_j, j = 1, 2$, defined in a heuristic way by

$$I_1 = E[(F(x_1, x_2) - a_1 x_1 - a_2 x_2 - b)^2], \tag{8.92}$$

$$I_2 = E[(2x_2 F(x_1, x_2) + \sigma^2(x_1, x_2) - (2a_1 + c_{12})x_1 x_2 - (2a_2 + c_{22})x_2^2$$

$$-(2b + d_2)x_2 - c_{11}(x_1)^2 - d_1 x_1 - e)^2], \tag{8.93}$$

i.e., from conditions

$$\frac{\partial I_1}{\partial a_i} = \frac{\partial I_1}{\partial b} = \frac{\partial I_2}{\partial c_{ij}} = \frac{\partial I_2}{\partial d_i} = \frac{\partial I_2}{\partial e} = 0, \quad i, j = 1, 2. \tag{8.94}$$

We note, that the proposed mean-square criteria $I_1$ and $I_2$ can be treated as mean-square errors corresponding to variables $x_i$ and $x_i x_j, i, j = 1, 2$, where only the drift terms of the Ito stochastic differential equations for variables $x_2$ and $x_2^2$ are linearized without taking into account the fact that the diffusion coefficients in the nonlinear equations and the corresponding ones in linearized equations are different.

Additionally, we note that conditions (8.94) give nonlinear algebraic equations for stationary moments of higher order. For instance, if functions $F(x_1, x_2)$ and $\sigma(x_1, x_2)$ are polynomials, then terms $E[x_2^2 F(x_1, x_2)]$ or $E[x_1 x_2 \sigma(x_1, x_2)]$ have to be replaced by some functions of lower-order moments. It can be obtained, for instance, by cumulant closure or by moments of the same order calculated for linearized system.

The discussed linearization method Brückner and Lin [2] extended to multi-degree-of-freedom systems described by

$$dx_1^{(j)}(t) = x_2^{(j)}(t)dt, \quad x_1^j(t_0) = x_{10}^j,$$

$$dx_2^{(j)}(t) = \mathbf{F}^{(j)}(\mathbf{x}_1, \mathbf{x}_2)dt + \sum_{k=1}^{M} \sigma_k^{(j)}(\mathbf{x}_1, \mathbf{x}_2)d\xi_k(t),$$

$$x_2^j(t_0) = x_{20}^j, \quad j = 1, \ldots, N, \tag{8.95}$$

where $\mathbf{x}_i = [x_i^{(1)}, \ldots, x_i^{(N)}]^T$, $i = 1, 2$, $F^{(j)}, \sigma_k^{(j)}$ are nonlinear scalar functions; $\xi_k$ are independent Wiener processes, the initial conditions $x_{i0}^j$, $i = 1, 2$, $j = 1, \ldots, N$, are random variable independent of $\xi_k$, $k = 1, \ldots, M$. The corresponding linearized equations have the following form:

$$dx_1^{(j)}(t) = x_2^{(j)}(t)dt, \quad x_1^{(j)}(t_0) = x_{10}^{(j)},$$

$$dx_2^{(j)}(t) = \sum_{k=1}^{M} [(a_{1_k}^{(j)} x_1^{(k)} + a_{2_k}^{(j)} x_2^{(k)}) + b^{(j)}]dt$$

$$+ \sum_{k=1}^{M} c_k^{(j)} d\xi_k(t), \quad x_2^j(t_0) = x_{20}^j, \quad j = 1, \ldots, N, \qquad (8.96)$$

where $a_{1_k}^{(j)}, a_{2_k}^{(j)}$, and $b^{(j)}$ are constant linearization coefficients, $\sum_{k=1}^{M} c_k^{(i)} c_k^{(j)}$ is a quadratic form with respect to $\mathbf{x_1}$ and $\mathbf{x_2}$. These terms are calculated from moment equations and necessary conditions for minimum of criteria $I_1^{(j)}$ and $I_2^{(jl)}$, $j = 1, \ldots, N$, $l = 1, \ldots, l_{max}$, defined by

$$I_1^{(j)} = E\left[\left(F^{(j)}(x_1, x_2) - \sum_{k=1}^{M}(a_{1_k}^{(j)} x_1^{(k)} - a_{2_k}^{(j)} x_2^{(k)}) - b^{(j)}\right)^2\right], \qquad (8.97)$$

$$I_2^{(jl)} = E\left[\left(x_2^{(l)} F^{(j)} + x_2^{(j)} F^{(l)} + \sum_{k=1}^{M} \sigma_k^{(i)} \sigma_k^{(j)}\right.\right.$$

$$- \sum_{k=1}^{M}(a_{1_k}^{(l)} x_2^{(j)} + a_{1_k}^{(j)} x_2^{(l)}) x_1^{(k)}$$

$$\left.\left. - \sum_{k=1}^{M}(a_{2_k}^{(l)} x_2^{(j)} + a_{2_k}^{(j)} x_2^{(l)}) x_2^{(k)} - b^{(l)} x_2^{(j)} - b^{(j)} x_2^{(l)} - \sum_{k=1}^{M} c_k^{(i)} c_k^{(j)}\right)^2\right].$$

$$(8.98)$$

where $l_{max} = max(p, q)$, $p$, and $q$ are degrees of polynomials of coordinates of functions $F^{(j)}$ and $\sigma_k^{(j)}$, respectively.

The further generalization of this approach Brückner and Lin presented in [1].

Another generalization has been proposed by Falsone [8]. He considered the general case of multidimensional nonlinear systems with parametric excitations described by the vector Ito stochastic differential equation

$$d\mathbf{x}(t) = \mathbf{\Phi}(\mathbf{x}, t)dt + \sum_{k=1}^{M} \boldsymbol{\sigma}_k(\mathbf{x}, t)d\xi_k, \quad \mathbf{x}(t_0) = \mathbf{x}_0, \qquad (8.99)$$

where $\mathbf{x} = [x_1, \ldots, x_n]^T$ is the state vector, $\mathbf{\Phi} = [\Phi_1, \ldots, \Phi_n]^T$ and $\boldsymbol{\sigma}_k = [\sigma_{k1}, \ldots, \sigma_{kn}]^T$ are vectors which coordinates are nonlinear deterministic functions, $\xi_k, k = 1, \ldots, M$ are mutually independent standard Wiener processes, and the initial condition $\mathbf{x}_0$ is a vector random variable independent of $\xi_k, k = 1, \ldots, M$.

The linearized system has the form

$$dx(t) = [\mathbf{A}(t)\mathbf{x}(t) + \mathbf{A}_0(t)]dt + \sum_{k=1}^{M}[\mathbf{D}_k(t)\mathbf{x}(t) + \mathbf{G}_k(t)]d\xi_k(t), \quad \mathbf{x}(t_0) = \mathbf{x}_0 ,$$

$$(8.100)$$

where $\mathbf{A}(t) = [a_{ij}(t)]$ and $\mathbf{D}_k(t) = [d_{ijk}(t)]$, $i, j = 1, \ldots, n$, $k = 1, \ldots, M$, and $\mathbf{A}_0(t) = [A_{01}(t), \ldots, A_{0n}(t)]^T$ and $\mathbf{G}_k(t) = [G_{k1}(t), \ldots, G_{kn}(t)]^T$ are time-dependent matrices and vectors of linearization coefficients.

Falsone proved that statistical linearization approach leads to the same results as Gaussian closure if the linearization is made on the coefficients of the Ito differential rule. Falsone [9] showed also that the proposed approach can be extended to systems with nonzero mean response and it can be treated as an extension of Kozin's *true linearization* for nonlinear dynamic systems under white noise external excitations to the case of nonlinear dynamic systems under white noise parametric and external excitations.

Using the idea of Young and Chang [21], we consider also for system (8.100) an equivalent linear system with external excitations

$$dx(t) = [\mathbf{A}(t)\mathbf{x}(t) + \mathbf{A}_0(t)]dt + \sum_{k=1}^{M} \mathbf{L}_k(t)d\xi_k(t), \quad \mathbf{x}(t_0) = \mathbf{x}_0 , \qquad (8.101)$$

where $\mathbf{L}_k(t) = [L_{k1}(t), \ldots, L_{kn}(t)]^T$, $k = 1, \ldots, M$, are new intensity of noise defined as follows:

$$\mathbf{L}_k(t)\mathbf{L}_k^T(t) = \mathbf{G}_k(t)\mathbf{G}_k^T(t) + \mathbf{D}_k(t)\mathbf{m_x}(t)\mathbf{A}_0^T(t) + \mathbf{A}_0(t)\mathbf{m_x}^T(t)\mathbf{D}_k^T(t)$$

$$+\mathbf{D}_k(t)\mathbf{\Gamma}_{Lp}(t)\mathbf{D}_k^T(t) , \qquad (8.102)$$

where $\mathbf{m_x}(t) = E[\mathbf{x}(t)]$, $\mathbf{\Gamma}_{Lp}(t) = E[\mathbf{x}(t)x^T(t)]$. Equivalence of systems (8.100) and (8.102) is understood in the sense that the moment equations of the first and second order for both systems are the same. They have the form

$$\frac{d\mathbf{m_x}(t)}{dt} = \mathbf{A}_0(t) + \mathbf{A}(t)\mathbf{m_x}(t), \quad \mathbf{m_x}(t_0) = E[\mathbf{x}_0] , \qquad (8.103)$$

$$\frac{d\mathbf{\Gamma}_{Lp}(t)}{dt} = \mathbf{m_x}(t)\mathbf{A}_0^T(t) + \mathbf{A}_0(t)\mathbf{m_x}^T(t) + \mathbf{\Gamma}_{Lp}(t)\mathbf{A}^T(t) + \mathbf{A}(t)\mathbf{\Gamma}_{Lp}(t)$$

$$+ \sum_{k=1}^{M} \left[ \mathbf{G}_k(t)\mathbf{G}_k^T(t) + \mathbf{D}_k(t)\mathbf{m_x}(t)\mathbf{A}_0^T(t) \right.$$

$$\left. +\mathbf{A}_0(t)\mathbf{m_x}^T(t)\mathbf{D}_k^T(t) + \mathbf{D}_k(t)\mathbf{\Gamma}_{Lp}(t)\mathbf{D}_k^T(t) \right] , \quad \mathbf{\Gamma}_{Lp}(t_0) = E[\mathbf{x}_0\mathbf{x}_0^T],$$

$$(8.104)$$

Similar considerations can be done in the case of energy criteria as well for statistical linearization as for equivalent linearization.

### 8.3.2 Criteria in Probability Density Functions Space

Similar to dynamic systems with additive excitations, there are two basic approaches of equivalent linearization in the space of probability density functions for dynamic systems with multiplicative excitations, namely, the direct method and an application of FPK equations.

### The Direct Method

In the case of the direct method for dynamic system described by (8.99) and approximated by (8.100) or by (8.101) the linearization coefficients and approximate characteristics of stationary solutions are determined by the application of a modified procedure $EL$–$PD$, where instead of stationary solutions of moment equations (6.123) and (6.124), we use the corresponding solutions of (8.103) and (8.104).

> **Procedure** $PEL$–$PD$
>
> *Step 1*: Substitute the initial values of linearization coefficients $a_{ij}, d_{kij}, C_i$, and $G_{kj}$, $i, j = 1, \ldots, n$, $k = 1, \ldots, M$, and calculate moments of stationary solutions of linearized system by solving (8.103) and (8.104).
>
> *Step 2*: Calculate approximate moments of the stationary response of the nonlinear system (8.99).
>
> *Step 3*: Calculate probability density functions $g_L(x_1, \ldots, x_n)$ and $g_N(x_1, \ldots, x_n)$ of the stationary response of linearized and nonlinear systems, respectively. They are defined by stationary moments obtained in Steps 1 and 2 and by the relations between the moments and the Hermite polynomials given in (6.119–6.121).
>
> *Step 4*: Consider a criterion, for instance $I_1$ and find the linearization coefficients $a_{ij_{\min}}, d_{kij_{\min}}, C_{i_{\min}}$ and $G_{kj_{\min}}$, $i, j = 1, \ldots, n$, $k = 1, \ldots, M$, which minimize jointly (for all nonlinear elements) criterion $I_{1e}$ given by (6.126). Next, substitute $a_{ij} = a_{ij_{\min}}$, $d_{kij} = d_{kij_{\min}}$, $C_i = C_{i_{\min}}$, and $G_{kj} = G_{kj_{\min}}$, $i, j = 1, \ldots, n$, $k = 1, \ldots, M$.
>
> *Step 5*: Calculate moments of the stationary solutions of linearized system (8.100) by solving (8.103) and (8.104).
>
> *Step 6*: Redefine the probability density functions for linearized systems by substituting the moments obtained in Step 5 relations defined by (6.118–6.121).
>
> *Step 7*: Repeat Steps 3–6 until $a_{ij}, d_{kij}, C_i, G_{kj}$, $i, j = 1, \ldots, n$, $k = 1, \ldots, M$ converge (with a given accuracy).

### The Application of Fokker–Planck–Kolmogorov Equations in Linearization

Another method of equivalent linearization with criteria in the probability density functions space is an approach where Fokker–Planck–Kolmogorov equations are used. It is a generalization of the method of application of FPK equations with the same criteria to dynamic systems with external excitations. The main idea is to replace the linearization of state equations by the linearization of the corresponding simplified FPK equations, i.e., instead of studying systems (8.99) and (8.100), we consider the corresponding FPK equations

$$\frac{\partial g_N}{\partial t} = -\sum_{i=1}^{n} \frac{\partial}{\partial x_i}[\Phi_i(\mathbf{x}, t)g_N] + \frac{1}{2}\sum_{i=1}^{n}\sum_{j=1}^{n}\frac{\partial^2}{\partial x_i \partial x_j}[b_{Nij}g_N] = 0 \qquad (8.105)$$

and

$$\frac{\partial g_L}{\partial t} = -\sum_{i=1}^{n} \frac{\partial}{\partial x_i}[(\mathbf{A}_i^T \mathbf{x} + A_{i0})g_{Lp}] + \frac{1}{2}\sum_{i=1}^{n}\sum_{j=1}^{n}\frac{\partial^2}{\partial x_i \partial x_j}[b_{Lpij}g_{Lp}] = 0 , \quad (8.106)$$

where $A_i^T$ is $i$th row of matrix $\mathbf{A}$, $\mathbf{B}_N = [b_{Nij}]$ and $\mathbf{B}_{Lp} = [b_{Lpij}]$ are diffusion matrices

$$b_{Nij} = b_{Lpij} = \sum_{k=1}^{M} G_{ki}G_{kj} . \qquad (8.107)$$

If we introduce the notations

$$p_1 = g_N, \quad p_{2i} = \frac{\partial g_N}{\partial x_i}, \quad q_1 = g_{Lp}, \quad q_{2i} = \frac{\partial g_{Lp}}{\partial x_i} , \qquad (8.108)$$

then (8.105) and (8.106) can be transformed to two systems of equations

$$\frac{\partial p_1}{\partial x_i} = p_{2i} , \ i = 1, \dots, n$$

$$\sum_{i=1}^{n}\left[\frac{\partial \phi_i}{\partial x_i}p_1 + \phi_i p_{2i}\right] - \frac{1}{2}\sum_{i=1}^{n}\sum_{j=1}^{n}\left[\frac{\partial^2 b_{ij}}{\partial x_i \partial x_j}p_1 + \frac{\partial b_{ij}}{\partial x_j}p_{2i} + \frac{\partial b_{ij}}{\partial x_i}p_{2j}\right.$$

$$\left. + b_{ij}\frac{\partial p_{2j}}{\partial x_i}\right] = 0 \qquad (8.109)$$

and

$$\frac{\partial q_1}{\partial x_i} = q_{2i} , \ i = 1, \dots, n$$

$$\sum_{i=1}^{n}\left[a_{ij}q_1 + (A_i^T \mathbf{x} + A_{i0})q_{2i}\right] - \frac{1}{2}\sum_{i=1}^{n}\sum_{j=1}^{n}\left[\frac{\partial^2 b_{ij}}{\partial x_i \partial x_j}q_1 + \frac{\partial b_{ij}}{\partial x_j}q_{2i} + \frac{\partial b_{ij}}{\partial x_i}q_{2j}\right.$$

$$\left. + b_{ij}\frac{\partial q_{2j}}{\partial x_i}\right] = 0$$

$$(8.110)$$

where $b_{ij} = b_{Nij}$ for system (8.109) and $b_{ij} = b_{Lpij}$ for system (8.110). Comparing the system equations (8.109) and (8.110), we find that $g_N$ and $\partial g_N/\partial x_i$ are approximated by $g_{Lp}$ and $\partial g_{Lp}/\partial x_i$, respectively. Then the following criterion has been proposed [16]:

$$I_3(\mathbf{A}, \mathbf{A}_0, \mathbf{D}_k, \mathbf{G}_k) = \int_{\mathbf{R}^n} [\epsilon_1^2(\mathbf{x}) + \epsilon_2^2(\mathbf{x})]dx , \qquad (8.111)$$

where

$$\epsilon_1(\mathbf{x}) = \sum_{i=1}^{n} \frac{\partial}{\partial x_i} \left[ (\Phi_i - \mathbf{A}_i^T\mathbf{x} - A_{i0})g_{Lp} \right] , \qquad (8.112)$$

$$\epsilon_2(\mathbf{x}) = \frac{1}{2} \sum_{i=1}^{n} \sum_{j=1}^{n} \frac{\partial^2}{\partial x_i \partial x_j} \left[ (b_{Nij} - b_{Lpij})g_{Lp} \right] . \qquad (8.113)$$

The necessary conditions of minimum of criterion $I_3$ have the form

$$\frac{\partial I_3}{\partial a_{ij}} = 0, \quad \frac{\partial I_3}{\partial d_{ijk}} = 0, \quad \frac{\partial I_3}{\partial A_{i0}} = 0, \quad \frac{\partial I_3}{\partial G_{kj}} = 0,$$

$$i, j = 1, \ldots, n, \quad k = 1, \ldots, M \qquad (8.114)$$

and linearization coefficients are determined by an iterative procedure, similar to Procedure $PEL$–$PD$. We illustrate the direct method and the application of FPK equations with an example of a single-degree-of-freedom system.

### Application to single-degree-of-freedom Systems

Consider a nonlinear oscillator described by the following vector Ito stochastic differential equation:

$$dx_1(t) = x_2(t)dt, \quad x_1(t_0) = x_{10},$$

$$dx_2(t) = -f(x_1, x_2)dt + \sum_{k=1}^{M} \sigma_k(x_1, x_2)d\xi_k(t), \quad x_2(t_0) = x_{20}, \quad (8.115)$$

where $f(x_1, x_2), \sigma k(x_1, x_2), k = 1, \ldots, M$ are nonlinear functions, $\xi_k(t)$ are mutually independent standard Wiener processes.

The corresponding linearized system has the form

$$dx_1(t) = x_2(t)dt, \quad x_1(t_0) = x_{10},$$

$$dx_2(t) = [-k_1x_1 - k_2x_2]dt + \delta_1x_1 \sum_{k=1}^{M} \beta_k d\xi_k(t)$$

$$+\delta_2x_2 \sum_{k=1}^{M} \gamma_k d\xi_k(t) + \delta_3 \sum_{k=1}^{M} \nu_k d\xi_k(t) ,$$

$$x_2(t_0) = x_{20}, \qquad (8.116)$$

where $k_1, k_2, \delta_1, \delta_2, \delta_3, \beta_k, \gamma_k,$ and $\nu_k, k = 1, \ldots, M,$ are linearization coefficients. Another equivalent linear system with only external excitations can also be considered

$$dx_1(t) = x_2(t)dt, \quad x_1(t_0) = x_{10},$$

$$dx_2(t) = [-k_1 x_1 - k_2 x_2]dt + \sum_{k=1}^{M} \sqrt{\delta_1^2 E[x_1^2]\beta_k^2 + \delta_2 E[x_2^2]\gamma_k^2 + \delta_3^2 \nu_k^2} d\xi_k(t),$$

$$x_2(t_0) = x_{20} . \tag{8.117}$$

To determine linearization coefficients and approximate stationary response characteristics for nonlinear system (8.115) by statistical linearization, a modified version of Procedure $SL–PD$ presented in Chap. 5 can be applied. The moment equations for linearized systems (8.116) and (8.117) are given by

$$\frac{dE[x_1^2]}{dt} = E[x_1 x_2], \quad E[x_1^2(t_0)] = x_{10}^2,$$

$$\frac{dE[x_1 x_2]}{dt} = E[x_2^2] - k_1 E[x_1^2] - k_2 E[x_1 x_2], E[(x_1 x_2)(t_0)] = x_{10} x_{20},$$

$$\frac{dE[x_2^2]}{dt} = -2k_1 E[x_1 x_2] - 2k_2 E[x_2^2] + \delta_1^2 E[x_1^2] \sum_{k=1}^{M} \beta_k^2$$

$$+\delta_2^2 E[x_2^2] \sum_{k=1}^{M} \gamma_k^2 + \delta_3^2 \sum_{k=1}^{M} \nu_k^2, \quad E[x_2^2(t_0)] = x_{20}^2. \tag{8.118}$$

The determination of linearization coefficients and approximate stationary response characteristics by the direct method of equivalent linearization requires the calculation of approximate probability density functions using the truncated Gram–Charlier expansion. As in the one-dimensional case, we show the application of criterion $I_{1e}$. For the two-dimensional system, it has the form

$$I_{1e} = \int_{-\infty}^{+\infty} \int_{-\infty}^{+\infty} (g_N(x_1, x_2) - g_{Lp}(x_1, x_2))^2 dx_1 dx_2 , \tag{8.119}$$

where $g_N(x_1, x_2)$ and $g_{Lp}(x_1, x_2)$ are defined by the truncated Gram–Charlier expansion, i.e.,

$$g_N(x_1, x_2) = \frac{1}{c_N} g_G(x_1, x_2) \left[1 + \sum_{k=3}^{N} \sum_{\sigma(\nu)=k} \frac{c_{\nu_1 \nu_2} H_{\nu_1 \nu_2}(x_1, x_2)}{\nu_1! \nu_2!}\right], \tag{8.120}$$

where

$$g_G(x_1, x_2) = \frac{1}{2\pi \sqrt{k_{11} k_{22} - k_{12}^2}} \exp \left\{ -\frac{k_{11}(x_2)^2 - 2k_{12} x_1 x_2 + k_{22}(x_1)^2}{2(k_{11} k_{22} - k_{12}^2)} \right\},$$

$$\tag{8.121}$$

where $c_N$ is a normalized constant, $c_{\nu_1\nu_2} = E[G_{\nu_1\nu_2}(x_1, x_2)]$ are quasi-moments and $\nu_1, \nu_2 = 0, 1, \ldots, N$, $\nu_1 + \nu_2 = 3, 4, \ldots N$, $k_{ij} = E[x_i x_j]$, $i, j = 1, 2$, $H_{\nu_1\nu_2}(x_1, x_2)$ and $G_{\nu_1\nu_2}(x_1, x_2)$ are Hermite polynomials of two variables. The stationary moments $k_{12} = k_{21} = 0$, $k_{11}$ and $k_{22}$ can be found from the moment equations (8.118).

The application of FPK equations approach leads to the following necessary conditions for a minimum of criterion $I_3$, namely:

$$\frac{\partial}{\partial k_i}(I_{p1} + I_{p2}) = 0, \quad i = 1, 2, \tag{8.122}$$

where

$$I_{p1} = \int_{-\infty}^{+\infty}\int_{-\infty}^{+\infty}\left\{\frac{\partial}{\partial x_2}\left([f(x_1, x_2) - k_1 x_1 - k_2 x_2]g_{Lp}(x_1, x_2)\right)\right\}^2 dx_1 dx_2, \tag{8.123}$$

$$I_{p2} = \int_{-\infty}^{+\infty}\int_{-\infty}^{+\infty}\frac{1}{4}\left\{\frac{\partial^2}{\partial x_2^2}\left(\sum_{k=1}^{M}[\sigma_k(x_1, x_2) - \delta_1^2 x_1^2 \beta_k^2\right.\right.$$
$$\left.\left. -\delta_2 x_2^2 \gamma_k^2 - \delta_3^2 \nu_k^2] g_{Lp}(x_1, x_2)\right)\right\}^2 dx_1 dx_2, \tag{8.124}$$

In the particular case when system (8.115) is approximated by (8.117), then criterion $I_{p2}$ defined by (8.124) is replaced by

$$I_{p2} = \int_{-\infty}^{+\infty}\int_{-\infty}^{+\infty}\frac{1}{4}\left\{\frac{\partial^2}{\partial x_2^2}\left(\sum_{k=1}^{M}[\sigma_k(x_1, x_2) - \delta_1^2 E[x_1^2]\beta_k^2\right.\right.$$
$$\left.\left. -\delta_2 E[x_2^2]\gamma_k^2 - \delta_3^2 \nu_k^2] g_{Lp}(x_1, x_2)\right)\right\}^2 dx_1 dx_2, \tag{8.125}$$

where $g_{Lp}(x_1, x_2) = g_{L_e}(x_1, x_2)$ appearing now in criteria (8.123) and (8.125) is defined by

$$g_{L_e}(x_1, x_2) = \frac{1}{c_{Le}\sqrt{k_{11}k_{22}}}\exp\left\{-\frac{1}{2}\left(\frac{(x_1)^2}{k_{11}} + \frac{(x_2)^2}{k_{22}}\right)\right\}, \tag{8.126}$$

where $c_{Le}$ is a normalized constant and

$$k_{11} = E[x_1^2] = \frac{\delta_3^2\sum_{k=1}^{M}\nu_k^2}{\left(2k_2 - \delta_2^2\sum_{k=1}^{M}\gamma_k^2\right)k_1 - \delta_1^2\sum_{k=1}^{M}\beta_k^2},$$

$$k_{12} = k_{21} = E[x_1 x_2] = 0, \quad k_{22} = E[x_2^2] = k_1 E[x_1^2]. \tag{8.127}$$

*Example 8.4.* Consider a simplified vector nonlinear Ito stochastic differential equation from Example 8.3., i.e.,

$$dx_1 = x_2 dt, \quad x_1(t_0) = x_{10},$$

$$dx_2 = [-\alpha x_1^3 - 2hx_2]dt - x_1^3 d\xi_1 + q_2 d\xi_0, \quad x_2(t_0) = x_{20}. \quad (8.128)$$

The linearized system is assumed in a special form of (8.116) or (8.117), i.e.,

$$dx = A_e x dt + q_1 D_{L_p} x d\xi_1 + Q d\xi_0, \quad x(t_0) = x_0, \quad (8.129)$$

or

$$dx = A_e x dt + q_1 D_{L_e} d\xi_1 + Q d\xi_0, \quad x(t_0) = x_0, \quad (8.130)$$

where

$$x = \begin{bmatrix} x_1 \\ x_2 \end{bmatrix}, A_e = \begin{bmatrix} 0 & 1 \\ -\alpha k_1 & -2h \end{bmatrix}, D_{L_p} = \begin{bmatrix} 0 & 0 \\ -k_1 & 0 \end{bmatrix}$$

$$D_{L_e} = \begin{bmatrix} 0 \\ -k_1\sqrt{E[x_1^2]} \end{bmatrix}, Q = \begin{bmatrix} 0 \\ q_2 \end{bmatrix}, x_0 = \begin{bmatrix} x_{10} \\ x_{20} \end{bmatrix}, \quad (8.131)$$

where $k_1$ is a linearization coefficient.

The moment equations for the nonlinear system (8.128) and linearized systems (8.129 or 8.130) are obtained by application the Ito formula to both the systems with the quasi-moment closure technique and without it, respectively.

When we apply the statistical linearization to the nonlinear functions defined by (8.128), i.e., $Y = x_1^3$, then one can show that the probability density function of the output variable $Y$ is given by

$$g_Y(y) = \frac{1}{3\sqrt{2\pi}\sigma_L y^{2/3}} \exp\left\{ -\frac{y^{2/3}}{2\sigma_L^2} \right\}, \quad (8.132)$$

where $\sigma_L^2 = E[x_1^2]$ is the variance of the input zero mean Gaussian variable and the probability density function of linearized variable $y = kx_1$ with the same input zero mean Gaussian variable has the form

$$g_L(y) = \frac{1}{\sqrt{2\pi}k\sigma_L} \exp\left\{ -\frac{y^2}{2k^2\sigma_L^2} \right\}. \quad (8.133)$$

To obtain the linearization coefficient and the approximate characteristics of the stationary solution of nonlinear system (8.128), we use a modified version of Procedure SL–PD presented in Chap. 5.

When we apply the direct equivalent linearization method to the nonlinear system (8.128) and linearized systems (8.129 or 8.130), then the approximate probability density function of the solution of system (8.128) and (8.129) are determined by formulas (8.120) and (8.121) for $N = 6$ and $k_2 = 2h$, while for linearized system (8.130) by a simplified form of (8.126), i.e.,

$$g_{Le}(x_1, x_2) = c_1\sqrt{k_1} \exp\left\{ -\frac{2h}{Q_L^2}(k_1 x_1^2 + x_2^2) \right\}, \quad (8.134)$$

where $c_1$ is a normalized constant and $Q_L^2 = q_1^2 k_1^2 E[x_1^2] + q_2^2$.

To obtain the linearization coefficient and the approximate characteristics of the stationary solution of nonlinear system (8.128), we use a modified version of Procedure *EL–PD* presented in Chap. 6.

The application of the FPK equations approach to nonlinear system (8.128) and linearized systems (8.129 or 8.130) leads to the necessary condition of minimum of the simplified criterion (8.111)

$$\frac{\partial}{\partial k_1}(I_{p1} + I_{p2}) = 0, i = 1, 2 , \tag{8.135}$$

where

$$I_{p1} = \int_{-\infty}^{+\infty} \int_{-\infty}^{+\infty} \left\{ \frac{\partial}{\partial x_2} [\alpha(x_1^3 - k_1 x_1) g_{Lp}(x_1, x_2)] \right\}^2 dx_1 dx_2 , \tag{8.136}$$

$$I_{p2} = \int_{-\infty}^{+\infty} \int_{-\infty}^{+\infty} \frac{1}{4} \left\{ \frac{\partial^2}{\partial x_2^2} [(q_1^2(x_1^6 - k_1^2 x_1^2) g_{Lp}(x_1, x_2)] \right\}^2 dx_1 dx_2 . \tag{8.137}$$

In the particular case when system (8.128) is approximated by a simplified version of system (8.117), then cirterion $I_{p1}$ is defined by (8.136) for $g_{Lp}(x_1, x_2) = g_{Le}(x_1, x_2)$ and criterion $I_{p2}$ defined by (8.137) is replaced by

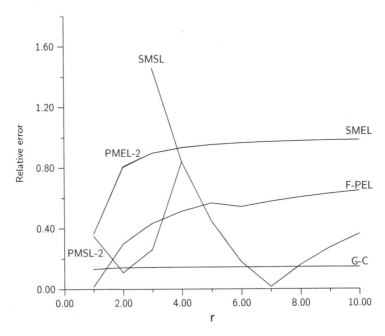

**Fig. 8.1.** Comparison of the relative errors of the response second–order moments $E[x_1^2]$ versus the ratio of parameters $r = \alpha/\lambda_0^2$ for $h = 0.5$, $q_1^2 = 0.2$, $q_2^2 = 0.2$, $\alpha = i \times 0.5$, $i = 1, ..., 10$, reproduced from [17]

$$I_{p2} = \int_{-\infty}^{+\infty} \int_{-\infty}^{+\infty} \frac{1}{4} \left\{ \frac{\partial^2}{\partial x_2^2} \left[ (q_1^2(x_1^6 - k_1^2 x_1^2) g_{Le}(x_1, x_2) \right] \right\}^2 dx_1 dx_2 , \quad (8.138)$$

where $g_{Lp}(x_1, x_2)$ is determined by formulas (8.120) and (8.121) for $N = 6$ and $k_2 = 2h$, while $g_{Le}(x_1, x_2)$ is defined by (8.134).

To illustrate the obtained results, we show a comparison of the relative errors of the second-order $E[x_1^2]$ and six-order $E[x_1^6]$ response moments versus the ratio of parameters $r = \alpha/\lambda_0^2$. In these comparisons we consider the moments obtained by second-order pseudomoment statistical linearization (PMSL-2), square metric probability density statistical linearization (SMSL), pseudomoment direct equivalent linearization (PMEL-2), Fokker–Planck–Kolmogorov equation linearization (F-PEL), and the Gaussian closure approach (G-C). We also use as an exact solution the stationary response of (8.128) obtained by the non-Gaussian approach proposed by Young and Chang [21] and discussed in Sect. 8.3. The numerical results for parameters $h = 0.5$, $q_1^2 = 0.2$, $\alpha = i \times 0.5$, $i = 1, \ldots, 10$, and $q_2^2 = 0.2$ for the relative errors of the second- and six-order response moments for the linearized model with parametric and external excitations are presented in Figs. 8.1 and 8.2, respectively.

Figures 8.1 and 8.2 show that there are no significant differences between the relative errors for the second-order moments obtained for the two

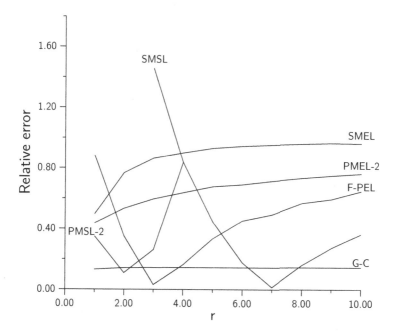

**Fig. 8.2.** Comparison of the relative errors of the response six–order moments $E[x_1^6]$ versus the ratio of parameters $r = \alpha/\lambda_0^2$ for $h = 0.5$, $q_1^2 = 0.2$, $q_2^2 = 0.2$, $alpha = i \times 0.5$, $i = 1, \ldots, 10$, reproduced from [17]

linearized models. Here, usually the Gaussian closure techniques gives smaller relative errors than the linearization methods in probability density functions space.

## Bibliography Notes

Further applications of statistical and equivalent linearization in response analysis of dynamic systems with stochastic parametric excitation were discussed by several authors.

Chang extended the joint approach with Young to SDOF system with hysteresis [5] and proposed a non-Gaussian linearization method [6].

Sobiechowski and Sperling [15] proposed an iterative statistical linearization method for MDOF nonlinear systems. The authors have followed an approach of Sperling [20] and applied the Gaussian closure to the differential equations for the generalized quasi-momentum functions. This procedure has the advantage that it leads to a linear system of equations for the quasi-momentum functions. These equations have to be solved in an iterative procedure. The proposed method can be applied to nonlinear dynamic systems as well for externally excited as for parametrically excited either by white noise or stationary colored noise. A wide study of application of statistical linearization to response analysis of dynamic systems under Poisson excitation was done by Proppe in his Ph.D. and published in [14].

Soize [18] has shown that the stochastic linearization method with random parameters is an efficient approach for identifying the following SDOF nonlinear second-order dynamical system driven by a broadband or a colored Gaussian noise:

$$M\ddot{x}(t) + 2\zeta\lambda_0 M\dot{x}(t) + M\lambda_0^2 x(t) + \varepsilon g(x(t), \dot{x}(t)) = q\eta(t) , \qquad (8.139)$$

where $M, \zeta, \lambda_0$, and $q$ are positive constants, $\varepsilon \geq 0$ is a small parameter, and $g(x, \dot{x})$ is a nonlinear function. The stochastic process $\eta(t)$ is either a real-valued second-order stationary Gaussian process, centered, mean-square continuous and having an integrable power spectral density function $S_\eta(\lambda)$, or a normalized Gaussian white noise with the power spectral density function $S_\eta(\lambda) = 1/2\pi$.

The linear stochastic differential equation with random parameters that approximate (8.139) has the form

$$M\ddot{y}(t) + 2\zeta\lambda_0 M(1 + \varepsilon\Theta)\dot{y}(t) + M\lambda_0^2(1 + \varepsilon\Lambda)y(t) = q\eta(t) , \qquad (8.140)$$

where $M, \zeta, \lambda_0, \varepsilon, q$, and $\eta$ are the previously defined quantities. The $R^2$-vector $(\Theta, \Lambda)$ is a random variable independent of the process $\eta$ such that its probability law $P_{\Theta,\Lambda}(d\theta, d\lambda)$ satisfies the following property concerning its support:

$$SuppP_{\Theta,\Lambda} \subset [\theta_0, +\infty] \times [\lambda_0, +\infty] , \qquad (8.141)$$

where $\theta_0$ and $\lambda_0$ are any real constants such that $1 + \varepsilon\theta_0 > 0$ and $1 + \varepsilon\lambda_0 > 0$.

The identification procedure proposed by Soize [18] consists of two main steps:

*Step 1*: Identification of constant coefficients $M$, $C = 2\zeta\lambda_0 M$, $K = \lambda_0^2 M$ using a modified Kozin's true stochastic linearization method with a mean-square criterion.

*Step 2*: Identification of probability law parameters $\theta_0$ and $\lambda_0$ in order to minimize a criterion for power spectral density functions $S_x(\lambda)$ and $S_y(\lambda)$. The first function is measured and the second one is given by the model defined by (8.140). This approach Soize [18] called "true stochastic linearization method with random parameters."

This approach was generalized by Soize and Lefur [19] for MDOF nonlinear systems

Naprstek and Fischer [13] studied spectral properties of a nonlinear oscillator with random perturbations of the parameters by a direct application of statistical linearization to nonlinear oscillator, i.e., in the equation of motion

$$\ddot{x}(t) - (2h + w_2(t))\dot{x}(t) + (\lambda_0^2 + w_1(t))x(t) + \beta^2 x^2(t)\dot{x}(t) = w_a(t) , \quad (8.142)$$

where $w_1(t), w_2(t)$, and $w_a(t)$ are Gaussian zero-mean correlated white noises, $h, \lambda_0$, and $\beta$ are constant parameters, the authors replaced the nonlinear element by a linear form obtained from statistical linearization.

Cai [3] has generalized the Brückner and Lin approach by considering an equivalent linear system with the same additive and parametric excitations.

# References

1. Brückner, A. and Lin, Y.: Application of complex stochastic averaging to nonlinear random vibration problems. Int. J. Non-Linear Mech. **22**, 237–250, 1987.
2. Brückner, A. and Lin, Y.: Generalization of the equivalent linearization method for nonlinear random vibration problems. Int. J. Non-Linear Mech. **22**, 227–235, 1987.
3. Cai, G.: Non-linear systems of multiple degrees of freedom under both additive and multiplicative random excitation. J. Sound Vib. **278**, 889–901, 2004.
4. Chang, R. and Young, G.: Methods and Gaussian criterion for statistical linearization of stochastic parametrically and externally excited nonlinear systems. Trans. ASME J. Appl. Mech. **56**, 179–185, 1989.
5. Chang, R.: A practical technique for spectral analysis of nonlinear systems under stochastic parametric and external excitations. Trans. ASME J. Vib. Acoust. **113**, 516–522, 1991.
6. Chang, R.: Non-Gaussian linearization method for stochastic parametrically and externally excited nonlinear systems. Trans. ASME J. Dyn. Syst. Meas. Control **114**, 20–26, 1992.
7. Evlanov, L. and Konstantinov, V.: Systems with Random Parameters. Nauka, Moskwa, 1976 (in Russian).

8. Falsone, G.: Stochastic linearization for the response of MDOF systems under parametric Gaussian excitation. Int. J. Non-Linear Mech. **27**, 1025–1037, 1992.

9. Falsone, G.: The true linearization coefficients for nonlinear systems under parametric white-noise excitations. Trans. ASME J. Appl. Mech. **74**, 161–163, 2007.

10. Ibrahim, R.: Parametric Random Vibration. Research Studies Press, Letchworth United Kingdom, 1985.

11. Kazakov, I. and Maltchikov, S.: Analysis of Stochastic Systems in State Space. Nauka, Moskwa, 1975 (in Russian).

12. Kottalam, J., Lindberg, K., and West, B.: Statistical replacement for systems with delta-correlated fluctuations. J. Stat. Phys. **42**, 979–1008, 1986.

13. Naprstek, J. and Fischer, O.: Spectral properties of a non-linear self-excited system with random perturbations of the parameters. Z. Angew. Math. Mech. **77**, 241–242, 1997 S1.

14. Proppe, C.: Statistical Linearization of Dynamical Systems Under Parametric Poisson White Noise Excitation. VDI Verlag, Düsseldorf, 2000 (in German).

15. Sobiechowski, C. and Sperling, L.: An iterative statistical linearization methods for MDOF systems. In: EUROMECH – 2nd European Nonlinear Oscillation Conference, Prague, pp. 419–422, September 9–13, 1996.

16. Socha, L.: Probability density equivalent linearization and non-linearization techniques. Arch. Mech. **51**, 487–507, 1999.

17. Socha, L.: Probability density statistical and equivalent linearization techniques. Int. J. System Sci. **33**, 107–127, 2002.

18. Soize, C.: Stochastic linearization method with random parameters for SDOF nonlinear dynamical systems: Prediction and identification procedures. Probab. Eng. Mech. **10**, 143–152, 1995.

19. Soize, C. and Le Fur, O.: Modal identification of weakly non-linear multidimensional dynamical systems using stochastic linearization method with random coefficients. Mech. Syst. Signal Process. **11**, 37–49, 1997.

20. Sperling, L.: Approximate analysis of nonlinear stochastic differential equations using certain generalized quasi-moment functions. Acta Mechanica **59**, 183–200, 1986.

21. Young, G. and Chang, R.: Prediction of the response of nonlinear oscillators under stochastic parametric and external excitations. Int. J. Non-Linear Mech. **22**, 151–160, 1987.

# Applications of Linearization Methods in Vibration Analysis of Stochastic Mechanical Structures

## 9.1 Introduction

The linearization methods presented in the previous sections constitute a relatively wide group of methods useful in approximation of some characteristics of stochastic dynamic systems. They are used as mathematical tools in other theoretical approaches and also in response analysis of mechanical and structural systems. The main areas of theoretical approaches are as follows:

- models of dynamic systems with hysteresis
- finite element method
- control of nonlinear stochastic dynamic systems
- sensitivity analysis of nonlinear stochastic dynamic systems
- reliability analysis of nonlinear stochastic dynamic systems

The main applications in mechanical and structural engineering:

- vibration of construction elements (beams and frames, shells, and plates)
- vibration of complex structures under:
  - earthquake excitations
  - sea wave excitations
  - wind excitations
- vehicle vibrations under road excitations

These applications and many other ones were discussed in books and review papers. See, for instance, [60, 74, 78, 79, 76]. In this chapter we concentrate only on a few main applications: two theoretical and three engineering. It includes models of dynamic systems with hysteresis and control problems of nonlinear stochastic dynamic systems, vibration of complex structures under earthquake excitations, the models of base isolation systems, and vibration of complex structures under sea wave or wind excitations.

L. Socha: *Applications of Linearization Methods in Vibration Analysis of Stochastic Mechanical Structures*, Lect. Notes Phys. **730**, 281–340 (2008)
DOI 10.1007/978-3-540-72997-6_9
© Springer-Verlag Berlin Heidelberg 2008

## 9.2 Applications in Hysteretic Systems

Many types of structural systems that undergo cycles of inelastic deformation under severe natural loads, such as strong ground motion, extreme wind, and wave action exhibit pinching of hysteresis loops. Various types of piecewise linear or smoothly varying hysteresis models have been developed by researchers interested in the response analysis of hysteretic systems to random excitations. Historically, the first smoothly varying hysteresis model was proposed by Bouc [10] and modified by Wen [84] further called the *Bouc–Wen model* or shortly the BW model. Because of the versatility and mathematical tractability of the BW model, it has been quickly extended and applied to a wide variety of engineering problems. A review of these works can be found in the survey papers by Wen [86] and Socha [78]. To obtain response characteristics of dynamic systems with hysteresis, several approximate methods have been applied such as statistical linearization, equivalent linearization, FPK equation solved by direct numerical integration or with application of cumulant closure technique, cumulant neglect closure, stochastic averaging, path integral solution technique, and a method based on balancing of the average energy dissipation. However, the particular applicability for multi-degree-of-freedom systems linearization methods makes these approaches the most popular in the response analysis of dynamic systems with hysteresis under random excitation. In what follows it will be shown the application of linearization methods for several hysteresis models. First we present the basic hysteresis models and next the moment equations for linearized systems. At the end of this subsection, we present an iterative procedure used as well in the determination of linearization coefficients as in calculations of response characteristics.

We start our considerations with the BW model. The equations of motion of an oscillator with a hysteretic element have the form

$$\dot{q}_1 = q_2 , \tag{9.1}$$

$$\dot{q}_2 = -\alpha\lambda_0^2 q_1 - 2\zeta_0\lambda_0 q_2 - (1-\alpha)\lambda_0^2 q_3 + \dot{\xi}(t) , \tag{9.2}$$

$$\dot{q}_3 = \alpha_1 q_2 - \beta q_3 |q_2| |q_3^{\alpha_2 - 1}| - \gamma q_2 |q_3|^{\alpha_2} , \tag{9.3}$$

where $q_1, q_2$, and $q_3$ are state variables, $q_3$ represents hysteresis variable, $\alpha, \zeta_0$, and $\lambda_0$, $\alpha \in [0, 1]$ are constant parameters, $\beta$ and $\gamma$ are the hysteresis shape parameters, $\alpha_1$ defines the tangent stiffness, and $\alpha_2$ is an odd number that represents the smoothness of the transition from elastic to plastic response; $\dot{\xi}(t)$ is a standard Gaussian white noise.

The linearization of the third equation in the BW model leads to the following approximate equation:

$$\dot{q}_3 = S_{e3} q_1 + C_{e3} q_2 + K_{e3} q_3 , \tag{9.4}$$

where the linearization coefficients $S_{e3}$, $C_{e3}$, and $K_{e3}$ according to [85] have the form

$$S_{e3} = E\left[\frac{\partial \dot{q}_3}{\partial q_1}\right] = 0 \,, \tag{9.5}$$

$$C_{e3} = E\left[\frac{\partial \dot{q}_3}{\partial q_2}\right] = \alpha_1 - \gamma\frac{2^{\alpha_2/2}}{\pi}\sum_{r=0}^{\alpha_2-1}\binom{\alpha_2}{r}\Gamma\left(\frac{r+1}{2}\right)\Gamma\left(\frac{\alpha_2-r+1}{2}\right)$$

$$\times (1 - \rho_{q_2 q_3}^2)^{r/2}\rho_{q_2 q_3}^{\alpha_2-r}\sigma_{q_3}^{\alpha_2} - \beta\frac{2^{\alpha_2/2}}{\sqrt{\pi}}\Gamma\left(\frac{\alpha_2+1}{2}\right)\sigma_{q_3}^{\alpha_2} \,, \tag{9.6}$$

$$K_{e3} = E\left[\frac{\partial \dot{q}_3}{\partial q_3}\right] = -\gamma\frac{2^{\alpha_2/2}}{\pi}\sum_{r=0}^{\alpha_2-1}\binom{\alpha_2-1}{r}\Gamma\left(\frac{r+1}{2}\right)\Gamma\left(\frac{\alpha_2-r+1}{2}\right)$$

$$\times (1 - \rho_{q_2 q_3}^2)^{r/2}\rho_{q_2 q_3}^{\alpha_2-r+1} - \beta\alpha_2\frac{2^{\alpha_2/2}}{\sqrt{\pi}}\Gamma\left(\frac{\alpha_2+1}{2}\right)\rho_{q_2 q_3}\sigma_{q_2}\sigma_{q_3}^{\alpha_2-1} \,,$$

$$\text{for} \quad \alpha_2 = odd \,, \tag{9.7}$$

where

$$\rho_{q_2 q_3} = \frac{E[q_2 q_3]}{\sigma_{q_2}\sigma_{q_3}}, \quad \sigma_{q_i} = \sqrt{E[q_i^2]} \quad i = 2,3 \,. \tag{9.8}$$

The expressions for the linearization coefficients (9.5–9.7) originally proposed by Wen [85] were simplified also under the assumption that the response joint probability distribution is Gaussian to the following form [34]:

$$S_{e3} = 0 \,,$$

$$C_{e3} = \alpha_1 - \gamma\frac{2^{\alpha_2/2}}{\pi}\Gamma\left(\frac{\alpha_2+2}{2}\right)\sigma_{q_3}^{\alpha_2}I_s - \beta\frac{2^{\alpha_2/2}}{\sqrt{\pi}}\Gamma\left(\frac{\alpha_2+1}{2}\right)\sigma_{q_3}^{\alpha_2} \,,$$

$$K_{e3} = -\gamma\frac{\alpha_2\sigma_{q_2}\sigma_{q_3}^{\alpha_2-1}}{\pi}\Gamma\left(\frac{\alpha_2+2}{2}\right)2^{\frac{\alpha_2}{2}}\left(2(1 - \rho_{q_2 q_3}^2)^{\frac{\alpha_2+1}{2}} + \rho_{q_2 q_3}I_s\right)$$

$$- \beta\alpha_2\frac{\rho_{q_2 q_3}\sigma_{q_2}\sigma_{q_3}^{\alpha_2-1}}{\sqrt{\pi}}\Gamma\left(\frac{\alpha_2+1}{2}\right)2^{\alpha_2/2} \,, \tag{9.9}$$

where

$$I_s = 2\int_l^{\pi/2} \sin^{\alpha_2}\theta d\theta \,, \tag{9.10}$$

$$l = arctan\left(\frac{\sqrt{1 - \rho_{q_2 q_3}^2}}{\rho_{q_2 q_3}}\right) \,. \tag{9.11}$$

In the particular case for $\alpha_2 = 1$, the linearization coefficients take the form

$$C_{e3} = \alpha_1 - \sqrt{\frac{2}{\pi}}(\gamma + \beta)\sigma_{q_3} \,,$$

$$K_{e3} = -\sqrt{\frac{2}{\pi}}(\gamma + \beta)\sigma_{q_2} \,. \tag{9.12}$$

Now we present the extended and improved models of hysteresis in single-degree-of-freedom systems using the notation of Ito stochastic differential equations and corresponding linearization coefficients.

### 9.2.1 Bouc, Babar, and Wen Model (BBW Model)[5, 6]

$$dq_1 = q_2 dt \, ,$$
$$dq_2 = [-\alpha\lambda_0^2 q_1 - 2\zeta_0\lambda_0 q_2 - (1-\alpha)\lambda_0^2 q_3]dt + d\xi(t) \, ,$$
$$dq_3 = \left[ \frac{1}{\zeta}\left[ \alpha_1 q_2 - \nu(\beta q_3|q_2||q_3^{\alpha_2-1}| + \gamma q_2|q_3|^{\alpha_2}) \right] \right] dt \, , \qquad (9.13)$$

where all parameters in BBW model have the same meaning as in BW model and $\zeta$ and $\nu$ are the deterioration control parameters. The linearization coefficient have the form (9.5), where $\alpha_1 := \frac{\alpha_1}{\zeta}$, $\beta := \frac{\beta\nu}{\zeta}$, and $\gamma := \frac{\gamma\nu}{\zeta}$.

The next two models are capable of generating asymmetric hysteresis patterns. The external excitation $\eta(t)$ is assumed to be a nonzero-mean process.

### 9.2.2 Babar Noori Model (BN Model)[4]

This model gives stiffness reduction in the unloading regimes of the stress–strain cycle.

$$dq_1 = q_2 dt \, ,$$
$$dq_2 = [-\alpha\lambda_0^2 q_1 - 2\zeta_0\lambda_0 q_2 - (1-\alpha)\lambda_0^2 q_3 + \eta(t)]dt,$$
$$dq_3 = \frac{\alpha_1}{4}\left\{ \left[ \left( \frac{1}{2} + \frac{1}{\pi}\arctan\frac{q_3 - q_{3+}}{q_{3_{ch}}} \right)\left( 1 - \left( \frac{q_3}{q_{3_{ult}}} \right)^{\alpha_2} \right) \right.\right.$$
$$\left. + \left( \frac{1}{2} - \frac{1}{\pi}\arctan\frac{q_3 - q_{3+}}{q_{3_{ch}}} \right)\mu \right] \times (q_2 + |q_2|)$$
$$+ \left[ \left( \frac{1}{2} - \frac{1}{\pi}\arctan\frac{q_3 - q_{3+}}{q_{3_{ch}}} \right)\left( 1 + \left| \frac{q_3}{q_{3_{ult}}} \right|^{\alpha_2-1}\left( \frac{q_3}{q_{3_{ult}}} \right) \right) \right.$$
$$\left.\left. + \left( \frac{1}{2} + \frac{1}{\pi}\arctan\frac{q_3 - q_{3-}}{q_{3_{ch}}} \right)\mu \right] \times (q_2 - |q_2|) \right\} dt \, , \qquad (9.14)$$

where $\alpha_1$, $\alpha_2$, $q_{3+}$, $q_{3-}$, $q_{ch}$, $q_{ult}$, and $\mu$ are the shape parameters.

### 9.2.3 Dobson Model (D Model) [21]

$$dq_1 = q_2 dt \, ,$$
$$dq_2 = [-\alpha\lambda_0^2 q_1 - 2\zeta_0\lambda_0 q_2 - (1-\alpha)\lambda_0^2 q_3 + \eta(t)]dt,$$
$$dq_3 = [\alpha_1 q_2 + \alpha_3(q_2 + |q_2|)(q_3 + |q_3|)^{\alpha_2} + \alpha_4(q_2 + |q_2|)(q_3 - |q_3|)^{\alpha_2}$$
$$+\alpha_5(q_2 - |q_2|)(q_3 + |q_3|)^{\alpha_2} + \alpha_6(q_2 - |q_2|)(q_3 - |q_3|)^{\alpha_2}] dt \, . \qquad (9.15)$$

where $\alpha_k$, $k = 3, 4, 5, 6$, are constant parameters.

The derivation of linearization coefficients by standard procedure is very complicated (for details, see [20]). Therefore Dobson et al. [21] proposed a different approach. The authors considered a single-degree-of-freedom system with nonzero-mean excitation in the form

$$dq_1 = q_2 dt \ ,$$
$$dq_2 = [-\alpha\lambda_0^2 q_1 - 2\zeta_0\lambda_0 q_2 - (1 - \alpha)\lambda_0^2 q_3 + \eta(t)]dt,$$
$$dq_3 = g(q_2, q_3)dt \ . \tag{9.16}$$

Dobson et al. [21] transformed system (9.16) to a zero-mean system by introducing new variables $y_i = q_i - E[q_i]$ and $\dot{\xi}(t) = \eta(t) - E[\eta(t)]$

$$dy_1 = y_2 dt \ ,$$
$$dy_2 = [-\alpha\lambda_0^2 y_1 - 2\eta_0\lambda_0 y_2 - (1 - \alpha)\lambda_0^2 y_3]dt + d\xi(t) \ ,$$
$$dy_3 = (g(q_2, q_3) - E[g(q_2, q_3)])dt \tag{9.17}$$

and proposed a new mean-square criterion $I = E[ee^T]$, where

$$e = [0 \ \ 0 \ \ g - E[g] - C_{e3}y_2 - K_{e3}y_3] \ , \tag{9.18}$$

where $C_{e3}$ and $K_{e3}$ are linearization coefficients.

Then the necessary conditions of minimum of $I$ are

$$\frac{\partial I}{\partial C_{e3}} = -2E[gy_2] + 2E[y_2 E[g]] + 2K_{e3}S_5 + 2C_{e3}S_4 = 0 \ , \tag{9.19}$$

$$\frac{\partial I}{\partial K_{e3}} = -2E[gy_3] + 2E[y_3 E[g]] + 2K_{e3}S_6 + 2C_{e3}S_5 = 0 \ . \tag{9.20}$$

Under assumptions that $E[g]$ is known and $E[y_i E[g]] = E[y_i]E[g] = 0$ , $E[gy_i] = E[gq_i] - E[g]E[q_i]$, $i = 2, 3$, the solutions of (9.19) and (9.20) are found in the form

$$C_{e3} = \frac{S_6(E[gq_2] - E[g]E[q_2]) - S_5(E[gq_3] - E[g]E[q_3])}{S_4 S_6 - S_5^2} \ , \tag{9.21}$$

$$K_{e3} = \frac{S_4(E[gq_3] - E[g]E[q_3]) - S_5(E[gq_2] - E[g]E[q_2])}{S_4 S_6 - S_5^2} \ , \tag{9.22}$$

where $S_4 = E[y_2^2], S_5 = E[y_2 y_3], and S_6 = E[y_3^2]$ are moments of linearized system, i.e., the first two equations of ((9.17)) and

$$dy_3 = (C_{e3}y_2 + K_{e3}y_3)dt \ . \tag{9.23}$$

## 9.2.4 Pinching Hysteresis Model (PH model) [25]

$$
\begin{aligned}
dq_1 &= q_2 dt, \\
dq_2 &= [-\alpha\lambda_0^2 q_1 - 2\zeta_0\lambda_0 q_2 - (1-\alpha)\lambda_0^2 q_3]dt + d\xi(t), \\
dq_3 &= \frac{1}{\zeta}\left[1 - \zeta_1 exp\left\{-(q_3 sgn(q_2) - \kappa z_u)^2/\zeta_2^2\right\}\right] \\
&\quad \times \left[q_2 - \nu(\beta|q_2||q_3|^{\alpha_2-1}q_3 - \gamma q_2|q_3|^{\alpha_2})\right]dt \,,
\end{aligned}
\tag{9.24}
$$

where

$$
z_u = \left[\frac{1}{\nu(\beta+\gamma)}\right]^{1/\alpha_2}, \tag{9.25}
$$

$\beta$, $\gamma$, and $\alpha_2$ are the hysteresis shape parameters, $\nu$, $\zeta$ are the strength and the stiffness parameters, respectively; and $\kappa$, $\zeta_1$, and $\zeta_2$ are constant parameters.

In further analysis it is assumed that $\nu$, $\zeta$, $\zeta_1$ and $\zeta_2$ are functions of $q_2$ and $q_3$. The strength and stiffness degradations are modeled, respectively, by

$$
\nu = \nu(\epsilon) = 1 + \delta_\nu\epsilon, \quad \zeta = \zeta(\epsilon) = 1 + \delta_\zeta\epsilon \,, \tag{9.26}
$$

where $\delta_\nu$ and $\delta_\zeta$ are constant parameters, $\epsilon$ is the hysteretic energy dissipation, given by

$$
\epsilon = (1-\alpha)\lambda_0^2 \int_{t_0}^{t_f} q_2(t)q_3(t)dt \tag{9.27}
$$

for given arbitrarily $t_0$ and $t_f$ and functions

$$
\zeta_1(\epsilon) = \zeta_{10}[1 - \exp\{-p\epsilon\}], \quad \zeta_2(\epsilon) = (\psi_0 + \delta_\psi\epsilon)(\lambda_1 + \zeta_1(\epsilon)) \,, \tag{9.28}
$$

where $p$ is the constant that controls the rate of initial drop in slope, $\zeta_{10}$ is the measure of total slip, $\psi_0$ is the parameter that contributes to the amount of pinching, $\delta_\psi$ is the constant specified for the desired rate of pinching spread, and $\lambda_1$ is the small parameter that controls the rate of change of $\zeta_2$ as $\zeta_1$ changes (for details, see [25]).

The third equation of (9.24) is replaced by the linearized form

$$
dq_3 = [C_{e3}q_2 + K_{e3}q_3]dt \,, \tag{9.29}
$$

where the linearization coefficients $C_{e3}$ and $K_{e3}$ are obtained such that the mean-square error $E[e^2]$ is minimized, where

$$
e = g_3(q_2, q_3) - C_{e3}q_2 - K_{e3}q_3 \tag{9.30}
$$

and $g_3(q_2, q_3)$ is the right-hand side of the third equation of (9.24). Assuming that responses $q_2$ and $q_3$ are jointly Gaussian and applying a direct linearization procedure proposed by Atalik and Utku [3], the authors have found the linearization coefficients in the form

$$C_{e3} = E\left[\frac{\partial g_3(q_2, q_3)}{\partial q_2}\right], \quad K_{e3} = E\left[\frac{\partial g_3(q_2, q_3)}{\partial q_3}\right]. \quad (9.31)$$

Since the degradation and pinching parameters $\nu, \eta, \zeta_1$, and $\zeta_2$ are functions of $q_2$ and $q_3$, the exact evaluation of the expected values is very difficult. Foliente et al. [25] proposed an approximate approach of the determination of linearization coefficients under the assumption that $\nu, \eta, \zeta_1$, and $\zeta_2$ are slowly varying functions of $q_2$ and $q_3$, it means that the partial derivatives involving $\nu, \eta, \zeta_1$, and $\zeta_2$ in both the equalities in (9.31) are neglected and by the first approximation, the parameters are replaced by their expected values $\mu_\nu, \mu_\eta, \mu_{\zeta_1}$, and $\mu_{\zeta_2}$, respectively. Then the approximated linearization coefficients are

$$C_{e3} = \frac{1}{\mu_\eta}\left[1 - \mu_\nu(\beta C_1 + \gamma C_2) - \mu_{\zeta_1}C_3 + \mu_\nu\mu_{\zeta_1}(\beta C_4 + \gamma C_5)\right], \quad (9.32)$$

$$K_{e3} = \frac{1}{\mu_\eta}\left[-\mu_\nu(\beta K_1 + \gamma K_2) + 2\frac{\mu_{\zeta_1}}{\mu_{\zeta_2}^2}(K_3 + \kappa z_u K_4) + 2\kappa z_u\frac{\mu_\nu\mu_{\zeta_1}}{\mu_{\zeta_2}^2}\right.$$

$$\times(\beta K_5 + \gamma K_6) - 2\kappa z_u\frac{\mu_\nu\mu_{\zeta_1}}{\mu_{\zeta_2}^2}(\beta K_7 + \gamma K_8)$$

$$\left. +\alpha_2\mu_\nu\mu_{\zeta_1}(\beta K_9 + \gamma K_{10})\right], \quad (9.33)$$

where the $C_i$'s $i = 1, \ldots, 5$ and the $K_j$'s $j = 1, \ldots, 10$ are constants determined by complex algebraic terms including special functions $\Gamma$, $erf$, $erfc$, and Gauss–Lauguerre quadrature (for details, see [25]).

### 9.2.5 Shape Memory Alloy Model (Yan and Nie [94])

$$\begin{aligned}
dq_1 &= q_2 dt, \\
dq_2 &= [-\alpha\lambda_0^2 q_1 - 2\zeta_0\lambda_0 q_2 - (1-\alpha)\lambda_0^2 q_3]dt + d\xi(t), \\
q_3 &= \{1 - sgn[sgn(|q_1| - a) + 1]\}q_1 \\
&\quad + \frac{sgn(|q_1| - a) + 1}{2}\left[\frac{sgn(q_1) + sgn(\dot{q}_2)}{2}(b-a) + a sgn(q_1)\right],
\end{aligned} \quad (9.34)$$

in which $sgn(x)$ gives $-1, 0$, or $1$ depending on whether $x$ is negative, zero, or positive, respectively; $a$ and $b$ are constant parameters of hysteresis.

Applying statistical linearization with the mean-square criterion to variable $q_3$ we find that

$$q_3 = S_{e3}q_1 + C_{e3}q_2, \quad (9.35)$$

where

$$S_{e3} = \frac{a+b}{\sqrt{2\pi}\sigma_{q_1}}\exp\left\{-\frac{a^2}{2\sigma_{q_1}^2}\right\}, \quad C_{e3} = \frac{a+b}{\sqrt{2\pi}\sigma_{q_2}}\left[1 - erf\left(\frac{a}{\sqrt{2}\sigma_{q_2}}\right)\right]. \quad (9.36)$$

$\sigma_{q_i}^2$, $i = 1, 2$ are variances of variables $q_i, i = 1, 2$, respectively.

Since the assumption of the Gaussian response joint probability distribution function is very strong particularly with respect to the hysteresis variable $q_3$, Hurtado and Barbat [34] adapted the idea of Kimura et al. [46] to smooth hysteresis and have applied to the mean-square criterion in equivalent linearization of hysteresis system described by the BBW model the mixed probability density function in the averaging operation.

### 9.2.6 Non-Gaussian Hysteresis Model (Hurtado and Barbat [34, 35])

It has been assumed that the probability density function for the $q_3$ variable in the BW model is defined by

$$f_{q_3}(q_3) = (1 - 2p)\phi_{q_3}(q_3) + p\delta(q_3 - z_x) + p\delta(q_3 + z_x) , \qquad (9.37)$$

where $\phi_3(z)$ is the univariate Gaussian probability density function, $\delta(.)$ is the Dirac delta function, $0 < p < 1$ is a weighting coefficient, and $z_x$ is the maximum attainable value of the nonlinear component of the restoring force given by

$$z_x = \left(\frac{\alpha_1}{\beta + \gamma}\right)^{1/\alpha_2} , \qquad (9.38)$$

where $\alpha_1$, $\alpha_2$, $\beta$, and $\gamma$ are the hysteresis parameters defined for the BW model. An illustration of the function $f_{q_3}(q_3)$ is given in Fig. 9.1.

The corresponding joint probability density function of variables $q_3$ and $q_1$ or $q_3$ and $q_2$ have the same form

$$f_{\tilde{x}q_3}(\tilde{x}, q_3) = (1 - 2p)\phi_{\tilde{x}q_3}(\tilde{x}, q_3) + p\delta(q_3 - z_x)\phi_{\tilde{x}}(\tilde{x}) + p\delta(q_3 + z_x)\phi_{\tilde{x}}(\tilde{x}) , \quad (9.39)$$

where $\tilde{x} = q_1$ or $\tilde{x} = q_2$. It means that variables $q_1$ and $q_2$ can be assumed to remain jointly Gaussian. $\phi_{q_i q_3}(q_i, q_3)$, $i = 1, 2$, are univariate Gaussian probability density functions.

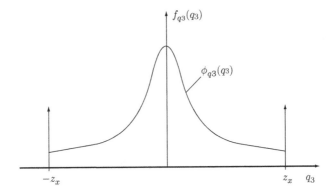

**Fig. 9.1.** An illustration of the non-Gaussian probability density function $f_{q_3}(q_3)$

Additionally, for separating the contributions of Gaussian and Dirac parts in calculation of the linearization coefficients, an assumption of simple splitting of their density functions is needed, i.e.,

$$f_{\tilde{x}}(\tilde{x}) = (1 - 2p)\phi_{\tilde{x}}(\tilde{x}) + 2p\phi_2(\tilde{x}) , \qquad (9.40)$$

$$f_{q_1 q_2}(q_2) = (1 - 2p)\phi_{q_1 q_2}(q_1, q_2) + 2p\phi_{q_1 q_2}(q_1, q_2) , \qquad (9.41)$$

where $\phi_{\tilde{x}}(\tilde{x})$ and $\phi_{q_1 q_2}(q_1, q_2)$ are univariate Gaussian probability density functions.

In the case of standard equivalent linearization, the linearized system has the form

$$\dot{q}_3 = \rho_e q_1 + c_e q_2 + k_e q_3 , \qquad (9.42)$$

where the vector of linearization coefficients consists of two groups of coefficients corresponding to the Gaussian and Dirac parts of the probability density function, namely,

$$[\rho_e c_e k_e] = (1 - 2p)[\rho_g c_g k_g] + 2p[\rho_d c_d k_d]V^{-1} , \qquad (9.43)$$

where $V$ is the covariance matrix defined by

$$V = E\left[\begin{bmatrix} q_1^2 & q_1 q_2 & q_1 q_3 \\ q_2 q_1 & q_2^2 & q_2 q_3 \\ q_3 q_1 & q_3 q_2 & q_3^2 \end{bmatrix}\right] , \qquad (9.44)$$

the subindices "$g$" and "$d$" denote the Gaussian and Dirac parts, respectively. The linearization coefficients corresponding to the Gaussian part are given by (9.9–9.11), where the parameters of the Gaussian probability density satisfy certain second-order constraints. Hurtado and Barbat [35] have shown that using the probability density function (9.37) the variance of the hysteretic variable $q_3$ is determined from the following equality:

$$E[q_3^2] = (1 - 2p)\sigma_{q_3}^2 + 2pz_x^2 . \qquad (9.45)$$

Hence

$$\sigma_{q_3}^2 = \frac{E[q_3^2] - 2pz_x^2}{(1 - 2p)} \qquad (9.46)$$

and similarly

$$\sigma_{\tilde{x}q_3} = \frac{E[\tilde{x}q_3]}{(1 - 2p)}, \quad \sigma_{\tilde{x}}^2 = E[\tilde{x}^2], \quad \sigma_{q_1 q_2} = E[q_1 q_2] , \qquad (9.47)$$

where again $\tilde{x}$ denotes either $q_1$ or $q_2$ and the expectations are obtained from the solution of the matrix covariance equation of the response under assumption that the linearized system is Gaussian.

Hurtado and Barbat [35] have shown that the linearization coefficients corresponding to the Dirac part are

$$\rho_d = \sigma_{q_1 q_2}(\alpha_1 - \beta z_x^{\alpha_2}) , \quad c_d = \sigma_{q_2}^2(\alpha_1 - \beta z_x^{\alpha_2}) , \quad k_d = -\sqrt{\frac{2}{\pi}}\gamma\sigma_{q_2}z_x^{\alpha_2+1} .$$
$$(9.48)$$

The second wide class of hysteresis are nonsmooth hysteresis with bilinear characteristics. First, the application of statistical linearization to dynamic systems with such hysteresis was shown by Kazakov [39] and next by Kazakov and Dostupov [40]. Some practical methods for the calculation of linearization coefficients for discontinuous and hysteretic elements were presented by Morosanov [58]. The idea of application of equivalent linearization to dynamic system with bilinear hysteresis appeared later, see, for instance, [2, 13, 45, 67]. For simplicity, we consider an oscillator with bilinear hysteresis under Gaussian stationary zero-mean colored noise with known correlation function described by

$$
\begin{aligned}
\dot{y}_1 &= y_2 , \\
\dot{y}_2 &= -\alpha y_1 - 2\zeta_0 y_2 - (1-\alpha)y_3 + \eta(t) , \\
\dot{y}_3 &= g(y_2, y_3) ,
\end{aligned}
$$
$$(9.49)$$

where $\alpha$ and $\zeta_0$ are constant positive parameters, $\eta(t)$ is a Gaussian stationary zero-mean colored noise with known correlation function,

$$g(y_2, y_3) = \begin{cases} y_2 \text{ for } |y_3| < 1; & y_2 \leq 0, y_3 = 1; & y_2 \geq 0, y_3 = -1 \\ 0 \text{ for } & y_2 \geq 0, y_3 = 1; & y_2 \leq 0, y_3 = -1 \end{cases} . \quad (9.50)$$

The equivalent linear system has the form

$$
\begin{aligned}
\dot{y}_1 &= y_2 , \\
\dot{y}_2 &= -\alpha y_1 - 2\zeta_0 y_2 - (1-\alpha)y_3 + \eta(t) , \\
\dot{y}_3 &= C_{e3}(t)y_2 + K_{e3}(t)y_3 ,
\end{aligned}
$$
$$(9.51)$$

where the linearization coefficients $C_{e3}(t)$ and $K_{e3}(t)$ are time-dependent functions determined by minimization of the mean-square criterion $I = E[ee^T]$, where

$$e = [0 \ \ 0 \ \ g(y_2, y_3) - C_{e3}y_2 - K_{e3}y_3] . \quad (9.52)$$

Then the necessary conditions of minimum of $I$ are

$$\frac{\partial I}{\partial C_{e3}} = -2E[g(y_2, y_3)y_2] + 2K_{e3}E[y_2 y_3] + 2C_{e3}E[y_2^2] = 0 , \quad (9.53)$$

$$\frac{\partial I}{\partial K_{e3}} = -2E[g(y_2, y_3)y_3] + 2K_{e3}E[y_3^2] + 2C_{e3}E[y_2 y_3] = 0 . \quad (9.54)$$

Then we find

$$C_e = \frac{E[y_3^2]E[g(y_2, y_3)y_2] - E[y_2 y_3]E[g(y_2, y_3)y_3]}{E[y_2^2]E[y_3^2] - (E[y_2 y_3])^2} , \quad (9.55)$$

$$K_e = \frac{E[y_2^2]E[g(y_2, y_3)q_3] - E[y_2 y_3]E[g(y_2, y_3)y_2]}{E[y_2^2]E[y_3^2] - (E[y_2 y_3])^2} , \qquad (9.56)$$

where $E[y_2^2]$, $E[y_2 y_3]$, and $E[y_3^2]$ are moments of linearized system (9.51). They can be determined directly under assumption that the response joint probability density function is Gaussian. However, this assumption is not realistic because the bilinear hysteresis represented by the state variable $y_3$ takes on the values $|y_3| \leq 1$. In fact, a non-Gaussian response joint probability density function should be considered.

Kimura et al. [46] have proposed to the mean-square criterion in equivalent linearization of hysteresis system described by a bilinear model, the truncated Gaussian probability density function in averaging operation.

For convenience of expression of such a function, three Gaussian distributed random variables $\{s_1, s_2, s_3\}$ are introduced. Their covariances are given by

$$\sigma_i^2 = E[s_i^2], \quad \sigma_{ij}^2 = E[s_i s_j], \quad i, j = 1, 2, 3 . \qquad (9.57)$$

The marginal probability density functions for variable $Y_i$, $i = 1, 2, 3$, and joint variables $Y_i Y_j$, $i, j = 1, 2, 3$, have the form of truncated Gaussian probability density function with a couple of delta functions, for instance,

$$p_{Y_3}(y_3) = \frac{1}{\sqrt{2\pi}\sigma_3} \exp\left\{-\frac{y_3^2}{2\sigma_3^2}\right\} rect(y_3) + [\delta(y_3 + 1) + \delta(y_3 - 1)]S , \quad (9.58)$$

where

$$S = \frac{1}{\sqrt{\pi}} erfc\left(\frac{1}{\sqrt{2}\sigma_3}\right), \quad erfc(x) = \int_x^\infty \exp\{-s^2\}ds, \quad rect(y_3)$$

$$= H(y_3 + 1) - H(y_3 - 1) , \qquad (9.59)$$

where $\delta(.)$ is the Dirac's delta function and $H(.)$ is the unit step function.

The joint probability density functions $p_{Y_1, Y_3}(y_1, y_3)$ and $p_{Y_2, Y_3}(y_2, y_3)$ can be derived from the relationship that $p_{Y_3}(y_3)$ has to be a marginal probability density function of the variables $Y_1, Y_3$ and $Y_2, Y_3$, respectively. These are given as follows:

$$f_{\tilde{X}Y_3}(\tilde{x}, y_3) = \phi_{\tilde{X}Y_3}(\tilde{x}, y_3)rect(y_3) + \delta(y_3 - 1)\phi_{\tilde{X}}(\tilde{x})erfc(v_{\tilde{X}})$$

$$+ \delta(y_3 + 1)\phi_{\tilde{X}}(\tilde{x})erfc(v_{\tilde{X}}) , \qquad (9.60)$$

where $\tilde{X} = Y_1$ or $\tilde{X} = Y_2$, $\tilde{x} = y_1$ or $\tilde{x} = y_2$, $\phi_{Y_i Y_3}(y_i, y_3)$, $i = 1, 2$, are univariate Gaussian probability density functions, i.e.,

$$\phi_{Y_i Y_j}(y_i, y_j) = \frac{1}{2\pi(\sigma_i^2 \sigma_j^2 - \sigma_{ij}^4)} \exp\left\{-\frac{\sigma_i^2 \sigma_j^2}{2(\sigma_i^2 \sigma_j^2 - \sigma_{ij}^4)}\left(\frac{y_i^2}{\sigma_i^2} - 2\frac{\sigma_{ij}^2 y_i y_j}{\sigma_i^2 \sigma_j^2} + \frac{y_j^2}{\sigma_j^2}\right)\right\} ,$$

$$i, j = 1, 2, 3, \quad i \neq j \qquad (9.61)$$

and

$$\phi_{Y_i}(y_i) = \frac{1}{\sqrt{2\pi}\sigma_i} \exp\left\{-\frac{y_i^2}{2\sigma_i^2}\right\}, \quad i = 1, 2, 3, \tag{9.62}$$

$$v_{Y_i} = \frac{\sigma_i}{\sqrt{2(\sigma_i^2\sigma_3^2 - \sigma_{i3}^4)}} - \frac{\sigma_{i3}^2 y_i}{\sqrt{2(\sigma_i^2\sigma_3^2 - \sigma_{i3}^4)}\sigma_i}, \quad i = 1, 2. \tag{9.63}$$

It means that variables $y_1$ and $y_2$ can be assumed to remain jointly Gaussian.

$$p_{Y_1}(y_1) = \phi_{Y_1}(y_1), p_{Y_2}(y_2) = \phi_{Y_2}(y_2), p_{Y_1Y_2}(y_1, y_2) = \phi_{Y_1Y_2}(y_1, y_2). \tag{9.64}$$

The differential equations for the second-order moments of the linearized system (9.51) are

$$\frac{dE[y_1^2]}{dt} = 2E[y_1y_2],$$

$$\frac{dE[y_1y_2]}{dt} = -\alpha E[y_1^2] + E[y_2^2] - 2\eta_0 E[y_1y_2] - (1-\alpha)E[y_1y_3] + E[y_1\eta(t)],$$

$$\frac{dE[y_1y_3]}{dt} = C_{e3}(t)E[y_1y_2] + K_{e3}(t)E[y_1y_3] + E[y_2y_3],$$

$$\frac{dE[y_2^2]}{dt} = -2\alpha E[y_1y_2] - 4\eta_0 E[y_2^2] - 2(1-\alpha)E[y_2y_3] + 2E[y_2\eta(t)],$$

$$\frac{dE[y_2y_3]}{dt} = C_{e3}(t)E[y_2^2] - \alpha E[y_1y_3]$$
$$\qquad + (K_{e3}(t) - 2\eta_0)E[y_2y_3] - (1-\alpha)E[y_3^2] + E[y_3\eta(t)],$$

$$\frac{dE[y_3^2]}{dt} = 2C_{e3}(t)E[y_2y_3] + 2K_{e3}(t)E[y_3^2]. \tag{9.65}$$

The relationship between the second-order moments of stationary solutions of nonlinear system (9.49) and (9.50) and the second-order moments of Gaussian random variables $\{s_1, s_2, s_3\}$ calculated by application of non-Gaussian probability density functions (9.58–9.63) are the following:

$$E[y_1^2] = \sigma_1^2,$$

$$E[y_1y_2] = \sigma_{12}^2,$$

$$E[y_1y_3] = \frac{2}{\sqrt{\pi}}\sigma_{13}^2 \, erf\left(\frac{1}{\sqrt{2}\sigma_3}\right),$$

$$E[y_2^2] = \sigma_2^2,$$

$$E[y_2y_3] = \frac{2}{\sqrt{\pi}}\sigma_{23}^2 \, erf\left(\frac{1}{\sqrt{2}\sigma_3}\right), \tag{9.66}$$

$$E[y_3^2] = \frac{2}{\sqrt{\pi}}\sigma_3^2 \, erf\left(\frac{1}{\sqrt{2}\sigma_3}\right) - \sqrt{\frac{2}{\pi}}\sigma_3 \exp\left\{-\frac{1}{2\sigma_3^2}\right\} + \frac{2}{\sqrt{\pi}}erfc\left(\frac{1}{\sqrt{2}\sigma_3}\right),$$

where

$$erf(x) = \int_0^x \exp\{-s^2\}ds \; . \tag{9.67}$$

These nonlinear relationships between the second-order moments can be represented in a form of nonlinear transformation

$$E[y_iy_j] = T(\sigma_{kl}), \quad i, j = 1, 2, 3, k, \quad l = 1, 2, 3 \;, \tag{9.68}$$

where $\sigma_k = \sigma_{kk}$. From (9.67) one can calculate numerically for given $E[y_iy_j]$ the moments $\sigma_k$ and $\sigma_{kl}$, $k, l = 1, 2, 3$, what is denoted by the inverse relation

$$\sigma_{kl} = T^{-1}(E[y_iy_j]), \quad i, j = 1, 2, 3, k, \quad l = 1, 2, 3 \;. \tag{9.69}$$

To obtain linearization coefficients and response characteristics, Kimura et al. [46] proposed the following iterative procedure.

**Procedure** *NG–HYST*

*Step 1:* Assume initial values for the second-order moments for the non-Gaussian variables $E[y_iy_j]$, $i, j = 1, 2, 3$ at $t = t_0$.

*Step 2:* Calculate values for the second-order moments $E[s_is_j]$, $i, j = 1, 2, 3$, at $t = t_0$ using the inverse transform $T^{-1}$ defined by (9.69).

*Step 3:* Determine the non-Gaussian probability density function for variables $\{y_1, y_2, y_3\}$ using the values for the second-order moments $E[s_is_j]$, $i, j = 1, 2, 3$, at $t = t_0$ obtained in Step 2 and formulas (9.58–9.64).

*Step 4:* Determine the linearization coefficients for variables $\{y_1, y_2, y_3\}$ using the values for the second-order moments $E[y_iy_j]$ and $E[y_ig(y_2y_3)]$, $i, j = 1, 2, 3$, at $t = t_0$ and relations (9.55) and (9.56).

*Step 5:* Assume the time step $\Delta t$ and solve the moment differential equations (9.65) and find the second-order moments $E[y_iy_j]$, $i, j = 1, 2, 3$, at $t = t_0 + \Delta t$. Substitute $t_0 = t_0 + \Delta t$ and go to Step 2.

*Step 6:* Repeat Steps 2–5. until $t \le t_f$ ($t_f$—given final time) or until characteristics converge.

At the end of the discussion of the application of non-Gaussian probability density function to the response study of bilinear model of hysteresis, we note that similar considerations can be done in the case of Gaussian white noise external excitation. Then the moment differential equations should be derived using the Ito formula.

# 9.3 Vibrations of Structures under Earthquake Excitations

A wide area of application of statistical and equivalent linearization is the response analysis of multistory structures under stochastic ground excitations.

Wen [85] and Baber and Wen [5, 6] considered a multi-degree-of-freedom shear beam model subjected to horizontal ground acceleration $\ddot{\xi}_g$ (see Fig. 9.2).

Noting that the quantities $v_i$ are the relative displacements of the $i$th and $(i+1)$th stories $v_i = x_{i+1} - x_i$, the equation displacements of the and stories, $\nu = X_{i+1} - Xi$, the equation of motion may be written in the form

$$m_i \left( \sum_{j=1}^{i} \ddot{v}_j + \ddot{\xi}_g \right) + q_i - q_{i+1} = 0 , \quad i = 1, \ldots, n , \tag{9.70}$$

where $m_i$ is the mass of the $i$th floor, $\ddot{\xi}_g$ is the ground acceleration, and $q_i$ is the $i$th restoring force, including viscous damping, given by

$$q_i = c_i \dot{v}_i + a_{1_i} k_i v_i + (1 - \alpha_{1_i}) k_i z_i , \quad i = 1, \ldots, n , \tag{9.71}$$

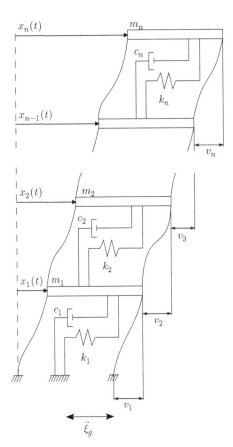

**Fig. 9.2.** Model of multi–degree of freedom shear beam subjected to horizontal ground acceleration, reproduced from [5]

where $c_i$ is the viscous damping, $k_i$ controls the initial tangent stiffness, $\alpha_{1_i}$, $0 \le \alpha_{1_i} \le 1$ controls the ratio of postyield stiffness, and $z_i$ is the $i$th hysteresis. Equation (9.70) can be rewritten with the accelerations decoupled if the $(i-1)$th equation is subtracted from the $i$th equation, for $i = 2, \ldots, n$. This result may be summarized by

$$\ddot{v}_i - (1-\delta_{i1})\frac{q_{i-1}}{m_{i-1}} + \frac{q_i}{m_i}\left[1 + (1-\delta_{i1})\frac{m_i}{m_{i-1}}\right] - (1-\delta_{in})\frac{q_{i+1}}{m_{i+1}}\left(\frac{m_{i+1}}{m_i}\right) = -\delta_{i1}\ddot{\xi}_g,$$
$$i = 1, \ldots, n ,  \tag{9.72}$$

in which $\delta_{i1}, \delta_{in}$ are Kronecker deltas.

The time history-dependent hysteretic restoring force is assumed in the form of the BWB model, i.e., modeled by the first-order nonlinear differential equation

$$\dot{z}_i = \frac{\alpha_{1_i}\dot{v}_i - \nu_i(\beta_i|\dot{v}_i||z_i|^{\alpha_{2_i}-1}z_i + \gamma_i\dot{v}_i|z_i|^{\alpha_{2_i}})}{\zeta_i} , \quad i = 1, \ldots, n , \tag{9.73}$$

in which $\alpha_{1_i}, \nu_i, \beta_i, \zeta_i, \gamma_i$, and $\alpha_{2_i}$ are parameters that control the hysteresis shape and model degradation.

Using equivalent linearization, Baber and Wen [5] looked for an equivalent linear equation in the form

$$\dot{z}_i = c_{e_i}\dot{v}_i + k_{e_i}z_i , \quad i = 1, \ldots, n . \tag{9.74}$$

It should be stressed that as seen above, equivalent linearization is not applied to the whole system (9.71–9.73) but only separately to each hysteresis.

Under the assumption that $\dot{v}_i$ and $z_i$ are zero-mean joint Gaussian processes, coefficients $c_{e_i}$ and $k_{e_i}$ can be found from a modification of conditions (9.5–9.8), i.e.,

$$c_{e_i} = E\left[\frac{\partial \dot{z}_i}{\partial \dot{v}_i}\right] , \quad k_{e_i} = E\left[\frac{\partial \dot{z}_i}{\partial z_i}\right] , \quad i = 1, \ldots, n \tag{9.75}$$

in closed form with the result

$$c_{e_i} = \frac{\mu_{\alpha_{1_i}} - \mu_{\nu_i}(\beta_i F_{1_i} + \gamma_i F_{2_i})}{\mu_{\zeta_i}} , \quad k_{e_i} = -\mu_{\nu_i}\frac{\beta_i F_{3_i} + \gamma_i F_{4_i}}{\mu_{\zeta_i}} , \quad i = 1, \ldots, n , \tag{9.76}$$

where

$$\mu_{\alpha_{1_i}} = E[\alpha_{1_i}], \; \mu_{\nu_i} = E[\nu_i], \; \mu_{\zeta_i} = E[\zeta_i],$$

$$F_{1_i} = \frac{\sigma_{z_i}^{\alpha_{2_i}}}{\pi}\Gamma\left(\frac{\alpha_{2_i}+2}{2}\right)2^{\frac{\alpha_{2_i}}{2}}(I_{s_{1i}} - I_{s_{2i}}), \; F_{2_i} = \frac{\sigma_{z_i}^{\alpha_{2_i}}}{\sqrt{\pi}}\Gamma\left(\frac{\alpha_{2_i}+1}{2}\right)2^{\frac{\alpha_{2_i}}{2}} ,$$

$$F_{3_i} = \frac{\alpha_{2_i}\,\sigma_{\dot{v}_i}\,\sigma_{z_i}^{\alpha_{2_i}-1}}{\pi}\Gamma\left(\frac{\alpha_{2_i}+2}{2}\right)2^{\alpha_{2_i}/2}\left[\frac{2(1-\rho_{\dot{v}_i z_i}^2)^{\frac{\alpha_{2_i}+1}{2}}}{\alpha_{2_i}} + \rho_{\dot{v}_i z_i}(I_{s_{1i}} - I_{s_{2i}})\right] ,$$

$$F_{4_i} = \frac{\alpha_{2_i}\rho_{\dot{v}_i z_i}\sigma_{\dot{v}_i}\sigma_{z_i}^{\alpha_{2_i}-1}}{\sqrt{\pi}}\Gamma\left(\frac{\alpha_{2_i}+1}{2}\right)2^{\frac{\alpha_{2_i}}{2}} , \quad i = 1,\dots,n. \qquad (9.77)$$

In the above, $\Gamma(.)$ is the gamma function, and

$$I_{s_{1i}} = \int_0^{\kappa_i} \sin^{\alpha_{2_i}}\theta d\theta, \quad I_{s_{2i}} = \int_{\kappa_i}^{\pi} \sin^{\alpha_{2_i}}\theta d\theta , \quad i = 1,\dots,n , \qquad (9.78)$$

where

$$\kappa_i = \arctan\left(-\frac{\sqrt{1-\rho_{\dot{v}_i z_i}^2}}{\rho_{\dot{v}_i z_i}}\right) , \quad i = 1,\dots,n . \qquad (9.79)$$

The integrals in equalities (9.78) are widely tabulated, see, for instance, [1, 36, 31]. The coefficients of (9.76) are response dependent since

$$\sigma_{z_i}^2 = E[z_i^2(t)], \quad \sigma_{\dot{v}_i}^2 = E[\dot{v}_i^2(t)], \quad \rho_{\dot{v}_i z_i} = \frac{E[\dot{v}_i(t)z_i(t)]}{\sigma_{\dot{v}_i}\sigma_{z_i}} , \quad i = 1,\dots,n . \qquad (9.80)$$

They can be found by application of standard iterative procedures.

Equivalent linearization was applied to various base isolation models by Su and Ahmadi [80] and Lin et al. [54]. A comparison of the performances of different base isolators for shear beam-type structures discussed in the literature was carried out by Ahmadi and his coworkers [81, 14]. We will shortly discuss the main base isolation systems including a laminated rubber bearing (LRB), with and without lead core, a resilient-friction base isolator (R-FBI), a friction-type base isolator developed under the auspices of Electricite de France (EDF), a sliding resilient-friction (SR-F), a high damping laminated rubber bearing (HD-LRB). Figures 9.3 and 9.4 show schematic diagrams of the mechanical behavior of these systems.

For an elastic nonuniform shear beam structural model with its base isolation system, the equation of vibration of the beam under earthquake ground acceleration $\ddot{u}_g$ is given by

$$\rho_0 A \frac{\partial^2 u(x,t)}{\partial t^2} + C_0 A \frac{\partial u(x,t)}{\partial t} - \frac{\partial}{\partial x}\left(KGA\frac{\partial u(x,t)}{\partial x}\right)$$
$$= -\rho_0 A(\ddot{u}_g(t) + \ddot{v}_b(t)) + F_s(t)\delta(x - H), 0 \le x \le l \qquad (9.81)$$

subject to boundary conditions

$$u(0,t) = 0, \quad \frac{\partial u(x,t)}{\partial x}\Big|_{x=L} = 0 . \qquad (9.82)$$

Here $u$ is the deflection of the beam relative to its base, $v_b$ is the displacement of the base raft relative to the ground, $G$, $K$, $A$, $C_0$, $\rho_0$, and $L$ are constant parameters. $F_s(t)$ is the force exerted by the secondary system on the structure, $\delta(x - H)$ is the Dirac delta function which indicates the concentrated force at

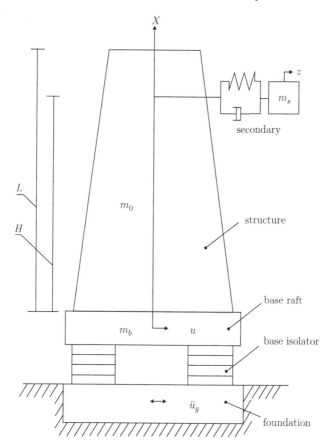

**Fig. 9.3.** Structural model of a base isolated nonuniform shear beam, from [14, 16]. Copyright John Wiley and Sons. Reproduced with permission

the attachment point $x = H$. The general solution of (9.81) with boundary conditions (9.82) in nondimensional form is assumed to be given as

$$u(x,t) = \sum_{n=1}^{N} q_n(t)\phi_n(x), \quad for \quad N \to \infty , \qquad (9.83)$$

where

$$\phi_n(x) = \frac{2}{\sqrt{2 - \frac{\sin 2\lambda_n}{\lambda_n}}} e^{ax} \sin(\lambda_n x) , \quad n = 1, \ldots, N, \qquad (9.84)$$

$N$ is the number of modes of vibration considered and the modal amplitudes $q_n(t)$ are determined by

$$\ddot{q}_n(t) + 2\zeta_n \lambda_n \dot{q}_n(t) + \lambda_n^2 q_n(t) = -\kappa_n(\ddot{u}_g(t) + \ddot{v}_b(t)) + (2\zeta_s \lambda_s \dot{z}(t)$$
$$+ \lambda_s^2 z(t))R_{s1} , \quad n = 1, \ldots, N , \qquad (9.85)$$

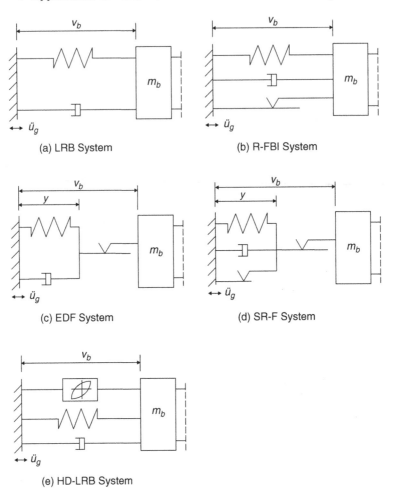

**Fig. 9.4.** Schematic diagrams of mechanical behavior of base isolation systems, from [14, 16]. Copyright John Wiley and Sons. Reproduced with permission

where $z(t)$ is the displacement of the secondary system, $\lambda_n$, $a$, $\zeta_n$, $\lambda_s$, $\zeta_s$, $R_{s1}$ and $\kappa_n$ are constant parameters defined by parameters of system (9.81).

The explicit equations governing the base raft displacement $v_b$ for various base isolation systems are described below.

### 9.3.1 LRB System

$$\ddot{v}_b + 2\zeta_0\lambda_0\dot{v}_b + \lambda_0^2 v_b = -\left(\ddot{u}_g + \sum_{n=1}^{N} B_n\ddot{q}_n + R_s\ddot{z}\right). \qquad (9.86)$$

Here, $\zeta_0$ is the effective damping coefficient of the rubber and $\lambda_0$ is the natural frequency of the base isolator, $R_s = m_s/m_0$ is a constant, ($m_0$ and

$m_s$ are masses of of structure and secondary system, respectively); $B_n$, $n = 1, ..., N$, are constant coefficients defined by parameters of system (9.81).

Since the LRB system has a linear structure, any statistical or equivalent linearization method is not used.

### 9.3.2 R-FBI System

$$\ddot{v}_b + 2\zeta_0\lambda_0\dot{v}_b + \lambda_0^2 v_b + \mu_1 g sgn(\dot{v}_b) = -(\ddot{u}_g + \sum_{n=1}^{N} B_n\ddot{q}_n + R_s\ddot{z}) . \quad (9.87)$$

Here, $\zeta_0$ is the damping coefficient, $\lambda_0$ is the natural frequency of the base isolator, $\mu_1$ is the friction coefficient, $g$ is the acceleration due to gravity, and $B_n$, $n = 1, \ldots, N$, are constant coefficients defined by parameters of system (9.81).

The equivalent linear system is given by

$$\ddot{v}_b + C_e\dot{v}_b + \lambda_0^2 v_b = -(\ddot{u}_g + \sum_{n=1}^{N} B_n\ddot{q}_n + R_s\ddot{z}) , \quad (9.88)$$

where $C_e$ is the equivalent damping coefficient. Applying statistical linearization and minimizing the mean-square error with respect to $C_e$ by the assumption that the excitation is Gaussian and $E[v_b] = 0$, one can show that

$$C_e = 2\zeta_0\lambda_0 + \mu_1 g\sqrt{\frac{2}{\pi E[\dot{v}_b^2]}} . \quad (9.89)$$

### 9.3.3 EDF System

$$\ddot{v}_b + \mu g sgn^*(\dot{v}_b - \dot{y}) = -(\ddot{u}_g + \sum_{n=1}^{N} B_n\ddot{q}_n + R_s\ddot{z}) , \quad (9.90)$$

$$2\zeta_0\lambda_0\dot{y} + \lambda_0^2 y = \mu g sgn^*(\dot{v}_b - \dot{y}) . \quad (9.91)$$

Here, $\zeta_0$ is the damping coefficient, $\lambda_0$ is the natural frequency of the elastomeric pad, $\mu$ is the friction coefficient, $g$ is the acceleration due to gravity, $y$ is a local displacement of the base raft, introduced in EDF and SR-F systems (see Fig. 9.4c and d), $B_n$ are constant coefficients defined by parameters of system (9.81), and

$$sgn^*(\dot{v}_b - \dot{y}) = \begin{cases} sgn(\dot{v}_b - \dot{y}) \ for \ \dot{v}_b \neq \dot{y} \ (sliding) , \\ -\frac{\ddot{z}_g + \ddot{y}}{\mu g} \quad for \ \dot{v}_b = \dot{y} \ (non\text{-}sliding) \end{cases} \quad (9.92)$$

With this definition for the $sgn^*$ function equations (9.90) and (9.91) include the sliding and nonsliding motion. According to [80] for parameters

$\lambda_0 \cong 2\pi$ rad/s, $\zeta_0 = 0.1$, and $\mu = 0.2$, the linearized form of the $sgn^*(\dot{v}_b - \dot{y})$ function is given as

$$sgn^*(\dot{v}_b - \dot{y}) \cong (1 - E[\gamma])\sqrt{\frac{2}{\pi E[(\dot{v}_b - \dot{y})^2]}}(\dot{v}_b - \dot{y})$$

$$- \frac{E[\gamma]}{\mu g}\left(\ddot{u}_g + \sum_{n=1}^{N} B_n \ddot{q}_n + \ddot{y} + R_s \ddot{z}\right), \qquad (9.93)$$

where a Gaussian closure approximation is used in the first term and $E[\gamma]$ is given by

$$E[\gamma] = 1 - \frac{\lambda_0^4}{\mu^2 g^2} E[y^2] \cong \frac{1}{1 + \frac{\lambda_0^4}{\mu^2 g^2} E[y^2]}. \qquad (9.94)$$

This approximation follows from Euler transformation and the assumption that $\frac{\lambda_0^4 E[y^2]}{\mu^2 g^2}$ is small.

Substituting (9.93) into (9.90) and (9.91), and linearizing the obtained equations, we find that the corresponding equivalent linear equations are given by

$$\ddot{v}_b + C_{e1}\dot{y} + \lambda_0^2 y = -\left(\ddot{u}_g + \sum_{n=1}^{N} B_n \ddot{q}_n + R_s \ddot{z}\right), \qquad (9.95)$$

$$\ddot{y} + C_{e2}\dot{y} + K_e y + C_e \dot{v}_b = -\left(\ddot{u}_g + \sum_{n=1}^{N} B_n \ddot{q}_n + R_s \ddot{z}\right), \qquad (9.96)$$

where the linearization coefficients are defined as

$$C_{e1} = 2\zeta_0\lambda_0, \quad C_{e2} = \frac{1}{E[\gamma]}\left[2\zeta_0\lambda_0 + \mu g(1 - E[\gamma])\sqrt{\frac{2}{\pi E[(\dot{v}_b - \dot{y})^2]}}\right],$$

$$C_e = -\frac{\mu g(1 - E[\gamma])}{E[\gamma]}\sqrt{\frac{2}{\pi E[(\dot{v}_b - \dot{y})^2]}}, \quad K_e = \frac{\lambda_0^2}{E[\gamma]}. \qquad (9.97)$$

### 9.3.4 SR-F System

$$\ddot{v}_b + \mu g\, sgn^*(\dot{v}_b - \dot{y}) = -\left(\ddot{u}_g + \sum_{n=1}^{N} B_n \ddot{q}_n + R_s \ddot{z}\right), \qquad (9.98)$$

$$2\zeta_0\lambda_0\dot{y} + \lambda_0^2 y + \mu_1 g\, sgn(\dot{y}) = \mu g\, sgn^*(\dot{v}_b - \dot{y}). \qquad (9.99)$$

Here, $\zeta_0$ is the damping coefficient, $\lambda_0$ is the natural frequency of the isolation system, $\mu$ and $\mu_1$ are the coefficients of friction of the upper plate and the isolator plate, respectively; $g$ is the acceleration due to gravity, and $B_n$, $n = 1, \ldots, N$, are constant coefficients defined by parameters of system (9.81).

With the $sgn^*$ function and its linearized form defined by (9.93) and $E[\gamma]$ given by (9.94) with $\mu$ being replaced by $(\mu - \mu_1)$, (9.98) and (9.99) govern motions of the base raft during both sliding and nonsliding phases in the upper friction plate. Chen and Ahmadi [14] assumed that the continuous sliding in the body friction plates of the isolator occurs and the stick duration can be neglected. Substituting (9.93) and (9.94) into (9.98) and (9.99) and linearizing the obtained equations, we find that the corresponding equivalent linear equations are

$$\ddot{v}_b + C_{e1}\dot{y} + \lambda_0^2 y = -\left(\ddot{u}_g + \sum_{n=1}^{N} B_n \ddot{q}_n + R_s \ddot{z}\right), \tag{9.100}$$

$$\ddot{y} + C_{e2}\dot{y} + K_e y + C_e \dot{v}_b = -\left(\ddot{u}_g + \sum_{n=1}^{N} B_n \ddot{q}_n + R_s \ddot{z}\right), \tag{9.101}$$

where the linearization coefficients are defined as

$$C_{e1} = 2\zeta_0 \lambda_0 + \mu_1 g \sqrt{\frac{2}{\pi E[\dot{y}^2]}},$$

$$C_{e2} = \frac{1}{E[\gamma]}\left[2\zeta_0 \lambda_0 + \mu g(1 - E[\gamma])\sqrt{\frac{2}{\pi E[(\dot{v}_b - \dot{y})^2]}} + \mu_1 g \sqrt{\frac{2}{\pi E[\dot{y}^2]}}\right],$$

$$C_e = -\frac{\mu g(1 - E[\gamma])}{E[\gamma]}\sqrt{\frac{2}{\pi E[(\dot{v}_b - \dot{y})^2]}}, \quad K_e = \frac{\lambda_0^2}{E[\gamma]}. \tag{9.102}$$

### 9.3.5 HD-LRB System

$$\ddot{v}_b + \frac{Mc}{m}\dot{v}_b + \frac{M}{m}\left[a_1 v_b(t) + a_2 |v_b(t)| v_b(t) + a_3 v_b^3(t) + b\left(1 - \frac{\beta}{A}|y|\right)y\right] =$$
$$-\left(\ddot{u}_g + \sum_{n=1}^{N} B_n \ddot{q}_n + R_s \ddot{z}\right), \tag{9.103}$$

$$\dot{y} = -\frac{\gamma}{\zeta}|\dot{v}_b(t)|y - \frac{\beta}{\zeta}\dot{v}_b(t)|y| + \frac{\alpha_1}{\zeta}\dot{v}_b(t). \tag{9.104}$$

Here, $M$ is the number of the aseismic bearings used, $c$ is the damping coefficient, $a_1, a_2,$ and $a_3$ are the stiffness parameters of the nonlinear spring, $y$ is the hysteretic variable (responsible for the hysteretic behavior of the bearing), $b, \zeta, \beta,$ and $\alpha_1$ are parameters that control the characteristics of the resulting hysteresis loop, and $B_n$, $n = 1, \dots, N$, are constant coefficients defined by parameters of system (9.81).

To obtain approximate mean-square responses, equivalent linearization was applied to the hysteresis element. The equivalent linear system is given by

$$\ddot{v}_b + C_e \dot{v}_b + D_e v_b + B_e y = -(\ddot{u}_g + \sum_{n=1}^{N} B_n \ddot{q}_n + R_s \ddot{z}) , \qquad (9.105)$$

$$\dot{y} + F_e \dot{v}_b + K_e y = 0 , \qquad (9.106)$$

where $D_e, B_e, F_e$, and $K_e$ are the linearization coefficients and $C_e = Mc/m$ is an effective damping constant. They were found from the mean-square criterion of linearization and have the form

$$D_e = \frac{M}{m} \left[ a_1 + 2\sqrt{\left(\frac{2}{\pi}\right)} a_2 \sigma_{v_b} + 3a_3 \sigma_{v_b}^2 \right] , \qquad (9.107)$$

$$B_e = \frac{Mb}{m} \left[ 1 - 2\sqrt{\left(\frac{2}{\pi}\right)} \frac{\beta}{A} \sigma_y \right] , \qquad (9.108)$$

and the coefficients $F_e$ and $K_e$ are determined by (9.12), i.e., $F_e = -C_{e3}$ and $G_e = -K_{e3}$.

In all the considered models of structures under earthquake excitation, it was assumed that the ground excitation are modeled by the generalized nonstationary Kanai–Tajimi model for the 1940 El Centro and the 1985 Mexico City earthquakes described by

$$\ddot{v}_g = -[2\zeta_g \lambda_g(t)\dot{\xi}_f(t) + \lambda_g^2(t)\xi_f(t)]e(t) , \qquad (9.109)$$

with

$$\ddot{\xi}_f + 2\zeta_g \lambda_g(t)\dot{\xi}_f(t) + \lambda_g^2(t)\xi_f(t) = w(t) , \qquad (9.110)$$

where $\xi_f$ is the filter response, $e(t)$ is a deterministic amplitude envelope, $\lambda_g(t)$ is the time dependent ground filter frequency, and $w(t)$ is a Gaussian white noise.

In all models of base isolation discussed by Ahmadi and his coworkers, the friction coefficients were assumed to be constant. Constantinou and Papageorgiou [17] considered sliding systems which utilize teflon sliding bearings. In the models of these systems, the coefficient of friction strongly depends on the velocity of sliding. The equation of motion for the slip displacement $x$ is

$$\ddot{x} + 2\zeta_0 \lambda_0 \dot{x} + \lambda_0^2 x + g\mu(\dot{x})z = -\ddot{u}_g , \qquad (9.111)$$

where $\zeta_0, \lambda_0$, and $g$ are constant parameters, $\mu(\dot{x})$ is the coefficient of sliding friction which depends on the slip velocity $\dot{x}$, and $\ddot{u}_g$ is the ground acceleration. $z$ is the hysteretic variable taking values in the interval $[-1, +1]$. This variable obeys equation (9.73), i.e.,

$$\dot{z} = \frac{\alpha_1 \dot{x} - \gamma|\dot{x}||z|^{\alpha_2 - 1}z - \beta\dot{x}|z|^{\alpha_2}}{\zeta} . \qquad (9.112)$$

The coefficient of sliding friction of teflon–steel interfaces closely approximates the following equation:

$$\mu(\dot{x}) = (f_{\max} - Df)\exp\{-a|\dot{x}|\}\,, \tag{9.113}$$

in which $f_{\max}$ represents the maximum value of friction coefficient attained at large velocities of sliding and $f_{\max} - Df$ represents the value at essentially zero velocity, and $a$ is a constant positive parameter.

Another method of reducing vibrations of structures under earthquake excitation is the application of tuned liquid column dampers (TLCD). They are a special type of tuned mass dampers (TMD) relying on the motion of a liquid mass in a tube-like container to counteract the external force, while a Bernoullian-type orifice forms damping forces that dissipate energy (see Fig. 9.5). This problem was studied, for instance, by Won et al. [88, 89] and Yalla and Kareem [93].

The equation of motion for the liquid oscillation subjected to the absolute acceleration $\ddot{y}_a$ of the container is given by

$$\rho A L \ddot{x}(t) + \frac{\rho A}{2}\gamma|\dot{x}(t)|\dot{x}(t) + 2\rho A g x(t) = -\rho A B \ddot{y}_a(t)\,, \tag{9.114}$$

where $\ddot{y}_a = \ddot{\xi}_g + \ddot{y}$ for SDOF structure ($\ddot{y}_a = \ddot{\xi}_g + \ddot{y}_n$ for MDOF structure), $x$ is the vertical elevation change of the liquid surface, and $\rho, L, B$, and $A$ are the density, length, width, and cross-sectional area of the liquid column, respectively. The constant $\gamma$ is the coefficient of head loss controlled by the opening ratio of the orifice at the center of the horizontal portion of the TLCD and $g$ is the acceleration due to gravity.

The application of statistical linearization gives the corresponding equivalent linear equation

$$\rho A L \ddot{x}(t) + c_e \dot{x}(t) + 2\rho A g x(t) = -\rho A B \ddot{y}_a(t)\,, \tag{9.115}$$

where $c_e$ is the equivalent linearization coefficient (damping coefficient) for the TLCD. Assuming Gaussian excitations and response, one can calculate using the mean-square criterion that

$$c_e = \sqrt{\frac{2}{\pi}}\rho A\gamma\sigma_{\dot{x}}\,, \tag{9.116}$$

where $\sigma_{\dot{x}}$ is the standard deviation of the liquid mass velocity $\dot{x}(t)$.

TLCD technique presented in [88] has been applied to flexible structures, while Ghosh and Basu [27] have applied to short period structures.

## 9.4 Vibrations of Structures Under Wave Excitations

The random nature of sea waves causes the necessity of random analysis of vibration of offshore structures and mooring cables. The physical modeling of the hydrodynamic forces acting on the fixed offshore structures or mooring cables is based on the Morison equation [57]. It can be presented in the general form [55] as the hydrodynamic loading on a single, uniform cylindrical member $e$ located between two nodes, namely,

$$f = \rho A (C_m - 1)(\ddot{U}_n - \ddot{u}_n) + \frac{1}{2}\rho D C_d \mid \dot{U}_n - \dot{u}_n \mid (\dot{U}_n - \dot{u}_n) + \rho A \ddot{U}_n \ , \quad (9.117)$$

where $f$ is the hydrodynamic force; $\rho$ is the density of surrounding water; $A$ is the cross-sectional area of member $e$; $D$ is the outside diameter; $C_m$ and $C_d$ are the inertia and drag coefficients, respectively; $\dot{u}_n$ and $\dot{U}_n$ are the normal velocity vectors of the member and the water particle, respectively.

Since $f$ in (9.117) is a nonlinear function of velocity $\dot{u}$, it was linearized by many authors in response analysis. The applications of such method for

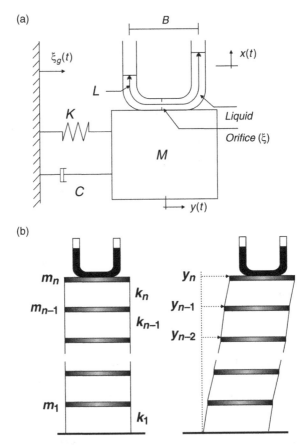

**Fig. 9.5.** Model of single-degree-of-freedom system with TLCD and shear beam structure with TLCD subjected to horizontal ground acceleration, from [89] with permission

analysis of general three-dimensional structures in the case of zero current may be found in [56, 26, 62, 61]. In these studies, a lumped mass model for a tower with $n$ degree of freedom is used, where the equations of motion are described in the matrix form

$$\mathbf{M}\ddot{\mathbf{U}}_t + \mathbf{C}\dot{\mathbf{U}} + \mathbf{K}\mathbf{U} = \rho(K_M - 1)\mathbf{V}(\dot{\mathbf{V}}_u - \ddot{\mathbf{U}}_t)$$

$$+\rho\mathbf{V}\dot{\mathbf{V}}_u + \rho K_D \mathbf{A}(\dot{\mathbf{V}}_u - \dot{\mathbf{U}}_t)|\dot{\mathbf{V}}_u - \dot{\mathbf{U}}_t| , \quad (9.118)$$

where vectors $\dot{\mathbf{V}}_u$ and $\ddot{\mathbf{V}}_u$ represent water particle velocities and accelerations, respectively, at the instantaneous deflected position of the structure; vectors $\dot{\mathbf{U}}_t$ and $\ddot{\mathbf{U}}_t$ represent the total structure velocities and accelerations, respectively, from a fixed reference; $\dot{\mathbf{U}}$ and $\mathbf{U}$ represent the structure velocities and displacements, respectively, as measured relative to its moving base; diagonal matrices $\mathbf{M}, \mathbf{V}$, and $\mathbf{A}$ represent structural lumped masses, volumes and areas, respectively; matrices $\mathbf{C}$ and $\mathbf{K}$ represent structural damping and stiffness coefficients, respectively (foundation effects may be included); and $\rho, K_M$, and $K_D$ are scalar quantities representing mass density of water, inertia coefficient, and drag coefficient, respectively; for $\mathbf{r} = \dot{\mathbf{V}}_u - \dot{\mathbf{U}}_t$ holds $\mathbf{r}|\mathbf{r}| = [r_1|r_1|, \ldots, r_n|r_n|]^T$.

In the above, coefficients $K_M$ and $K_D$ are assumed to be constants with $\dot{V}_o = \dot{V}_u$ and $\ddot{V}_o = \ddot{V}_u$ where subscript "$o$" refers to the undeflected structure coordinates. Using this approximation, substituting the relation $\mathbf{U}_t = \mathbf{U}_g + \mathbf{U}$, where each component in the vector $\mathbf{U}_g$ is the same time history of horizontal ground displacement, and substituting the relation $\mathbf{r} \equiv \mathbf{V}_o - \mathbf{U}_t$, (9.118) becomes

$$[\mathbf{M} + \rho(K_M - 1)\mathbf{V}]\ddot{\mathbf{r}} + \mathbf{C}\dot{\mathbf{r}} + \mathbf{K}\mathbf{r} + \rho K_D \mathbf{A}\dot{\mathbf{r}}|\dot{\mathbf{r}}| = [\mathbf{M} - \rho\mathbf{V}]\ddot{\mathbf{V}}_o + \mathbf{C}\dot{\mathbf{V}}_o$$

$$+\mathbf{K}\mathbf{V}_o - \mathbf{C}\ddot{\mathbf{U}}_g - \mathbf{K}\mathbf{U}_g .$$

$$(9.119)$$

Linearizing each element $\dot{r}_i|\dot{r}_i|$ by $a_i\dot{r}_i$, $i = 1,\ldots,n$, where $a_i$ are the linearization coefficients, one can show that, under the assumption of Gaussian response

$$a_i = \frac{E[\dot{r}_i^2|\dot{r}_i|]}{E[\dot{r}_i^2]} = \sqrt{\frac{8}{\pi}}E[\dot{r}_i^2] , \quad i = 1,\ldots,n . \quad (9.120)$$

The vector $|\dot{\mathbf{r}}|\dot{\mathbf{r}} = [\dot{r}_1|\dot{r}_1|\ldots\dot{r}_n|\dot{r}_n|]^T$ in (9.119) can now be replaced by vector $\mathbf{a}\dot{\mathbf{r}}$, where $\mathbf{a} = diag\{a_i\}$.

Thus, one obtains the desired linearized equations of motion, namely

$$[\mathbf{M}+\rho(K_M-1)\mathbf{V}]\ddot{\mathbf{r}}+\mathbf{C}_e\dot{\mathbf{r}}+\mathbf{K}\mathbf{r} = [\mathbf{M}-\rho\mathbf{V}]\ddot{\mathbf{V}}_o+\mathbf{C}\dot{\mathbf{V}}_o+\mathbf{K}\mathbf{V}_o-\mathbf{C}\ddot{\mathbf{U}}_g-\mathbf{K}\mathbf{U}_g ,$$

$$(9.121)$$

where $\mathbf{C}_e = \mathbf{C} + \rho K_D \mathbf{a}$.

Penzien and his coauthors [62, 61] considered such structures under both earthquake and random wave excitation.

Some models of guyed offshore towers were considered by Brynjolfsson and Leonard [12, 11]. A conceptual model of such a tower is one where the governing equation has two main sources of nonlinearities: fluid–structure interaction and the restoring force of the cable system. Therefore, it has both nonlinear stiffness and damping.

Following Penzien and Kaul [61], Brynjolfsson and Leonard [11] proposed a nonlinear model in the form

$$\mathbf{M}_v \ddot{\mathbf{x}} + \mathbf{C}_{tw} \dot{\mathbf{x}} + \mathbf{K}_{tw} \mathbf{x} = -\mathbf{M}_v \mathbf{1} \ddot{x}_g - \mathbf{K}_{nl}(\mathbf{x})\mathbf{x} + \mathbf{C}_h |\mathbf{V} - \dot{\mathbf{x}} - \mathbf{1}\dot{x}_g|(\mathbf{V} - \dot{\mathbf{x}} - \mathbf{1}\dot{x}_g) \,, \tag{9.122}$$

where

$$\mathbf{M}_v = \mathbf{M}_{tw} + \mathbf{M}_a, \quad \mathbf{M}_a = \rho C_a \mathbf{V}_e$$

$$\mathbf{C}_h = 0.5 \rho C_d \mathbf{A}_e$$

$$\mathbf{K}_{nl} = [K_{nl,ij}], = \begin{cases} 0 & \text{for } i,j \neq n \\ c_{2,nn}(1 - \exp\{-c_{1,nn}|x_n|\}) & \text{for } i = j = n \end{cases} . \tag{9.123}$$

In the above $\mathbf{x}, \dot{\mathbf{x}}$, and $\ddot{\mathbf{x}}$ are the horizontal tower displacement, velocity, and acceleration, respectively; $\mathbf{M}_{tw}, \mathbf{C}_{tw}$, and $\mathbf{K}_{tw}$ are linear portions of the structural mass, damping, and stiffness matrices of the tower, respectively; $\mathbf{M}_a$ is the diagonal added mass matrix from the Morison equation; $\mathbf{1} = [1, 1 \ldots 1]^T$; $\dot{x}_g$ and $\ddot{x}_g$ are the ground velocity and acceleration, respectively; $\mathbf{K}_{nl}(\mathbf{x})$ is the diagonal nonlinear cable stiffness (one element in present study); $n$ denotes the node where the cable is connected to the tower; $\mathbf{C}_h$ is the diagonal drag matrix and $\mathbf{V}$ is the steady current; $\rho$ is the density of seawater; $\mathbf{V}_e$ and $\mathbf{A}_e$ are the equivalent inertial volume and projected area of the tower, respectively; $C_a$ and $C_d$ are added mass and the drag coefficients, respectively; $c_1$ and $c_2$ are constants that determine the cable nonlinearities.

Applying statistical linearization to every nonlinear element in (9.122), the element $(\mathbf{V} - \dot{\boldsymbol{\beta}})|\mathbf{V} - \dot{\boldsymbol{\beta}}|$ is replaced by $\mathbf{a} + \mathbf{B}\dot{\boldsymbol{\beta}}$ and $\mathbf{K}_{nl}(\mathbf{x})$ by $\mathbf{h} + \mathbf{D}\bar{\mathbf{x}}$, where $\bar{\mathbf{x}} = \mathbf{x} - \mathbf{x}_0$ is a zero-mean random process, $\mathbf{x}_0 = E[\mathbf{x}]$ and $\dot{\boldsymbol{\beta}} = \dot{\mathbf{x}} + \mathbf{1}\dot{x}_g$ is the absolute velocity of the structure.

The elements of the linearization vectors $\mathbf{a} = [a_1, \ldots, a_n]^T$, $\mathbf{h} = [h_1, \ldots, h_n]^T$ and matrices $\mathbf{B} = [b_{ij}]$, $\mathbf{D} = [d_{ij}], i, j = 1, ..., n$ can be found from the relations [11]

$$a_k = E[|V_k - \dot{\beta}_k|(V_k - \dot{\beta}_k)] \quad k = 1, \ldots, n \,, \, \tag{9.124}$$

$$b_{kk} = \frac{E[|V_k - \dot{\beta}_k|(V_k - \dot{\beta}_k)\dot{\beta}_k]}{E[\dot{\beta}_k^2]} \,, \quad k = 1, \ldots, n \,, \tag{9.125}$$

$$h_n = E[K_{nl,nn}(x_n)x_n] \,, \tag{9.126}$$

$$d_{nn} = \frac{E[K_{nl,nn}(x_n)\bar{x}_n]}{E[\bar{x}_n^2]} \,, \tag{9.127}$$

where $n$ is the node of the cable attachment to the tower.

## 9.5 Vibrations of Structures Under Wind Excitations

Besides the earthquakes, winds effects on structures are the most important sources of uncertainty in predicting the response of structures. Due to the random nature of wind-induced load, a probabilistic approach based on random vibration analysis was proposed in the literature [19, 18, 69, 71]. Recently, the problem of the determination probabilistic characteristics of wind, stochastic models of wind, and responses of structures to wind excitation were presented in Nigam and Narayanan's book [60]. Usually the models of structures are assumed to be linear. However, if we want to reduce undesired vibrations caused by wind excitation, we use some additional dampers and then nonlinear models of considered structures appear. In this case, in response analysis, statistical linearization is used. We review some recent results in this field shortly.

Xu et al. [92] and Xu et al. [91] considered along- and cross-wind response of tall/slender structures, respectively. The structure is modeled as a lumped mass multi-degree-of-freedom system taking into account both bending and shear. The wind excitation is modeled as a stochastic process, which is stationary in time and nonhomogeneous in space. To reduce the vibration of structure, three types of passive damping devices connected to the $n$th mass are considered, namely tuned mass dampers (TMD), tuned liquid column damper (TLCD) and tuned liquid column/mass damper (TLCMD). When a damper is connected to the $n$th mass of the main structure its governing equations of motions are given below

### 9.5.1 Tuned Mass Damper (TMD)

$$M_d \ddot{y}_d + C_d \dot{Z} + K_d Z = 0 , \tag{9.128}$$

$$y_d = Z + y_n , \tag{9.129}$$

where $M_d, K_d, C_d$, and $y_d$ are the mass, spring, damping parameters, and absolute displacement of the damper, respectively; $Z$ is the relative displacement of the damper with respect to the main structure; $y_n$ is the absolute displacement of the $n$th mass of the main structure.

For the main structure, the resultant external force, $P_{nt}$ applied to the $n$th mass is determined by

$$P_{nt} = P_n + C_d \dot{Z} + K_d Z , \tag{9.130}$$

where $P_n$ is the wind force acting at the $n$th mass.

### 9.5.2 Tuned Liquid Column Damper (TLCD)

$$y_n^+ = y_n^- = y_n, \quad \phi_n^+ = \phi_n^-, \quad M_n^+ = M_n^- , \tag{9.131}$$

$$Q_n^+ = Q_n^- - (\rho AL + m_n)\ddot{y}_n - \beta_n \dot{y}_n - \rho AB\ddot{x} + P_n \,, \tag{9.132}$$

$$\rho AL\ddot{x} + \frac{1}{2}p\rho A \,|\, \dot{x} \,|\, \dot{x} + 2\rho Agx = \rho AB\ddot{y}_n \,, \tag{9.133}$$

where $y_n^+, \phi_n^+, M_n^+$, and $Q_n^+$ are the displacement, angular displacement, bending moment, and shear force at the top end of the $n$th mass, respectively; $y_n^-, \phi_n^- M_n^-$, and $Q_n^-$ are the same quantities at the bottom end of the $n$th mass, respectively; $m_n, \beta_n, P_n$ are the mass, damping coefficients and the wind force acting at the $n$th mass, respectively. $\rho, A, B, L$, and $p$ are constant parameters; $x$ is the elevation of the liquid; $g$ is the gravitational constant.

Since only the third equation is nonlinear, the corresponding linearized equation has the form

$$\rho AL\ddot{x} + \frac{1}{2}p\rho AC_p\dot{x} + 2\rho Agx = \rho AB\ddot{y}_n \,, \tag{9.134}$$

where $C_p = \sqrt{\frac{E[\dot{x}^2]}{2\pi}}$ is the linearization coefficient.

### 9.5.3 Tuned Liquid Column/Mass Damper (TLCMD)

$$\rho AL\ddot{x} + \frac{1}{2}p\rho A \,|\, \dot{x} \,|\, \dot{x} + 2\rho Agx = \rho AB\ddot{y}_d \,, \tag{9.135}$$

$$\rho AB\ddot{x} + (\rho AL + M_d)\ddot{y}_d + C_d\dot{Z} + K_dZ = 0 \,, \tag{9.136}$$

$$y_d = Z + y_n \,, \tag{9.137}$$

where $M_d$ and $y_d$ are the container mass exclusive of liquid column mass and the absolute displacement of the container, respectively; other parameters are the same as before.

To obtain response characteristics, Xu et al. [92] and Xu et al. [91] used statistical linearization with mean-square criterion for displacements (standard equivalent linearization).

## 9.6 Applications in Control Problems

Stochastic linearization was also used as mathematical tool in control theory. For the first time, it was done by Wonham and Cashman [90], who combined statistical linearization and LQG theory to the determination of a quasioptimal control for a nonlinear stochastic system. This idea was later developed by many authors, for instance, Beaman [8], Kazakov [41], and Yoshida [96]. In these papers, systems with nonlinear plants, linear actuators, and nonlinear criteria were studied and the linearization coefficients were determined from the mean-square criterion. The quasioptimal control was obtained from an iterative procedure, where the linearization coefficients were derived using the mean-square linearization criterion and the optimal control for linear stochastic system with the mean-square control criterion was calculated. In what follows, we present this approach and we show a comparison between the applicability of statistical linearization methods with moment criteria and criteria in probability density space to the determination of quasioptimal external control for a nonlinear dynamic system excited by a colored Gaussian noise and the mean-square criterion. Also, a comparison between the applicability of statistical and equivalent linearization methods with criteria in probability density space to the determination of quasioptimal control proposed by Socha and Błachuta [73] will be presented.

## 9.6.1 Linear Quadratic Gaussian Problems

Consider the following optimal control problem. The dynamic system is described by the following vector Ito stochastic differential equation:

$$dx(t) = [\mathbf{Ax(t)} + \mathbf{Bu(t)}]dt + \sum_{k=1}^{M} \mathbf{G_k} d\xi_k(t), \quad \mathbf{x(t_0)} = \mathbf{x_0}, \qquad (9.138)$$

where $\mathbf{x}(t)$ is the state vector $\mathbf{x} \in \mathbf{R}^n$, $\mathbf{u}(t)$ is the control vector $\mathbf{u} \in \mathbf{R}^m$, $\mathbf{G}_k$ are constant vectors of intensity of noise, $\mathbf{G}_k \in \mathbf{R}^n$, $\xi_k$, $k = 1, \dots, M$, are correlated standard Wiener processes with covariance matrix $\mathbf{Q}_\xi$; the initial condition $\mathbf{x}_0$ is a vector random variable independent of $\xi_k$, $k = 1, \dots, M$; $\mathbf{A}$, $\mathbf{B}$, and $\mathbf{Q}_\xi$ are constant matrices of $n \times n$, $n \times m$, and $M \times M$ dimensions, respectively.

The mean-square criterion is defined by

$$I = \lim_{t_f \to \infty} E \left\{ \frac{1}{t_f - t_0} \int_{t_0}^{t_f} \left[ \mathbf{x}^T(t)\mathbf{Qx}(t) + 2\mathbf{x}^T(t)\mathbf{Nu}(t) + \mathbf{u}^T(t)\mathbf{Ru}(t) \right] dt \right\},$$
$$(9.139)$$

where $t_f$ is a finite moment of control process, $\mathbf{Q}$, $\mathbf{N}$, and $\mathbf{R}$ are matrices of $n \times n$, $n \times m$, and $m \times m$ dimensions, respectively. $\mathbf{Q} \geq \mathbf{0}$ and $\mathbf{R} > \mathbf{0}$ are symmetric $(\mathbf{A}, \mathbf{B})$ is stabilizable and $(\mathbf{A}, \mathbf{Q})$ is detectable [52]. It is assumed that the state vector $\mathbf{x}$ is complete measurable.

Using results of LQG theory, the optimal control [52] is determined by

$$\mathbf{u}(t) = -\mathbf{Kx}(t), \qquad (9.140)$$

where the control gain matrix $\mathbf{K}$ is constant and given by

$$\mathbf{K} = \mathbf{R}^{-1}(\mathbf{N}^T + \mathbf{B}^T \mathbf{P}), \qquad (9.141)$$

where $\mathbf{P}$ is a symmetric, positive-definite solution of the algebraic Riccati equation

$$\mathbf{PA}_N + \mathbf{A}_N^T \mathbf{P} - \mathbf{PBR}^{-1}\mathbf{B}^T\mathbf{P} + \mathbf{Q}_N = \mathbf{0} , \qquad (9.142)$$

where

$$\mathbf{A}_N = \mathbf{A} - \mathbf{BR}^{-1}\mathbf{N}^T \leq \mathbf{0}, \quad \mathbf{Q}_N = \mathbf{Q} - \mathbf{NR}^{-1}\mathbf{N}^T \geq \mathbf{0} . \qquad (9.143)$$

The mean-square criterion for optimal control is determined by

$$I = \text{tr}(\mathbf{PGQ}_\xi \mathbf{G}^T) \qquad (9.144)$$

or alternatively can be calculated from the algebraic Lyapunov equation

$$(\mathbf{A} - \mathbf{BK})\mathbf{V} + \mathbf{V}(\mathbf{A} - \mathbf{BK}) + \mathbf{GQ}_\xi \mathbf{G}^T = \mathbf{0} , \qquad (9.145)$$

where $\mathbf{G} = [\mathbf{G}_1, \ldots, \mathbf{G}_M]$, $\mathbf{V}$ is the covariance matrix of the vector state $\mathbf{x}$, i.e.,

$$\mathbf{V}(t) = E[\mathbf{x}(t)\mathbf{x}^T(t)] \qquad (9.146)$$

### 9.6.2 Algorithms for Quasioptimal Control Problems for Nonlinear System

Consider the following optimal control problem. The nonlinear stochastic system is described by

$$dx(t) = [\mathbf{Ax}(t) + \mathbf{\Phi}(\mathbf{x}) + \mathbf{Bu}(t)]dt + \sum_{k=1}^{M} \mathbf{G}_k d\xi_k, \quad \mathbf{x}(t_0) = \mathbf{x}_0 , \quad (9.147)$$

where $\mathbf{x} \in \mathbf{R}^n$ and $\mathbf{u} \in \mathbf{R}^m$ are the state vector and the control vector, respectively. $\mathbf{A}$ and $\mathbf{B}$ are time-invariant matrices of appropriate dimensions, $\mathbf{\Phi} = [\Phi_1, \ldots, \Phi_n]^T$ is a vector nonlinear function such that $\mathbf{\Phi}(\mathbf{0}) = \mathbf{0}$, $\mathbf{G}_k$ are time- invariant deterministic vectors, $\xi_k$, $k = 1, \ldots, M$, are correlated standard Wiener processes with covariance matrix $\mathbf{Q}_\xi$; the initial condition $\mathbf{x}_0$ is a vector random variable independent of $\xi_k$, $k = 1, \ldots, M$. We assume that the unique solution of (9.147) exists. The control strategy is designed to minimize the criterion (9.139). This problem is called nonlinear quadratic Gaussian problem NLQG.

We assume that the nonlinear vector $\mathbf{\Phi}(\mathbf{x})$ can be substituted by a linearized form

$$\mathbf{\Phi}(\mathbf{x}) \cong \mathbf{A_e x} , \qquad (9.148)$$

where $\mathbf{A_e}$ is an $n \times n$ matrix of linearization coefficients such that $(\mathbf{A} + \mathbf{A_e}, \mathbf{B})$ is stabilizable and $(\mathbf{A} + \mathbf{A_e}, \mathbf{Q})$ is detectable [52].

Then the optimal control for the linearized system

$$dx_L(t) = [(\mathbf{A} + \mathbf{A_e})\mathbf{x}_L(t) + \mathbf{Bu}(t)]dt + \sum_{k=1}^{M} \mathbf{G}_k d\xi_k, \quad \mathbf{x}_L(t_0) = \mathbf{x}_0 \quad (9.149)$$

where $\mathbf{x}_L(t)$ is the state vector of the linearized system, can be found by the LQG standard method presented in previous subsection in the linear feedback form (9.140), i.e.,

$$\mathbf{u} = -\mathbf{K}\mathbf{x}_L , \tag{9.150}$$

where the gain matrix $\mathbf{K}$ is equal to

$$\mathbf{K} = \mathbf{R}^{-1}(\mathbf{N}^T + \mathbf{B}^T\mathbf{P}) , \tag{9.151}$$

and $\mathbf{P}$ is a positive solution of the algebraic Riccati equation

$$\mathbf{P}(\mathbf{A}_N + \mathbf{A}_e) + (\mathbf{A}_N + \mathbf{A}_e)^T\mathbf{P} - \mathbf{P}\mathbf{B}\mathbf{R}^{-1}\mathbf{B}^T\mathbf{P} + \mathbf{Q}_N = \mathbf{0} , \tag{9.152}$$

where

$$\mathbf{A}_N + \mathbf{A}_e = \mathbf{A} + \mathbf{A}_e - \mathbf{B}\mathbf{R}^{-1}\mathbf{N}^T \leq 0, \quad \mathbf{Q}_N = \mathbf{Q} - \mathbf{N}\mathbf{R}^{-1}\mathbf{N}^T \geq 0 . \tag{9.153}$$

Substituting (9.150) into (9.149) we find

$$d\mathbf{x}_L(t) = [(\mathbf{A} + \mathbf{A_e} - \mathbf{BK})\mathbf{x}_L(t)]dt + \sum_{k=1}^{M} \mathbf{G}_k d\xi_k, \quad \mathbf{x}_L(t_0) = \mathbf{x}_0 . \tag{9.154}$$

The corresponding covariance matrix equation has the form

$$(\mathbf{A} + \mathbf{A_e} - \mathbf{BK})\mathbf{V}_L + \mathbf{V}_L(\mathbf{A} + \mathbf{A_e} - \mathbf{BK})^T + \sum_{k=1}^{M} \mathbf{G}_k\mathbf{Q}_\xi\mathbf{G}_k^T = \mathbf{0}$$

$$\tag{9.155}$$

and the criterion is equal

$$I_L = E[\mathbf{x}_L^T(\mathbf{Q} + \mathbf{K}^T\mathbf{R}\mathbf{K})\mathbf{x}_L] = tr[(\mathbf{Q} + \mathbf{K}^T\mathbf{R}\mathbf{K})\mathbf{V}_L] , \tag{9.156}$$

where the subindex $L$ corresponds to the linearized problem, "$tr$" denotes the trace of matrix and

$$\mathbf{V}_L = E[\mathbf{x}_L\mathbf{x}_L^T] . \tag{9.157}$$

To obtain the quasioptimal control usually the authors have used the mean-square criterion in statistical linearization. Since in the vibration analysis of stochastic systems in mechanical and structural engineering many criteria of linearization were considered, and we show a comparison of application of two groups of criteria, namely moment criteria and criteria in probability density functions space.

**Moment Criteria**

We consider four criteria of statistical linearization: the mean-square error of displacement, equality of the second-order moments of nonlinear and linearized elements, the mean-square error of potential energy of displacement, and true linearization method.

In statistical linearization the elements of the nonlinear vector have to be replaced by corresponding equivalent elements "in the sense of a given criterion" in a linear form. The following criteria for a scalar nonlinear function $\varphi(x)$ are considered.

**Criterion of Linearization** 1 − *cont.* Mean-square error of displacements.

$$E\left[(c_1 x - \varphi(x))^2\right] \to \min . \tag{9.158}$$

**Criterion of Linearization** 2 − *cont.* Equality of the second-order moments of nonlinear and linearized elements.

$$E\left[(c_2 x)^2\right] = E[(\varphi(x))^2] . \tag{9.159}$$

**Criterion of Linearization** 3 − *cont.* Mean-square error of potential energies.

$$E\left[\left(\int_0^x (c_3 x - \varphi(x))dx\right)^2\right] \to \min . \tag{9.160}$$

where $c_i, i = 1, 2, 3$ are linearization coefficients.

In true linearization [51], the linearization coefficients are determined from the equality of covariance matrices of responses of nonlinear and linearized systems to random excitations. The main idea of this approach for multidimensional system can be formulated as a next criterion.

**Criterion of Linearization** 4 − *cont.* True linearization.

$$E\left[\mathbf{x}_N \mathbf{x}_N^T\right] = E[\mathbf{x}_L \mathbf{x}_L^T] , \tag{9.161}$$

where $\mathbf{x}_N$ and $\mathbf{x}_L$ are stationary solutions of nonlinear and linearized dynamic systems, respectively.

In the application of linear feedback gain obtained for linearized system to the nonlinear system, the state equation and the corresponding mean-square control criterion have the form

$$d\mathbf{x}_N(t) = [(\mathbf{A} - \mathbf{B}\mathbf{K})\mathbf{x}_N(t) + \mathbf{\Phi}(\mathbf{x}_N(t))]dt + \sum_{k=1}^{M} \mathbf{G}_k d\xi_k, \quad \mathbf{x}_N(t_0) = \mathbf{x}_0 , \tag{9.162}$$

$$I_N = E[\mathbf{x}_N^T(\mathbf{Q} + \mathbf{K}^T\mathbf{R}\mathbf{K})\mathbf{x}_N] = tr[(\mathbf{Q} + \mathbf{K}^T\mathbf{R}\mathbf{K})\mathbf{V}_N] , \tag{9.163}$$

where the subindex $N$ corresponds to the original nonlinear problem and

$$\mathbf{V}_N = E[\mathbf{x}_N \mathbf{x}_N^T] . \qquad (9.164)$$

In a general case, the covariance matrix $\mathbf{V}_N$ can be found approximately. To obtain the linearization matrix $\mathbf{A}_e$ and quasioptimal control, one of the four proposed linearization criteria in an iterative procedure can be applied. For the first three criteria, the modified procedure given in [96] can be used

**Procedure** SL–CONT

*Step 1:* First put $\mathbf{A}_e = 0$ in (9.149) and then solve (9.152) and (9.155). The solutions of (9.152) and (9.155) are $\mathbf{P}$ and $\mathbf{V}_L$, respectively.

*Step 2:* Substitute $\mathbf{P}$ obtained in Step 1 into (9.151) and find $\mathbf{K}$. Next, substitute $\mathbf{K}$ into (9.150) and (9.156) find $\mathbf{u}$ and $I_L$, respectively.

*Step 3:* Find linearization coefficients from one of criteria of linearization: Criteria of linearization $1 - cont - 3 - cont$ to create the new linearization matrix $\mathbf{A}_e = [\mathbf{A}_{eij}]$.

*Step 4:* If the accuracy is greater than a given parameter $\varepsilon_1$, then repeat solving equations (9.152) and (9.155) and Steps 2–4 until $\mathbf{V}_L$ and $\mathbf{P}$ converge.

In the case of true linearization, the determination of linearization coefficients from condition (9.161) for multidimensional systems is not always unique and additional conditions are required, for instance, the equality of even higher-order linearized systems, i.e.,

$$E\left[\mathbf{x}_N^{[p]}\right] = E\left[\mathbf{x}_L^{[p]}\right] , \qquad (9.165)$$

where $\mathbf{x}^{[p]}$ denotes $p$th forms of the of the components of vector $\mathbf{x}$, i.e.,

$$\mathbf{x}^{[p]} = x_1^{p_1} x_2^{p_2} \dots x_n^{p_n} , \quad \sum_{j=1}^{n} p_j = p , \quad p = 4, 6, \dots, 2k \qquad (9.166)$$

or to condition (9.161) could be added the following one:

$$I_A = \min_{\mathbf{A}_{eij}} \left| E\left[\mathbf{x}_N^{[p]}(\mathbf{A}_e)\right] - E\left[\mathbf{x}_L^{[p]}(\mathbf{A}_e)\right] \right| . \qquad (9.167)$$

The following procedure for Criterion of linearization $4 - cont$ is a modified version of a standard one given in [96].

**Procedure** TRU–CONT

*Step 1:* First put $\mathbf{A}_e = 0$ in (9.149) and then solve (9.152) and (9.155). The solutions of (9.152) and (9.155) are $\mathbf{P}$ and $\mathbf{V}_L$, respectively.

*Step 2:* Substitute $\mathbf{P}$ obtained in Step 1 into (9.151) and find $\mathbf{K}$. Next, substitute $\mathbf{K}$ into (9.150) and (9.156) find $\mathbf{u}$ and $I_L$, respectively.

*Step 3:* Substitute **K** obtained in Step 2 into (9.163) and (9.164) and then find exactly or approximately $I_N$ and $V_N$ .

*Step 4:* Find linearization coefficients from Criterion of linearization 4 – *cont* and if necessary from additional conditions (9.165) or apply a minimization procedure with respect to coefficients $\mathbf{A}_e = [\mathbf{A}_{eij}]$.

*Step 5:* If the accuracy is greater than a given parameter $\varepsilon_1$, then repeat solving (9.152) and (9.155) and Steps 2–4 until $\mathbf{V}_L$ and **P** converge.

From a numerical point of view, because of the weak convergence of both the procedures, sometimes it convenient to split the determination of linearization coefficients from other steps of the procedure. For example, the modified version of Procedure SL–CONT denoted by SL–CONTa has the form

**Procedure** SL–CONTa

*Step 1:* Select $\mathbf{A}_e = 0$ in (9.149) and then solve (9.152). The solution of (9.152) is **P**.

*Step 2:* Substitute **P** obtained in Step 1 into (9.151) and find the matrix **K**. Next, substitute **K** and $\mathbf{A}_e = 0$ into (9.155) and solve the equation. The solution of (9.155) is $\mathbf{V}_L$.

*Step 3:* For each nonlinear element, find the linearization coefficient which minimize the selected criterion (9.158) or (9.159) or (9.160) and create the matrix of linearization coefficients $\mathbf{A}_e(\mathbf{V}_L)$.

*Step 4:* Substitute the matrix of linearization coefficients $\mathbf{A}_e(\mathbf{V}_L)$ obtained in Step 3 into (9.155) and then solve the equation.

*Step 5:* If the error is greater than a given parameter $\varepsilon_1$, then repeat Steps 3 and 4 until $\mathbf{V}_L$ converges.

*Step 6:* Substitute the matrix of linearization coefficients $\mathbf{A}_e(\mathbf{V}_L)$ into Riccati equation (9.152) and then solve the equation.

*Step 7:* Substitute the matrix **P** obtained in Step 6 into the covariance equation (9.155) and then solve the equation.

*Step 8:* If the error is greater than a given $\varepsilon_2$, then repeat Steps 3–7 until $\mathbf{V}_L$ and **P** converge.

*Step 9:* Calculate criteria $I_L$ and $I_N$ given by (9.156) and (9.163), respectively.

We illustrate the main results of this subsection by two examples.

*Example 9.1.* [75] (one dimensional) Consider the nonlinear scalar stochastic system

$$dx(t) = [-ax(t) - \alpha x^3(t) + bu(t)]dt + gd\xi(t), \quad x(t_0) = x_0 , \quad (9.168)$$

where $a, \alpha, b$, and $g$ are positive constants, $\xi(t)$ is the standard Wiener process, and the mean-square criterion is

$$I = E[q\bar{x}^2 + r\bar{u}^2] , \quad (9.169)$$

where $\bar{x}$ is the stationary solution of (9.168) and $\bar{u}$ is the corresponding stationary control, and $q$ and $r$ are positive constants. The linearized system has the following form:

$$dx_L(t) = [-ax_L(t) - \alpha k x_L(t) + bu(t)]dt + gd\xi(t) , \qquad (9.170)$$

where $k$ is a linearization coefficient. The optimal control for linearized system (9.170) has the feedback form

$$\bar{u} = -\frac{Pb}{r}\bar{x}_L , \qquad (9.171)$$

where $\bar{x}_L$ is the stationary solution of (9.170) and $\bar{u}$ is the corresponding stationary control, and $P$ is a scalar positive solution of the algebraic Riccati equation

$$\beta P^2 + 2\gamma P - q = 0 , \qquad (9.172)$$

where $\beta = b^2/r$ and $\gamma = a + \alpha k$ . Hence

$$P = \frac{1}{\beta}(-\gamma + \sqrt{\gamma^2 + q\beta}) . \qquad (9.173)$$

The optimal value of the criterion for linearized system is equal to

$$I = E[q\bar{x}_L^2 + r\bar{u}^2] = (q + \beta P^2)E[\bar{x}_L^2] . \qquad (9.174)$$

The second moment can be found from the Lyapunov equation

$$-2(a + \alpha k + \beta P)E[\bar{x}_L^2] = 0 . \qquad (9.175)$$

Hence

$$E[\bar{x}_L^2] = \frac{g^2}{2(a + \alpha k + \beta P)} = \frac{g^2}{2\sqrt{(a + \alpha k)^2 + q\beta}} . \qquad (9.176)$$

If we apply the obtained linear feedback control to nonlinear system, we obtain the state equation and the corresponding criterion

$$dx_N(t) = [-ax_N(t) - \alpha x_N^3(t) - \beta P x_N(t)]dt + gd\xi(t) , \qquad (9.177)$$

$$I_{N_{opt}} = (q + \beta P^2)E[\bar{x}_N^2] , \qquad (9.178)$$

where the second-order moment $E[\bar{x}_N^2]$ can be found in the analytical form

$$E[\bar{x}_N^2] = \int_{-\infty}^{+\infty} \bar{x}_N^2 g_N(\bar{x}_N)d\bar{x}_N , \qquad (9.179)$$

where

$$g_N(\bar{x}_N) = \frac{1}{c_N} exp\left\{-\frac{2}{g^2}(a + \beta P)\bar{x}_N^2 + \alpha \frac{\bar{x}_N^4}{2}\right\} , \tag{9.180}$$

$c_N$ is a normalized constant. One can show [39, 23] that linearization coefficients for Criteria of linearization $1 - cont$, $2 - cont$, and $3 - cont$ have the form

$$k_1 = 3E[\bar{x}_L^2], \quad k_2 = \sqrt{15}E[\bar{x}_L^2], \quad k_3 = 2.5E[\bar{x}_L^2] , \tag{9.181}$$

respectively. In the case of Criterion of linearization $4 - cont$, it can be found from condition $E[\bar{x}_N^2] = E[\bar{x}_L^2]$, where $E[\bar{x}_N^2]$ and $E[\bar{x}_L^2]$ are defined by (9.179) and (9.176), respectively.

To illustrate the obtained results, a comparison of the criterion $I_{N_{opt}}$ defined by (9.178) and relative errors based on criteria defined by (9.178) and (9.169) for different linearization methods versus parameters of system (9.168) has been shown. In this comparison, four criteria of linearization, namely, mean-square error of displacements, equality of second-order moments, mean-square error of potential energies, and true linearization were considered. The numerical results denoted by circles, diamond, cross lines, and dots, respectively, are presented in Figs. 9.6–9.8. The direct comparison of the criterion $I_{N_{opt}}$ versus parameter $\alpha$ has been shown in Fig. 9.6. Since for the scalar system the optimal feedback control and the corresponding mean-square criterion in exact forms do not exist and they exist only for the linear feedback control, we use as a measure of accuracy the relative error defined by

$$\Delta_{opt} = \Delta_{opt}(par) = \frac{|I_{N_{opt}}(par) - I_{lin}(par)|}{I_{lin}(par)} , \tag{9.182}$$

where $I_{N_{opt}}(par)$ and $I_{lin}(par)$ are criteria for nonlinear system with linear feedback and linear system with linear feedback, respectively. The argument "$par$" denotes a parameter of system (9.168) or criterion (9.169). The relative errors are presented in Figs. 9.7 and 9.8.

*Example 9.2.* [75] (Duffing oscillator)

Consider the Duffing oscillator described by

$$dx_1(t) = x_2(t)dt, \quad x_1(t_0) = x_{10},$$
$$dx_2(t) = [-\lambda_0^2 x_1(t) - 2hx_2(t) - \alpha x_1^3(t)$$
$$+bu(t)]dt + gd\xi(t), x_2(t_0) = x_{20}, \tag{9.183}$$

where $\lambda_0, h, \alpha, b$, and $g$ are positive constants, $\xi(t)$ is the standard Wiener process, and the initial conditions $x_{10}$ and $x_{20}$ are independent random variables and independent of $\xi(t)$. The mean-square control criterion is

$$I = E[\bar{\mathbf{x}}^T \mathbf{Q} \bar{\mathbf{x}} + r\bar{u}^2] , \tag{9.184}$$

where $\bar{\mathbf{x}} = [x_1(+\infty) \ x_2(+\infty)]^T$ is the stationary vector state, $\bar{u} = u(+\infty)$ is the corresponding stationary control, $\mathbf{Q} = diag\{Q_i\}$; $Q_i$ and $r$ are positive constants.

The linearized system has the following form:

$$dx_1(t) = x_2(t)dt, \quad x_1(t_0) = x_{10} ,$$
$$dx_2(t) = [-\lambda_0^2 x_1(t) - 2hx_2(t) - \alpha k x_1(t)$$
$$+ bu(t)]dt + gd\xi(t), \quad x_2(t_0) = x_{20}, \tag{9.185}$$

where $k$ is a linearization coefficient. The coordinates of the stationary solutions of algebraic Riccati and covariance equations denoted by $\mathbf{P} = [p_{ij}]$ and $\mathbf{V}_L = [v_{L_{ij}}]$, respectively, for $i, j = 1, 2$, are the following:

$$p_{12} = \frac{1}{\beta}\left(-\gamma + \sqrt{\gamma^2 + Q_1\beta}\right), \quad p_{22} = \frac{1}{\beta}\left(2h + \sqrt{4h^2 + \beta(Q_2 + 2p_{12})}\right),$$
$$\tag{9.186}$$

$$p_{11} = 2hp_{12} + \gamma p_{22} + \beta p_{12}p_{22},$$

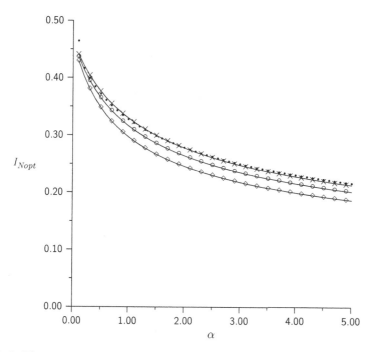

Fig. 9.6. The comparison of optimization criteria obtained by application of different statistical linearization methods versus parameter $\alpha$ with $a = 1$, $b = 1$, $g = 1$, $q = 1$, $r = 1$

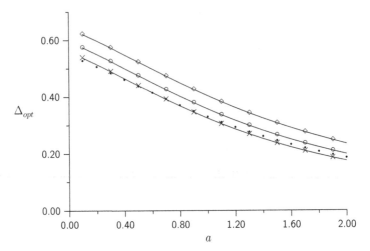

**Fig. 9.7.** The comparison of relative errors of optimal criteria obtained by application of different statistical linearization methods versus parameter $a$ with $\alpha = 1$, $b = 1$, $g = 1$, $q = 1$, $r = 1$

$$v_{L_{22}} = \frac{g^2}{2(2h + \beta p_{22})}, \quad v_{L_{12}} = 0, \quad v_{L_{11}} = \frac{v_{L_{22}}}{\gamma + \beta p_{12}}, \tag{9.187}$$

where $\beta = b^2/r$ and $\gamma = \lambda_0^2 + \alpha k$ . The optimal value of the criterion for linearized system is equal to

$$I_L = (Q_1 + \beta p_{12}^2)v_{L_{11}} + (Q_2 + \beta p_{22}^2)v_{L_{22}} . \tag{9.188}$$

If we apply the obtained linear feedback control to the nonlinear problem (9.183), (9.184) we obtain the following state equation and the corresponding criterion:

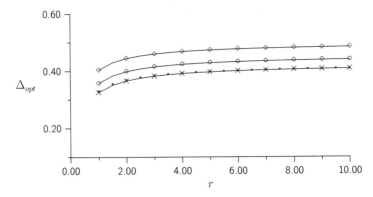

**Fig. 9.8.** The comparison of relative errors of optimal criteria obtained by application of different statistical linearization methods versus parameter $r$ with $a = 1$, $\alpha = 1$, $b = 1$, $g = 1$, $q = 1$

$$dx_1 = x_2 dt, \qquad x_1(t_0) = x_{10},$$

$$dx_2 = [-\lambda_0^2 x_1 - 2hx_2 - \alpha x_1^3 - \beta(x_1 p_{12} + x_2 p_{22})]dt$$

$$+gd\xi, \qquad x_2(t_0) = x_{20}, \tag{9.189}$$

$$I_{N_{opt}} = (Q_1 + \beta p_{12}^2)v_{N_{11}} + (Q_2 + \beta p_{22}^2)v_{N_{22}}, \tag{9.190}$$

where the second-order moments $v_{N_{11}}$ and $v_{N_{22}}$ can be found in the analytical form.

$$v_{N_{ii}} = \int_{-\infty}^{+\infty} \int_{-\infty}^{+\infty} x_i^2 g_N(x_1, x_2) dx_1 dx_2, \quad i = 1, 2, \tag{9.191}$$

where

$$g_N(x_1, x_2) = \frac{1}{c_N} \exp\left\{-\frac{2h + \beta p_{22}}{g^2}\left[(\lambda_0^2 + \beta p_{12})x_1^2 + \alpha\frac{x_1^4}{2} + x_2^2\right]\right\}, \tag{9.192}$$

where $c_N$ is a normalized constant. The linearization coefficients for for Criteria of linearization $1 - cont$, $2 - cont$, and $3 - cont$ have the form (9.181)

$$k_1 = 3E[x_1^2], \quad k_2 = \sqrt{15}E[x_1^2], \quad k_3 = 2.5E[x_1^2] \tag{9.193}$$

For Criterion of linearization $4 - cont$ it can be found from the equality $v_{N_{ii}} = v_{L_{ii}}$, where $v_{N_{ii}}$ and $v_{L_{ii}}$ are defined by (9.191) and (9.187), respectively.

To obtain the quasioptimal control for the Duffing oscillator the iterative procedures SL–CONT or SL–CONTa and TRU–CONT can be used. To illustrate the obtained results a comparison of the criterion $I_{N_{opt}}$ defined by (9.190) and criterion (9.184) for different linearization methods versus parameters of system (9.183) has been shown. In this comparison as for the one-dimensional system, four criterion of linearization, namely, mean-square error of displacements, equality of second-order moments, mean-square error of potential energies, and true linearization were considered. Also, the same characteristics and relative errors are calculated. The numerical results denoted by circles, diamond, cross lines, and dots, respectively, are presented in Figs. 9.9–9.11.

### Statistical Linearization with Criteria in Probability Density Functions Space

In this subsection we discuss an application of statistical and equivalent linearization methods linearization with criteria in probability density functions space to the determination of quasioptimal control for the nonlinear dynamic systems under external Gaussian excitations with the mean-square control criterion. It has been first considered in [73]. First we study statistical linearization. We assume that the elements of nonlinear vector $\mathbf{\Phi} = [\Phi_1, \ldots, \Phi_n]^T$ in (9.147) denoted by

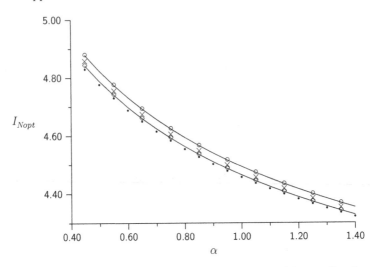

**Fig. 9.9.** The comparison of optimization criteria obtained by application of different statistical linearization methods versus parameter $\alpha$ with $\lambda_0^2 = 1$, $h = 0.05$, $b = 1$, $g = 1$, $Q_1 = Q_2 = 1$, $r = 100$

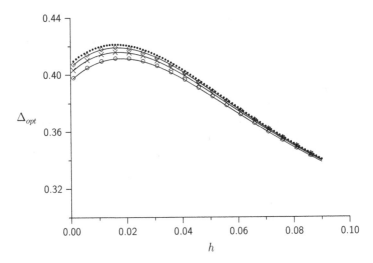

**Fig. 9.10.** The comparison of relative errors of optimal criteria obtained by application of different statistical linearization methods versus parameter $h$ with $\lambda_0^2 = 1$, $\alpha = 1$, $b = 1$, $g = 1$, $Q_1 = Q_2 = 1$, $r = 100$

$$Y_j = \Phi_j(x_j), \quad j = 1, \ldots n \qquad (9.194)$$

are replaced by linearized forms

$$Y_j = k_j x_j, \quad j = 1, \ldots, n . \qquad (9.195)$$

Two equivalence criteria in probability density functions space for scalar functions $\Phi_j(x_j)$ for $j = 1, \ldots, n$ are considered

Criterion of linearization $1 - SLPD$. Square probability metric

$$I_{1j} = \int_{-\infty}^{+\infty} (g_N(y_j) - g_L(y_j))^2 dy_j , \quad j = 1, \ldots, n , \tag{9.196}$$

where $g_N(y_j)$ and $g_L(y_j)$ are probability density functions of variables defined by (9.194) and (9.195), respectively.

Criterion of linearization $2 - SLPD$. Pseudomoment metric

$$I_{2j} = \int_{-\infty}^{+\infty} |y_j|^{2l} |g_N(y_j) - g_L(y_j)| dy_j , \quad j = 1, \ldots, n . \tag{9.197}$$

If we assume that the input processes are Gaussian processes with mean values $m_{x_j}$ for $j = 1, \ldots, n$ and the probability density functions

$$g_I(x_j) = \frac{1}{\sqrt{2\pi}\sigma_{x_j}} \exp\left\{-\frac{x_j^2}{2\sigma_{x_j}^2}\right\} , \tag{9.198}$$

where $\sigma_{x_j}^2 = E[x_j^2]$, then the output processes $Y_j$ for $j = 1, \ldots, n$ from the static linear elements defined by equality (9.195) are also Gaussian and their corresponding probability density functions are given by

$$g_I(x_j) = \frac{1}{\sqrt{2\pi}k_j\sigma_{x_j}} \exp\left\{-\frac{x_j^2}{2k_j^2\sigma_{x_j}^2}\right\} . \tag{9.199}$$

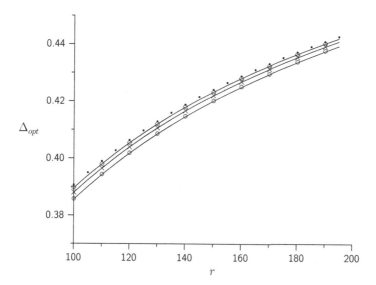

**Fig. 9.11.** The comparison of relative errors of optimal criteria obtained by application of different statistical linearization methods versus parameter $r$ with $\lambda_0^2 = 1$, $h = 0.05$, $\alpha = 1$, $b = 1$, $g = 1$, $Q_1 = Q_2 = 1$

To apply the proposed criteria (9.196) and (9.197), we have to find the probability density functions $g_N(y_j)$. This problem was discussed in Sect. 5.3. It was shown that except some special cases when the scalar functions are strictly monotonically increasing or decreasing, functions of the form given by (9.194) with continuous derivatives $\Phi'_j(x_j)$ for all $x_j \in R$ approximation methods have to be used.

In a general case, in order to obtain approximate probability density functions of scalar nonlinear random variables $Y_j = \phi_j(x_j)$, $j = 1, \ldots, n$, one can use, for instance, the Gram–Charlier expansion [63].

$$g_{Y_j}(y_j) = \frac{1}{\sqrt{2\pi}c_j\sigma_{y_j}} \exp\left\{-\frac{(y_j - m_{y_j})^2}{2\sigma_{y_j}^2}\right\} \left[1 + \sum_{\nu=3}^{N} \frac{c_{\nu j}}{\nu!} H_\nu\left(\frac{y_j - m_{y_j}}{\sigma_{y_j}}\right)\right],$$

(9.200)

where $m_{y_j} = E[y_j]$, $\sigma_{y_j}^2 = E[(y_j - m_{y_j})^2]$; $c_{\nu j} = E[G_\nu(y_j - m_{y_j})]$, $\nu = 3, 4, \ldots, N$, are quasi-moments, $c_j$ are normalized constants, and $H_\nu(x)$ and $G_\nu(x)$ are Hermite polynomials of one variable. For instance, a few first polynomials are given below

$$H_3(x) = \frac{1}{\sigma_x^6}(x^3 - 3\sigma_x x), \quad H_4(x) = \frac{1}{\sigma_x^8}(x^4 - 6\sigma_x^2 x^2 + 3\sigma_x^4),$$

$$H_5(x) = \frac{1}{\sigma_x^{10}}(x^5 - 10\sigma_x^2 x^3 + 15\sigma_x^4 x),$$

$$H_6(x) = \frac{1}{\sigma_x^{12}}(x^6 - 15\sigma_x^2 x^4 + 45\sigma_x^4 x^2 - 15\sigma_x^6).$$

(9.201)

In contrast to standard statistical linearization with criteria in state space, one cannot find expressions for linearization coefficients in an analytical form. However, in some particular cases, some analytical considerations can be done. For instance, for criterion $I_{1_j}$ defined by (9.196) and for an input zero-mean Gaussian process, the necessary condition of minimum can be derived in the following form for $j = 1, \ldots, n$:

$$\frac{\partial I_{1_j}}{\partial k_j} = 2\int_{-\infty}^{+\infty} (g_N(y_j) - g_L(y_j))\frac{1}{k_j}\left(1 - \frac{y_j^2}{k_j^2\sigma_{x_j}^2}\right)g_L(y_j)dy_j = 0. \quad (9.202)$$

**Equivalent Linearization with Criteria in the Probability Density Functions Space**

As was mentioned in Sect. 6, the objective of equivalent linearization with criterion in probability density space is to find for nonlinear dynamic system (9.147) for $\mathbf{B} = \mathbf{0}$ an equivalent linear dynamic system in the form

$$d\mathbf{x}(t) = \mathbf{A}\mathbf{x}dt + \sum_{k=1}^{M} \mathbf{G}_k d\xi_k(t), \quad \mathbf{x}(t_0) = \mathbf{x}_0, \quad (9.203)$$

where $A = [a_{ij}]$ is a matrix of constant linearization coefficients. As in the previous case, two criteria of linearization are proposed

Criterion of linearization $1 - ELPD$. Square probability metric

$$I_3 = \int_{-\infty}^{+\infty} \cdots \int_{-\infty}^{+\infty} (g_N(\mathbf{y}) - g_L(\mathbf{y}))^2 dy_1 \ldots dy_n . \tag{9.204}$$

Criterion of linearization $2 - ELPD$. Pseudomoment metric

$$I_4 = \int_{-\infty}^{+\infty} \cdots \int_{-\infty}^{+\infty} |\mathbf{y}|^{[2l]} |g_N(\mathbf{y}) - g_L(\mathbf{y})| dy_1 \ldots dy_n , \tag{9.205}$$

where $\mathbf{y} = [y_1, \ldots, y_n]^T$, $\mathbf{y}^{[2l]} = y_1^{q_1} y_2^{q_2} \cdots y_n^{q_n}$, $\sum_{i=1}^{n} q_i = 2l$, and $g_N(\mathbf{y})$ and $g_L(\mathbf{y})$ are probability density functions of solutions of nonlinear (9.147) for $\mathbf{B} = \mathbf{0}$ and linearized systems (9.203), respectively.

In the case of linearized system the probability density of the solution of system (9.203) is known and can be expressed as follows [63]:

$$g_L(\mathbf{y}) = \frac{1}{\sqrt{(2\pi)^n |\mathbf{K}_L|}} \exp\{-\frac{1}{2}\mathbf{y}^T \mathbf{K}_L^{-1} \mathbf{y}\} , \tag{9.206}$$

where $\mathbf{K}_L = \mathbf{K}_L(t) = E[\mathbf{x}(t)\mathbf{x}(t)^T]$ is the covariance matrix of the solution $\mathbf{x} = \mathbf{x}(t)$ of system (9.203), $|\mathbf{K}_L|$ denotes the determinant of the matrix $\mathbf{K}_L$. The matrix $\mathbf{K}_L$ satisfies the following equation:

$$\frac{d\mathbf{K}_L(t)}{dt} = \mathbf{K}_L \mathbf{A}^T + \mathbf{A}\mathbf{K}_L + \sum_{k=1}^{M} \mathbf{G}_k \mathbf{G}_k^T . \tag{9.207}$$

To apply the proposed criteria we have to find the probability density $g_N(\mathbf{x})$. Unfortunately, except some special cases, it is impossible to find the function $g_N(\mathbf{x})$ in an analytical form. As it was discussed in Sect. 6.3 the approximate probability density function of the stationary solution of the nonlinear dynamic system can be found, for instance, by Gram–Charlier expansion [63].

## Iterative Procedures

The difference between both the approaches, statistical, and equivalent linearization methods with criteria in probability density space, implies the difference in iterative procedures for the determination of the quasioptimal controls for the NLQG problem. In the case of statistical linearization, the determination of linearization coefficients requires two loops in an iterative procedure, while in the case of equivalent linearization the direct minimization of the linearization criterion can be applied and the only one iteration is used. The following proposed procedures uses Criteria of linearization $1 - SLPD$

and $2 - SLPD$ for statistical linearization and $1 - ELPD$ and $2 - ELPD$ for equivalent linearization, respectively.

**Procedure** PDF–ST–LIN–CONT (for statistical linearization)

*Step 1:* Select $\mathbf{A}_e = 0$ in (9.147) and then solve (9.152). The solution of (9.152) is $\mathbf{P}$.

*Step 2:* Substitute $\mathbf{P}$ obtained in Step 1 into (9.151) and find the matrix $\mathbf{K}$. Next, substitute $\mathbf{K}$ and $\mathbf{A}_e = \mathbf{0}$ into (9.155) and solve the equation. The solution of (9.155) is $\mathbf{V}_L$.

*Step 3:* For each nonlinear and linearized elements, calculate the corresponding probability density functions given by (5.109) and (9.199), respectively. If formula (5.109) cannot be applied, then instead of (5.109) we may use (9.200).

*Step 4:* For each nonlinear element find the linearization coefficient, which minimizes the selected criterion (9.196) or (9.197).

*Step 5:* Substitute the matrix of linearization coefficients $\mathbf{A}_e(\mathbf{V}_L)$ obtained in Step 4 into (9.155), and then solve the equation.

*Step 6:* If the error is greater than a given parameter $\varepsilon_1$, then repeat Steps 3–6 until $\mathbf{V}_L$ converges.

*Step 7:* Substitute the matrix of linearization coefficients $\mathbf{A}_e(\mathbf{V}_L)$ into Riccati equation (9.152), and then solve the equation.

*Step 8:* Substitute the matrix $\mathbf{P}$ obtained in Step 7 into covariance equation (9.155), and then solve the equation.

*Step 9:* If the error is greater than a given $\varepsilon_2$, then repeat Steps 3–9 until $\mathbf{V}_L$ and $\mathbf{P}$ converge.

*Step 10:* Calculate criteria $I_L$ and $I_N$ given by (9.156) and (9.163), respectively.

To determine the probability density function from (9.203), we have to solve the moment equations of higher order for the whole nonlinear dynamic system using one of the closure techniques and then to calculate higher-order moments for variables from the domain of nonlinear elements.

**Procedure** PDF–EQ–LIN–CONT (for equivalent linearization)

*Step 1: and 2:* From previous procedure.

*Step 3:* For $\mathbf{K}$ obtained in Step 2 calculate moment equations for the nonlinear system (9.162) and the covariance matrix $\mathbf{V}_L$ for the linearized system (9.154). Next, define the probability density functions of stationary solutions of the nonlinear and linearized systems.

*Step 4:* Calculate the matrix of linearization coefficients $\mathbf{A}_e(\mathbf{V}_L)$, which minimizes criterion (9.204) or (9.205).

*Step 5:* Substitute the matrix of linearization coefficients $\mathbf{A}_e(\mathbf{V}_L)$ obtained in Step 4 into Riccati equation (9.152), and then solve the equation.

*Step 6:* Substitute $\mathbf{P}$ obtained in Step 5 into (9.151) and find the matrix $\mathbf{K}$. Next, substitute $\mathbf{K}$ and $\mathbf{A}_e(\mathbf{V}_L)$ into equation (9.155) and solve the equation.

*Step 7:* For $\mathbf{K}$ obtained in Step 6 calculate moment equations for nonlinear system (9.162) and the covariance matrix $\mathbf{V}_L$ for the linearized system (9.149). Next, redefine the probability density functions of stationary solutions of nonlinear and linearized systems.

*Step 8:* If the error is greater than a given parameter $\varepsilon$, then repeat Steps 4–7 until $\mathbf{V}_L$ converges.

*Example 9.3.* (Duffing oscillator) Consider again the Duffing oscillator described in a modified form to (9.183)

$$dx_1(t) = x_2(t)dt, \qquad x_1(t_0) = x_{10} \,,$$
$$dx_2(t) = [-2hx_2(t) - f(x_1(t)) + bu(t)]\,dt + gd\xi(t), \quad x_2(t_0) = x_{20}, \ (9.208)$$

where $f(x_1)$ is a nonlinear function defined by

$$f(x_1) = Y_1 = \lambda_0^2 x_1 + \alpha x_1^3 \,, \tag{9.209}$$

where $\lambda_0, h, \alpha, b$, and $g$ are positive constants; $\xi(t)$ is the standard Wiener process and the mean-square criterion is

$$I = E[\bar{\mathbf{x}}^T \mathbf{Q}\bar{\mathbf{x}} + r\bar{u}^2] \,, \tag{9.210}$$

where $\bar{x} = [\bar{x}_1, \bar{x}_2]^T$ is the stationary solution of (9.208) $\mathbf{Q} = diag(Q_i), \ i = 1, 2; \ Q_i, r$ are positive constants.

The linearized system has the following form:

$$dx_1(t) = x_2(t)dt, \qquad x_1(t_0) = x_{10} \,,$$
$$dx_2(t) = [-2hx_2(t) - kx_1(t) + bu(t)]dt + gd\xi(t), \quad x_2(t_0) = x_{20}, \ (9.211)$$

where $k$ is a linearization coefficient. In this example, we will linearize directly $f(x_1)$, i.e., we replace $f(x_1)$ by $cx_1$. The reason of this modification is the following. If we denote by $k$ and $k_p$ the linearization coefficients for a criterion in probability density functions space for functions $f(x_1)$ and $\alpha x_1^3$, respectively, then one can show that $c \neq \lambda_0^2 + k_p$.

The coordinates of the solutions of algebraic Riccati and covariance equations denoted by $\mathbf{P} = [p_{ij}]$ and $\mathbf{V}_L = [v_{L_{ij}}]$, respectively, for $i, j = 1, 2$, are the following:

$$p_{12} = \frac{1}{\beta}(-c + \sqrt{c^2 + Q_1\beta}), \qquad p_{22} = \frac{1}{\beta}\left(2h + \sqrt{4h^2 + \beta(Q_2 + 2p_{12})}\right) \,,$$
$$p_{11} = 2hp_{12} + cp_{22} + \beta p_{12}p_{22}, \tag{9.212}$$

and

$$v_{L_{22}} = \frac{g^2}{2(2h + \beta p_{22})}, \quad v_{L_{12}} = 0, \quad v_{L_{11}} = \frac{v_{22}}{\gamma + \beta p_{12}}, \quad (9.213)$$

where $\beta = b^2/r$. The optimal value of the criterion for linearized system is

$$I_L = (Q_1 + \beta p_{12}^2)v_{L_{11}} + (Q_2 + \beta p_{22}^2)v_{L_{22}} . \quad (9.214)$$

Applying the obtained linear feedback control to nonlinear system, we obtain the state equation and the corresponding criterion

$$dx_1(t) = x_2(t)dt, \quad x_1(t_0) = x_{10} ,$$
$$dx_2(t) = [-2hx_2(t) - \lambda_0^2 x_1(t) - \alpha x_1^3(t)$$
$$- \beta(x_1(t)p_{12} + x_2(t)p_{22})]dt + gd\xi(t) , x_2(t_0) = x_{20} \quad (9.215)$$

and

$$I_{N_{opt}} = (Q_1 + \beta p_{12}^2)v_{N_{11}} + (Q_2 + \beta p_{22}^2)v_{N_{22}} , \quad (9.216)$$

where the second-order moments $v_{N_{11}}$ and $v_{N_{22}}$ can be found in the analytical form by direct averaging, i.e.,

$$v_{N_{ii}} = \int_{-\infty}^{+\infty} \int_{-\infty}^{+\infty} x_i^2 g_N(x_1, x_2)dx_1 dx_2, i = 1, 2 , \quad (9.217)$$

where $g_N(x_1, x_2)$ is the probability density function of the stationary solution of equation (9.208)

$$g_N(x_1, x_2) = \frac{1}{c_N} exp \left\{ -\frac{2h + \beta p_{22}}{g^2} \left[ (\lambda_0^2 + \beta p_{12})x_1^2 + \alpha \frac{x_1^4}{2} + x_2^2 \right] \right\} \quad (9.218)$$

and $c_N$ is a normalized constant.

To obtain the linearization coefficient in the stationary case of statistical linearization, we find the probability density function of the output variable $y$ defined by (9.209)

$$g_Y(y) = \frac{1}{\sqrt{2\pi}\sigma_L} exp \left\{ -\frac{(v_1 + v_2)^2}{2\sigma_L^2} \right\} \frac{1}{6a\alpha} \left[ \frac{a+y}{v_1^2} + \frac{a-y}{v_2^2} \right] , \quad (9.219)$$

where

$$v_1 = \left( \frac{a+y}{2\alpha} \right)^{\frac{1}{3}}, \quad v_2 = \left( \frac{y-a}{2\alpha} \right)^{\frac{1}{3}}, \quad a = \sqrt{y^2 + 4\lambda_0^6/27\alpha} . \quad (9.220)$$

The probability density of the linearized variable,

$$Y_1 = kx_1 , \quad (9.221)$$

has the form

$$g_L(y) = \frac{1}{\sqrt{2\pi}\sigma_L} \exp\left\{-\frac{y^2}{2k^2\sigma_L^2}\right\} , \qquad (9.222)$$

where $\sigma_L^2 = E[x_1^2]$ is the variance of the input Gaussian variable.

The linearization coefficient can be found from the minimization of criterion (9.196) or (9.197). To obtain the quasioptimal control, use the iterative Procedure PDF–ST–LIN–CONT proposed in this section.

In the case of equivalent linearization, the probability density function of the stationary solution of system (9.208) with $f$ given by (9.209) is defined by (9.218), while the probability density function of the stationary solution of the linearized system (9.211) has the form

$$g_L(x_1, x_2) = \frac{4h\sqrt{c}}{c_L g^2} \exp\left\{-\frac{2h + \beta p_{22}}{g^2}\left[(c + \beta p_{12})x_1^2 + x_2^2\right]\right\} . \qquad (9.223)$$

To obtain the quasioptimal control the iterative procedures proposed in this section are used. To illustrate these methods, a comparison of criterion $I_{N_{opt}}$ defined by (9.216) and the relative error based on criteria (9.209) and (9.210) for five different linearization methods versus parameters of system (9.208) is considered. In this comparison, the following criteria are considered: mean-square error of the displacement, square metric and pseudomoment metric for statistical linearization in probability density functions space, and square metric and pseudomoment metric for equivalent linearization in probability density functions space. The numerical results denoted by lines with circles, stars, squares, triangles, and crosses, respectively are presented in Figs. 9.12 and 9.13. The parameters selected for calculations are $\lambda_0^2 = 1, \alpha = 1, b = 1, g = 1, Q_1 = Q_2 = 1$, and $r = 100$. The direct comparison of criterion $I_{N_{opt}}$ versus parameter $h$ has been shown in Fig. 9.12.

Since for the Duffing oscillator the optimal feedback control and the corresponding mean-square criterion in exact forms do not exist and they exist for the linear feedback control, we use as a measure accuracy the relative error defined by

$$\Delta_{opt} = \Delta_{opt}(par) = \frac{|I_{N_{opt}}(par) - I_{lin}(par)|}{I_{lin}(par)} , \qquad (9.224)$$

where $I_{N_{opt}}(par)$ and $I_{lin}(par)$ are the criteria for nonlinear system with linear feedback and linear system with linear feedback, respectively. The argument "$par$" denotes a parameter of system (9.208) or criterion (9.210). The relative errors are presented in Fig. 9.12 for par=$h$.

Numerical studies show that for a given mean-square criterion of minimization (9.210), there are no significant differences between the considered linearization methods. However, the relative errors obtained for linearization methods with criteria in probability density functions space are smaller than the corresponding ones for standard statistical linearization.

The conclusion that there are no significant differences between the considered linearization methods applied in control problem was confirmed by the

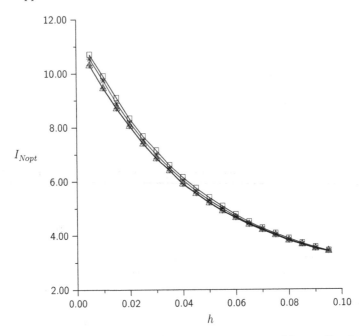

**Fig. 9.12.** The comparison of optimization criteria obtained by application of different linearization methods versus parameter $h$

author [78], where the multicriteria were used. The set of dominating points in the example of the Duffing oscillator for the two criteria $I_1$ and $I_2$ defined by mean-square error of displacement and mean-square error of energy for parameters $\lambda_0^2 = 1$, $h = 0.05$, $b = 1$, $\alpha = 1$, $g = 1$, $Q_1 = Q_2 = 1$, $r = 100$ is presented in Fig. 9.14.

The obtained numerical results show that the considered mean-square criteria $I_1$ and $I_2$ are linearly dependent and there are no dominating points.

## Bibliography Notes

### Ref.9.1.

One can find further applications of statistical and equivalent linearization in several papers. Grundmann et al. [32] have shown an application of an evolutionary algorithm in an iterative procedure with statistical linearization to the determination of response characteristics of linear system with one-dimensional model of hysteresis given by Suzuki and Minai [82]. For a 1-DOF system, the equations of motion and hysteresis have the form

$$\ddot{x} + 2h\dot{x} + \lambda_0^2[\alpha x + (1 - \alpha)x_f z] = F(t) , \qquad (9.225)$$

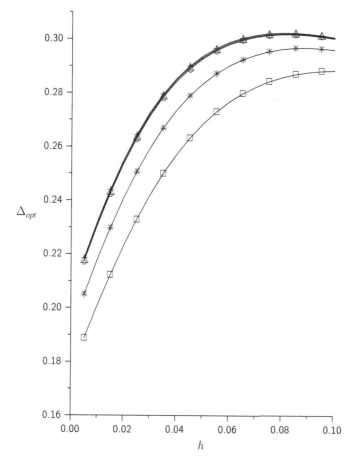

**Fig. 9.13.** The comparison of relative errors of optimal criteria obtained by application of different linearization methods versus parameter $h$

$$\dot{z} = \dot{x}/x_f[1 - H(\dot{x}/x_f)H(z - 1) - H(-\dot{x}/x_f)H(-z - 1)] , \qquad (9.226)$$

where $h, \lambda_0^2, \alpha$ and $x_f$ are positive constant parameters, $F(t)$ is a force at the elastoplastic spring, $H(.)$ is the Heaviside function.

A response analysis of a simple single-degree-of-freedom system with a hysteretic element subjected to narrowband seismic motion was considered by Silva and Ruiz [70]. The equation of motion and hysteretic element were described by the BBW model and the response characteristics were obtained by direct application of statistical linearization.

Basili and De Angelis [7] considered two adjacent structures each modeled as a SDOF system under ground excitation and connected by a hysteretic damper. Using the Bouc–Wen model statistically linearized, the authors proposed a method of the determination of optimal parameters of considered damper. This method is based on an optimization procedure with an energy

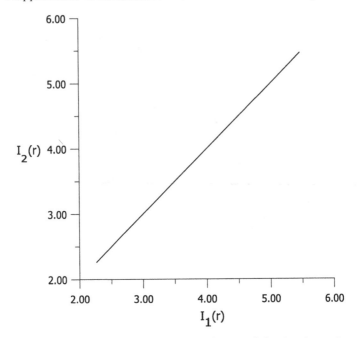

**Fig. 9.14.** A graphical illustration of the set of dominating points

criterion. Two typical external excitations were used: Gaussian white noise and Kanai–Tajimi filtered white noise.

Duval et al. [22] have applied the approach by Dobson [21] for obtaining the linearization coefficients in the response study of MDOF systems with hysteresis under Gaussian and non-Gaussian and zero or nonzero mean excitations by statistical linearization. Duval et al. [22] considered a 2-DOF system with two hysteresis that were linearized separately.

The vibration analysis for an oscillator with the Biot hysteretic element was presented by Spanos and Tsavachidis [77]. The authors considered the following integrodifferential equation of motion in dimensionless form:

$$\ddot{x}(t) + 4\pi^2(\dot{x}(t)\varepsilon x) + 8\pi\zeta \int_{t_0}^t E_1[h(t-\tau)]\dot{x}(\tau)d\tau = \sigma w(t) , \qquad (9.227)$$

where $\varepsilon, \zeta$ and $\sigma$ are positive constant parameters, $h(t)$ is the impulse transfer function of a linear first order dynamic system, $w(t)$ is a standard Gaussian white noise, and $E_1[x]$ is defined by

$$E_1[a] = \int_a^\infty \frac{e^{-s}}{s}ds , \qquad (9.228)$$

where $a \in R$. The authors obtained response characteristics by application of statistical linearization to the cubic term and the spectral analysis to the linearized integrodifferential equation.

**Ref.9.2.**

Several other applications of statistical or equivalent linearization to the response analysis of structures under earthquake excitation was shown in the literature, for instance,

Koliopulos and Chandler [50] have adopted equivalent linearization for the analysis of hysteretic torsionally coupled oscillators under biearthquake excitations. To obtain linearization coefficients, the authors used polynomial expansions of three groups of nonlinear elements and implemented properties of Gaussian variables. These groups involves expectations of the following forms:

$$(a)\ E[sgn(v_1)v_2\mid v_2\mid], \qquad (b)E[\mid v_2\mid\mid v_2\mid], \qquad (c)E[sgn(v_1)\mid v_2\mid v_3]\ .$$
$$(9.229)$$

After applying the linearization procedure, Koliopulos and Chandler [50] have obtained approximations

$$(a)\quad E[sgn(v_1)v_2\mid v_2\mid] \cong E[v_2^2]\left(1.25E[v_1v_2]-0.25(E[v_1v_2])^3\right)\ ,$$

$$(b)\quad E[\mid v_2\mid\mid v_2\mid] \cong \sqrt{E[v_1^2]E[v_2^2]}(0.64+0.36(E[v_1v_2])^2)\ ,$$

$$(c)\quad E[sgn(v_1)\mid v_2\mid v_3] \cong \sqrt{E[v_2^2]E[v_3^2]}\left(0.64E[v_1v_3]+0.61E[v_1v_2]E[v_2v_3]\right.$$
$$\left. -0.25E[v_1v_3](E[v_1v_2])^3\right) \qquad (9.230)$$

where $v_i$, $i = 1, 2, 3$ are stochastic processes.

The main advantage of the proposed method is that it is able to provide useful insights in a computationally efficient manner. We note that this approach according to Koliopulos and Chandler [50], has been already earlier successfully implemented in the context of marine and wind engineering by Koliopulos [47] and Koliopulos [48].

One can find some further results of application of equivalent linearization in response analysis of base-isolated (R-FBI) multistory structures in [33].

The response analysis of 1-DOF system with tuned mass damper with nonlinear viscous damping was presented by Rüdinger [68]. In his study, he applied the statistical linearization to a few models of nonlinear viscous damping, for instance, to the functions

$$(a)\ \alpha_0 sgn(v_2)\mid v_2\mid^\nu, \qquad (b)\ \alpha_1 sgn(v_2)+\alpha_2 v_2\ , \qquad (9.231)$$

where $\alpha_i$, $i = 0, 1, 2$, and $\nu$, are positive constants and $v_2$ is the relative velocity of the considered mass damper.

**Ref.9.3.**

In the case of wave excitation, some other authors have also used statistical linearization and nonlinearization in vibration analysis of offshore structures. The Morison equation was used by Quian and Wang [64] to calculate the excitation of the wave forces on the structure. As an example they considered three-dimensional model of jacket-type offshore platforms with the fluid–structure interaction and the coupling of the motion in different directions at each node. To obtain the response characteristics, statistical linearization was applied.

Liu and Bergdahl [55] used the Morison equation to model the hydrodynamic forces acting on the mooring cable. Two example structures, one flexible riser and one deep water mooring were presented. The normal and tangential drag forces acting on a unit length of the cable were linearized by statistical linearization.

Benfratello and Falsone [9] proposed to approximate the nonlinearity $P|P|$ appearing in the Morison equation by an equivalent polynomial having the form

$$P|P| = \sum_{i=0}^{N} \alpha_i P^i \,, \tag{9.232}$$

where the coefficients $\alpha_i$, $i = 0, 1, \ldots, N$ are determined by minimizing the mean-square errors between the nonlinear element $P|P|$ and the two terms of series (9.232). Benfratello and Falsone [9] used this approximation to obtain the wave excitations in the form of non-Gaussian process (a polynomial of a Gaussian-filtered process) and next studied a SDOF offshore structure subjected to the actions due to the horizontal velocity of the waves.

An application of the methods of statistical quadratization and cubization in approximation of wave and wind excitations acting on offshore platforms was shown by Quek et al. and Koh [65], Li et al. [53], and Kareem et al. [38]. In all the papers, the offshore platforms were modeled by SDOF systems and the approximate solutions are based on the Volterra series approach and computation is done in the frequency domain. We note that the theoretical aspects of statistical quadratization and cubization were discussed in [72].

Wolfram [87] proposed an alternative approach to statistical linearization with an example of the Morison equation. The main idea of this method is to find an equivalent linear system which yields the same characteristics, for instance, expected value of peak force as for nonlinear system.

**Ref.9.4.**

Application of statistical linearization in the response analysis of structures under wind excitation has been shown in some papers. Zhang et al. [97] studied an $n$-story structure with TLD subject to random wind excitation. To obtain response characteristics, the authors used statistical linearization with mean-square criterion for energy of displacements.

Chen and Ahmadi [15] considered the probabilistic responses of base-isolated structures to random wind loadings. The structure was modeled as a rigid block and the models of isolators are the same as in [14], namely, LRB, R-FBI, and HD-LRB.

Kareem and Li [37] studied a tension-leg platform under wind loading in the presence of waves and currents. The nonlinearities of hydrodynamic origin and those resulting from tether-restoring force were quadratized. An example was utilized to demonstrate that the frequency-domain analysis gives an excellent agreement with the time-domain simulation results.

Floris [24] discussed a linear structure subjected to wind excitations in the form of nonlinear terms with Gaussian excitations that make non-Gaussian excitations. He found the response characteristics by considering the corresponding nonlinear system with an equivalent Gaussian excitations.

### Ref.9.5.

The idea of combining statistical linearization and LQG approach to the determination of quasioptimal control in nonlinear stochastic systems was explored and developed by many authors. Recently, the problem of the determination of the quasioptimal control for linear system with constraints on control variable was considered by Gokcek et al. [28, 29, 30]. They have studied two groups of systems by two approaches called in the literature as LQR and LQG theories, respectively. The corresponding synthesis equations have the form

(a) for LQR theory

$$\begin{aligned}
\dot{\mathbf{x}} &= \mathbf{A}\mathbf{x} + \mathbf{B}_2 sat(\mathbf{K}\mathbf{x}) + \mathbf{B}_1\mathbf{w} , \\
\mathbf{z} &= (\mathbf{C}_1 + \mathbf{D}_{12}\mathbf{K})\mathbf{x} , \\
\mathbf{u} &= \mathbf{K}\mathbf{x}.
\end{aligned} \tag{9.233}$$

(b) for LQG theory

$$\begin{aligned}
\dot{\mathbf{x}}_G &= \mathbf{A}\mathbf{x}_G + \mathbf{B}_2 sat(\mathbf{K}_C\mathbf{x}) + \mathbf{B}_1\mathbf{w} , \\
\dot{\mathbf{x}}_C &= \mathbf{M}\mathbf{x}_C - \mathbf{L}\mathbf{C}_2\mathbf{x}_G - \mathbf{L}_{21}\mathbf{w} , \\
\mathbf{z} &= \mathbf{C}_1\mathbf{x}_G + \mathbf{D}_{12}\mathbf{K}\mathbf{x}_C , \\
\mathbf{u} &= \mathbf{K}\mathbf{x}_C,
\end{aligned} \tag{9.234}$$

where $\mathbf{x}, \mathbf{x}_G$, and $\mathbf{x}_C$ are state vectors, $\mathbf{u}$ is the control vector, $\mathbf{w}$ is a standard vector uncorrelated white noise process, $\mathbf{z}$ is a vector controlled output, and $\mathbf{A}, \mathbf{B}_1, \mathbf{B}_2, \mathbf{C}_1, \mathbf{C}_2, \mathbf{M}, \mathbf{L}, \mathbf{K}, \mathbf{K}_C, \mathbf{D}_{12}$, and $\mathbf{L}_{21}$ are constant matrices.

$$sat(u) = \begin{cases} +1, & \text{for} \quad u > +1 \\ u & \text{for} \quad -1 \leq u \leq +1 \\ -1, & \text{for} \quad u < -1. \end{cases} \tag{9.235}$$

Using statistical linearization to $sat(x)$ function and suitable assumptions regarding the control system Gokcek et al. have found sufficient conditions for optimal control that minimizes the variance of the vector $\mathbf{z}$, for corresponding linearized systems, it means that the authors have shown that the LQR/LQG design methodology admits an extension to systems with saturating actuators called by SLQR/SLQG theory, respectively.

Another extensions of the Wonham and Cashman's approach was proposed by Kazakov [42], who considered a stochastic nonlinear vector equation

$$\dot{\mathbf{y}} = \mathbf{A}(\mathbf{y}, t) + \mathbf{B}(t)\mathbf{u} + \mathbf{N}(t)\mathbf{w}(t), \quad \mathbf{y}(t_0) = \mathbf{y}_0 , \qquad (9.236)$$

where $t_0$ and $t_n$ are the initial and final moments of control, respectively; $\mathbf{y}$, $\mathbf{u}$, and $\mathbf{w}$ are state, control, and white noise disturbances vectors, respectively; $\mathbf{A}(\mathbf{y}, t)$ is a deterministic nonlinear vector function, $\mathbf{B}(t)$ and $\mathbf{N}(t)$ are deterministic time-dependent matrices of appropriate dimensions. The observation equation is a nonlinear vector algebraic stochastic equation

$$\mathbf{z} = \mathbf{C}(\mathbf{y}, t) + \mathbf{v} , \qquad (9.237)$$

where $\mathbf{z}$ is an vector observation variable, $\mathbf{A}(\mathbf{y}, t)$ is a nonlinear vector function, and $\mathbf{v}$ is a white noise disturbance vector. The objective of the considered problem was to find the optimal control $\mathbf{u} \in U_0 = \{\mathbf{u} : |u_i| \leq u_{i0}\}, i = 1, \ldots, r$, of the feedback form that minimizes the mean-square criterion

$$I_k(\mathbf{y}, \mathbf{u}, \mathbf{w}, t_0, t_k) = E\left[\mathbf{y}^T(t_k)\mathbf{\Gamma}\mathbf{y}(t_k) + \int_{t_0}^{t_k} (\mathbf{y}^T(t)\mathbf{L}\mathbf{y}(t) + \mathbf{u}^T(t)\mathbf{R}^{-1}\mathbf{u}(t))dt\right],$$
$$(9.238)$$

where $t_0$ and $t_k$ are the initial and final moments of control, respectively; $\mathbf{\Gamma}$ and $\mathbf{L}$ are positive-definite matrices and $\mathbf{R}$ is a diagonal matrix of the positive coefficients.

This problem was generalized by Kazakov in his next papers. In [44] the state equation has the form

$$\dot{\mathbf{y}} = \mathbf{A}(\mathbf{y}, t) + \mathbf{B}(\mathbf{y}, t)\mathbf{u} + \mathbf{S}(t) + \mathbf{N}(\mathbf{y}, t)\mathbf{w}(t), \quad \mathbf{y}(t_0) = \mathbf{y}_0 , \qquad (9.239)$$

where $\mathbf{S}(t)$ is a deterministic vector function. The theoretical equation of motion is described by

$$\dot{\mathbf{y}}_\tau = \mathbf{A}(\mathbf{y}_\tau, t) + \mathbf{S}(t), \quad \mathbf{y}_\tau(t_0) = \mathbf{y}_0 . \qquad (9.240)$$

The observation equation is a linear vector algebraic stochastic equation

$$\mathbf{z} = \mathbf{C}(t) + \mathbf{v}(t) . \qquad (9.241)$$

The objective of the considered problem is to find the optimal control $\mathbf{u} \in U_0 = \{\mathbf{u} : |u_i| \leq u_{i0}\}, i = 1, \ldots, r$, of the feedback form that minimizes the mean-square criterion

$$I_k(\mathbf{y}, \mathbf{u}, \mathbf{w}, t_0, t_k) = E\left[\mathbf{e}^T(t_k)\mathbf{\Gamma}\mathbf{e}(t_k) + \int_{t_0}^{t_k} (\mathbf{e}^T(t)\mathbf{L}\mathbf{e}(t) + \mathbf{u}^T(t)\mathbf{R}^{-1}\mathbf{u}(t))dt\right].$$
$$(9.242)$$

The error equation has the form

$$\dot{\mathbf{e}} = \mathbf{A}(\mathbf{y}, t) - \mathbf{A}(\mathbf{y}_\tau, t) + \mathbf{B}(\mathbf{y}, t)\mathbf{u} + \mathbf{N}(\mathbf{y}, t), \mathbf{W}(t), \quad \mathbf{e}(t_0), = \mathbf{0} , \qquad (9.243)$$

Several other nonlinear nondifferentiable criteria of optimization were considered by Kazakov [43].

An extension of Wonham and Cashman's idea for quasioptimal control for seismic-excited hysteretic structural systems without and with observation equation was done by Yang et al. [95] and Suzuki [83], respectively. Raju and Narayanan [66] and Narayanan and Senthil [59] applied this approach to the two-degree-of-freedom vehicle suspension model with fully observed state vector. In both cases, equivalent linearization is applied to hysteretic parts of considered systems and LQG theory with or without filtering equation is used.

A comparison between the applicability of statistical linearization methods with moment criteria and criteria in probability density space to the determination of quasioptimal external control for the nonlinear dynamic system excited by a colored Gaussian noise and the mean-square criterion was presented by the author [?]. Also, a comparison between the applicability of statistical and equivalent linearization methods with criteria in probability density space to the determination of quasioptimal external control was done by Socha and Błachuta [73].

To obtain sufficient conditions for almost sure asymptotic stability of the Duffing oscillator under filtered white noise external excitations, Koliopulos and Langley [49] have used equivalent linearization.

# References

1. Abramowitz, M. and Stegun, I.: Handbook of Mathematical Functions. Dover, New York, 1972.
2. Asano, K. and Iwan, W.: An alternative approach to the random response of bilinear hysteretic systems. Earthquake Eng. Struct. Dyn. **12**, 229–236, 1984.
3. Atalik, T. and Utku, S.: Stochastic linearization of multi-degree-of-freedom nonlinear systems. Earthquake Eng. Struct. Dyn. **4**, 411–420, 1976.
4. Baber, T. and Noori, M.: Modeling general hysteresis behaviour and random vibration application, J. Vibration, Acoustics, Stres and Reliability in Design, 108, 411420, 1986.
5. Baber, T. and Wen, Y.: Random vibration of hysteretic degrading systems. ASCE J. Eng. Mech. Div. **107**, 1069–1087, 1981.
6. Baber, T. and Wen, Y.: Stochastic response of multi-story yielding frames random vibration of hysteretic degrading systems. Earthquake Eng. Struct. Dyn. **10**, 403–416, 1982.
7. Basili, M. and De Angelis, M.: Optimal passive control of adjacent structures interconnected with nonlinear hysteretic devices. J. Sound Vib. **301**, 106–125, 2007.
8. Beaman, J. J.: Non-linear quadratic Gaussian control. Int. J. Control **39**(2), 343–361, 1984.
9. Benfratello, S. and Falsone, G.: Non-Gaussian approach for stochastic analysis of offshore structures. ASCE J. Eng. Mech. **121**, 1173–1180, 1995.
10. Bouc, R.: Forced vibration of mechanical system with hysteresis. In: Proc. of 4th Conference on Nonlinear Oscillation, Prague, 1967.

11. Brynjolfsson, S. and Leonard, J.: Effects of currents on the stochastic response to earthquakes of multiple degree of freedom models of guyed offshore towers. Eng. Struct. **10**, 194–203, 1988.
12. Brynjolfsson, S. and Leonard, J.: Response of guyed offshore towers to stochastic loads, time domain vs. frequency domain. Eng. Struct. **10**, 106–116, 1988.
13. Caughey, T.: Random excitation of a loaded nonlinear string. Trans. ASME J. Appl. Mech. **27**, 575–578, 1960.
14. Chen, Y. and Ahmadi, G.: Stochastic earthquake response of secondary systems in base-isolated structures. Earthquake Eng. Struct. Dyn. **21**, 1039–1057, 1992.
15. Chen, Y. and Ahmadi, G.: Wind effects on base-isolated structures. ASCE J. Eng. Mech. **118**, 1708–1727, 1992.
16. Chen, Y. and Ahmadi, G.: Performance of a high damping rubber bearing base-isolation system for a shear beam structure. Earthquake Eng. Struct. Dyn. **23**, 729–744, 1994.
17. Constantinou, M. and Papageorgiou, A.: Stochastic response of practical sliding isolation systems. Probab. Eng. Mech. **5**, 27–34, 1990.
18. Davenport, A.: The application of statistical concept to the wind loading of structures. Proc. Inst. Civil. Engrs. **19**, 449–473, 1961.
19. Davenport, A.: The spectrum of horizontal gustiness near the ground in high winds. Q. J. R. Meterd Soc. **87**, 194–211, 1961.
20. Dobson, S., Noori, M., Hou, Z., Dimentberg, M., and Baber, T.: Modeling and random vibration analysis of SDOF systems with asymmetric hysteresis. Int. J. Non-Linear Mech. **32**, 669–680, 1997.
21. Dobson, S., Noori, M., Hou, Z., and Dimentberg, M.: Direct implementation of stochastic linearization for SDOF systems with general hysteresis. Struct. Eng. Mech. **6**, 473–484, 1998.
22. Duval L. et al.: Zero and non-zero mean analysis of MDOF hysteretic systems via direct linearization. In: B. Spencer and E. Johnson (eds.), Stochastic Structural Dynamics, 77–84. Balkema, Rotterdam, 1999.
23. Elishakoff, I.: Method of stochastic linearization: Revisited and improved. In: S. P. D. and C. Brebbia (eds.), Computational Stochastic Mechanics, 101–111. Computational Mechanics Publication and Elsevier Applied Science, London, 1991.
24. Floris, C.: Equivalent Gaussian process in stochastic dynamics with application to along-wind response of structures. Int. J. Non-Linear Mech. **31**, 779–794, 1996.
25. Foliente, G., Singh, M., and Noori, M.: Equivalent linearization of generally pinching hysteretic, degrading systems. Earthquake Eng. Struct. Dyn. **25**, 611–629, 1996.
26. Foster, E.: Model for nonlinear dynamics of offshore structures. ASCE J. Struct. Eng. Div. **96**, 41–67, 1970.
27. Ghosh, A. and Basu, B.: Seismic vibration control of short period structures using the liquid column damper. Eng. Struct. **26**, 1905–1913, 2004.
28. Gokcek, C., Kabamba, P., and Meerkov, S.: Disturbance rejection in control systems with saturating actuators. Nonlinear Anal. **40**, 213–226, 2000.
29. Gokcek, C., Kabamba, P., and Meerkov, S.: Optimization of disturbance rejection in systems with saturating actuators. J. Math. Anal. Appl. **249**, 135–159, 2000.
30. Gokcek, C., Kabamba, P., and Meerkov, S.: An LQR/LQG theory for systems with saturating actuators. IEEE Trans. Autom. Control **46**, 1529–1542, 2001.

31. Gradshteyn, I. and Ryzhnik, I.: Table of Integrals, Series and Products, 4th edn. Academic Press, London, 1965.
32. Grundmann, H., Hartmann, C., and Waubke, H.: Structures subjected to stationary stochastic loadings. Preliminary assessment by statistical linearization combined with an evolutionary algorithm. Comput. Struct. **67**, 53–64, 1998.
33. Hong, W. and Kim, H.: Performance of multi-story structure with a resilient-friction base isolation system. Comput. Struct. **82**, 2271–2283, 2004.
34. Hurtado, J. and Barbat, A.: Improved stochastic linearization method using mixed distributions. Struct. Safety **18**, 49–62, 1996.
35. Hurtado, J. and Barbat, A.: Equivalent linearization of the Bouc–Wen hysteretic model. Eng. Struct. **22**, 1121–1132, 2000.
36. Janke, E., Emde, F., and Lösch, F.: Tafeln Höherer Funktionen, 6th edn. B. G. Teubner, Stuttgart, 1960.
37. Kareem, A. and Li, Y.: Wind-excited surge response of tension-leg platform-frequency-domain approach. ASCE J. Eng. Mech. **119**, 161–183, 1993.
38. Kareem, A., Zhao, J., and Tognarelli, M.: Surge response statistics of tension leg platforms under wind and wave loads: Statistical quadratization approach. Probab. Eng. Mech. **10**, 225–240, 1995.
39. Kazakov, I. E.: Approximate probabilistic analysis of the accuracy of the operation of essentially nonlinear systems. Avtomatika i Telemekhanika **17**, 423–450, 1956.
40. Kazakov, I. and Dostupov, B.: Statistical Dynamic of Nonlinear Automatic Systems. Fizmatgiz, Moskwa, 1962 (in Russian).
41. Kazakov, I.: Analytical synthesis of a quasi-optimal additive control in a nonlinear stochastic system. Avtomatika i Telemekhanika **45**, 34–46, 1984.
42. Kazakov, I.: Gaussian approximation when optimizing a control in a stochastic nonlinear system. J. Comp. Syst. Sci. Int. **32**, 118–122, 1994.
43. Kazakov, I.: Analytical construction of a conditionally optimal control in a nonlinear stochastic system by a complex local functional. Avtomatika i Telemekhanika **56**, 34–46, 1995.
44. Kazakov, I.: Optimization of control in a nonlinear stochastic system by a local criterion. J. Comput. Syst. Sci. Int. **35**, 940–947, 1996.
45. Kimura, K., Yagasaki, K., and Sakata, M.: Non-stationary response of a system with bilinear hysteresis subjected to non-white random excitation. J. Sound Vib. **91**, 181–194, 1983.
46. Kimura, K., Yasumuro, H., and Sakata, M.: Non-Gaussian equivalent linearization for non-stationary random vibration of hysteretic systems. Probab. Eng. Mech. **9**, 15–22, 1994.
47. Koliopulos, P.: Quasi-static and dynamic response statistics of linear SDOF systems under Morison-type wave forces. Eng. Struct. **10**, 24–36, 1988.
48. Koliopulos, P.: Application of the separability assumption on the statistics of linear SDOF systems under square-Gaussian excitation. Appl. Math. Model. **14**, 184–198, 1990.
49. Koliopulos, P. and Langley, R.: Improved stability analysis of the response of a Duffing oscillator under filtered white noise. Int. J. Non-Linear Mech. **28**, 145–155, 1993.
50. Koliopulos, P. and Chandler, A.: Stochastic linearization of inelastic seismic torsional response – formulation and case-studies. Eng. Struct. **17**, 494–504, 1995.

51. Kozin, F.: The method of statistical linearization for non-linear stochastic vibration. In: F. Ziegler and G. Schueller (eds.), Nonlinear Stochastic Dynamic Engineering Systems, 45–56. Springer, Berlin, 1989.

52. Kwakernaak, H. and Sivan, R.: Linear Optimal Control Systems. J. Wiley, New York, 1972.

53. Li, X., Quek, S., and Koh, C.: Stochastic response of offshore platform by statistical cubization. ASCE J. Eng. Mech. **121**, 1056–1068, 1995.

54. Lin, B., et al.: Response of base isolated buildings to random excitation with a Clough- Penzien model. Earthquake Eng. Struct. Dyn. **18**, 49–62, 1989.

55. Liu, Y. and Bergdahl, L.: Influence of current and seabed friction on mooring cable response: Comparison between time-domain and frequency-domain analysis. Eng. Struct. **19**, 945–953, 1997.

56. Malhotra, A. and Penzien, J.: Response of offshore structures to random wave forces. ASCE J. Struct. Eng. Div. **107**, 2155–2173, 1970.

57. Morison, J., O'Brien, M., Johnson, J., and Shaaf, S.: The force exerted by surface waves on piles. AIME Petrol Trans. **189**, 149–154, 1950.

58. Morosanov, J.: Practical methods for the calculation of the linearization factors for arbitrary nonlinearities. Avtomatika i Telemekhanika **29**, 81–86, 1968.

59. Narayanan, S. and Senthil, S.: Stochastic optimal active control of a 2-DOF quarter car model with non-linear passive suspension elements. J. Sound Vib. **211**, 495–506, 1998.

60. Nigam, N. and Narayanan, S.: Applications of Random Vibrations. Narosa Publishing House, New Delhi, 1994.

61. Penzien, J. and Kaul, M.: Response of offshore towers to strong motion earthquakes. Earthquake Eng. Struct. Dyn. **1**, 55–68, 1972.

62. Penzien, J., Kaul, M., and Berge, B.: Stochastic response of offshore towers to random sea waves and strong motion earthquakes. Comput. Struct. **2**, 733–756, 1972.

63. Pugachev, V. and Sinitsyn, I.: Stochastic Differential Systems. Wiley, Chichester, 1987.

64. Qian, J. and Wang, X.: 3-dimensional stochastic response of offshore towers to random sea waves. Comput. Struct. **2**, 385–390, 1992.

65. Quek, S., Li, X., and Koh, C.: Stochastic response of jack-up platform by the method for statistical quadratization. Appl. Ocean Res. **16**, 113–122, 1994.

66. Raju, G. and Narayanan, S.: Active control of nonstationary response of a 2-degree of freedom vehicle model with nonlinear suspension. Sadhana-Acad. P. Eng. S. **20**, 489–499, 1995.

67. Roberts, J. and Spanos, P.: Random Vibration and Statistical Linearization. John Wiley and Sons, Chichester, 1990.

68. Rüdinger, F.: Tuned mass damper with nonlinear viscous damping. J. Sound Vib. **300**, 932–948, 2007.

69. Sachs, P.: Wind Forces in Engineering. Pergamon Press, Oxford, 1978.

70. Silva, F. and Ruiz, S.: Calibration of the equivalent linearization Gaussian approach applied to simple for hysteretic systems subjected to narrow band seismic motions. Struct. Safety **22**, 211–231, 2000.

71. Simiu, E. and Scanian, R.: Wind Effects on Structures. John Wiley, New York, 1978.

72. Sobiechowski, C. and Socha, L.: Statistical linearization of the Duffing oscillator under non-Gaussian external excitation. J. Sound Vib. **231**, 19–35, 2000.

73. Socha, L. and Błachuta, M.: Application of linearization methods with probability density criteria in control problems. In: Proc. American Control Conference, Chicago, 2775–2779, CD–ROM Vol. 4, Danvers, MA USA, 2000.

74. Socha, L. and Soong, T.: Linearization in analysis of nonlinear stochastic systems. Appl. Mech. Rev. **44**, 399–422, 1991.

75. Socha, L.: Application of true linearization in stochastic quasi-optimal control problems. J. Struct. cont. **7**, 219–230, 2000.

76. Spanos, P.: A method of analysis of nonlinear vibrations caused by modulated random excitation. Int. J. Non-Linear Mech. **16**, 1–11, 1981.

77. Spanos, P. and Tsavachidis, S.: Deterministic and stochastic analyses of nonlinear system with a Biot visco-elastic elements. Earthquake Eng. Struct. Dyn. **30**, 595–612, 2001.

78. Socha, L.: Linearization in analysis of nonlinear stochastic systems – recent results. Part I. Theory. ASME Appl. Mech. Rev. **58**, 178–205, 2005.

79. Socha, L.: Linearization in analysis of nonlinear stochastic systems – recent results. Part II. Applications. ASME Appl. Mech. Rev. **58**, 303–315, 2005.

80. Su, L. and Ahmadi, G.: Response of frictional base isolation systems to horizontal-vertical random earthquake excitations. Probab. Eng. Mech. **3**, 12–21, 1988.

81. Su, L. Ahmadi, G. and Tadjbakhsh, I.: A comparative study of performances of various base isolation systems – Part I: Shear beam structures. Earthquake Eng. Struct. Dyn. **18**, 11–32, 1989.

82. Suzuki, Y. and Minai, R.: Application of stochastic differential equations to seismic reliability analysis of hysteretic structures. Lecture Notes Eng. **32**, 1987.

83. Suzuki, Y.: Stochastic control of hysteretic structural systems. Sadhana **20**, 475–488, 1995.

84. Wen, Y.: Method for random vibration of hysteretic systems. ASCE J. Eng. Mech. **102**, 249–263, 1976.

85. Wen, Y.: Equivalent linearization for hysteretic systems under random excitation. Trans. ASME J. Appl. Mech. **47**, 150–154, 1980.

86. Wen, Y.: Methods of random vibration for inelastic structures. Appl. Mech. Rev. **42**, 39–52, 1989.

87. Wolfram, J.: On alternative approaches to linearization and Morison's equation for wave forces. Proc. Roy. Soc. Lond. A. Mat. **455**, 2957–2974, 1999.

88. Won, A., Pires, J., and Haroun, M.: Stochastic seismic performance evaluation of tuned liquid column dampers. Earthquake Eng. Struct. Dyn. **25**, 1259–1274, 1996.

89. Won, A., Pires, J., and Haroun, M.: Performance assessment of tuned liquid column dampers under random seismic loading. Int. J. Non-Liner Mech. **32**, 745–758, 1997.

90. Wonham, W. and Cashman, W. F.: A computational approach to optimal control of stochastic saturating systems. Int. J. Control **10**(1), 77–98, 1969.

91. Xu, Y., Kwok, K., and Samali, B.: The effect of tuned mass dampers and liquid dampers on corrs-wind response of tall slender structures. J. Wind Eng. Ind. Aerodyn. **40**, 33–54, 1992.

92. Xu, Y., Samali, B., and Kwok, K.: Control of along-wind response of structures by mass and liquid dampers. ASCE. J. Eng. Mech. **118**, 20–39, 1992.

93. Yalla, S. and Kareem, A.: Optimum absorber parameters for tuned liquid column dampers. ASCE J. Struct. Eng. **126**, 906–915, 2000.

94. Yan, X. and Nie, J.: Response of SMA superelastic systems under random excitation. J. Sound Vib. **238**, 893–901, 2000.

95. Yang, J., Li, Z., and Vongchavalitkul, S.: Stochastic hybrid control of hysteretic structures. Probab. Eng. Mech. **9**, 125–133, 1994.

96. Yoshida, K.: A method of optimal control of non-linear stochastic systems with non-quadratic criteria. Int. J. Control **39**(2), 279–291, 1984.

97. Zhang, X., Zhang, R., and Xu, Y.: Analysis on control of flow-induced vibration by tuned liquid damper with crossed tube-like containers. J. Wind Eng. Ind. Aerodyn. **50**, 351–360, 1993.

# 10

# Accuracy of Linearization Methods

Since linearization methods are approximate methods in the study of nonlinear stochastic systems, the analysis of their accuracy is one of the most important problems. This analysis is done in a few fields such as a theoretical study of the accuracy, a comparison of approximated characteristics with corresponding exact ones, or with simulations or with experiments. We will discuss all these fields below.

## 10.1 Theoretical Study of the Accuracy of Linearization Methods

The first theoretical accuracy analysis of statistical linearization methods was proposed by the "father" of statistical linearization Kazakov [62, 61]. We present the main idea of this study.

Consider again a dynamic system (5.1) with external excitation described by the nonlinear vector Ito stochastic differential equation

$$dx(t) = \Phi(x, t)dt + \sum_{k=1}^{M} G_{k0}(t)d\xi_k(t), \quad x(t_0) = x_0 , \qquad (10.1)$$

where $x = [x_1, \ldots, x_n]^T$ is the vector state, $\Phi = [\Phi_1, \ldots, \Phi_n]^T$ is a nonlinear vector function, $G_{k0} = [\sigma_{k0}^1, \ldots, \sigma_{k0}^n]^T$, $k = 1, \ldots, M$, are deterministic vectors of intensities of noise, $\xi_k$, $k = 1, \ldots, M$, are independent standard Wiener processes; the initial condition $x_0$ is a vector random variable independent of $\xi_k(t)$, $k = 1, \ldots, M$. We assume that the solution of (10.1) exists.

In the derivation of statistical linearization, the nonlinear function $\Phi(x, t)$ is approximated by $\Phi_0(m_x, \Theta_x, t) + K(m_x, \Theta_x, t)x^0$, where $\Phi_0 = [\Phi_0^1, \ldots, \Phi_0^n]^T$ is a nonlinear vector function of the moments of variables $x_j$, $x^0 = [x_1^0, \ldots, x_n^0]^T$, $K = [k_{ij}]$ is an $n \times n$ matrix of statistical linearization coefficients.

L. Socha: *Accuracy of Linearization Methods*, Lect. Notes Phys. **730**, 341–376 (2008)
DOI 10.1007/978-3-540-72997-6_10                © Springer-Verlag Berlin Heidelberg 2008

$$\mathbf{m_x} = E[\mathbf{x}] = [m_{x_1}, \ldots, m_{x_n}]^T, \; m_{x_i} = E[x_i], \; \mathbf{\Theta_x} = [\Theta_{ij}] = \left[E[x_i^0 x_j^0]\right],$$

$i, j = 1, \ldots, n$, and $x_i^0$ is the centralized $i$th coordinate of the vector state $\mathbf{x}$, $x_i^0 = x_i - m_{x_i}$.

Then (10.1) can be approximated by the linearized equation

$$d\mathbf{x}(t) = \left[\mathbf{\Phi}_0(\mathbf{m}_x, \mathbf{\Theta_x}, t) + \mathbf{K}(\mathbf{m}_x, \mathbf{\Theta_x}, t)\mathbf{x}^0\right] dt + \sum_{k=1}^{M} \mathbf{G}_{k0}(t) d\xi_k(t), \quad \mathbf{x}(t_0) = \mathbf{x_0},$$

(10.2)

where the coordinates of the vector $\mathbf{m}_x$ and the matrix $\mathbf{\Theta_x}$ satisfy the differential equations

$$\frac{d\mathbf{m}_{xi}(t)}{dt} = \Phi_0^i(\mathbf{m_x}, \mathbf{\Theta_x}, t), \quad \mathbf{m}_{xi}(t_0) = E[\mathbf{x}_{i0}], i = 1, \ldots, n, \qquad (10.3)$$

$$\frac{d\Theta_{ij}(t)}{dt} = \sum_{l=1}^{n}(k_{il}\Theta_{lj}(t) + k_{jl}\Theta_{il}(t)) + \sum_{k=1}^{M} \sigma_{k0}^i(t)\sigma_{k0}^j(t),$$

$$\Theta_{ij}(t_0) = E[x_{i0}^0 x_{j0}^0] = \Theta_{ij0}, i, j = 1, \ldots, n, \qquad (10.4)$$

where $m_{xi}$ are elements (components) of vector $\mathbf{m_x}$, i.e., $\mathbf{m_x} = [m_{x1}, \ldots, m_{xn}]^T$ and $\Theta_{ij}$ are elements of the matrix $\mathbf{\Theta_x}$, i.e., $\mathbf{\Theta_x} = [\Theta_{ij}]$.

The exact formula for the nonlinear function $\mathbf{\Phi}(\mathbf{x}, t)$ is

$$\mathbf{\Phi}(\mathbf{x}, t) = \mathbf{\Phi}_0(\mathbf{m}_x, \mathbf{\Theta_x}, t) + \mathbf{K}(\mathbf{m}_x, \mathbf{\Theta_x}, t)\mathbf{x}^0 + \mathbf{\Psi}(\mathbf{x}, t), \qquad (10.5)$$

where $\mathbf{\Psi}(\mathbf{x}, t)$ is a vector nonlinear function determining the error of approximation by a linearization procedure. We denote by $\mathbf{x}(t)$ and $\mathbf{x}(t) + \mathbf{\Delta}(t)$ the approximate and exact solution of (10.2) and (10.1), respectively.

Substitution of equality (10.5) into (10.1) and taking into account that the exact solution of (10.1) is $\mathbf{x}(t) + \mathbf{\Delta}(t)$, we obtain

$$d(\mathbf{x}(t) + \mathbf{\Delta}(t)) = \left[\mathbf{\Phi}_0(\mathbf{m}_x, \mathbf{\Theta_x}, t) + \mathbf{K}(\mathbf{m}_x, \mathbf{\Theta_x}, t)(\mathbf{x}^0 + \mathbf{\Delta}^0) + \mathbf{\Psi}(\mathbf{x} + \mathbf{\Delta}, t)\right] dt$$

$$+ \sum_{k=1}^{M} \mathbf{G}_{k0}(t) d\xi_k(t), \quad \mathbf{x}(t_0) + \mathbf{\Delta}(t_0) = \mathbf{x_0}, \qquad (10.6)$$

where $\mathbf{\Delta}^0$ is the centralized error of linearization, i.e.,

$$\mathbf{\Delta}^0(t) = \mathbf{\Delta} - \mathbf{m}_\Delta, \qquad \mathbf{m}_\Delta = \mathbf{m}_\Delta(t) = E[\mathbf{\Delta}(t)]. \qquad (10.7)$$

Taking into account in (10.6) equality (10.5), we find

$$d(\mathbf{\Delta}(t)) = [\mathbf{K}(\mathbf{m}_x, \mathbf{\Theta_x}, t)\mathbf{\Delta}^0 + \mathbf{\Psi}(\mathbf{x} + \mathbf{\Delta}, t)] dt, \quad \mathbf{\Delta}(t_0) = \mathbf{0}, \qquad (10.8)$$

where

$$\Psi(\mathbf{x}+\Delta,t) = \Phi(\mathbf{x}+\Delta,t) - \Phi_0(\mathbf{m}_x,\Theta_\mathbf{x},t) - \mathbf{K}(\mathbf{m}_x,\Theta_\mathbf{x},t)(\mathbf{x}^0+\Delta^0) \,. \quad (10.9)$$

We note that although (10.9) looks like deterministic, in fact it is stochastic, because the vector $\mathbf{x}(t) + \Delta(t)$ is the solution of stochastic differential equation (10.1).

To estimate the parameters of the error process $\Delta(t)$, one can use the estimates of the mean value $\mathbf{m}_\Delta$ and correlation variables $\Theta_\Delta$.

From (10.8) we find

$$\dot{\mathbf{m}}_\Delta(t) = E_\Delta\left[E_x[\Psi(\mathbf{x}+\Delta,t)|\Delta]\right] = E_\Delta[\hat{\Psi}(\mathbf{x}+\Delta,t)|\Delta], \quad \mathbf{m}_\Delta(t_0) = \mathbf{0} \,,$$
$$(10.10)$$

where $E_\Delta[.]$ and $E_x[.]$ are the expectations with respect to $\Delta$ and $x$, respectively.

We assume that the function $\hat{\Psi}(\mathbf{x}+\Delta,t)$ is twice differentiable and can be presented in Taylor series

$$\hat{\Psi}(\Delta,t) = \hat{\Psi}_0(t) + \mathbf{P}\Delta + \mathbf{R} \,, \quad (10.11)$$

where $\hat{\Psi}_0(t) = \hat{\Psi}(\mathbf{0},t)$, $\mathbf{P} = [P_{kj}]$ is a matrix whose elements are defined by

$$P_{kj} = \frac{\partial \hat{\Psi}_k(\Delta,t)}{\partial \Delta_j}\Big|_{\Delta=0}, \quad j,k=1,\ldots,n \,, \quad (10.12)$$

$\mathbf{R} = [\mathbf{R}_1,\ldots,\mathbf{R}_n]^T$ is a vector whose elements $\mathbf{R}_k = \Delta^T\mathbf{C}_k\Delta$ are defined by matrices $\mathbf{C}_k = [c_{ijk}]$ whose elements are

$$c_{kij} = \frac{1}{2}\frac{\partial^2 \hat{\Psi}_k(\Delta,t)}{\partial\Delta_i\partial\Delta_j}\Big|_{\Delta=0}, \quad i,j,k=1,\ldots,n \,. \quad (10.13)$$

Taking into account that $\hat{\Psi}_0(t) = 0$ and equality (10.11), (10.10) for coordinates can be presented in the form

$$\dot{\mathbf{m}}_{\Delta_k}(t) = \sum_{l=1}^{n} P_{kl}\mathbf{m}_{\Delta_l} + \sum_{l=1}^{n}\sum_{j=1}^{n} c_{klj}(\mathbf{m}_{\Delta_l}\mathbf{m}_{\Delta_j} + \theta_{\Delta_l\Delta_j}),$$
$$\mathbf{m}_{\Delta_k}(t_0) = 0, \quad k=1,\ldots,n \,, \quad (10.14)$$

where $\theta_{\Delta_l\Delta_j}(t) = E[\Delta_l^0(t)\Delta_j^0(t)], \quad l,j=1,\ldots,n.$

Similarly using (10.8) and (10.10), one can derive the differential equation for centralized variables $\Delta_k^0$

$$d(\Delta_k^0(t)) = \left[\sum_{l=1}^{n} K_{kl}\Delta_l^0 + \Psi_k^0(\mathbf{x}+\Delta,t)\right]dt, \quad \Delta_k^0(t_0) = 0, \quad k=1,\ldots,n \,,$$
$$(10.15)$$

where

$$\Psi_k^0(\mathbf{x} + \boldsymbol{\Delta}, t) = \Phi_k^0(\mathbf{x} + \boldsymbol{\Delta}, t) - m_{\Psi_k}, \quad k = 1, \ldots, n. \tag{10.16}$$

To estimate the accuracy of statistical linearization, we have to solve (10.14) and from (10.15) to estimate the correlation functions (second-order moments).

The solution of (10.15) can be formally presented as follows:

$$\Delta_k{}^0(t) = \sum_{l=1}^{n} \int_{t_0}^{t} g_{kl}(t, \tau) \Psi_l^0(\mathbf{x}(\tau) + \boldsymbol{\Delta}(\tau), \tau) d\tau, \quad k = 1, \ldots, n, \tag{10.17}$$

where $g_{kl}(t, \tau)$ is the impulse function for system (10.15).

Hence, one can calculate the second-order moments $\theta_{\Delta_k{}^0 \Delta_l{}^0} = E[\Delta_k{}^0(t) \Delta_l{}^0(t)]$ as follows:

$$\theta_{\Delta_k{}^0 \Delta_l{}^0}(t) = \sum_{i=1}^{n} \sum_{j=1}^{n} \int_{t_0}^{t} \int_{t_0}^{t} g_{ki}(t, \tau_1) g_{lj}(t, \tau_2) K_{\Psi_i^0 \Psi_j^0}(\tau_1, \tau_2) d\tau_1 d\tau_2,$$

$$k, l = 1, \ldots, n, \tag{10.18}$$

where $K_{\Psi_i^0 \Psi_j^0}(\tau_1, \tau_2) = E[\Psi_i^0(\mathbf{x}(\tau_1) + \boldsymbol{\Delta}(\tau_1), \tau_1) \Psi_j^0(\mathbf{x}(\tau_2) + \boldsymbol{\Delta}(\tau_2), \tau_2)]$ is the correlation function of variables $\Psi_i^0$ and $\Psi_j^0$.

For stationary solutions, the variables $\theta_{\Delta_k{}^0 \Delta_l{}^0}$ are constant and according to (3.61–3.64) defined in frequency domain by

$$\theta_{\Delta_k{}^0 \Delta_l{}^0} = \sum_{r=1}^{n} \sum_{j=1}^{n} \int_{-\infty}^{+\infty} \Phi_{kr}(i\lambda) \Phi_{lj}(-i\lambda) S_{\Psi_r^0 \Psi_j^0}(\lambda) d\lambda, \quad k, l = 1, \ldots, n,$$

$$\tag{10.19}$$

where

$$\Phi_{kr}(i\lambda) = \int_{-\infty}^{+\infty} g_{kr}(\tau) e^{-i\lambda\tau} d\tau, \tag{10.20}$$

$$S_{\Psi_r^0 \Psi_j^0}(\lambda) = \frac{1}{2\pi} \int_{-\infty}^{+\infty} K_{\Psi_r^0 \Psi_j^0}(\tau) e^{-i\lambda\tau} d\tau. \tag{10.21}$$

By calculations of estimations of the variables $\theta_{\Delta_k{}^0 \Delta_l{}^0}$, the practical important system according to Kazakov [61] is wideband system. In this case, one can use an approximation $K_{\Psi_r^0 \Psi_j^0}(\tau_1, \tau_2) \approx \theta_{\Delta_k{}^0 \Delta_l{}^0}(t)$ and for stationary systems $\Phi_{kr}(i\lambda) \approx \Phi_{kr}(0)$.

Then equalities (10.18) and (10.19) can be replaced by

$$\theta_{\Delta_k{}^0 \Delta_l{}^0}(t) \approx \sum_{r=1}^{n} \sum_{j=1}^{n} \theta_{\Psi_r^0 \Psi_j^0}(t) \int_{t_0}^{t} g_{ki}(t, \tau_1) d\tau_1 \int_{t_0}^{t} g_{lj}(t, \tau_2) d\tau_2, \quad k, l = 1, \ldots, n,$$

$$\tag{10.22}$$

$$\theta_{\Delta_k{}^0 \Delta_l{}^0} \approx \sum_{r=1}^{n} \sum_{j=1}^{n} \theta_{\Psi_r^0 \Psi_j^0} \Phi_{kr}(0) \Phi_{lj}(0), \quad k, l = 1, \ldots, n, \tag{10.23}$$

respectively, where

$$\theta_{\Psi_r^0 \Psi_j^0} \approx \int_{-\infty}^{+\infty} S_{\Psi_r^0 \Psi_j^0}(\lambda)d\lambda, \quad r,j = 1,\dots,n .\qquad (10.24)$$

In the case of narrowband systems, these approximations have to be replaced by the corresponding inequalities

$$\theta_{\Delta_k^0 \Delta_l^0}(t) \le \sum_{r=1}^{n}\sum_{j=1}^{n} \theta_{\Psi_r^0 \Psi_j^0}(t) \int_{t_0}^{t} g_{ki}(t,\tau_1)d\tau_1 \int_{t_0}^{t} g_{lj}(t,\tau_2)d\tau_2, \quad k,l = 1,\dots,n ,$$

$$(10.25)$$

$$\theta_{\Delta_k^0 \Delta_l^0} \le \sum_{r=1}^{n}\sum_{j=1}^{n} \theta_{\Psi_r^0 \Psi_j^0} \Phi_{kr}(0)\Phi_{lj}(0), \quad k,l = 1,\dots,n .\qquad (10.26)$$

Similarly, using (10.14), one can calculate the estimates for variables $m_{\Delta_k}$ for nonstationary and stationary systems as follows:

$$m_{\Delta_k}(t) \le \sum_{r=1}^{n} \int_0^t \tilde{g}_{kr}(t,\tau_1) \left[ \sum_{\alpha=1}^{n}\sum_{j=1}^{n} c_{k\alpha j}(\tau_1) \sum_{\beta=1}^{n}\sum_{\gamma=1}^{n} \theta_{\Psi_\beta^0 \Psi_\gamma^0}(\tau_1) \right.$$

$$\left. \times \int_0^t g_{\alpha\beta}(t,\tau_2)d\tau_2 \int_0^t g_{j\gamma}(t,\tau_3)d\tau_3 \right] d\tau_1, \quad k = 1,\dots,n , \quad (10.27)$$

$$m_{\Delta_k} \le \sum_{r=1}^{n} \tilde{\Phi}_{kr}(0) \sum_{\alpha=1}^{n}\sum_{j=1}^{n} c_{k\alpha j} \sum_{\beta=1}^{n}\sum_{\gamma=1}^{n} \theta_{\Psi_\beta^0 \Psi_\gamma^0} \Phi_{\alpha\beta}(0)\Phi_{j\gamma}(0), \quad k = 1,\dots,n ,$$

$$(10.28)$$

where $\tilde{g}_{kr}(t,\tau_1)$ and $\tilde{\Phi}_{kr}(\lambda)$, $k,r = 1,\dots,n$ are the impulse transfer function and its Fourier transform for system (10.14), respectively.

In particular case for nonlinearities depending on single variables, the obtained formulas for estimates simplify to the following forms:

for the nonstationary system

$$\theta_{\Delta_k^0 \Delta_l^0}(t) \le \sum_{r=1}^{n} \int_{-\infty}^{+\infty} \Psi_r^2(x_r,t)p(x_r,t)dx_r \int_{t_0}^{t} g_{kr}(t,\tau_1)d\tau_1 \int_{t_0}^{t} g_{lr}(t,\tau_2)d\tau_2,$$

$$k,l = 1,\dots,n ,$$

$$(10.29)$$

$$m_{\Delta_k}(t) \le \sum_{r=1}^{n} \int_0^t \tilde{g}_{kr}(t,\tau_1) \left[ c_{krr}(\tau_1)\theta_{\Psi_r^0 \Psi_r^0}(\tau_1) \left( \int_0^t g_{kr}(t,\tau_2)d\tau_2 \right)^2 \right] d\tau_1,$$

$$k = 1,\dots,n$$

$$(10.30)$$

and for the stationary system

$$\theta_{\Delta_k 0 \Delta_l 0} \leq \sum_{r=1}^{n} \int_{-\infty}^{+\infty} \Psi_r^2(x_r) p(x_r) dx_r \Phi_{kr}(0) \Phi_{lr}(0), \quad k, l = 1, \ldots, n , \quad (10.31)$$

$$m_{\Delta_k} \leq \sum_{r=1}^{n} \tilde{\Phi}_{kr}(0) c_{krr} \theta_{\Psi_r^0 \Psi_r^0} [\Phi_{rr}(0)]^2, \quad k = 1, \ldots, n . \quad (10.32)$$

where $p(x_r, t)$ and $p(x_r)$ are the probability density functions of a normal stochastic process and a normal variable, respectively; $c_{krr}$ are defined by (10.13).

An illustration of the considered estimates is given in the Kazakov example.

*Example 10.1.* [61] Consider the one-dimensional nonlinear dynamic system

$$dx(t) = [-ax(t) - ka\phi(x) + kam_v]dt + \sqrt{2\pi s_0} kad\xi(t) , \quad (10.33)$$

where
$$\phi(x) = \varepsilon sign(x(t)) , \quad (10.34)$$

$a, k, \varepsilon, m_v$, and $s_0$ are positive constants.

Applying statistical linearization we rewrite (10.33) in the form

$$dx(t) = [-ax(t) - ka\phi_0 - kak_1 x^0 - ka\Psi(x) + kam_v]dt + \sqrt{2\pi s_0} kad\xi(t) , \quad (10.35)$$

where $\phi_0$ and $k_1$ are linearization coefficients defined by

$$\phi_0 = 2\varepsilon \Phi \left( \frac{m_x}{\sqrt{\theta_x}} \right), k_1 = \frac{2\varepsilon}{\sqrt{2\pi\theta_x}} exp \left\{ -\frac{m_x^2}{2\theta_x} \right\},$$

$$\theta_x = E[(x - m_x)^2], m_x = E[x] ,$$

$$\Psi(x) = \varepsilon sign(x(t)) - \phi_0 - k_1(x - m_x),$$

$$\Phi(x) = \frac{1}{\sqrt{2\pi}} \int_0^x exp\{-\frac{\tau^2}{2}\} d\tau. \quad (10.36)$$

The differential equations for the mean value and the variance for the linearized system have the form

$$\dot{m}_x = -am_x - ka\phi_0(m_x, \theta_x) + kam_v , \quad (10.37)$$

$$\dot{\theta}_x = -2a\theta_x - 2kak_1\theta_x + 2\pi s_0 k^2 a^2 . \quad (10.38)$$

In the stationary case, the solutions satisfy the following algebraic equations:

$$m_x + 2\varepsilon k\Phi \left( \frac{m_x}{\sqrt{\theta_x}} \right) = km_v, \quad \theta_x \left[ 1 + \frac{2\varepsilon}{\sqrt{2\pi\theta_x}} exp \left\{ -\frac{m_x^2}{2\theta_x} \right\} \right] = \pi s_0 k^2 a . \quad (10.39)$$

The differential equations for the mean value of the error process $m_\Delta$ and the centralized error process $\Delta^0(t)$ have the form

$$\dot{m}_\Delta = -a m_\Delta + \frac{1}{2}\frac{m_x}{\theta_x} k_1 (m_\Delta^2 + \theta_{\Delta^0\Delta^0}) , \qquad (10.40)$$

$$d\Delta^0 = [-a\Delta^0 - k a k_1 \Delta^0 - ka\Psi(x + \Delta)]dt . \qquad (10.41)$$

From inequalities (10.31) and (10.32) follow the estimations for characteristics $m_\Delta$ and $\theta_{\Delta^0\Delta^0}$

$$m_\Delta \le \frac{k k_1 m_x}{2\theta_x} \theta_{\Delta^0\Delta^0}, \quad \theta_{\Delta^0\Delta^0} \le \frac{k^2}{(1 + kk_1)^2}\theta_\Psi , \qquad (10.42)$$

where

$$\theta_\Psi = \int_{-\infty}^{+\infty} \Psi^2(x)\frac{1}{\sqrt{2\pi\theta_x}}exp\left\{-\frac{(x - m_x)^2}{2\theta_x}\right\} dx = \varepsilon^2 \left[1 - 4\Phi^2\left(\frac{m_x}{\sqrt{\theta_x}}\right) - \frac{2}{\pi}\right] . \qquad (10.43)$$

Kazakov [61] compared the relative errors $\frac{m_\Delta}{m_x}$ and $\frac{\theta_\Delta}{\theta_x}$ in this example for five sets of parameters. The numerical calculations gave the errors in the intervals 0.03–0.40 and 0.07–0.35, respectively.

Another approach of the estimating of the accuracy of the solutions of linearized equations was proposed by Kolovskii [68]. He considered the nonlinear dynamic system described by an operator equation in the form

$$Q(p)x + R(p)f(x, \dot{x}) = \eta(t) , \qquad (10.44)$$

where $Q(p)$ and $R(p)$ are polynomials of the differential operator where the degree of $Q(p)$ is higher than the degree of $R(p)$, $f(x, \dot{x})$ is one-dimensional nonlinear function, and $\eta(t)$ is a stationary wideband process.

We approximate the nonlinear function by a linear one in the form

$$f(x, \dot{x}) \approx f_0 + k_x x + k_v \dot{x} \qquad (10.45)$$

and we introduce the nonlinear error of the approximation

$$\phi(x, \dot{x}) = f(x, \dot{x}) - f_0 - k_x x - k_v \dot{x} , \qquad (10.46)$$

where $f_0$, $k_x$, and $k_v$ are the linearization coefficients.

Next, we rewrite (10.44) using the function $\phi(x, \dot{x})$

$$[Q(p)x + R(p)(k_x + pk_v)]x + R(0)f_0 \doteq -R(p)\phi[x(t), \dot{x}(t)] + \eta(t) . \qquad (10.47)$$

The linearized equation is sought in the form

$$[Q(p)x + R(p)(k_x + pk_v)]x + R(0)f_0 = \eta(t) . \qquad (10.48)$$

The linearization coefficients one can find by minimization of the mean-square criterion

$$I = \theta_\phi = E[\phi^2(x, \dot{x})] - m_\phi^2$$

$$= E[(f(x, \dot{x}) - f_0 - k_x x - k_v \dot{x})^2] - (E[(f(x, \dot{x})$$

$$- f_0 - k_x x - k_v \dot{x}])^2 , \tag{10.49}$$

$$k_x = \frac{1}{\theta_x} E[f(x, \dot{x})(x - m_x)], \, k_v = \frac{1}{\theta_v} E[f(x, \dot{x})\dot{x}], \, f_0 = E[f(x, \dot{x})] - k_x m_x ,$$
$$\tag{10.50}$$

where $\theta_\phi = E[(\phi(x, \dot{x}) - m_\phi)^2], \ \theta_x = E[(x - m_x)^2], \ \theta_v = E[(\dot{x})^2],$ $m_x = E[x], \ m_\phi = E[\phi(x, \dot{x})]$. We note that in stationary case $E[\dot{x}] = 0$ and $E[(x\dot{x})] = 0$.

We denote the solutions of (10.47) and (10.48) by $x_N(t)$ and $x_L(t)$, respectively, and the error process and its derivative by

$$\Delta(t) = x_N(t) - x_L(t), \quad \Delta_v(t) = \dot{\Delta}(t) . \tag{10.51}$$

One can derive using (10.47) and (10.48) the differential equation for the error process

$$[Q(p)x(t) + R(p)(k_x + pk_v)]\Delta = -R(p)\phi[x(t), \dot{x}(t)] . \tag{10.52}$$

Then one can derive estimates for the mean value and variances of the processes $\Delta(t)$ and $\dot{\Delta}(t)$

$$m_\Delta = E[\Delta] = \frac{-R(0)}{Q(0) + k_v R(0)} m_\phi, \quad m_{\Delta_v} = E[\Delta_v] = 0 , \tag{10.53}$$

$$\theta_\Delta = \int_{-\infty}^{+\infty} S_\Delta(\lambda)d\lambda = \int_{-\infty}^{+\infty} \left| \frac{R(i\lambda)}{Q(i\lambda) + R(i\lambda)(k_x + k_v i\lambda)} \right|^2 S_\phi(\lambda)d\lambda , \tag{10.54}$$

$$\theta_{\Delta_v} = \int_{-\infty}^{+\infty} S_{\Delta_v}(\lambda)d\lambda = \int_{-\infty}^{+\infty} \left| \frac{R(i\lambda)i\lambda}{Q(i\lambda) + R(i\lambda)(k_x + k_v i\lambda)} \right|^2 S_\phi(\lambda)d\lambda ,$$
$$\tag{10.55}$$

where $\theta_\Delta = E[(\Delta - m_\Delta)^2], \ \theta_{\Delta_v} = E[(\Delta_v - m_{\Delta_v})^2], \ S_\phi(\lambda), \ S_\Delta(\lambda)$ and $S_{\Delta_v}(\lambda)$ are power spectral density functions of processes $\phi(x(t), \dot{x}(t)), \ \Delta(t)$ and $\dot{\Delta}(t)$, respectively.

The estimates presented in (10.54) and (10.55) are inconvenient in applications, because the power spectral density function $S_\phi$ is usually unknown (the solution appearing in the nonlinear error $\phi(x, \dot{x})$ remains unknown). Some simplifications can be done the function $\phi(x, \dot{x})$ satisfies Lipshitz condition or when we consider separately the cases of wideband and narrowband filter systems.

In the case when the operational transfer function $L(p)$ defined by

$$L(p) = \frac{R(p)}{Q(p) + R(p)(k_x + k_v p)} \tag{10.56}$$

represents a wideband filter with a frequency border $\lambda_b$, i.e., $S_\phi(\lambda)$ is sufficient small for $|\lambda| < \lambda_b$, then an approximation can be used

$$\theta_\Delta \approx \int_{-\lambda_b}^{+\lambda_b} |L(i\lambda)| S_\phi(\lambda) d\lambda \approx L^2(0) \int_{-\lambda_b}^{+\lambda_b} S_\phi(\lambda) d\lambda \approx L^2(0) \theta_\phi \tag{10.57}$$

and in a similar way one can approximate

$$\mu_\Delta^2 = m_\Delta^2 + \theta_\Delta \approx L^2(0) \mu_\phi^2 , \tag{10.58}$$

where $\mu_\phi$ is the root mean-square value of $\phi(x(t), \dot{x}(t))$.

In the case of the dynamic system with a weak nonlinearity, the dependence of $|L(i\lambda)|$ on the parameters $k_x$ and $k_v$ is weak and it can be assumed that $|R(i\lambda)(k_x + k_v i\lambda)| << |Q(i\lambda)|$ for $|\lambda| < \lambda_b$. Therefore, for partially linearized system (10.48) ($k_x = k_v = 0$), the above characteristics have the form

$$\theta_{\Delta_0} \approx \int_{-\lambda_b}^{+\lambda_b} \left| \frac{R(i\lambda)}{Q(i\lambda)} \right| S_\phi(\lambda) d\lambda \approx L^2(0) \theta_f \tag{10.59}$$

and

$$\mu_{\Delta_0}^2 \approx L^2(0) \mu_f^2 \tag{10.60}$$

where $\mu_f$ is the root mean square value of $f(x(t), \dot{x}(t))$, $\theta_f = E[(f(x, \dot{x}) - m_f)^2]$, $m_f = E[f(x, \dot{x})]$.

The estimations (10.59) and (10.60) follow from the fact that the inequality $R(0) << Q(0)$ implies the approximation $\frac{R(0)}{Q(0)} \approx L(0)$.

As a criterion of the possible effectiveness of statistical linearization Kolovskii proposed [68] the ratio of $\mu_\Delta$ to the value $\mu_{\Delta_0}$. From (10.57–10.60), we obtain

$$\frac{\mu_\Delta}{\mu_{\Delta_0}} = \frac{\mu_\phi}{\mu_f} \approx \sqrt{\frac{\theta_\phi}{\theta_f}} . \tag{10.61}$$

The substitution of linearization coefficients determined by (10.50) to the criterion (10.49) yields

$$\theta_{\phi \min} = \theta_f - k_x^2 \theta_x - k_v^2 \theta_v - m_\phi^2$$
$$= \theta_f - \frac{1}{\theta_x} \left( E[f(x, \dot{x}) x] - E[f(x, \dot{x})] m_x \right)^2 - \frac{1}{\theta_v} \left( E[(fx, \dot{x}) \dot{x}] \right)$$
$$- \left( E[f(x, \dot{x}) - f_0 - k_x x - k_v \dot{x}] \right)^2 . \tag{10.62}$$

To illustrate the obtained results, Kolovskii [68] considered a few examples.

For instance, for the nonlinear function from Example 10.1, i.e., $f(x, \dot{x}) = \varepsilon sign(x)$ $m_\eta = 0$, $m_x = 0$, $m_\phi = 0$, $k_v = 0$, and

$$E[(f(x, \dot{x})x] = \varepsilon \int_{-\infty}^{+\infty} \tau sgn(\tau) \frac{1}{\sqrt{2\pi\theta_x}} \exp\left\{-\frac{\tau^2}{2\theta_x}\right\} d\tau = \varepsilon\sqrt{\frac{2\theta_x}{\pi}} . \qquad (10.63)$$

Since in our example $\theta_f = \varepsilon^2$ we obtain from (10.61)

$$\sqrt{\frac{\theta_{\phi min}}{\theta_f}} = \sqrt{1 - \frac{2}{\pi}} \approx 0.602 . \qquad (10.64)$$

Now we discuss this criterion of effectiveness of statistical linearization on a simplified scalar dynamic system considered in Example 5.1.

*Example 10.2.* Consider the nonlinear scalar dynamic system

$$dx = -Ax^3 dt + \sqrt{q}d\xi, \quad x(t_0) = x_0 , \qquad (10.65)$$

where $A > 0$ and $q > 0$ are constant coefficients and $\xi$ is a standard Wiener process, and the initial condition $x_0$ is a random variable independent of $\xi$.

We assume that the stationary solution $x(t)$ is approximately a Gaussian process $N(0, \sigma_x)$, and the probability density function is given by (5.16) for $\sigma_x^2 = \theta_x$ and $m_x = E[x] = 0$.

The linearization coefficients for the nonlinear function $Y = Ax^3$ are defined by (5.80–5.83) for $m_x = E[x] = 0$, i.e.,

Criterion 1 − SL

$$k_1^{(1)} = \sqrt{\frac{D\Phi}{\sigma_x^2}} = A\sqrt{15}\sigma_x^2 . \qquad (10.66)$$

Criterion 2 − SL

$$k_1^{(2)} = 3A\sigma_x^2 . \qquad (10.67)$$

Criterion 3 − SL

$$k_1^{(3)} = 2\sqrt{\frac{E[U^2(x)]}{E[x^4]}} = \frac{\sqrt{35}}{2}A\sigma_x^2 . \qquad (10.68)$$

Criterion 4 − SL

$$k_1^{(4)} = 2\frac{E[U(x)x^2]}{E[x^4]} = \frac{5}{2}A\sigma_x^2 . \qquad (10.69)$$

where $U(x) = A\frac{x^4}{4}$ is the potential energy of the nonlinear element.

Hence, we find that all linearization coefficients are linearly dependent on $\sigma_x^2$, i.e.,

$$k_1^{(j)} = \alpha_j A \sigma_x^2, \quad j = 1, 2, 3, 4, \tag{10.70}$$

where

$$\alpha_1 = \sqrt{15}, \quad \alpha_2 = 3, \quad \alpha_3 = \frac{\sqrt{35}}{2}, \quad \alpha_4 = \frac{5}{2}. \tag{10.71}$$

Since the nonlinear function is an odd function and the mean zero excitations $m_\eta = 0$, we also have $m_x = 0$, $m_\phi = 0$, $k_v = 0$, and $\theta_f = A^2 E[x^6]$.

To calculate the square of the ratio $\sqrt{\frac{\theta_{\phi \min}}{\theta_f}}$ for all linearization coefficients, we use the simplified mean square criterion defined by (10.49), i.e.,

$$I = \theta_\phi = E[\phi^2(x, \dot{x})] - m_\phi^2 = E[(Ax^3 - k_x x)^2]. \tag{10.72}$$

Then we obtain

$$\theta_{\phi_j \min} = E[(Ax^3 - \alpha_j A \sigma_x^2 x)^2]$$

$$= \theta_f - 2\alpha_j A \sigma_x^2 E[x^4] + \alpha_j^2 A^2 \sigma_x^4 E[x^2], \quad j = 1, 2, 3, 4. \tag{10.73}$$

Hence we check the square of the ratio $\sqrt{\frac{\theta_{\phi \min}}{\theta_f}}$

$$\sqrt{\frac{\theta_{\phi_j \min}}{\theta_f}} = \sqrt{1 - 2\alpha_j \frac{\sigma_x^2 E[x^4]}{E[x^6]} + \alpha_j^2 \frac{\sigma_x^4 E[x^2]}{E[x^6]}}$$

$$= \sqrt{1 - \frac{2}{5}\alpha_j + \frac{1}{15}\alpha_j^2}, \quad j = 1, 2, 3, 4. \tag{10.74}$$

The calculations yield

$$\frac{\theta_{\phi_1 \min}}{\theta_f} = -\frac{2}{5}3.87, \quad \frac{\theta_{\phi_2 \min}}{\theta_f} = 1 - \frac{9}{15}, \quad \frac{\theta_{\phi_3 \min}}{\theta_f} = 1 - \frac{\sqrt{35}}{5} + \frac{7}{12}, \quad \frac{\theta_{\phi_4 \min}}{\theta_f} = \frac{6.25}{15}, \tag{10.75}$$

Hence it follows that the estimation for Criterion $1^0$ is incorrect while for Criteria $2^0$–$4^0$ are equal

$$\sqrt{\frac{\theta_{\phi_2 \min}}{\theta_f}} = \sqrt{1 - \frac{9}{15}} \approx 0.632, \quad \sqrt{\frac{\theta_{\phi_3 \min}}{\theta_f}} = \sqrt{\frac{19}{12} - \frac{\sqrt{35}}{5}} \approx 0.632,$$

$$\sqrt{\frac{\theta_{\phi_4 \min}}{\theta_f}} = \sqrt{\frac{6.25}{15}} \approx 0.645. \tag{10.76}$$

Now we extend the Kolovskii's approach to the case when instead of the mean-square criterion for displacements (10.49), we will minimize the mean-square criterion for potential energies of the nonlinear element $f(x)$ and its linearized form $k_x x$, i.e.,

$$I_U = \theta_{U_\phi} = E\left[\left(U(f(x)) - k_x\frac{x^2}{2}\right)^2\right].\tag{10.77}$$

After transformation we obtain

$$I_U = \theta_{U_\phi} = E\left[(U^2(f(x))] - k_x E\left[U(f(x))x^2\right] + \frac{k_x^2}{4}E[x^4]\right.$$

$$= \theta_{U_f} - k_x E[U(f(x))x^2] + \frac{k_x^2}{4}E[x^4].\tag{10.78}$$

In the case of nonlinear function $f(x) = Ax^3$ and four linearization coefficients
$$k_1^{(j)} = \alpha_j A\sigma_x^2, \quad j = 1,2,3,4,\text{ we find}$$

$$\theta_{U_{\phi j\,\min}} = E\left[\left(\left(A\frac{x^4}{4}\right)^2 - \alpha_j A\sigma_x^2\frac{x^2}{2}\right)^2\right]$$

$$= \theta_{U_f} - \frac{1}{4}\alpha_j A\sigma_x^2 E[x^6] + \alpha_j^2 A^2\sigma_x^4 E[x^4], \quad j = 1,2,3,4\tag{10.79}$$

Hence, we calculate the ratio $\sqrt{\dfrac{\theta_{U_{\phi j\,\min}}}{\theta_{U_f}}}$

$$\sqrt{\frac{\theta_{U_{\phi j\,\min}}}{\theta_{U_f}}} = \sqrt{1 - \alpha_j\frac{\sigma_x^2 E[\frac{x^6}{4}]}{E[\frac{x^8}{16}]} + \alpha_j^2\frac{\sigma_x^4 E[x^4]}{4E[\frac{x^8}{16}]}}$$

$$= \sqrt{1 - \frac{4}{7}\alpha_j + \frac{4}{35}\alpha_j^2}, \quad j = 1,2,3,4.\tag{10.80}$$

The calculations yield

$$\sqrt{\frac{\theta_{U_{\phi 1\,\min}}}{\theta_{U_f}}} = \sqrt{1 - \frac{4\sqrt{15}}{7} + \frac{12}{7}} \approx 0.707, \quad \sqrt{\frac{\theta_{U_{\phi 2\,\min}}}{\theta_{U_f}}} = \sqrt{1 - \frac{12}{7} + \frac{36}{35}} \approx 0.56,$$

$$\sqrt{\frac{\theta_{U_{\phi 3\,\min}}}{\theta_{U_f}}} = \sqrt{1 - \frac{4\sqrt{35}}{14} + 1} \approx 0.556, \quad \sqrt{\frac{\theta_{U_{\phi 4\,\min}}}{\theta_{U_f}}} = \sqrt{1 - \frac{10}{7} + \frac{5}{7}} \approx 0.534.$$

$$\tag{10.81}$$

In the case of the dynamic system with the same wide band filter operator $L(p)$ and a nonlinear "not weak" function $f(x, \dot{x})$, the linearization coefficients are derived from the minimization of $\mu_\Delta^2$ defined also in this case by (10.58)

$$\mu_\Delta^2 \approx L^2(0)\mu_\phi^2 = \frac{R^2(0)}{[Q(0) + k_x R(0)]^2}\mu_\phi^2.\tag{10.82}$$

Kolovskii [68] has shown using the conditions

$$\frac{\partial\mu_\Delta^2}{\partial k_x} = 0, \quad \frac{\partial\mu_\Delta^2}{\partial k_v} = 0,\tag{10.83}$$

that the linearization coefficients $f_0$ and $k_v$ have the form as in (10.58), while $k_x$ is defined by

$$k_x = \frac{Q(0)E[f(x,\dot{x})(x - m_x)] + R(0)(\theta_f - k_v^2\theta_v)}{Q(0)\theta_x + R(0)E[f(x,\dot{x})(x - m_x)]}. \qquad (10.84)$$

Kolovskii [68] has also considered the case of the dynamic system with the narrowband filter operator $L(p)$ for two cases when $L(i\lambda)$ depends weakly and strongly on the method of the determination of linearization coefficients (for details, see [68]). We note only that the error of approximation for the cubic nonlinear function was smaller in the case of the narrowband filter operator $L(p)$ than in the case of the wideband filter.

## 10.2 Comparison of Linearized and Exact Response Characteristics

Unfortunately, the theoretical study of the accuracy of linearization methods are complicated and practically limited to one- and two-dimensional dynamic systems. Usually, the authors compare the response characteristic obtained by their new linearization methods with the corresponding exact one. We also show such a comparison for a group of nonlinear stochastic dynamic systems. The are one-degree-of-freedom systems with nonlinear spring functions modeled by odd order polynomials under Gaussian white noise external excitation and described by the Ito stochastic differential equations, i.e.,

$$dx_1(t) = x_2(t)dt$$
$$dx_2(t) = [-2h_1x_2(t) - \lambda_{01}^2 x_1(t) - \varepsilon_1 f(x_1(t))]dt + q_1 d\xi_1(t) , \quad (10.85)$$

where $h_1, \lambda_{01}, \varepsilon_1$, and $q_1$ are constant parameters and the nonlinear function $f(x_1)$ is considered as one of the fifth functions $f(x_1) = f_i(x_1)$, $i = 1, \ldots, 5$, modeled by odd order polynomials, i.e.,

$$f_1(x_1) = sign(x_1), \quad f_2(x_1) = x_1^2 sign(x_1), \quad f_3(x_1) = x_1^3,$$
$$f_4(x_1) = x_1^4 sign(x_1), \quad f_5(x_1) = x_1^5. \qquad (10.86)$$

The Ito equations for linearized system are

$$dx_1(t) = x_2(t)dt$$
$$dx_2(t) = [-2h_1x_2(t) - \lambda_{01}^2 x_1(t) - \varepsilon_1 k(E[x_1^2(t)])x_1(t)]dt + q_1 d\xi_1(t),$$
$$\qquad (10.87)$$

where $k$ is a linearization coefficient.

The moment equation for linearized system are

$$\frac{dE[x_1^2(t)]}{dt} = 2E[x_1(t)x_2(t)] ,$$

$$\frac{dE[x_1(t)x_2(t)]}{dt} = E[x_2^2(t)] - 2h_1 E[x_1(t)x_2(t)]$$
$$- \lambda_{01}^2 E[x_1^2(t)] - \varepsilon_1 k_i(E[x_1^2(t)])E[x_1^2(t)] ,$$

$$\frac{dE[x_2^2(t)]}{dt} = -4h_1 E[x_2^2(t)] - 2\lambda_{01}^2 E[x_1(t)x_2(t)]$$
$$- 2\varepsilon_1 k_i(E[x_1^2(t)])E[x_1(t)x_2(t)] + q_1^2. \tag{10.88}$$

We find the stationary moments for linearized system from equations

$$2E[x_1(t)x_2(t)] = 0 ,$$

$$E[x_2^2(t)] \; -2h_1 E[x_1(t)x_2(t)] - \lambda_{01}^2 E[x_1^2(t)] - \varepsilon_1 k_i(E[x_1^2(t)])E[x_1^2(t)] = 0 ,$$

$$-4h_1 E[x_2^2(t)] - 2\lambda_{01}^2 E[x_1(t)x_2(t)]$$
$$-2\varepsilon_1 k_i(E[x_1^2(t)])E[x_1(t)x_2(t)] + q_1^2 = 0. \tag{10.89}$$

In our accuracy analysis, we compare four basic moment criteria introduced in Sect. 5 and also considered in previous subsection, i.e., criteria $1^0$–$4^0$ defined by (5.5) and (5.6), (5.17) and (5.18), (5.61), and (5.64), respectively.

They have the form

$$
\begin{aligned}
&\text{for} \quad sgn(x_1) \quad k_i = \alpha_i \sigma_{x_1}^{-1} \\
&\text{for} \quad x_1^2 sgn(x_1) \quad k_i = \beta_i \sigma_{x_1} \\
&\text{for} \quad\quad x_1^3 \quad\quad k_i = \gamma_i \sigma_{x_1}^2 \quad i = 1,2,3,4 . \\
&\text{for} \quad x_1^4 sgn(x_1) \quad k_i = \delta_i \sigma_{x_1}^3 \\
&\text{for} \quad\quad x_1^5 \quad\quad k_i = \rho_i \sigma_{x_1}^4
\end{aligned}
\tag{10.90}
$$

**Table 10.1.** The linearization coefficients $k_i$ that determine $\alpha_i$, $\beta_i$, $\gamma_i$, $\delta_i$ and $\rho_i$

| Function | Potential energy | $k_1$ | $k_2$ | $k_3$ | $k_4$ |
|---|---|---|---|---|---|
| $sgn(x_1)$ | $|x_1|$ | $\sigma_{x_1}^{-1}$ | $\sqrt{\frac{2}{\pi}}\,\sigma_{x_1}^{-1}$ | $\frac{2}{3}\sqrt{3}\,\sigma_{x_1}^{-1}$ | $\frac{4}{3}\sqrt{\frac{2}{\pi}}\,\sigma_{x_1}^{-1}$ |
| $x_1^2 sgn(x_1)$ | $\frac{|x_1|^3}{3}$ | $\sqrt{3}\,\sigma_{x_1}$ | $2\sqrt{\frac{2}{\pi}}\,\sigma_{x_1}$ | $\frac{2}{3}\sqrt{5}\,\sigma_{x_1}$ | $\frac{16}{9}\sqrt{\frac{2}{\pi}}\,\sigma_{x_1}$ |
| $x_1^3$ | $\frac{x_1^4}{4}$ | $\sqrt{15}\,\sigma_{x_1}^2$ | $3\,\sigma_{x_1}^2$ | $\frac{\sqrt{35}}{2}\,\sigma_{x_1}^2$ | $\frac{5}{2}\,\sigma_{x_1}^2$ |
| $x_1^4 sgn(x_1)$ | $\frac{|x_1|^5}{5}$ | $\sqrt{105}\,\sigma_{x_1}^3$ | $8\sqrt{\frac{2}{\pi}}\,\sigma_{x_1}^3$ | $3\sqrt{\frac{7}{5}}\,\sigma_{x_1}^3$ | $\frac{32}{5}\sqrt{\frac{2}{\pi}}\,\sigma_{x_1}^3$ |
| $x_1^5$ | $\frac{x_1^6}{6}$ | $\sqrt{945}\,\sigma_{x_1}^4$ | $15\,\sigma_{x_1}^4$ | $\sqrt{385}\,\sigma_{x_1}^4$ | $\frac{35}{3}\,\sigma_{x_1}^4$ |

**Table 10.2.** The algebraic equations for stationary solutions of linearized systems

| Function | Algebraic moment equation |
|---|---|
| $sgn(x_1)$ | $\frac{q_1^2}{4h_1} - \left(\lambda_{01}^2 + \varepsilon_1\alpha_i\frac{1}{\sigma_{x_1}}\right)\sigma_{x_1}^2 = 0$ |
| $x_1^2 sgn(x_1)$ | $\frac{q_1^2}{4h_1} - \left(\lambda_{01}^2 + \varepsilon_1\beta_i\sigma_{x_1}\right)\sigma_{x_1}^2 = 0$ |
| $x_1^3$ | $\frac{q_1^2}{4h_1} - \left(\lambda_{01}^2 + \varepsilon_1\gamma_i\sigma_{x_1}^2\right)\sigma_{x_1}^2 = 0$ |
| $x_1^4 sgn(x_1)$ | $\frac{q_1^2}{4h_1} - \left(\lambda_{01}^2 + \varepsilon_1\delta_i\sigma_{x_1}^3\right)\sigma_{x_1}^2 = 0$ |
| $x_1^5$ | $\frac{q_1^2}{4h_1} - \left(\lambda_{01}^2 + \varepsilon_1\rho_i\sigma_{x_1}^4\right)\sigma_{x_1}^2 = 0$ |

The coefficients $\alpha_i, \beta_i, \gamma_i, \delta_i$, and $\rho_i$ are defined by the values given in Table 10.1.

The stationary solutions of linearized systems satisfy the corresponding algebraic equations as shown in Table 10.2.

Solving the algebraic equations given in Table 10.2 with coefficients $\alpha_i, \beta_i, \gamma_i, \delta_i$, and $\rho_i$ determined in Table 10.1 we find the corresponding values for $\sigma_{x_1 lin}^2$.

The exact values of the second-order moments for the stationary solution of (10.85) $\sigma_{x_1 exact}^2$ we find using the exact probability density function, i.e.,

$$\sigma_{x_1 exact}^2 = E[x_1^2]_{exact} = \int_{-\infty}^{+\infty}\int_{-\infty}^{+\infty} x_1^2 g_{N1}(x_1, x_2)dx_1 dx_2 , \qquad (10.91)$$

where

$$g_{N1}(x_1, x_2) = c_1 \exp\left\{-\frac{2h_1}{q_1^2}\left[x_2^2 + \lambda_{01}^2 x_1^2 + 2\varepsilon_1\int_0^{x_1} f(s)ds\right]\right\} , \qquad (10.92)$$

$$c_1^{-1} = \int_{-\infty}^{+\infty}\int_{-\infty}^{+\infty} \exp\left\{-\frac{2h_1}{q_1^2}\left[x_2^2 + \lambda_{01}^2 x_1^2 + 2\varepsilon_1\int_0^{x_1} f(s)ds\right]\right\} dx_1 dx_2 \qquad (10.93)$$

For all groups of linearization coefficients $k_i$, $i = 1, 2, 3, 4$, we calculate the relative errors defined by

$$\Delta_{f(x_1)} = \frac{|\sigma_{x_1 exact}^2 - \sigma_{x_1 lin}^2|}{\sigma_{x_1 exact}^2} . \qquad (10.94)$$

The results for parameters $h_1 = 0.05$, $\lambda_{01} = 1$, $\varepsilon_1 = 0.1$, and $q_1 = 1$ are summarized in Table 10.3.

**Table 10.3.** The numerical results for parameters $h_1 = 0.05$, $\lambda_{01} = 1$, $\varepsilon_1 = 0.1$ and $q_1 = 1$

| Relative error | $k_1$ | $k_2$ | $k_3$ | $k_4$ |
|---|---|---|---|---|
| $\Delta_{sgn(x_1)}$ | 0.0089 | 0.0001 | 0.0158 | 0.0118 |
| $\Delta_{x_1^2 sgn(x_1)}$ | 0.0232 | 0.0056 | 0.0086 | 0.0187 |
| $\Delta_{x_1^3}$ | 0.1274 | 0.0529 | 0.0487 | 0.0010 |
| $\Delta_{x_1^4 sgn(x_1)}$ | 0.2620 | 0.1375 | 0.0356 | 0.0740 |
| $\Delta_{x_1^5}$ | 0.3735 | 0.2224 | 0.2823 | 0.1625 |

One can observe from the obtained results that the relative errors are growing for higher-order polynomial spring functions and the criteria for potential energies give better approximations than criteria for displacements. As it was mentioned earlier, the relative errors are in the interval $[1 - 40]$ percentage.

## 10.3 Comparison of Linearized and Simulated Response Characteristics

The comparison presented in Sect. 10.2 is limited to simple dynamic systems for which the probability density function of the stationary solution of nonlinear system is known in an exact form. Also, the calculation of multiple integrals, which is necessary in the determination of exact response characteristics by numerical integration is very slow and inaccurate in higher dimension systems. Therefore, the verification of linearization methods for multi-degree-of-freedom systems was usually done by a comparison of their response characteristics with their corresponding ones obtained by simulation. The literature results one can collect into two groups depending on the type of simulated excitation.

The first group consist of artificial excitations usually wideband stationary processes generated by computer in the form of piecewise constant realizations in the form [30]

$$x_r(t) = \lim_{N \to \infty} x_r^N(t) = \lim_{N \to \infty} \sum_{k=-N}^{N-1} a_{kr} \gamma(t - k\Delta t - \varepsilon_r), \quad r = 1, 2, \ldots, \quad (10.95)$$

where $x_r^N$ is a realization of $2N\Delta t$ length, $a_{kr}$ are independent zero-mean random variables with given probability distribution, $\Delta t$ is a constant time interval for a given order of approximation $N$, $\varepsilon_r$ is a random variable uniformly distributed on the interval $[0, \Delta t]$ independent of variables $a_{kr}$, and $\gamma(t) = 1(t) - 1(t - \Delta t)$ is the deterministic window function, i.e.,

$$1(t) = \begin{cases} 1 \text{ for } t > 0 \\ 0 \text{ for } t \leq 0. \end{cases}$$

A modification of this type of the generator of pseudorandom sample paths is defined by

$$x_r(t) = \left(1 - \frac{t - \varepsilon_r - k\Delta t}{\Delta t}\right) a_{kr} + \frac{t - \varepsilon_r - k\Delta t}{\Delta t} a_{k+1r}, \quad r = 1, 2, \ldots \quad (10.96)$$

for $\varepsilon_r + k\Delta t \leq t \leq \varepsilon_r + (k+1)\Delta t$. The realizations of random variables $a_k = cG_k$ appear in equal time intervals $\Delta t$, where $G_k$, $k = 1, 2, \ldots$ are standard Gaussian random variables $N(0, 1)$. The parameter $c$ is the scale coefficient defined by

$$c = \sqrt{\frac{S_0}{\Delta t}}, \qquad (10.97)$$

where $S_0$ is the power spectral density of the approximated Gaussian white noise.

Clough and Penzien [30] have shown that the power spectral density of the process (10.96) has the form

$$S_x(\lambda, \Delta t) = S_0 \frac{6 - 8\cos\lambda\Delta t + 2\cos 2\lambda\Delta t}{\lambda^4(\Delta t)^4}. \qquad (10.98)$$

Another type of the generator of pseudorandom sample paths is a sum of harmonics with random phases and deterministic or random amplitudes defined by

$$x_r(t) = \lim_{N\to\infty} x_r^N(t) = \lim_{N\to\infty} \sum_{k=1}^{N} A_k \cos(\lambda_k t + \phi_k), \quad r = 1, 2, \ldots, \quad (10.99)$$

where $\phi_k$ are independent random variables uniformly distributed in the interval $[0, 2\pi]$, $A_k$ and $\lambda_k$ are deterministic or random coefficients depending on the proposed approach, for instance, Shinozuka and Jan [91] and Shinozuka [89] proposed

$$A_k = \sqrt{2S_0(\lambda_k)\Delta\lambda},$$

$$\lambda_k = \left(k - \frac{1}{2}\right)\Delta\lambda + \delta\lambda_k + \lambda_{\min},$$

$$\Delta\lambda = \frac{\lambda_{\max} - \lambda_{\min}}{N}, \qquad (10.100)$$

where $\delta\lambda_k$ are independent random variables uniformly distributed in the interval $[-\frac{\delta\lambda'}{2}, \frac{\delta\lambda'}{2}]$, for $\delta\lambda' \ll \delta\lambda$, $\lambda_{min}$ and $\lambda_{max}$ are the minimal and the maximal frequency in the considered frequency interval, respectively. In [89] it is suggested to use $\delta\lambda' = 0.05\,\delta\lambda$.

One can find the details regarding parameters of random generators and correlation functions and spectral characteristics in many monographs and survey papers, for instance, [30, 90, 89]. These two basic approaches were developed in order to increase the efficiency of Monte Carlo simulation. In particular, the methods of increasing the efficiency of generating samples, reducing the necessary sample size for a specified accuracy and controlling Monte Carlo simulation. In this last case, samples are divided and in different intervals there are different densities of random points. One can find a review of these methodologies, for instance, in [83].

The simulation of wideband and narrowband processes is usually done using the proper filters as it was discussed in previous chapters. Usually, the parameters of filters are identified from a proposed power spectral density of the response of a stochastic dynamic system. In this case as well the structure of the filter is identified as the parameters for a given structure are estimated. Usually, the literature methods are based on identification procedures in frequency domain, see, for instance, [86] or ARMA approach [35]. It was discussed partially in Sect. 6.4. The methods of simulation of nonstationary and non-Gaussian processes were discussed partially in Sects. 5.4. and 5.5, respectively.

As was mentioned earlier in many papers, the authors have verified their linearization methods by simulation. The following examples of SDOF systems were considered: in the plate model [24, 46], in the beam model with potential energy criteria [39, 44], with friction dampers [42, 56], with tuned mass damper with nonlinear viscous damping [85], for the squeeze film system [9], for the Duffing oscillator under narrowband external excitation [58], the Duffing oscillator under colored noise external excitation [64, 87], with the power spectral density criteria in [10, 11], with polynomial nonlinearities and parametric excitations [16, 20, 43, 98, 118], with polynomial nonlinearities and continuous non-Gaussian excitations [96], with polynomial nonlinearities and Poisson white noise external excitations [96, 47] and parametric excitations [95, 82], with hysteresis under external excitations in [3, 4, 5, 6, 12, 36, 45, 48, 53, 54, 65, 76, 77, 117, 112], with hysteresis and parametric excitations [19], and with polynomial nonlinearities and nonstationary external excitations [57, 94].

There are also examples of MDOF systems where simulation were used to verify the accuracy of linearization methods. The following examples of MDOF systems were considered: with polynomial nonlinearities [71, 72, 108], for MDOF systems of construction elements [23, 25, 29, 37, 48, 63, 101, 106], with friction dampers [18, 55], with friction dampers and energy criteria [119], with the power spectral density criteria in [99], with polynomial nonlinearities and parametric excitations [16], for MDOF system by applying finite element models [41, 73, 88], and with friction under Poisson white noise excitation [81].

The second group of excitations used in the comparison study of linearized models are records of natural excitations registered during earthquake or wind excitations, for instance, the 1940 El Centro and the 1985 Mexico City earthquakes. They were modeled by Kanai–Tajimi filter model for the 1940 El Centro and the 1985 Mexico City earthquakes described by (9.109) and (9.110). The following examples of SDOF systems were considered: [8, 59, 67], SDOF primary system with attached SDOF secondary system [52, 122], for one-story model with a base isolation system [59].

In the case of MDOF systems with tuned liquid damper [114, 115], with hysteresis usually used in the response analysis of multistory buildings in [7, 27, 28, 33, 40, 66, 78, 111, 105, 104].

The simulations were also used in statistical tests of accuracy of linearization methods. Chang and Yong [22] proposed to apply chi-square Gaussian goodness–of–fit test as a measure of the discrepancy between the output non-Gaussian process of a nonlinear parametrically and externally excited dynamic system and the Gaussian probability density function.

The accuracy of the nonlinearization methods discussed in some papers and reviewed in Chap. 7 was verified by simulations. For instance, in SDOF system [17, 38, 70, 80, 84, 102, 109, 110, 121, 123, 124, 125] and in MDOF system [60, 103].

## 10.4 Validation of Linearization Method by Experiments

The usability of linearization methods in modeling and response analysis was also verified by experiments. Unfortunately, the number of verifications by experiments is significantly less than by simulation methods. In this subsection we describe a few of them.

Hanawa and Shimizu [50] have studied a shaking table experiment for the piping system with a friction support in two directions. It was done on the large scale electric–hydraulic shaking table ($15[m] \times 10[m]$) in the National Research Institute for Earth Science and Disaster Prevention in Japan. The experimental response characteristics were compared with corresponding theoretical ones obtained by application of statistical linearization. Detailed explanation on the vibration experiment is given [51] and a brief summary in [50]. Figure 10.1 shows the piping model with 2-dimensional in-plane sliding motion. The important point of the model are such as restrains, friction support, anchors and displacement sensors are indicated by symbols R, F, A and D, respectively. The excitations in the form of narrow band stochastic process were synthesized by summation of harmonics.

The piping system has been so designed that the contact point of the piping moves in-plane two-dimensional motion on the teflon bearing plate supported by the shaking table. The modal equations of motion are given by

$$\ddot{x}_i(t) + 2\zeta_{0i}\lambda_{0i}\dot{x}_i(t) + \lambda_{0i}^2 x_i(t) = -\beta_i\ddot{\xi}_g - \frac{F_0}{M_i}\frac{\dot{x}_i(t)}{\sqrt{\dot{x}_1^2(t) + \dot{x}_2^2(t)}} , \qquad (10.101)$$

for the first and second modes of vibration, where $i = 1, 2$, $\lambda_{0i}$, $\zeta_{0i}$, and $M_i$ denote the $i$th natural angular frequency, modal damping ratio, and modal

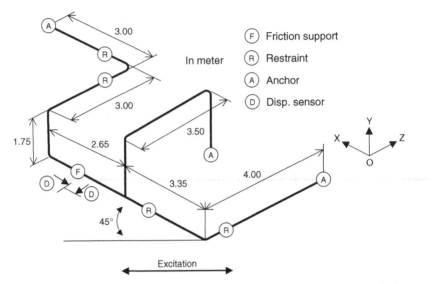

**Fig. 10.1.** Experimental set–up of piping system, reproduced from [50]

mass, respectively; $\beta_i = m_i/M_i$ and $F_0 = \mu_d N$, $m_i$, $\mu_d$, and $N$ are modal excitation mass, frictional coefficient, and normal load, respectively; $\ddot{\xi}_g$ is a narrowband stochastic process.

Displacements of the piping relative to the friction bearing at the friction support $U_x, U_y$, and $U_z$ can be approximated according to [50] by

$$U_x \simeq x_1 + \rho_1 x_2, \quad U_y \simeq 0, \quad U_z \simeq \rho_2 x_1 + x_2 , \tag{10.102}$$

where $\rho_i \ll 1$, $i = 1, 2$ are constants. Hence, it follows that relative displacements and relative velocities $V_x$ and $V_z$ are

$$U_x \simeq x_1, \quad U_z \simeq x_2,$$
$$V_x = \dot{U}_x \simeq \dot{x}_1, \quad V_z = \dot{U}_z \simeq \dot{x}_2. \tag{10.103}$$

The frictional forces in (10.101) are approximated by the equivalent linearized viscous damping coefficients as follows:

$$\frac{F_0}{M_i} \frac{\dot{x}_i}{\sqrt{\dot{x}_1^2 + \dot{x}_2^2}} \cong 2\zeta_{ei1}\lambda_{01}\dot{x}_1 + 2\zeta_{ei2}\lambda_{02}\dot{x}_2, \quad i = 1, 2 . \tag{10.104}$$

Using relation (10.101), one can define the difference $\varepsilon_i$ between the outputs of the nonlinear element and the linearized element for the $i$th mode

$$\varepsilon_i = \frac{F_0}{M_i} \frac{\dot{x}_i}{\sqrt{\dot{x}_1^2 + \dot{x}_2^2}} - 2\zeta_{ei1}\lambda_{01}\dot{x}_1 + 2\zeta_{ei2}\lambda_{02}\dot{x}_2, \quad i = 1, 2 , \tag{10.105}$$

where $\zeta_{ei1}, \zeta_{ei2}, i = 1, 2$, are linearization coefficients.

To obtain linearization coefficients, Hanawa and Shimizu [50] proposed to use the mean-square criterion

$$I = E[\varepsilon_1^2 + \varepsilon_2^2] \tag{10.106}$$

and the necessary conditions for minimum of $I$ are

$$\frac{\partial I}{\partial \zeta_{e11}} = \frac{\partial I}{\partial \zeta_{e12}} = \frac{\partial I}{\partial \zeta_{e21}} = \frac{\partial I}{\partial \zeta_{e22}} = 0 . \tag{10.107}$$

Then, the linearization coefficients are equal to

$$\zeta_{e12} = \zeta_{e21} = 0, \quad \zeta_{e11} = \frac{F_0}{\sqrt{2\pi}\lambda_{01}M_1} \frac{r}{(1-r^2)}[K(\rho) - L(\rho)]\frac{1}{\sigma_{\dot{x}_1}} ,$$

$$\zeta_{e22} = \frac{F_0}{\sqrt{2\pi}\lambda_{02}M_2} \frac{1}{(1-r^2)}[-r^2 K(\rho) + L(\rho)]\frac{1}{\sigma_{\dot{x}_2}}, \tag{10.108}$$

where $K(\rho)$ and $L(\rho)$ are the complete elliptic functions of the first and second kind defined by

$$K(\rho) = \int_0^{\frac{\pi}{2}} \frac{1}{\sqrt{1 - \rho \sin^2 \alpha}} d\alpha, \quad L(\rho) = \int_0^{\frac{\pi}{2}} \sqrt{1 - \rho \sin^2 \alpha} \, d\alpha \tag{10.109}$$

and

$$\rho = \frac{\sigma_{\dot{x}_2}^2 - \sigma_{\dot{x}_1}^2}{\sigma_{\dot{x}_2}^2} = 1 - r^2, \quad r = \frac{\sigma_{\dot{x}_1}}{\sigma_{\dot{x}_2}}, \quad 0 < \rho < 1 . \tag{10.110}$$

Hanawa and Shimizu [50] have compared several response characteristics obtained by numerical calculations and from experiments. From (10.103), it follows that the standard deviations of the velocity response in the $X$- and $Z$-direction of the piping at the friction support are approximated by the standard deviations of the velocity response for the first two modes $\sigma_{\dot{x}_1}$ and $\sigma_{\dot{x}_2}$. Hence it follows that the standard deviations of the relative displacements $\sigma_{U_x}$ and $\sigma_{U_z}$ in the $X$- and $Z$-direction are approximated by $\sigma_{U_x} \simeq \sigma_{\dot{x}_1}/\lambda_{01}$ and $\sigma_{U_z} \simeq \sigma_{\dot{x}_2}/\lambda_{02}$.

The comparison of the standard deviations of the relative displacements $\sigma_{U_x}$ and $\sigma_{U_z}$ against the maximum input acceleration $a$ is shown in Figure 10.2 and 10.3.

The marks $\bullet$ and $\circ$ denotes the experimental values for the piping with and without friction, respectively. Solids lines indicate the calculate values for values $F_0 = 191[N]$ and $F_0 = 0[N]$ in equation (10.104), respectively. The obtained computational and experimental results are in a good agreement.

The obtained computational and experimental results are in a good agreement. Similar agreement the authors have also observed in other characteristics. However, for some values of parameters, statistical linearization cannot be applied, for instance, for the maximum input acceleration $a$ being lower than

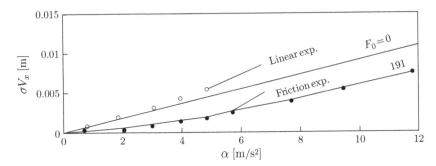

**Fig. 10.2.** Standard deviation of displacement against acceleration input in X–direction, reproduced from [50]

2.5–3.0 m/s². This indicates that the linearized model (by applying statistical linearization) itself does not stably exists lower than the value of 2.5–3.0 m/s² of maximum input acceleration. One can find the further interesting observations and conclusions regarding linearized model in [50].

Another comparison of theoretical computations obtained by the finite element method and the statistical linearization with experimental results was presented by Chen et al. [26]. The authors studied the random response of thermally buckled simply supported beam, clamped beam, and simply supported–clamped beam. The experiments were done for a clamped–clamped (C–C) aluminum beam 405 mm × 20 mm × 2.2 mm at room temperature with 100 microstrain prestretching. A lumped mass (exciter coil) of 26.4 g was attached to the center of the beam. The details of equipments as well as realization of random excitations are given in [26].

The mathematical model based on the principal of virtual work after separation time-independent or static components from time-dependent or dynamic components is given by the two vector equations

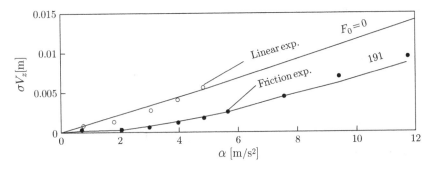

**Fig. 10.3.** Standard deviation of displacement against acceleration input in Z–direction, reproduced from [50]

$$(\mathbf{K} + \mathbf{K}_{N0} - \mathbf{K}_{N\Delta T})\{W\}_s + \frac{1}{2}\mathbf{N}_{1_s}\{W\}_s + \frac{1}{3}\mathbf{N}_{2_s}\{W_s\}_s = \mathbf{P}_{\Delta T} + E[\mathbf{P}_p(t)]$$

$$(10.111)$$

$$\mathbf{M}\{\ddot{W}\}_t + (\mathbf{K} + \mathbf{K}_{N0} - \mathbf{K}_{N\Delta T} + \mathbf{N}_{1_s} + \mathbf{N}_{2_{ss}})\{W\}_t$$
$$+ (\frac{1}{2}\mathbf{N}_{1_t} + \mathbf{N}_{2_{st}})\{W\}_t + \frac{1}{3}\mathbf{N}_{2_{tt}}\{W\}_s$$
$$= \mathbf{P}_p(t) - E[\mathbf{P}_p(t)],\qquad (10.112)$$

where $\{W\}_s$ and $\{W\}_t$ denotes the time-independent or static components and time-dependent or dynamic components of the constrained nodal displacements of the assembled system, respectively; $\mathbf{M}$ is a mass matrix, $\mathbf{K}$ is the linear elastic stiffness matrix, $\mathbf{K}_{N0}$ is the linear stiffness matrix due to initial prestretching axial force $N_0$; $\mathbf{N}_1$ and $\mathbf{N}_2$ are the nonlinear stiffness matrices which depend linearly and quadratically upon displacements, respectively; $\mathbf{P}_p(t)$ and $\mathbf{P}_{\Delta T}$ are the random excitation and the thermal load vectors; the subscripts $\{.\}_s$ and $\{.\}_t$ in nonlinear stiffness matrix $\mathbf{N}_1$ or $\mathbf{N}_2$ denotes that the matrix is evaluated with the static deflection $\{W\}_s$ and dynamic deflection $\{W\}_t$, respectively.

After solving $\{W\}_s$ from (10.111) and the matrices $\mathbf{N}_{1_s}$ and $\mathbf{N}_{2_{ss}}$ are evaluated, (10.112) is approximately solved. The procedure consists with three main steps. First, the linear frequencies and mode shapes of the deformed structure are obtained by solving the eigenvalue problem

$$(\mathbf{K} + \mathbf{K}_{N0} - \mathbf{K}_{N\Delta T} + \mathbf{N}_{1_s} + \mathbf{N}_{2_{ss}})\{\phi\}_n = \lambda_n^2 \mathbf{M}\{\phi\}_n . \qquad (10.113)$$

Then, we approximate $\{W\}_t$ by a finite sum

$$\{W\}_t \simeq [\Phi]\mathbf{q} = \sum_{n=1}^{N}\{\phi\}_n q_n . \qquad (10.114)$$

Using the finite series, the following approximations are assumed:

$$\mathbf{N}_{1_t} \simeq \sum_{n=1}^{N} q_n \mathbf{N}_{1_t}^{(n)}, \quad \mathbf{N}_{1_{st}} \simeq \sum_{n=1}^{N} q_n \mathbf{N}_{2_t}^{(n)}, \quad \mathbf{N}_{2_{tt}} \simeq \sum_{n=1}^{N}\sum_{r=1}^{N} q_n q_r \mathbf{N}_{2_{tt}}^{(nr)},$$

$$(10.115)$$

where the super scripts $(n)$ and $(nr)$ denote the corresponding nonlinear modal stiffness matrix evaluated with the modes $\{\phi\}_n$ and $\{\phi\}_n, \{\phi\}_r$, respectively.

Substituting these nonlinear modal stiffness matrices into (10.112) and multiplying both sides by $[\Phi]^T$ we obtain a set of nonlinear coupled modal equations of motion

$$\mathbf{M}_d\ddot{\mathbf{q}} + \mathbf{K}_{linear}\dot{\mathbf{q}} + [\Phi]^T \sum_{n=1}^{N} q_n \left(\frac{1}{2}\mathbf{N}_{1_t}^{(n)} + \mathbf{N}_{2_{st}}^{(n)}\right)[\Phi]\mathbf{q}$$

$$+ [\Phi]^T \left(\frac{1}{3}\sum_{n=1}^{N}\sum_{r=1}^{N} q_n q_r \mathbf{N}_{2_{tt}}^{(nr)}\right)[\Phi]\mathbf{q}$$

$$-[\Phi]^T(\mathbf{P}_p(t) - E[\mathbf{P}_p(t)]) = 0, \tag{10.116}$$

where

$$\mathbf{M}_d = [\Phi]^T\mathbf{M}[\Phi] = diag\{m_i\},$$

$$\mathbf{K}_{linear} = [\Phi]^T(\mathbf{K} + \mathbf{K}_{N0} - \mathbf{K}_{N\Delta T} + \mathbf{N}_{1_s} + \mathbf{N}_{2_{ss}})[\Phi]$$

$$= diag\{\lambda_i^2 m_i\}, \quad i = 1,\ldots,N. \tag{10.117}$$

After rewriting (10.116) in the form

$$\mathbf{M}_d\ddot{\mathbf{q}} + \mathbf{g}(\mathbf{q}) - [\Phi]^T(\mathbf{P}_p(t) - E[\mathbf{P}_p(t)]) = 0 \tag{10.118}$$

it is linearized to the following form:

$$\mathbf{M}_d\ddot{\mathbf{q}} + \tilde{\mathbf{K}}\mathbf{q} - \Phi^T(\mathbf{P}_p(t) - E[\mathbf{P}_p(t)]) = 0 , \tag{10.119}$$

where $\tilde{\mathbf{K}} = [\tilde{K}_{ij}]$, $i,j = 1,\ldots,N$ is an equivalent linear modal stiffness matrix of linearization coefficients, that can be determined by minimizing the mean-square criterion defined by

$$I = E[\mathbf{e}^T\mathbf{e}]$$

with respect to $\tilde{K}_{ij}$, for $i,j = 1,\ldots,N$, where $\mathbf{e} = \mathbf{g}(\mathbf{q}) - \tilde{\mathbf{K}}\mathbf{q}$ and the averaging operation does not depend on linearization coefficients.

Next, in order to uncouple the linearized equations of motion the modal transformation is used once more. Then we obtain

$$\tilde{\mathbf{K}}\{\tilde{\phi}\} = \Lambda^2\mathbf{M}_d\{\tilde{\phi}\} , \tag{10.120}$$

where $\Lambda^2 = diag\{\lambda_j^2\}$, $j = 1,\ldots,N$ and the modal transformation matrix is defined as

$$\mathbf{q} = [\{\tilde{\phi}\}_1, \{\tilde{\phi}\}_2,\ldots,\{\tilde{\phi}\}_N]\,\boldsymbol{\eta} = \tilde{\Phi}\,\boldsymbol{\eta} . \tag{10.121}$$

Then (10.119) is transformed to a set of uncoupled modal equations

$$\ddot{\eta}_j + h\dot{\eta}_j + \Lambda_j^2\eta_j = \tilde{f}_j, \quad j = 1,\ldots,N , \tag{10.122}$$

where

$$\tilde{f}_j = \frac{1}{\tilde{m}_j}\tilde{\Phi}^T(\mathbf{P}_p(t) - E[\mathbf{P}_p(t)]), \quad \tilde{m}_j = \tilde{\Phi}^T\mathbf{M}_d\tilde{\Phi}_{|j}, \quad j = 1,\ldots,N \tag{10.123}$$

and the modal damping term $h\dot{\eta}_j = 2\zeta\lambda_1\dot{\eta}_j$ has been added to (10.122), where $\lambda_1$ is the first linear frequency of the system (10.116) without nonlinear terms, i.e., $(\mathbf{K} + \mathbf{K}_{N0})\{\phi\}_1 = \lambda_1^2\mathbf{M}\{\phi\}_1$

Then the covariance matrix of the displacement vector can be calculated by application of an iterative procedure and is approximated by

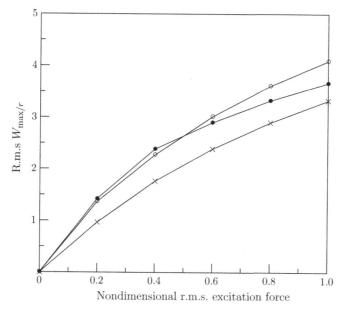

**Fig. 10.4.** The comparison of r.m.s. $W_{\max}/r$ for a clamped aluminum beam: $-\circ-$, FE $h = 0.005$; $- \times -$, FE $h = 0.01$; $- \bullet -$, experiments, from [26] with permission

$$E[\{W\}\{W\}^T] \simeq [\Phi]E[\mathbf{q}\mathbf{q}^T][\Phi]^T \ . \tag{10.124}$$

Chen et al. [26] have compared some characteristics obtained by numerical calculations and experiments. For example, the comparison of the root mean square for $W_{\max}/r$ for a clamped aluminum beam is presented in Fig. 10.4, where $r = \sqrt{I/A}$ is the radius of gyration of the cross-sectional area of the beam.

## 10.5 Limitations of Applicability of Linearization Methods

The careful analysis of linearization methods shows as well the wide spectrum of fields of applications as some important limitations. In this subsection we briefly discuss the main limitations.

The first one is the accuracy of linearization methods. From the discussion in previous subsections, it follows that the accuracy of response characteristics is acceptable for simple weakly nonlinear systems (one or two degree of freedom). For strongly nonlinear systems, the error of the predicted response characteristics may be up to several 10 percentages [113]. For higher-order degree, the accuracy decreases and only in some special cases of weakly nonlinear functions is acceptable. However, we have to remember that stochastic linearization is the only one practical analytical mathematical tool in approximate analysis of MDOF stochastic dynamic systems. It is clear that the accuracy depends on criteria of linearization. In the case if we compare response characteristics, the energy criteria give better results than the other ones. The direct comparison of moment criteria, probability density criteria, and spectral criteria is not reasonable, because of qualitative different aims of these groups of criteria (they are not comparable).

The mentioned qualitative differences appear if we consider the spectral response characteristics. In this case we have to extend the dimension of equivalent linearized system in order to obtain the similar response power spectral density. For example, the response power spectral density for the Duffing oscillator contains three peaks, while the corresponding power spectral density for linearized oscillator only one peak. It was discussed in detail in Sect. 6.4. Similar qualitative difference appears when we consider a class of nonlinear dynamic systems having multimodal response probability density functions. Then the standard linearization methods fail and a special treatment is required. This problem was partially discussed in Sect. 6.6.4.

The main qualitative difference is connected with stability analysis of original nonlinear system and a linearized one. We show it with a simple example.

*Example 10.3.* Consider the nonlinear scalar dynamic system

$$dx = -[\lambda x + Ax^3 + (\gamma - \frac{3}{2}\beta^2)x^5]dt + \beta x^3 d\xi, \quad x(t_0) = x_0 , \quad (10.125)$$

where $\lambda > 0, A > 0, \gamma > 0$, and $\beta > 0$ are constant coefficients and $\xi$ is a standard Wiener process, the initial condition $x_0$ is a random variable independent of $\xi$.

The linearized system has the form

$$dx = -[\lambda x + Ak_1 + (\gamma - \frac{3}{2}\beta^2)k_2]xdt + \beta k_1 xd\xi, \quad x(t_0) = x_0 , \quad (10.126)$$

where $k_1$ and $k_2$ are linearization coefficients defined by the mean-square criterion, i.e.,

$$k_1 = 3E[x^2] = 3\sigma^2, \quad k_2 = 15(E[x^2])^2 = 15\sigma^4 . \quad (10.127)$$

Assuming the Lyapunov function in the form of a quadratic function

$$V = ax^2 , \quad (10.128)$$

where $a > 0$ is a constant, we find the sufficient condition for the asymptotic stability with probability 1 of trivial solution of nonlinear system (10.125)

$$LV = -2a[\lambda x^2 + Ax^4 + (\gamma - \frac{3}{2}\beta^2)x^6] + a\beta^2 x^6 \le 0, \quad x(t_0) = x_0 . \quad (10.129)$$

Hence, we find that the sufficient condition is

$$\gamma > 2\beta^2 . \quad (10.130)$$

Similar considerations for linearized system leads to the following inequality:

$$LV = -2[\lambda x + Ak_1 x + (\gamma - \frac{3}{2}\beta^2)k_2 x]x + 9\beta^2 k_1^2 x^2$$
$$= -2\lambda x^2 - 18A\sigma^2 - (30\gamma - 45\beta^2)\sigma^4 + 81\beta^2\sigma^4 < 0 . \quad (10.131)$$

Hence, we find that the sufficient condition for the asymptotic stability of linearized system (10.126) is

$$\gamma > 4.2\beta^2 . \quad (10.132)$$

The comparison of conditions (10.130) and (10.132) leads to the expected conclusion, that the sufficient conditions of asymptotic stability for nonlinear and linearized systems are not equivalent. It means that from stability of nonlinear system does not follow stability of linearized system. This is the important fact that should be taken into account in the linearization procedure of nonlinear stochastic dynamic systems with parametric excitations.

## Bibliography Notes

### Ref.10.1.

The theoretical accuracy analysis is not much developed in the literature. Usually, the authors discussed the accuracy of their proposed linearization approaches on numerical examples, comparing with standard equivalent linearization even if in the title the words "error analysis of statistical linearization" are used.

Similar to Kazakov [61] and Kolovskii [68], in only a few papers, a theoretical justification of some linearization methods was carried out, see, for instance, [15, 14].

Recently, Skrzypczyk [92, 93] has proposed theoretical bounds for the standard deviation errors based on characteristics of linearized system. He has considered both nonlinear and linearized dynamic systems described by stochastic integral equation in the form

$$x(t) = \eta(t) + \int_0^t k(t,s)f(s,x(s))ds , \quad (10.133)$$

$$y(t) = \eta(t) + \int_0^t k(t,s)l(s,y(s))ds , \quad (10.134)$$

respectively, where $\eta$ is a stochastic process, $k(t,s)$ is a kernel function, $f(t,x)$ is a nonlinear vector function, $l(t,y)$ is a linearized vector function defined by $l(t,y) = C_n(t)y + c(t)$, and $y = [y_1^T, y_2^T]^T \in R^n$, $y_1 \in R^p, p \leq n$. It is assumed that $C_n y = C y_1$, $C : R^p \to R^n$ and $c = 0$.

Skrzypczyk has proved under suitable assumptions that if we denote the solutions of (10.133) and (10.134) by $\bar{x}(t)$ and $\bar{y}(t)$, respectively, then

$$\sup_{t\in R^1} |\sigma_x - \sigma_y| \le \sup_{t\in R^1} \left(E[|\overline{x}(t) - \overline{y}(t)|^2]\right)^{1/2} \le \gamma \sup_{t\in R^1} \rho(l) , \qquad (10.135)$$

where $\sigma_x = \sqrt{var[x(t)]}, \sigma_y = \sqrt{var[y(t)]}$, $\gamma$ is a constant determined by assumption parameters, and

$$\rho(l) = \rho(l(t,\overline{y}(t))) = \left(E[|f(t,\overline{y}(t)) - C_n\overline{y}(t) - c(t)|^2]\right)^{1/2} . \qquad (10.136)$$

Skrzypczyk [92, 93] has also considered an illustrative example of nonlinear oscillator under stationary colored noise excitation. He has shown that the absolute standard deviation errors $|\sigma_x - \sigma_y|$ for two standard linearization methods obtained by numerical calculations evaluated exactly, along with theoretical bounds determined by (10.135). This statement confirms the applicability of the theoretical bounds.

**Ref.10.2.**

A similar comparison of relative errors for an oscillator with odd order polynomial spring and Van der Pol oscillator was given in [1, 2], for nonlinear oscillators with energy criteria [120] and for MDOF systems with third-order polynomial springs relative errors of stationary autocovariance functions were calculated in [97]. However, as exact solutions a "good approximation" by quasi-momentum functions was applied.

**Ref.10.3.**

Some authors proposed new approximate methods for response analysis of nonlinear stochastic systems. To show the advantage of their approaches, they have compared examples of response characteristics obtained by their approaches with corresponding ones obtained by statistical linearization and simulations. This gives an indirect comparison of statistical linearization with "better" approximation methods and with simulations. The two main groups of such methods are cumulant closure and quasi-moments closure methods. The comparisons of response characteristics obtained by these approaches with the exact ones and statistical linearization were given, for instance, in [13, 31, 32, 34, 69, 97, 107, 116, 113].

**Ref.10.4.**

In several further papers the mathematical models of dynamic systems with application stochastic linearization methods were verified by experiments. Han and Kim [49] have applied the combined methods of statistical linearization and the LQG theory for the determination of suboptimal control for a timing belt driving cart system.

Ng [75] analyzed random thermal acoustic responses of aircraft structures by two theoretical approaches, namely, equivalent linearization and Fokker–Planck–Kolmogorov equation, and by physical experiments. For theoretical consideration, the author used von Karman equation and Galerkin's method. The obtained simplified model for the first mode was the nonlinear oscillator with a third-order polynomial term. Experimental results were obtained for an aluminum plate with combined radiant heating and base excitation. The comparison of the acoustic response characteristics obtained by these approaches confirmed that the Fokker–Planck–Kolmogorov approach gives accurate results while equivalent linearization gives good results only for some types of response.

Pires [79] applied equivalent linearization for the stochastic response analysis in the time domain of horizontally layered soil deposits under vertically propagating random shear waves. The author proposed a new model of soil that is an extension of the well-known Bouc–Wen smooth hysteretic model. The Pires model represents the nonlinear hysteretic shearing stress–strain behavior of soils observed under cyclic loading with either fixed or variable limits. The main idea of the Pires model is to consider a finite number of nonlinear hysteretic springs described by the Bouc–Wen model with a somewhat large value of $n$ (say $M = 2$) are used in parallel, i.e., if the shearing stress $\tau$ is given by

$$\tau = (1 - \alpha)G_m\gamma + (1 - \alpha)\sum_{k=1}^{M} G_{m_k} z_k \, , \tag{10.137}$$

where each of the dummy variables $z_k$ is given by

$$\dot{z}_k = \alpha_{1k}\dot{\gamma} - \beta_k \mid \dot{\gamma} \mid\mid z_k \mid^{\alpha_{2k}-1} z_k - \delta_k \mid \dot{\gamma} \mid\mid z_k \mid^{\alpha_{2k}} \, , \tag{10.138}$$

with the following two conditions:

$$\sum_{k=1}^{M} G_{m_k} z_k = G_m, \qquad \sum_{k=1}^{M} \tau_{m_k} z_k = \tau_m \, , \tag{10.139}$$

where

$$\tau_{m_k} = G_{m_k} \left( \frac{\alpha_{1k}}{\beta_k + \gamma_k} \right)^{1/\alpha_{2k}} , \tag{10.140}$$

$\alpha_{1k}, \beta_k, \gamma_k$, and $\delta_k$ are constant parameters, $\alpha_{2k}$ is an odd number.

As an illustration, Pires has shown a comparison of the surface acceleration versus expected maximum base acceleration for soft soil obtained by the proposed method and experiments.

Song and Der Kiureghian [100] used experimental results to determine a generalized Bouc–Wen model. The obtained model successfully described the asymmetric hysteresis loops of existing flexible strap connectors as observed in laboratory experiments.

## Ref.10.5.

The problem of stability of linearized systems was considered by Naumov [74] in the context of control systems. Chang and Lin [21] proposed a new linearization model of stochastic parametrically excited nonlinear system for the prediction of accurate response under robust stability boundary.

# References

1. Anh, N. and Schiehlen, W.: New criterion for Gaussian equivalent linearization. Eur. J. Mech. A/Solids **16**, 1025–1039, 1997.
2. Anh, N. and Schiehlen, W.: A technique for obtaining approximate solutions in Gaussian equivalent linearization. Comput. Methods Appl. Mech. Eng. **168**, 113–119, 1999.
3. Asano, K. and Iwan, W.: An alternative approach to the random response of bilinear hysteretic systems. Earthquake Eng. Struct. Dyn. **12**, 229–236, 1984.
4. Baber, T.: Nonzero mean random vibration of hysteretic systems. ASCE J. Eng. Mech. **110**, 1036–1049, 1983.
5. Baber, T. and Noori, M.: Random vibration of degrading, pinching systems. ASCE J. Eng. Mech. **111**, 1010–1026, 1985.
6. Baber, T. and Noori, M.: Modeling general hysteresis behavior and random vibration application. Trans. ASME J. Vib. Acoust. Stress Rel. Des. **108**, 411–420, 1986.
7. Baber, T. and Wen, Y.: Random vibration of hysteretic degrading systems. ASCE J. Eng. Mech. Div. **107**, 1069–1087, 1981.
8. Basu, B. and Gupta, V.: A note on damage-based inelastic spectra. Earthquake Eng. Struct. **25**, 421–433, 1996.
9. Bellizzi, S., Bouc, R., and Defilippi, M.: Response spectral densities and identification of a randomly excited non-linear squeeze film oscillator. Mech. Syst. Signal Proc. **12**, 693–711, 1998.
10. Bernard, P. and Taazount, M.: Random dynamics of structures with gaps: Simulation and spectral linearization. Nonlinear Dyn. **5**, 313–335, 1994.
11. Bouc, R.: The power spectral density of response for a strongly non-linear random oscillator. J. Sound Vib. **175**, 317–331, 1994.
12. Bouc, R. and Boussaa, D.: Drifting response of hysteretic oscillators to stochastic excitation. Int. J. Non-Linear Mech. **37**, 1397–1406, 2002.
13. Bover, D.: Moment equation methods for nonlinear stochastic systems. J. Math. Anal. Appl. **65**, 306–320, 1978.
14. Budgor, A.: Studies in nonlinear stochastic processes I. Approximate solutions of nonlinear stochastic differential equations by the method of statistical linearization. J. Stat. Phys. **15**, 355–374, 1976.
15. Bunke, H.: Statistical linearization. Z. Angew. Math. Mech. **52**, 79–84, 1972 (in German).
16. Cai, G.: Non-linear systems of multiple degrees of freedom under both additive and multiplicative random excitation. J. Sound Vib. **278**, 889–901, 2004.
17. Cavaleri, L. and Di-Paola, M.: Statistic moments of the total energy of potential systems and application to equivalent non-linearization. Int. J. Non-Linear Mech. **35**, 573–587, 2000.
18. Cha, D. and Sinha, A.: Statistics of responses of a mistuned and frictionally damped bladed disk assembly subjected to white noise and narrow band excitations. Probab. Eng. Mech. **21**, 384–396, 2006.
19. Chang, R.: A practical technique for spectral analysis of nonlinear systems under stochastic parametric and external excitations. Trans. ASME J. Vib. Acoust. **113**, 516–522, 1991.
20. Chang, R.: Non-Gaussian linearization method for stochastic parametrically and externally excited nonlinear systems. Trans. ASME J. Dyn. Syst. Meas. Control **114**, 20–26, 1992.

21. Chang, R. and Lin, S.: Statistical linearization model for the response prediction of nonlinear stochastic systemsthrough information closure method. Trans. ASME J. Appl. Mech. **126**, 438–448, 2004.

22. Chang, R. and Young, G.: Methods and Gaussian criterion for statistical linearization of stochastic parametrically and externally excited nonlinear systems. Trans. ASME J. Appl. Mech. **56**, 179–185, 1989.

23. Chang, T., Chang, H., and Liu, M.: A finite element analysis on random vibration of nonlinear shell structures. J. Sound Vib. **291**, 240–257, 2006.

24. Chang, T. and Ke, J.: Nonlinear dynamic response of a nonuniform orthotropic circular plate under random excitation. Comput. Struct. **60**, 113–123, 1996.

25. Chen, C. and Yang, H.: Flexible thin shell elements under nonwhite and nonzero mean loads. ASCE J.Eng. Mech. **119**, 1680–1697, 1993.

26. Chen, R., Mei, C., and Wolfe, H.: Comparison of finite element non-linear beam random response with experimental results. J. Sound Vib. **195**, 719–737, 1996.

27. Chen, Y. and Ahmadi, G.: Stochastic earthquake response of secondary systems in base-isolated structures. Earthquake Eng. Struct. Dyn. **21**, 1039–1057, 1992.

28. Chen, Y. and Ahmadi, G.: Performance of a high damping rubber bearing base-isolation system for a shear beam structure. Earthquake Eng. Struct. Dyn. **23**, 729–744, 1994.

29. Cheng, G., Lee, Y., and Mei, C.: Nonlinear random response of internally hinged beams. Finite Elem. Anal. Des. **39**, 487–504, 2003.

30. Clough, R. and Penzien, J.: Dynamics of Structures. IV Random Vibrations. McGraw-Hill, New York, 1972.

31. Crandall, S.: Non-Gaussian closure for random vibration of nonlinear oscillators. Int. J. Non-Linear Mech. **15**, 303–313, 1980.

32. Crandall, S.: Non-Gaussian closure for stationary random vibration. Int. J. Non-Linear Mech. **20**, 1–8, 1985.

33. Cunha, A.: The role of the stochastic equivalent linearization method in the analysis of the nonlinear seismic response of building structures. Earthquake Eng. Struct. **23**, 837–857, 1994.

34. Dashevski, M.: Approximate analysis of the accuracy of nonstationary nonlinear systems using the method of semiinvariants. Avtomatika i Telemekhanika **28**, 1673–1690, 1967.

35. Daucher, D., Fogli, M., and Clair, D.: Modelling of complex dynamical behaviours using a state representation technique based on a vector ARMA approach. Probab. Eng. Mech. **21**, 73–80, 2006.

36. Dobson, S., Noori, M., Hou, Z., and Dimentberg, M.: Direct implementation of stochastic linearization for SDOF systems with general hysteresis. Struct. Eng. Mech. **6**, 473–484, 1998.

37. Duval L. et al.: Zero and non-zero mean analysis of MDOF hysteretic systems via direct linearization. In: B. Spencer and E. Johnson (eds.), Stochastic Structural Dynamics, pp. 77–84. Balkema, Rotterdam, 1999.

38. Elishakoff, I. and Cai, G.: Approximate solution for nonlinear random vibration problems by partial stochastic linearization. Probab. Eng. Mech. **8**, 233–237, 1993.

39. Elishakoff, I., Fang, J., and Caimi, R.: Random vibration of a nonlinearly deformed beam by a new stochastic linearization method. Int. J. Solids Struct. **32**, 1571–1584, 1995.

40. Ellison, J., Ahmadi, G., and Grodsinsky, C.: Stochastic response of pasive vibration control systems to g-jitter excitation. Microgravity Sci. Tec. **10**, 2–12, 1997.

41. Emam, H.H. Pradlwarter, H., and Schueller, G.: On the computational implementation of EQL in FE-analysis. In: B. Spencer and E. Johnson (eds.), Stochastic Structural Dynamics, pp. 85–91. Balkema, Rotterdam, 1999.

42. Er, G. and Iu, V.: Stochastic response of base-excited coulomb oscillator. J. Sound Vib. **233**, 81–92, 2000.

43. Falsone, G.: Stochastic linearization for the response of MDOF systems under parametric Gaussian excitation. Int. J. Non-Linear Mech. **27**, 1025–1037, 1992.

44. Fang, J., Elishakoff, I., and Caimi, R.: Nonlinear response of a beam under stationary random excitation by improved stochastic linearization method. Appl. Math. Modeling **19**, 106–111, 1995.

45. Foliente, G., Singh, M., and Noori, M.: Equivalent linearization of generally pinching hysteretic, degrading systems. Earthquake Eng. Struct. Dyn. **25**, 611–629, 1996.

46. Ghazarian, N. and Locke, J.: Nonlinear random response of antisymmetric angle-ply laminates under thermal-acoustic loading. J. Sound Vib. **186**, 291–309, 1995.

47. Grigoriu, M.: Equivalent linearization for Poisson white noise input. Probab. Eng. Mech. **10**, 45–51, 1995.

48. Grundmann, H., Hartmann, C., and Waubke, H.: Structures subjected to stationary stochastic loadings. Preliminary assessment by statistical linearization combined with an evolutionary algorithm. Comput. Struct. **67**, 53–64, 1998.

49. Han, S. I. and Kim, J.: Nonlinear quadratic Gaussian control with loop transfer recovery. Mechatronics **13**, 273–293, 2003.

50. Hanawa, Y. and Shimizu, N.: Statistical seismic response analysis of piping system with a teflon friction support. JSME Int. J. C-Mech. Syst. **45**, 393–401, 2002.

51. Hanawa, Y., Shimizu, N., Kobayashi, H., and Ogawa, N.: Vibration test and analysis of responses of piping system with friction support. Trans. Jpn. Soc. Mech. Eng. **65**, 4611–4617, 1999(in Japanese).

52. Huang, C., Zhu, W., and Soong, T.: Nonlinear stochastic response and reliability of secondary systems. J. Eng. Mech.-ASCE **120**, 177–196, 1994.

53. Hurtado, J. and Barbat, A.: Improved stochastic linearization method using mixed distributions. Struct. Safety **18**, 49–62, 1996.

54. Hurtado, J. and Barbat, A.: Equivalent linearization of the Bouc–Wen hysteretic model. Eng. Struct. **22**, 1121–1132, 2000.

55. Inaudi, J. and Kelly, J.: Mass damper using friction-dissipating devices. J. Eng. Mech. **121**, 142–149, 1995.

56. Inaudi, J., Leitmann, G., and Kelly, J.: Single-degree-of-freedom nonlinear homogeneous systems. J. Eng. Mech. **120**, 1543–1562, 1994.

57. Iwan, W. and Mason, A.: Equivalent linearization for systems subjected to nonstationary random excitation. Int. J. Non-Linear Mech. **15**, 71–82, 1980.

58. Iyengar, R. and Roy, D.: Conditional linearization in nonlinear random vibration. ASCE J. Eng. Mech. **119**, 197–200, 1996.

59. Jangid, R. and Datta, T.: Performance of base-isolation systems for asymmetric building subject to random-excitation. Eng. Struct. **17**, 443–454, 1995.

60. Kareem, A., Zhao, J., and Tognarelli, M.: Surge response statistics of tension leg platforms under wind and wave loads: Statistical quadratization approach. Probab. Eng. Mech. **10**, 225–240, 1995.

61. Kazakov, I.: Statistical Theory of Control Systems in State Space. Nauka, Moskwa, 1975 (in Russian).

62. Kazakov, I. and Dostupov, B.: Statistical Dynamic of Nonlinear Automatic Systems. Fizmatgiz, Moskwa, 1962 (in Russian).

63. Köylüoglu, H., Nielsen, S., and Cakmak, A.: Stochastic dynamics of geometrically nonlinear structures with random properties subject to stationary random excitation. J. Sound Vib. **190**, 821–841, 1996.

64. Kimura, K. and Sakata, M.: Non-stationary responses of a non-symmetric nonlinear system subjected to a wide class of random excitation. J. Sound Vib. **76**, 261–272, 1981.

65. Kimura, K., Yasumuro, H., and Sakata, M.: Non-Gaussian equivalent linearization for non-stationary random vibration of hysteretic systems. Probab. Eng. Mech. **9**, 15–22, 1994.

66. Koliopulos, P. and Chandler, A.: Stochastic linearization of inelastic seismic torsional response – formulation and case-studies. Eng. Struct. **17**, 494–504, 1995.

67. Koliopulos, P., Nichol, E., and Stefanou, G.: Comparative performance of equivalent linearization methods for inelastic seismic design. Eng. Struct. **16**, 5–10, 1994.

68. Kolovskii, M.: Estimating of the accuracy of solutions obtained by the method of statistical linearization. Avtomatika i Telemekhanika **27**, 1692–1701, 1966.

69. Lee, J.: Improving the equivalent linearization method for stochastic Duffing oscillators. J. Sound Vib. **186**, 846–855, 1995.

70. Li, X., Quek, S., and Koh, C.: Stochastic response of offshore platform by statistical cubization. ASCE J. Eng. Mech. **121**, 1056–1068, 1995.

71. Lin, J., Wang, J., and Zhang, Y.: Nonstationary random response of MDOF Duffing systems. Shock Vib. **11**, 615–624, 2004.

72. Micaletti, R., Cakmak, A., Nielsen, S., and Koyluoglu, H.: Error analysis of statistical linearization with Gaussian closure for large-degree-of-freedom systems. Probab. Eng. Mech. **13**, 77–84, 1998.

73. Muravyov, A. and Rizzi, S.: Determination of nonlinear stiffness with application to random vibration of geometrically nonlinear structures. Comput. Struct. **81**, 1513–1523, 2003.

74. Naumov, B.: The Theory of Nonlinear Automatic Control Systems. Frequency Methods. Nauka, Moscow, 1972 (in Russian).

75. Ng, C.: The nonlinear acoustic response of thermally buckled plates. Appl. Acoust. **59**, 237–254, 2000.

76. Ni, Y., Ying, Z., Ko, J., and Zhu, W.: Random response of integrable Dunhem hysteretic systems under non-white excitations. Int. J. Non-Linear Mech. **37**, 1407–1419, 2002.

77. Noori, M. and Davoodi, H.: Zero and nonzero mean random vibration analysis of a new general hysteresis model. Probab. Eng. Mech. **1**, 192–201, 1986.

78. Pagnini, L. and Solari, G.: Stochastic analysis of the linear equivalent response of bridge piers with aseismic devices. Earthquake Eng. Struct. **28**, 543–560, 1999.

79. Pires, J.: Stochastic seismic response analysis of soft soil sites. Nuclear Eng. Design **160**, 363–377, 1996.

80. Polidori, D. C. Beck, J., and Papadimitriou, C.: A new stationary PDF approximation for non-linear oscillators. Int. J. Non-Linear Mech. **35**, 657–673, 2000.

81. Proppe, C.: Equivalent linearization of MDOF systems under external poisson white noise excitation. Probab. Eng. Mech. **17**, 393–399, 2002.

82. Proppe, C.: Stochastic linearization of dynamical systems under parameteric poisson white noise excitation. Int. J. Non-Linear Mech. **38**, 543–555, 2003.

83. Proppe, C., Pradlwarter, H., and Schueller, G.: Equivalent linearization and Monte Carlo simulation in stochastic dynamics. Probab. Eng. Mech. **18**, 1–15, 2003.

84. Quek, S., Li, X., and Koh, C.: Stochastic response of jack-up platform by the method for statistical quadratization. Appl. Ocean Res. **16**, 113–122, 1994.

85. Rüdinger, F.: Tuned mass damper with nonlinear viscous damping. J. Sound Vib. **300**, 932–948, 2007.

86. Roberts, J., Dunne, J., and Debonos, A.: A spectral method for estimation of nonlinear parameters from measured response. Probab. Eng. Mech. **10**, 199–207, 1995.

87. Sakata, M. and Kimura, K.: Calculation of the non-stationary mean square response of a non-linear system subjected to non-white excitation. J. Sound Vib. **73**, 333–343, 1980.

88. Schueller, G. and Pradlwarter, H.: On the stochastic response of nonlinear FE models. Arch. Appl. Mech. **69**, 765–784, 1999.

89. Shinozuka, M.: Simulation of multivariate and multidimensional random processes. J. Acoust. Soc. Am. **49**, 357–367, 1977.

90. Shinozuka, M. and Deodatis, G.: Simulation of stochastic processes by spectral representation. Appl. Mech. Rev. **44**, 191–203, 1991.

91. Shinozuka, M. and Jan, C.: Digital simulation of random processes and its applications. J. Sound Vib. **25**, 111–128, 1972.

92. Skrzypczyk, J.: Accuracy analysis of statistical linearization methods applied to nonlinear continuous systems described by random integral equations. J. Theor. Appl. Mech. **32**, 841–865, 1994.

93. Skrzypczyk, J.: Accuracy analysis of statistical linearization methods applied to nonlinear dynamical systems. Rep. Math. Phys. **36**, 1–20, 1995.

94. Smyth, A. and Masri, S.: Nonstationary response of nonlinear systems using equivalent linearization with a compact analytical form of the excitation process. Probab. Eng. Mech. **17**, 97–108, 2002.

95. Sobiechowski, C.: Statistical linearization of dynamical systems under parametric delta-correlated excitation. Z. Angew. Math. Mech. **79**, 315–316, 1999 S2.

96. Sobiechowski, C. and Socha, L.: Statistical linearization of the Duffing oscillator under non-Gaussian external excitation. J. Sound Vib. **231**, 19–35, 2000.

97. Sobiechowski, C. and Sperling, L.: An iterative statistical linearization method for MDOF systems. In: EUROMECH – 2nd European Nonlinear Oscillation Conference, Prague, pp. 419–422, September 9–13, 1996.

98. Soize, C.: Stochastic linearization method with random parameters for SDOF nonlinear dynamical systems: Prediction and identification procedures. Probab. Eng. Mech. **10**, 143–152, 1995.

99. Soize, C. and Le Fur, O.: Modal identification of weakly non-linear multidimensional dynamical systems using stochastic linearization method with random coefficients. Mech. Syst. Signal Process. **11**, 37–49, 1997.

100. Song, J. and DerKiureghian, A.: Generalized Bouc-Wen model highly asymmetric hysteresis. ASCE J. Eng. Mech. **132**, 610–618, 2006.

101. Spanos, P., Chevallier, A., and Politis, N.: Nonlinear stochastic drill-string vibrations. Trans. ASME J. Appl. Mech. **124**, 512–518, 2002.

102. Spanos, P. and Donley, M.: Equivalent statistical quadratization for nonlinear system. ASCE J. Eng. Mech. **117**, 1289–1310, 1991.

103. Spanos, P. and Donley, M.: Non-linear multi-degree-of-freedom system random vibration by equivalent statistical quadratization. Int. J. Non-Linear Mech. **27**, 735–748, 1992.

104. Su, L. Ahmadi, G. and Tadjbakhsh, I.: A comparative study of performances of various base isolation systems – Part I: Shear beam structures. Earthquake Eng. Struct. Dyn. 1989.

105. Su, L. and Ahmadi, G.: Response of frictional base isolation systems to horizontal-vertical random earthquake excitations. Probab. Eng. Mech. **3**, 12–21, 1988.

106. Sun, J., Bao, W., and Miles, R.: Fatique life prediction of nonlinear plates under random excitations. J. Vib. Acoust. **120**, 353–360, 1998.

107. Sun, J. and Hsu, C.: Cumulant-neglect closure method for nonlinear systems under random excitations. Trans. ASME J. Appl. Mech. **54**, 649–655, 1987.

108. To, C.: Recursive expression for random response of nonlinear systems. Comput. Struct. **29**, 451–457, 1988.

109. Tognarelli, M., Zhao, J., Rao, K., and Kareem, A.: Equivalent statistical quadratization and qubicization for nonlinear system. ASCE J. Eng. Mech. **123**, 512–523, 1997.

110. Tognarelli, M., Zhao, J., and Kareem, A.: Equivalent statistical qubicization for system and forcing nonlinearization. ASCE J. Eng. Mech. **123**, 890–893, 1997.

111. Wang, J. and Lin, J.: Seismic random response analysis of hysteretic systems with Pseudo Excitation Method. Acta Mech. Solida Sin. **13**, 246–253, 2000.

112. Wen, Y.: Equivalent linearization for hysteretic systems under random excitation. Trans. ASME J. Appl. Mech. **47**, 150–154, 1980.

113. Wojtkiewicz, S., Spencer, B., and Bergman, I.: On the cumulant–neglect closure method in stochastic dynamics. Int. J. Non-Linear. Mech. **31**, 657–684, 1996.

114. Won, A., Pires, J., and Haroun, M.: Stochastic seismic performance evaluation of tuned liquid column dampers. Earthquake Eng. Struct. Dyn. **25**, 1259–1274, 1996.

115. Won, A., Pires, J., and Haroun, M.: Performance assessment of tuned liquid column dampers under random seismic loading. Int. J. Non-Liner Mech. **32**, 745–758, 1997.

116. Wu, W. F. and Lin, Y.: Cumulant-neglect closure for nonlinear oscillators under random parametric and external excitations. Int. J. Non-Linear Mech. **19**, 349–362, 1984.

117. Yan, X. and Nie, J.: Response of SMA superelastic systems under random excitation. J. Sound Vib. **238**, 893–901, 2000.

118. Young, G. and Chang, R.: Prediction of the response of nonlinear oscillators under stochastic parametric and external excitations. Int. J. Non-Linear Mech. **22**, 151–160, 1987.

119. Zhang, R., Elishakoff, I., and Shinozuka, M.: Analysis of nonlinear sliding structures by modified stochastic linearization methods. Nonlinear Dyn. **5**, 299–312, 1994.

120. Zhang, X., Zhang, R., and Xu, Y.: Analysis on control of flow-induced vibration by tuned liquid damper with crossed tube-like containers. J. Wind Eng. Ind. Aerodyn. **50**, 351–360, 1993.
121. Zhu, W. and Deng, M.: Equivalent non-linear system method for stochastically excited and dissipated integrable Hamiltonian systems-resonant case. J. Sound Vib. **274**, 1110–1122, 2004.
122. Zhu, W., Huang, C., and Soong, T.: Response and reliability of secondary systems in yielding structures. Probab. Eng. Mech. **9**, 145–155, 1994.
123. Zhu, W. and Lei, Y.: Equivalent nonlinear system method for stochastically excited and dissipated integrable Hamiltonian systems. Trans. ASME J. Appl. Mech. **64**, 209–216, 1997.
124. Zhu, W., T.T., S., and Lei, Y.: Equivalent nonlinear system method for stochastically excited integrable Hamiltonian systems. Trans. ASME J. Appl. Mech. **61**, 618–623, 1994.
125. Zhu, W. and Yu, J.: The equivalent non-linear system method. J. Sound Vib. **129**, 385–395, 1989.

# Index